Biology of Adventitious
Root Formation

BASIC LIFE SCIENCES

Ernest H. Y. Chu, Series Editor

The University of Michigan Medical School
Ann Arbor, Michigan

Alexander Hollaender, Founding Editor

A Continuation Order Plan is available for this series. A continuation order will bring delivery of each new volume immediately upon publication. Volumes are billed only upon actual shipment. For further information please contact the publisher.

Biology of Adventitious Root Formation

Edited by

Tim D. Davis

Texas A&M University Research and Extension Center
The Texas Agricultural Experiment Station
Dallas, Texas

and

Bruce E. Haissig

USDA Forest Service
North Central Forest Experiment Station
Rhinelander, Wisconsin

Plenum Press ● New York and London

Library of Congress Cataloging-in-Publication Data

Biology of adventitious root formation / edited by Tim D. Davis and
  Bruce E. Haissig.
        p.   cm. -- (Basic life sciences ; v. 62)
     "Proceedings of the First International Symposium on the Biology
  of Adventitious Root Formation, held April 18-22, 1993, in Dallas,
  Texas"--T.p. verso.
   Includes bibliographical references and index.
   ISBN 0-306-44627-8
   1. Plant cuttings--Rooting--Congresses. 2. Roots (Botany)-
  -Congresses.   I. Davis, Tim D. II. Haissig, Bruce E.
  III. International Symposium on the Biology of Adventitious Root
  Formation (1st : 1993 : Dallas, Tex.) IV. Series.
  SB123.75.B56 1994
  581.1'0428--dc20   93-46186
                                                              CIP

Proceedings of the First International Symposium on the Biology of Adventitious Root Formation, held April 18–22, 1993, in Dallas, Texas

The opinions expressed herein reflect the views of the authors, and mention of any trade names or commercial products does not necessarily constitute endorsement by the funding sources.

Cover illustration: Adventitious roots and shoots regenerate in different places when a *Bryophyllum* leaf is suspended entirely in air (upper, right), compared to dipping the apex into water (upper, left). Adventitious roots and shoots regenerate only in the notches of the lower side of a *Bryophyllum* leaf suspended in air (lower, left), and a red pigment develops only in such leaves. This "directed influence of gravity" does not exist when a leaf is wholly suspended in water (lower, right). [From Loeb J. (1924), *Regeneration from a Physico-Chemical Viewpoint*, New York, NY, McGraw-Hill Book Co., pp. 21 and 27].

ISBN 0-306-44627-8

©1994 Plenum Press, New York
A Division of Plenum Publishing Corporation
233 Spring Street, New York, N.Y. 10013

Printed in the United States of America

Host and Sponsors

Texas A&M University Research and Extension Center, Dallas, Texas (Host)

International Plant Propagators' Society
- Southern Region
- Western Region

Texas Forest Service

Texas Utilities Company (TU Services)

U.S. Department of Energy Biofuels Feedstock Development Program

U.S. Department of Agriculture Forest Service
- Forest Management Research
- North Central Forest Experiment Station
- Southern Forest Experiment Station

Executive Committee

Tim D. Davis
Texas A&M University
Research and Extension Center
17360 Coit Road
Dallas TX 75252-6599 U.S.A.

R. Daniel Lineberger
Texas A&M University
Department of Horticultural Science
Horticulture/Forestry Building, Room 202
College Station TX 77843-2133 U.S.A.

Bruce E. Haissig (Chairman)
USDA-Forest Service
North Central Forest Experiment Station
Forestry Sciences Laboratory
P.O. Box 898
Rhinelander WI 54501 U.S.A.

James A. Reinert
Texas A&M University
Research and Extension Center
17360 Coit Road
Dallas TX 75252-6599 U.S.A.

Stanley L. Krugman
USDA Forest Service
14th and Independence S.W.
P.O. Box 96090
Washington DC 20090-0090 U.S.A.

Narendra Sankhla
J.N. Vyas University
Department of Botany
Jodhpur 342001
India

Executive Secretary
Edith Franson
USDA Forest Service
North Central Forest Experiment Station
Forestry Sciences Laboratory
P.O. Box 898
Rhinelander WI 54501 U.S.A.

FOREWORD

Charles E. Hess
Department of Environmental Horticulture
University of California
Davis, CA 95616

Research in the biology of adventitious root formation has a special place in science. It provides an excellent forum in which to pursue fundamental research on the regulation of plant growth and development. At the same time the results of the research have been quickly applied by commercial plant propagators, agronomists, foresters and horticulturists (see the chapter by Kovar and Kuchenbuch, by Ritchie, and by Davies and coworkers in this volume). In an era when there is great interest in speeding technology transfer, the experiences gained in research in adventitious root formation may provide useful examples for other areas of science. Interaction between the fundamental and the applied have been and continue to be facilitated by the establishment, in 1951, of the Plant Propagators' Society, which has evolved into the International Plant Propagators' Society, with active programs in six regions around the world. It is a unique organization which brings together researchers in universities, botanical gardens and arboreta, and commercial plant propagators. In this synergistic environment new knowledge is rapidly transferred and new ideas for fundamental research evolve from the presentations and discussions by experienced plant propagators.

In the past 50 years, based on research related to the biology of adventitious root formation, advances in plant propagation have been made on two major fronts. Perhaps the most successful has been the development of systems to provide a favorable environment for maintaining cuttings during the process of root formation. Once glasshouses had been developed in the late 1800s, the next major, original approach was to further restrict the volume of air surrounding the cuttings and to allow the humidity to increase by evaporation from the medium and the cuttings until an equilibrium inside and outside of the cutting was reached. The confined space did present a problem, however, because it was necessary to reduce the amount of light energy entering therein. Without shading, the area surrounding the cuttings became a heat trap and the cuttings were killed. Also, high humidity and warm temperatures provided a very favorable environment for the growth of pathogenic fungi. The introduction of plastic film made construction of propagation beds less expensive and easier, but the principles remained the same.

In the 1940s, humidification systems were introduced. The addition of moisture in the form of water vapor to the cutting environment made it possible to enlarge the confinement area and thereby provide a larger air volume which acted as a buffer to temperature fluctuation. This meant less shading was necessary. Humidification did not eliminate the need for shading nor was the potential for fungal disease substantially reduced.

In the search for nozzles that could raise the humidity, some were selected that emitted a fine spray rather than water vapor. As a result, a film of free water developed on the leaves. Subsequently, it was found that a direct spray of water on the leaves permitted rooting without confinement, and propagation could be done in full sunlight. Mist propagation was immediately adapted by the nursery industry. Refinements followed in the development of inexpensive and non-clogging nozzles, and it was found that intermittent mist was superior to constant mist. The incidence of disease also was reduced, possibly by the "washing" action of the mist and/or by the presence of free water on the surface of the leaf as compared to high humidity. Cells were also fully turgid under mist, which may have provided physical resistance to invasion by pathogens. The introduction of mist propagation made possible the use of cuttings taken from new growth of plants whose cells had not fully matured. The presence of cells, which were still dividing and differentiating, extended the range of plant materials that could be propagated from cuttings as well as shortened the time required for root initiation. Comparative measurement of the environments under mist and closed propagation units showed higher light intensity and lower leaf temperatures under mist. Cuttings propagated under mist had greater rates of photosynthesis, lower rates of respiration and greater accumulated carbohydrates.

The most sophisticated form of environmental control came with the development of plant tissue culture under aseptic conditions (see chapter by Haissig and Davis in this volume). Defined media were developed to support the growth and development of stem segments, meristems and, eventually, single cells. The technique has been used commercially to produce virus-free plants and to clonally propagate horticultural and forest species. It is a valuable tool in root formation research, and it is also used as a technique to produce transgenic plants and to screen for somaclonal (genetic and epigenic) variation. Associated with each of the techniques used for environmental regulation of moisture loss is the necessity to harden-off the new plants or plantlets during the transition from a controlled to a more open environment in the growth chamber, glasshouse or field.

The second major contribution to plant propagation, based on studies of the biology of adventitious root formation, arose from the search for substances which might specifically determine rooting (see the chapter by Blakesley, and by Haissig and Davis in this volume). The concept that there may be a specific root-forming substance was suggested in the late 1800s: If a green plant were maintained in conditions suitable for growth, it would synthesize a specific root-forming substance in the leaves which moved towards the root system. Then, if a shoot were cut from the plant, the cut surface would prevent further movement of the substance which would accumulate at the base of the cutting, promoting root initiation. In the late 1920s and early 1930s, physiological evidence was obtained which suggested that an acidic root-promoting substance, termed rhizocaline, was produced by leaves and transported basipetally in the phloem. In this time period auxin (indole-3-acetic acid, IAA) was discovered and when a careful comparison of the properties of rhizocaline and auxin was made, they were found to be very similar. Synthetic IAA was found to be as active in stimulating root formation as any extract that had been tested, and it was suggested that the term rhizocaline could be eliminated. Nonetheless, the search for rhizocaline has continued (see the chapter by Haissig and Davis in this volume).

The discovery of auxin and its role in stimulating rooting had a major effect upon the root formation research agenda and the field of plant propagation. From the applied side of the world, it is interesting to note that early Dutch propagators inserted grain into the base of cuttings to enhance root formation. Presumably, the auxin released by the germinating seed was the source of enhancement. Other forms of root-stimulating compounds were synthesized and found more effective than IAA. Indolebutyric acid and naphthaleneacetic acid became standard tools of the plant propagator. Extensive applied research was conducted to determine the best method of applying the root-promoting substances.

Although a primary focus of the research agenda was devoted to the role of auxins in root formation, some workers continued to provide evidence that other special substances were essential for rooting in addition to auxin—a renewal of the rhizocaline concept. Improved rooting was demonstrated by grafting an easy-to-root scion onto a difficult-to-root cutting when auxin was also provided at the base of the cutting. Girdling below the graft or the removal of leaves from the easy-to-root scion eliminated the stimulatory effect. A mixture of auxin, sucrose and nitrogen, the latter in the form of amino acids, could substitute for the leaves. Other substances, such as ethylene, were shown to stimulate root formation. By the early 1950s it was widely accepted that some essential substance or substances were synthesized in the leaves and/or buds, accumulated at the base of the cutting and, in combination with auxin, determined the rooting response.

Subsequent research comparing substances in easy- and difficult-to-root cuttings provided additional evidence for the role of special substances in root formation in addition to auxin (see the chapter by Haissig and Davis in this volume). A bioassay specific for root formation was developed and at least four root-promoting substances were demonstrated in easy-to-root cuttings which acted synergistically with auxin in stimulating rooting. It was suggested that the substances may serve as "cofactors" with auxin. One cofactor was identified as chlorogenic acid. It has been suggested throughout this time period that phenolic compounds play a role in root formation. In addition to the presence or absence of substances involved in root formation, the relative supply of one substance to another can be important as can be seen in the "balance" of auxin and cytokinin. If the balance is in favor of auxin, root formation may take place. If the balance is in favor of cytokinin, cell division is predominant.

Anatomical studies showed that cells in the endodermis and pericycle are involved in root initiation. The presence of phloem fibers are not believed to be a barrier to root formation. However, the dedifferentiation of mature cells which allows cell division and primordium initiation is believed to be one of the crucial first steps in the process of root initiation. The primary stimulus is yet unknown (see the chapter by Haissig and Davis in this volume).

Although a lot of knowledge and practical application of the knowledge have been achieved over the past 50 years, the challenges and opportunities facing the researcher and plant propagator are even greater. The development of the tools of molecular biology are providing us with the ability to ask questions and find answers that could not be considered 10 or more years ago (see the chapter by Hamill and Chandler, by Hand, by Riemen-schneider, and by Tepfer and coworkers in this volume). The purpose of The First International Symposium on the Biology of Adventitious Root formation has been to assess where we are now, to identify new research opportunities and to determine how they may be funded. A complete understanding of the regulation of adventitious root formation is yet to be found, so agronomists, foresters and horticulturists still have many plants which are difficult to root. The scientific community and society will benefit when the answers are found.

PREFACE

It is our pleasure to present you with the Keynote Address and 19 of the 20 invited research papers from "The First International Symposium on the Biology of Adventitious Root Formation," which was held at Dallas, Texas, from April 18-22, 1993. In addition, the Editors have contributed a chapter, "A historical assessment of adventitious rooting research to 1993," which was to a great extent the theme of the symposium. Summaries of volunteer poster presentations were previously published in *Abstracts of Presentations at the First International Symposium on the Biology of Adventitious Root Formation*, B.E. Haissig and T.D. Davis, eds., USDA Forest Service, North Central Forest Experiment Station, General Technical Report No. 154, 1993.

The symposium provided the first international forum for the discussion of past research progress and future research directions in botany, quantitative and molecular genetics, biochemistry, biophysics, and plant physiology concerning *the biology* of adventitious root formation. Historically, interests in the biology of adventitious root formation have ranged from academic studies within plant morphogenesis and, somewhat, in ecology to empirical trials to improve cloning of special genotypes for agronomy, forestry, and horticultural research and practice. Thus, this volume serves all such special interests in one way or another and, as such, should prove useful to students, teachers, and researchers.

Yet, no single volume can comprehensively describe the wealth of information that has been amassed through research on adventitious root formation during the last 150 and more years. In counterpoint, and as a detraction to our reporting abilities, there are many critical voids in specific subjects that are relevant to adventitious rooting. In that regard, biology means "life-reckoning," including aspects of structure, function, growth, origin, evolution, and distribution. In that broad biological sense much about adventitious rooting has not been researched, even to the extent of complete observation.

All chapters were technically and anonymously reviewed by several qualified scientists, including members of the Advisory Board and the editors, and, indeed, many scientists who were not even associated with the symposium. The chapter authors responded to the reviews by revising their manuscripts, often extensively, so this volume contains information that goes beyond what was originally presented at the symposium. Indeed, titles of some chapters even differ from the corresponding symposium presentations. All-in-all, however, the ultimate responsibilities for chapter content, and precision and accuracy in presentation, interpretations, conclusions and opinions reside with the authors.

The editors and members of the Advisory Board have enjoyed working on this volume because it will foster research on the internationally important subject of adventitious root formation. Thus, we thank our host, our financial sponsors, and Plenum Press for their willing and generous support. We also thank all those individuals who, in the background and without due recompense, made this volume possible through their diligent efforts. In particular, we recognize the special, diverse contributions of Ms. Patti Davis and Ms. Karin Haissig. We also gratefully acknowledge the assistance of Paul Graff and Betty Knight.

Tim D. Davis
Bruce E. Haissig

CONTENTS

INDUCTION OF ROOTING

ROOT SYSTEM DEVELOPMENT AND PLANT GROWTH

SPECIAL CHAPTER

EPILOGUE

THE ORIGIN, DIVERSITY AND BIOLOGY OF SHOOT-BORNE ROOTS

Peter W. Barlow

Department of Agricultural Sciences
University of Bristol
AFRC Institute of Arable Crops Research
Long Ashton Research Station
Bristol BS18 9AF, UK

INTRODUCTION

The term "adventitious root" is widely used to designate a root that arises either on an already lateralized root axis or at a site on the plant that is not itself a root (e.g., on a shoot or leaf) (Esau, 1953). In the latter case, the root, strictly speaking, need not be adventitious since, etymologically, this refers to a root located at an unusual site on the plant whereas such a root might be developing at a site in a way that is entirely consistent with the normal ontogenic pattern of the plant. It would be more exact to designate such a root as "shoot-borne." This, in turn, leads to the idea that there are two types of roots: one is the shoot-borne type whose origin is self-defining, the other is the pole-borne root whose origin is from one of the poles of the embryo. According to Guédès, (Guédès et al, 1979) there is only one type of root: all roots are shoot-borne because even the embryonic root derives from the shoot-pole of the embryo. The same argument has been made for grasses (Tillich, 1977). Although this view may seem a reasonable argument from an evolutionary perspective, it is not one accepted by all morphologists. Whatever view is taken, it does seem that the term "adventitious root" can be restricted to its "true" meaning as referring to a root which develops out of the normal temporal sequence and/or at an unusual location. In most if not all cases, this would apply to a root that develops as a result of wounding and is thus evidence of a regenerative response. Adventitious roots, therefore, are simply a class of shoot-borne (or root-borne) roots developed under rather special circumstances (see the chapter by Haissig and Davis in this volume).

Although the present chapter sets out to examine the diversity of shoot-borne roots, (Barlow et al., 1986) and will consider them as a type of root distinct from the embryonically derived pole-borne type, there will be discussion, particularly in the next section, of both types of root/shoot-borne roots (including the truly adventitious roots) and pole-borne roots, as well as their counterparts in the shoot system—that is, root-borne and pole-borne shoots. No special pleading should be necessary for this liberty since ultimately it is impossible not to take a holistic view of the plant and its organs and thus consider all of them together.

Biology of Adventitious Root Formation, Edited by
T.D. Davis and B.E. Haissig, Plenum Press, New York, 1994

CLASSIFICATION OF ROOTING TYPES

Heeding the above-mentioned criticisms surrounding the terminology of rooting, especially those voiced by Groff and Kaplan (1988), it is evidently necessary to try and clarify not only the types of root (and shoot), but also the interrelationships between root and shoot systems. A useful starting point in this last-mentioned area is the synthesis attempted by Aeschimann and Bocquet (1980). However, the terminology presented in their paper clearly derives from the concepts of allorhizy and homorhizy introduced by Karl von Goebel and Wilhelm Troll. For these authors, allorhizy describes cases where the root and shoot systems are of embryonic origin and are joined at the hypocotyl; homorhizy describes cases where the embryonic root seems to have been lost and all the roots of the plant arise from the shoot. Aeschimann and Bocquet (1980) introduced a third term, homocauly, to describe root-borne shoots and thus draw attention both to the equal status of the shoot system and to its symmetrical relationship with the root in classifications of either system. Their analysis draws upon the useful distinction between pole- and shoot-borne roots (and their cauline counterparts) mentioned earlier. Although aspects of their terminology seem difficult to apply in practice, a more serious failing, as pointed out by Groff and Kaplan (1988), is that it omits a very widespread type of root/shoot system. To remedy this, Groff and Kaplan amended their classification and jettisoned the terms allorhizy, homorhizy and homocauly. In their place they proposed the neutral terms class 1, 2, 3, and 4, the latter describing the root/shoot system that had been omitted by Aeschimann and Bocquet.

While the proposals of Groff and Kaplan are sound as far as they go, they do not go far enough. This is largely because the root/shoot systems which their classes illustrate indicate the possibility of developing a logical and comprehensive terminology that reflects the interrelationships both within and between root and shoot systems. The basis of this logical system, outlined in Fig. 1, is that: a) root or shoot can produce a member of the contrasting class, and b) the two systems are cumulative (i.e., no members are lost, or, if they are, the loss is registered). The system only categorizes the interrelationships between roots and shoots; that roots and shoots can branch to produce additional members of the same class is not included.

The scheme shown in Fig. 1 indicates the possible types of root/shoot inter-relationships, derived from the alphabet $[S_1 R_1, S_2, R_2]$ and how one type may be derived from another, indicated by the arrows. In this respect, it is an example of a structuralist solution to a defined problem of biological relationships (Piaget et al., 1971). The scheme may be regarded from three aspects. First, it is classificatory; the branching pattern of a given root/shoot system is described by the letters within the bracket [ ]. Second, the scheme can be ontogenetic; the arrows denote a developmental sequence of branching. Third, the scheme might be read as a phylogenetic sequence; that is, the arrows denote an evolutionary sequence. The first and third aspects of the scheme fulfill, respectively, the classificatory aims of Groff and Kaplan (1988) and the phylogenetic aims of Aeschimann and Bocquet (1980). The correspondence between the classifications of root/shoot system types due to the two pairs of authors and that presently proposed is demonstrated in Table 1. However, the inherent logic of the scheme predicts root/shoot branching types that are not described by either pair of authors. Also, it will be evident that one example of adventitious rooting could be designated by this scheme as $[S_1, R_1^\dagger, {}_{S1}R_2]$; it could then develop further to $[S_1, R_1^\dagger, {}_{S1}R_2, {}_{R2}S_2]$, this representing regeneration. Likewise, *in vitro* root cultures could be designated $[S_1^\dagger, R_1]$. In some circumstances, adventitious shoots will sprout from such roots [(e.g., the *Convolvulus* root cultures studied by Bonnett and Torrey (1966)]; they would be notated as $[S_1^\dagger, R_1, {}_{R1}S_2]$. Examples of the various kinds of root/shoot system types throughout the plant kingdom are listed in Table 2.

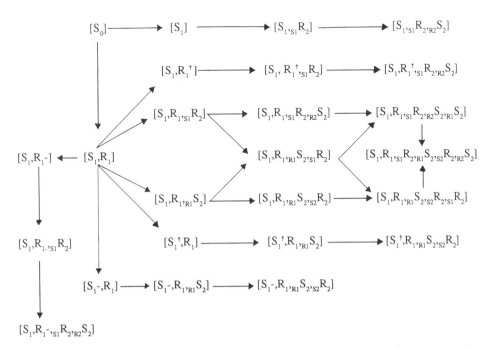

Figure 1. Scheme illustrating the interconnections between the various types of root and shoot systems. For further explanation see the text and Table 1. S_0, a primitive shoot; S_1, a pole-borne shoot, R_1, a pole-borne root; S_2, a root-borne shoot; R_2, a shoot-borne root; †, death of the root or shoot meristem; R- or S- indicate the absence or extreme reduction (vestigiality) of the root or shoot.

Table 1. Classifications of the types of root and shoot system interconnections according to Aeschimann and Bocquet (1980), and Groff and Kaplan (1988).

Type	Terminology According To:	
	Aeschimann & Bocquet	Groff & Kaplan
$[S_1,R_1]$	Allorhizy	Class 1
$[S_{1,S1}R_2]$	Primitive homorhizy	--------
$[S_1,R_{1,S1}R_2]$	Homorhizy	Class 2
$[S_1,R_{1,R1}S_2]$	Homocauly	Class 3
$[S_1,R_{1,R1}S_{2,S2}R_2]$	--------	Class 4
$[S_1,R_1{}^{\dagger}{}_{,S1}R_2]$	Vicariant homorhizy	Class 2
$[S_1{}^{\dagger},R_{1,R1}S_2]$	Vicariant homocauly	Class 3

3

The root/shoot system classes discussed by Groff and Kaplan (1988) are essentially those readily observed in the field. They do not necessarily reflect the full potentiality of roots or shoots to produce the other kind of organ. For example, in their illustration of the class 4 system, shoot-borne roots $[R_2]$ are carried only on root-borne shoots (i.e., class 4 is thus $[S_1,R_{1,R1}S_{2,S2}R_2]$) and not on pole-borne shoots (i.e., they are not $[S_1,R_{1,R1}S_{2,S1}R_2]$). It could be asked whether the pole-borne shoot $[S_1]$ of class 4 is inherently unable to produce roots $[_{S1}R_2]$, or whether it does not do so because, unlike the root-borne shoot $[_{R1}S_2]$, none of the pole-borne shoot is in an environment favoring rooting. Thus, some experimental testing is necessary to elucidate not only which types of root/shoot systems actually exist in natural environments, but also whether those that do exist show their full branching potential. Similarly, it could be questioned whether the shoot-borne roots $[R_2]$ of Groff and Kaplan's class 2 are unable to bear shoots $[_{R2}S_2]$ because of the former's linkage with the pole-borne shoot $[S_1]$. That is, the type $[S_1,R_{1,S1}R_2]$ precludes further development to $[S_1,R_{1,S1}R_{2,R2}S_2]$. If this linkage were broken, it would be interesting to see whether $[_{R2}S_2]$ could be formed from $[R_2]$, or whether a shoot system and a new shoot-borne root system could regenerate from the remaining pole-borne shoot $[S_1]$; that is, $[S_{1,S1}R_2]$ develops into $[S_{1,S1}R_{2,R2}S_2]$.

Table 2. Distribution of root and shoot system types in species belonging to the Pteridophytes, Gymnosperms and Angiosperms. [Many of the species mentioned below are referred to by Groff and Kaplan (1988); others are mentioned in the text.]

Type	Pteridophytes	Gymnosperms	Angiosperms	
			Monocots	Dicots
$[S_1,$	None	Many	None	Annuals
$[S_1,R_{1,S1}R_2]$	Most	*Sequoia sempervirens*[1]	Most	Most
$[S_1,R_{1,R1}S_2]$	None	*Stangeria*	None	*Monotropa* (adult)
$[S_1,R_{1,R1}S_{2,S2}R_2]$	*Ophioglossum petiolatum*	None	Orchidaceae	Many
$[S_1-,R_1]$	------	------	------	Monotropa (juvenile)
$[S_1{}^\dagger,R_1]$	------	------	------	Root cultures
$[S_1-,R_{1,R1}S_2]$	------	------	------	Podostemaceae
$[S_1,R_1-]$	*Salvinia*	------	*Wolffia*	*Utricularia*
$[S_1,R_1{}^\dagger,_{S1}R_2]$	------	------	*Taeniophyllum*	------

[1]A species named in the table means that it is a representative of the system-type which is otherwise infrequent in the taxonomic class or subclass in question; ------ means that a species has not been specifically identified in the relevant category.

GENETIC DIVERSITY OR PHENOTYPIC PLASTICITY OF ROOT TYPES?

The classificatory scheme shown in Fig. 1 raises a number of questions about the integrity of the elements within the root- and shoot-system classes. For example, is there a fundamental difference between the pole-borne roots $[R_1]$ and the shoot-borne roots $[R_2]$? This question is particularly pertinent in the light of Zobel's (1986, 1989) proposal that there are four, or possibly five, different categories of root in both monocotyledons and dicotyledons—pole-borne, lateral, basal, shoot-borne and, possibly, collateral. Obviously there are morphological and anatomical differences between them, but there is evidence which also points to genetical and physiological differences. One extreme interpretation would be that each of the five categories is specified by a distinct group of genes. A more moderate position would be that there is a core set of genes which determines a basic root *Bauplan*, and that this *Bauplan* is then modified by the action of additional root determining genes. At the other extreme is a situation where the core set of genes again specifies the root *Bauplan*, but the phenotype of the root is modified during its development by influences from the cellular milieu and not directly by changes in genetic control. That is, the root phenotype is plastic due to its response to influences that impinge upon it during the primordium stage. The phenotype then stabilizes at some early point in development unless the meristem again is caused to respond to influences from the external environment that further modify its development. For example, roots of *Holcus lanatus* vary in their appearance (altered radius, root hair density, etc.) depending on external nitrate levels (Robinson, 1989). Moreover, such plastic responses are also modulated by the genotype as evidenced by the response of adventitious roots of different clones of *Betula pendula* to mycorrhizal association (Lavender et al., 1992). The extent to which genotype and environment modify root diversity could be tested through the use of cDNA libraries constructed from the putatively different categories of root borne on the same plant. In the present context, it would be of interest to know whether the shoot-borne roots (both adventitious and natural, as well as the basal roots formed at the pole-borne root/hypocotyl boundary) are fundamentally different from each other and from the pole-borne set of roots. A second question is whether shoots of different origins (i.e., $[S_1]$ or $[S_2]$) are, by their genetic constitution, predisposed to form different types of shoot-borne roots. In other words, is $[_{s1}R_2]$ different from $[_{s2}R_2]$? This would clarify the distinct behavior of the two types of shoots ($[S_1]$ and $[S_2]$) in Groff and Kaplan's class 4 root/shoot system.

THE EVOLUTIONARY ORIGIN OF SHOOT-BORNE AND POLE-BORNE ROOTS

The scheme in Fig. 1 shows an entity designated $[S_0]$ which, with respect to the phylogenetic reading of the figure, can be regarded as a primitive plant consisting of only a shoot. It is generally agreed that the root is a later evolutionary addition, but whether it has a mono- or polyphyletic origin is not clear. The distinction already made between pole-borne and shoot-borne roots suggests it originated on at least two independent occasions. But what, if anything, caused roots to develop? I propose that the evolution of the root in early land plants was related to their increasing photosynthetic efficiency. Teleologically, the root can be viewed as a device to rid the plant of excess carbon derived from photosynthesis. Furthermore, the excretion of carbon from roots could have been a means whereby the quality of early soils was improved with the consequent promotion of an increasingly varied land flora.

Early land plants of the lower Devonian period (400 MY BP), such as *Rhynia major* and *Horneophyton lignieri*, possessed both erect and prostrate axes [equivalent to shoots and rhizomes, respectively (Stewart, 1983)]. There were no leaves. Rhizoids were present on the ventral side of the prostrate axes, where they joined with the erect axes. These rhizoids seem already to be foreshadowing the shoot-borne roots, not only in their location but also in their probable involvement in solute uptake; they are located where surplus photosynthates might be expected to accumulate. Later in evolutionary time, microphylls developed on the erect axes. In *Asteroxylon*, for example, the microphylls were photosynthetic as evidenced by the presence of stomata. The density of microphylls on the stems indicates that the shoot organ as a whole would possess an increased photosynthetic capacity. Significantly, *A. mackei* was one of the first plants to exhibit roots. These organs also developed on the ventral side of the junction between the prostrate and erect axes suggesting some peculiarity of this site, as does this region in the earlier rhizoid-bearing plants. Other fossilized plants of the Lower Devonian have also been discovered possessing roots: these are either on presumed subterranean stems, as in the Drepanophycopsid, *Drepanophycus spinaeformis*, (Rayner, 1983) or on aerial stems, as in the Zosterophyll, *Sawdonia ornata* (Rayner, 1984).

Suppose that the increased surface area of photosynthetically active tissue produced more carbohydrate than could be incorporated into the growing shoot system. Suppose, further, that this carbohydrate accumulated at the base of a photosynthesizing shoot. Then, the excess carbohydrate in this region might trigger renewed cell growth, division and cellulose synthesis. Due to an inherent polarization of growth in the new cells, a new axis would spring from the preexisting one carrying forward the growing and dividing cells at its distal end. Growth of this axis would continue for as long as excess carbohydrate was available. An anatomist contemporary with these early land plants might call the group of cells whose growth and division was triggered by the excess carbohydrate a "primordium" or "meristem," and the out-growing, neo-formed axis a "shoot-borne root." What better way, then, to eliminate this carbohydrate, which might otherwise interfere with photosynthesis, (Sonnewald and Willmitzer, 1992) than by harmlessly sequestering it as cellulose, particularly in a ventrally located organ so that it would neither shade nor entangle the erect parts of the plant. Moreover, roots of present-day plants seem to possess a respiratory activity in excess of the requirements for growth and function (Caldwell et al., 1979) suggesting that respiration might be an additional means of removing excess carbon from the plant. Particularly interesting in this regard is cyanide-insensitive respiration (Lambers, 1980) which represents a process for "incinerating" excess carbon. Moreover, excess carbon can be excreted from roots and thereby serve as an attractive bait for beneficial (as well as pathogenic) microorganisms, and it also facilitates the establishment of mycorrhizal associations (Whipps, 1990)

With the development of the seed habit during the Carboniferous period (325 MY BP) (Stewart, 1983) a similar recapitulation of events, again brought about by carbohydrate excess, might now apply within the embryo. Insoluble carbohydrates stored within the seed and later metabolized to a soluble form, may have induced a "meristem" at the hitherto unoccupied anti-apical pole (McLean and Ivimey-Cook, 1951) of the embryo located opposite the embryonic, apical shoot-pole. As the shoot-pole continued to grow and photosynthesize, the excess carbohydrate which it produced would be sequestered in the emerging, neo-formed pole-borne root structure.

So far, the proposed pathway of root development has been viewed as an adaptive response to a particular set of internal physiological conditions. However, there are no known ways in which acquired endogenous adaptive features can be directly integrated into the normal ontogenetic program of the organism. This is the dilemma of Lammarckism (Waddington, 1957; Jablonka et al., 1992). Therefore, the next necessary step in root evolution would be a means of regularizing the presence of this new organ through the acquisition of a genetic program that would reproducibly determine its development. It is possible that mutation in the ancient plant genome and the generation of root-forming genes could solve the problem. An alternative possibility (Harper et al., 1991) is that an early land plant(s) was (were) genetically transformed by microbe-borne DNA sequences which were root-determining (cf. the Ri plasmid of *Agrobacterium rhizogenes*). However, this explanation requires more assumptions (hence violating the principle of the *Doctor invicibilis*, William of Occam, that *Entia non sunt multiplicanda praeter necessitatem*) than the simpler hypothesis that analogous mutations occurred within the plant's own genome. Both hypotheses are untestable, unfortunately. However, it would be interesting to know whether roots can be induced to form on the rootless rhizomes of *Psilotum nudum,* for example, following infection with *A. rhizogenes* since *Psilotum* is believed to be derived from an extinct ancestral form that resembles members of the Devonian Rhyniopsida (Stewart, 1983; Rouffa, 1978).

Evolution appears to be adept at "tinkering" both with already-formed parts (Jacob, 1977) and with the metabolic units from which they are made (Ugolev, 1990). As a result, neo-formed structures are invested with a functional significance of their own. Thus, it seems a relatively straight-forward step to transform the cellular cellulosic dump that was the proto-root into the elegant, gravity-responsive, load-bearing, nutrient-capturing organ that is the present-day root.

Whatever the exact origin of roots, if the premise is correct that they were, and still are, a means of removing excess carbon from the plant body, it might explain their characteristic sites on present-day plants since they should still continue to form where the excess accumulates. The frequent location of shoot-borne roots at the leaf base or node suggests that these sites could be where carbohydrates naturally accumulate. It is known, for example, that both the nodes of stems and the junction between pole-borne shoot and root are associated with restrictions on the upward flow of water in the xylem (Luxová, 1986; Salleo et al., 1990). The same may also be true of carbohydrate fluxes through the phloem, though this problem has been studied less, partly because the techniques involve detailed anatomical investigation (O'Brien and Zee, 1982) and the use of radioactive tracers (Wardlaw, 1965).

DIVERSITY IN THE POSITIONING OF SHOOT-BORNE ROOTS

Diversity in Space

The development during evolution, of leaves, nodes, internodes and axillary buds has resulted in a branched system of axes (Barlow, 1989) which can even be seen in fossil plants. The relationship of roots with defined locations on the shoot axis is fluid. In some fossil species [e.g., *Sawdonia ornata*, (Rayner, 1983)] the roots occur where shoot axes branch, while in the Lygnopterid, *Lygnopteris oldhamia*, they arise freely on the stem, apparently with no preferred location (Scott, 1962). In present-day plants, the location of root primordia with respect to the node and internode varies in a way that is characteristic of the species. Primordia locations vary not only around the circumference of the stem, but also along the intervening length of internode. Some examples have been described by Weber (1936, 1953) and Gill (1969), a few of which from both monocotyledonous and dicotyledonous species are included in Fig. 2. All the diagrams show the position of the shoot-borne root with reference to the nearest leaf and node. In the Labiateae, (Weber, 1936) the node is more than just a reference point for classifying root positions: it is also an important anatomical site that presumably specifies whether a root primordium will form, and if so, where. The most frequent situation in this family is that shoot-borne roots form either immediately above or below a node. Less frequently, roots form at both locations (e.g., in yellow archangel, *Lamiastrum galeobdolon*, Fig. 2F). However, the presence of roots in the middle of an internode is quite common, though when this occurs roots are also very frequently found at the associated node. The sequence of precedence of rooting sites has not been systematically investigated in any species whose shoots show such multiple sites. Neither have detailed anatomical or developmental studies been made to clarify the role of the node in root initiation. The Labiateae, for example, which, because of their square stems and nodes, have many clear internal reference points for such a study, would furnish excellent material for discovering why some species in the family form their roots at the distal end of the internode, while others form them at the proximal end. This material would also be favorable for a correlation of root positions with local fluxes of hormones, carbohydrates, etc., along the length of the node and internode.

The shoot-borne roots of shoots which grow horizontally often develop on the ventral side. Whether this is a gravimorphic or thigmomorphic effect is obviously open to question (Steinitz et al., 1992), just as is the role of leaves and buds in the production of shoot-borne roots. Moreover, many species exhibit anisophylly, while some others, particularly members of the Acanthaceae, exhibit anisoclady (i.e., leaves and buds, respectively, show different forms or have different growth potentials.) These natural variables of leaf and bud development, together with possibilities for reorienting root-bearing stems with respect to the gravity vector, as well as the readiness with which adventitious roots will form on cuttings, give a wealth of experimental possibilities to discover the positional controls that normally regulate root initiation and out-growth. Indeed, work on cut stems of *Sambucus nigra* (Warren-Wilson and Warren-Wilson, 1977) has already shown that adventitious root production is self-inhibiting; removing already formed-roots induces a new flush of such roots. Probably each root sets up an inhibitory field that prevents new roots from forming in their immediate vicinity. The same situation may apply to the induction of adventitious root-borne roots (Blakely et al., 1982).

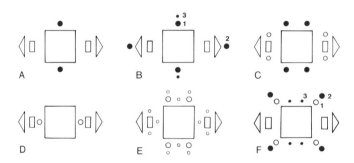

Figure 2. Diagrams illustrating the diversity, within species of the family Labiateae, of the position of shoot-borne roots in relation to the site of the node (represented as □; O —root above the node, and ● —root below the node), leaf (c), and axillary bud ([]) (where present). The order of emergence of the roots (1, 2, 3) is shown in two cases (B and F). The species are: A, *Scutellaria galericulata*; B, *Glechoma hederacea*; C, *Mentha arvensis*; D, *Mentha rotundifolia*; E, *Lycopus europea*; F, *Lamiastrum galeobdolon*. [A, C, E, redrawn from Weber (1953); B, D, F, author's observations]

The site of origin and orientation of shoot-borne roots within the canopy of trees and shrubs is a variable which may have implications for the general ecology of the species. Nadkarni (1981) drew attention to the extensive network of shoot-borne roots of epiphytes which penetrate the detritus accumulating in the canopy of forest trees growing in Olympic Peninsula, Washington State (USA). More recently, Lüttge (1991) has reviewed the ecophysiology of hemi-epiphytic species of *Clusia* from forests in Central America and The Antilles whose shoot-borne roots play an important role in defining the form of the plant; for example, some species become epiphytic by losing contact with the soil, whereas others start as epiphytes and then become self-supporting trees after sending their roots down into the soil. These examples indicate that the ability of species to manipulate the distribution of their shoot-borne roots may be important in defining the type of community in which these species exist. In a further example, Gill (1969) examined the sites of emergence of aerial roots from trees, shrubs and vines comprising an elfin forest in Puerto Rico. His findings are summarized in Fig. 3. Although anatomical features peculiar to each species may determine the precise sites where their roots are formed, the distribution of sites may have an influence on the structure, both physical and taxonomic, of this distinctive plant community.

Diversity in Space and Time

Diversity also extends to the temporal distribution of roots (i.e., their longevity) in relation to the overall energy balance within the community. In an Alaskan tundra community (Shaver and Billings, 1975) all shoot-borne roots of the sedge *Eriophorum angustifolium* die each year, whereas other graminoid species in the same community, such as *Carex aquatilis* or *Dupontia fischeri*, possess shoot-borne roots with slower turn-over times (their roots live for four to six and five to eight years, respectively). This may bear on how the ecosystem processes scarce or fluctuating levels of nutrients in the soil, and on how the growth pattern of the plant community is regulated. Other aspects of how the temporal pattern of shoot-borne root growth is adjusted to the growth strategy of the shoot and is linked with seasonal changes are described for four Chinese species (Liu and Liu, 1993). These authors were able to identify three patterns of shoot-borne root behavior: a renewal type where all roots are lost each year, a type where root growth is continuous throughout the year, and an intermediate type where some roots die but others continue growing on a seasonal basis. This work might be considered as the starting point of a complex project that aims to relate the structure and evolution of plant communities with the evolution of the forms of the plants that compose them.

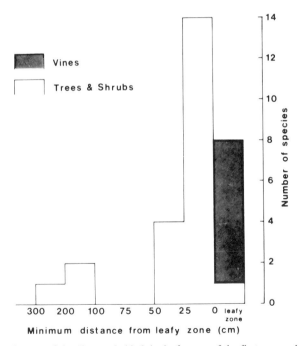

Figure 3. Frequency diagram of the distance behind the leafy zone of the first emerged shoot-borne roots on 22 species (from 17 families) of trees and shrubs (open bars) and 8 species (from 8 families) of vines (shaded bar) growing in an elfin forest in Puerto Rico. [data from Gill (1969)].

The introduction of a temporal aspect to the development of diverse types of shoot-borne roots inevitably leads to thoughts about the way in which the distributions of root types of contemporary plants have arisen during evolutionary time. The phylogeny of root types, and the positions of these roots on the plant, can be approached by examining the current diversity of these two features within a closed taxonomic grouping. Such a study has already been made by Hoshikawa (1969) for seedlings within the Gramineae. He examined 219 species spanning 88 genera. His intention was to use morphological features of the seedling root system as a means of verifying systematic relationships within the family and for proposing a phylogeny of the grass subfamilies. However, his data can also be used to schematize possible ontogenetic transformations that achieve increasingly complex patterns of root system morphogenesis. Because evolution seems to be concerned with developing increasingly more complex relationships both within and between species, such ontogenetic transformations probably occur within an evolutionary timescale.

Hoshikawa's findings and proposed combinations of four genotypic determinants (α, β, γ, δ) that govern the rooting pattern in the Gramineae are summarized in Table 3. Multiple copies of three of the determinants increase the numbers of mesocotylar roots (β), basal roots (γ) and cotyledonary nodal roots (δ), respectively, whereas it is assumed that multiple copies of α, which determines the pole-borne roots, have as much (or as little) effect on this root type as one copy (there is, after all, only one site for such a root in the embryo). However, as the scheme stands, the only cumulative phenotypic trait of this system is that for the mesocotylar roots (phenotype class 11 in Table 3). An additional phenotypic character that is also noted is the height of the mesocotyl; this is either dwarf (d) (in phenotype classes 1 and 6) or tall (D).

Table 3. Summary of the phenotypic and genotypic states that contribute to various shoot-borne rooting patterns in the Gramineae. The icons under the phenotype state visually represent the rooting patterns on a shoot (which can be either tall or short as shown by the height of the vertical line). Also listed are examples of the taxonomic classes to which these patterns correspond.

State			Taxon		
Phenotype		Genotype[1]			
Number	Icon		Sub-family	Tribe	Genus
1		d,α	Bambusoideae	Bambuseae	*Bambusa*
2		D,α	Oryzoideae	Oryzeae	*Chikushicloa*
3		D,α,β	Arundinoideae	Arundineae	*Phragmites*
4		D,α,δ	Festucoideae	Stipeae	*Stipa*
5		D,α,γ	Apparently undiscovered, though it could be		
			Festucoideae	Meliceae	*Strebochaeta*
6		d,α,γ	Festucoideae	Triticeae	*Triticum*
7		D,α,α,β,δ	Festucoideae	Aveneae	*Beckmannia*
8		D,α,α,γ,δ	Festucoideae	Festuceae	*Bromus*
9		D,α,α,β,γ	Eragrostoideae	Chlorideae	*Eleusine*
10		D,α,α,α,β,γ,δ	Oryzoideae	Ehrharteae	*Ehrharta*
11		D,α,α,α,β,β,γ	Eragrostoideae	Eragrosteae	*Muhlenbergia*

[1] Key to root genotypes: α, pole root; β, mesocotylar root; γ, basal root; δ, cotyledonary nodal root.

Schemes can be devised to describe the morphological transformations in the shoot-borne system and which might even serve as a framework for experimentation. For example, the transformations can be represented using the notion of "morphogenetic distance," whereby it is possible to calculate the number of mutational steps needed to achieve certain morphological states (Lück and Lück, 1993).

ANATOMICAL DIVERSITY OF SHOOT-BORNE ROOTS

Shoot-borne roots within large but related taxonomic groupings (e.g., families and subfamilies) show anatomical diversity. The Gramineae have been quite extensively studied in this respect. For example, Goller (1977) examined transverse sections of nodal roots of 257 species and was able to identify a number of variable cellular characters associated with all tissues: epidermis, exodermis, cortex, endodermis, pericycle, phloem, xylem and pith. Nevertheless, certain characters were correlated, thus enabling three broad anatomical syndromes to be identified that corresponded to the subfamily Festucoideae and Panicoideae, and to an Intermediate taxonomic grouping (including the Oryzoideae). The endodermis shows conspicuous variation between taxa. Differential wall thickening on the inner surface of its cells characterizes the "U" type of cell, while uniform thickening characterizes the "O" cell type. The shape of the cell, i.e., whether the major diameter is in the radial or circumferential plane, leads to further differentiation between taxa. Even studies of the nucleolus of epidermal cells (Brown and Emery, 1957) and of the cells from which root hairs originate (Row and Reeder, 1957) have led to correlations with taxonomic status: Panicoid grasses tend to have persistent nucleoli at mitosis and equal-sized hair and non-hair cells whereas Festucoid grasses tend to have no persistent nucleoli and hairs originate from the shorter cell of an unequal epidermal division.

Goller (1977) studied the roots of only relatively few members of the subfamily Bambusoideae and classified them all, on the basis of cellular anatomy, as belonging to the Panicoid group. A more recent survey of roots of 15 species of Bambusoideae (Raechal and Curtis, 1990) has shown more variability than Goller encountered. In fact, anatomical features varied so widely that it was impossible to differentiate, using these criteria, taxa above the tribe level in this subfamily.

Such variation is of interest not only from a taxonomic point of view, where it can be used to distinguish genera and even species, (Sobotik and Kutschera-Mitter, 1991) but also from an ontogenetic point of view. Here, it would be interesting to establish interrelationships between characters and also to discover how much plasticity each character displays as a response to environmental variation. The aerenchyma of grass roots is one such plastic response of the root cortex, and variation in the form of this tissue in roots is also quite striking. Further examples of aerenchyma are mentioned later.

DIVERSITY IN THE FORM AND FUNCTION OF SHOOT-BORNE ROOTS

Just as structuralist concepts (Piaget, 1971) can help deduce the basic units and relationships that contribute to a system such as root and shoot, so the concepts of general systems theory can help analyze and isolate the fundamental functional properties of living systems. In the present context, the two concepts appear to be quite closely related. Hierarchies are a feature of general-systems thinking and the whole plant is a good example of a "level" in a hierachically organized living system. Its component elements at the next lower level are organs, an element class which includes roots. Each level of a living system is constituted of a canonical set of 20 subsystems, (Miller and Miller, 1990) although there may be fewer in plants (Miller, 1978; Barlow, 1987). Six of these subsystems are relevant in the present context. They are: extruder, supporter, motor, ingestor, distributor and converter, and energy store. All have to do with the processing of matter-energy. The reproducer subsystem will also be mentioned; its significance is self-evident.

A consequence of evolution is the increasing complexity of living forms. Complexity arises by the optimizing of the subsystems which support (and define) each level in the hierarchy within the context of a restrictive, or selective, environment. In the scenario of root evolution put forward earlier, the first-formed root (whether shoot-borne or pole-borne does not matter) was regarded as an extruder subsystem, developed for the removal of superfluous carbon. However, if it is true as Jacob (1977) suggests that evolution makes use of tinkering or "bricolage," then the emergence of the root (both figuratively, in evolutionary terms, and physically as a projection into the surrounding environment) immediately provided an opportunity to develop, at the level of the whole plant, two other subsystems, the supporter and the ingestor, which in the early land plants must have been quite rudimentary. To varying extents, roots retain these three subsystem properties to the present day, the two last-mentioned ones (supporter and ingestor) being those most immediately associated with roots. Moreover, much of the diversity of forms of shoot-borne roots can be seen as part of a stratagem to optimize these and other subsystems to enable plants to grow in widely differing environments. The fact that different names (e.g., stilt root, air root, root tuber, etc.) are given to different forms of shoot-borne roots suggests that each form "emphasizes" a particular root-based subsystem function that relates to the whole plant level. Thus, stilt roots emphasize the supporter subsystem, air roots the distributor subsystem, root tubers the storage subsystem, and so on. Although this view obviously has a strong element of subjectivity, it seems nevertheless a viable base upon which not only to classify the diversity of root forms, but also to aid in understanding their significance. This correspondence between shoot-borne root form and subsystem emphasis will be explored under the headings which follow.

Supporter Subsystem

Stilt Roots. Good examples of stilt, or prop, roots are found in the Palmeae. Here, they are thick, shoot-borne roots which emerge from the stem up to one meter or more above the ground. During his travels in the Amazon, the naturalist Alfred Russell Wallace noted being able to walk beneath arches of such roots produced from 20-meter-tall trees of *Iriartea exorrhiza* (Wallace, 1853). Observations of the early development of palms (Schatz et al., 1985) illustrate some of the advantages which stilt roots confer. Stilt roots of young specimens of *Socratea durissima* and *Iriartea gigantea* develop early and stabilize the upward growth of the stem permitting it to gain rapid penetration of gaps in the canopy. In these early stages, *S. durissima*, in contrast to *I. gigantea*, invests more resources into the stilt roots than into the stem, enabling it to achieve a correspondingly greater height per shoot weight. This stratagem differs from that of *Welfia georgii*, a species that grows alongside, where no stilt roots are produced and the young plant remains as a squat rosette within the understory of the forest. It is also worth remarking that, in some species, e.g., [*Piper auritum* (Piperaceae)], the stilt root is capable of bearing shoots after it has penetrated the soil (Greig and Mauseth, 1991). In this respect, the shoot-borne roots are participating in the reproducer subsystem. Stilt roots are often associated with palm trees that live in areas subject to flooding. The potentially lethal effects of flooding are mitigated by the presence of stilt roots high on the trunk, often above the flood level, and by the presence of pneumatorhizae that facilitate gas exchange (Frangi and Ponce, 1985). However, stilt roots generally seem to be an intrinsic property of the plant and are not developed as a response to flooding.

Internal Roots. Somewhat similar in appearance to stilt roots are the "internal" roots of the Velloziaceae and Xanthorreaceae. The roots are internal in that they are obscured from view by the marcescent leaves which ensheathe the stem. When the leaf bases are removed from stems of *Kingia australis* (Xanthorreaceae), an Australian species of "grass tree," numerous shoot-borne roots are seen to encircle the stem (Staff and Waterhouse, 1981; Lamont, 1981a). These roots are initiated about 10 cm from the shoot apex and increase in number with increasing distance down the stem. After their production ceases in the proximal zone of the stem, branching of the roots further increases their number (Fig. 4). Because the base of the shoot dies along with its associated roots quite early in the life of the plant (when the plants are about 100 years old!), it is obvious that tall stems—which may reach up to eight meters in height when ca. 550 years old—are entirely supported by their internal shoot-borne roots. This support is not from the roots acting as struts, as stilt roots tend to be, but through their forming a rigid column of pseudo-stem, known as a caudex.

The veritable cascade of roots down the stem of *Kingia*, which resembles a tight bundle of cellulosic stalactites, also emphasizes the roots' role (as an extruder subsystem) in excreting all the excess carbon that cannot be utilized because of the slow growth and simple form of the shoot. Although roots of old plants of *Kingia* may spend hundreds of years in an arid, aerial environment, their ingesting function is not impaired. Experiments show (Lamont, 1981b) that they retain the ability to import water and minerals, and presumably do this during the infrequent rainfalls in the Australian bush.

Clasping Roots. The "strangling" figs (*Ficus* spp.) take the pseudo-stem function of shoot-borne roots a step further. The seeds of many epiphytic species of *Ficus* germinate in a crotch high in the canopy of a support tree. The pole-borne root and its laterals, as well as shoot-borne roots, descend the trunk of the support tree, fusing with one another as they grow down to the ground. A similar fusion occurs between roots borne on axes of the shoot system far removed from the original germination site. These form massive pillar roots which are able to support separate regions of the canopy. When the tree is old and its integrity begins to break up, these roots then support new independent trees which are clones (ramets) of the original tree. Roots of this species thus combine both the supporter and ingestor subsystems. In the early stages of growth another class of shoot-borne root that are apogravitropic, extends upwards penetrating the nutrient-rich detritus that collects in the leaf bases and branch angles of the support tree (Putz and Holbook, 1989).

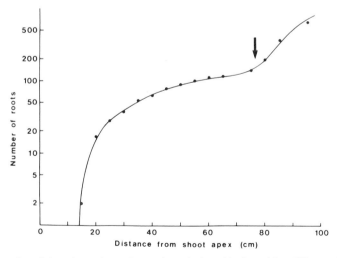

Figure 4. The number of shoot-borne internal roots (note the logarithmic scale) at different distances from the shoot apex, down a stem of a mature specimen of *Kingia australis*. The arrow indicates the distance at which the roots begin to branch. [Drawn from data kindly supplied by Dr I.A. Staff, which had also been published in Fig. 8.20 of Staff and Waterhouse (1981).]

In hemiepiphytic lianes, the two functional attributes of supporter and ingestor are clearly separated within a dimorphic root system (Barlow, 1986). For example, in *Philodendron melanochrysum* there are two types of root, one known as "clasping," the other as "feeding." Both types develop at the same nodes of the stem, but whereas clasping roots are thinner, more numerous and grow horizontally to girdle the stem of the support plant, the feeding roots are fatter, less numerous (one per node) and grow vertically downward and enter the soil. Feeding roots have a well developed vascular system but a poorly developed sclerenchyma (a supporter subsystem at the organ level), whereas the converse holds for clasping roots.

The dimorphism of the shoot-borne root system in *P. melanochrysum* may be a result of the internal environment of the shoot within which the root primordia are initiated— feeding roots arise below a leaf scar, whereas clasping roots arise to one side of this location. However, it cannot be ruled out that the different roots arise as a result of different genetic programs activated by different internal environments.

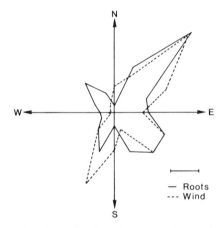

Figure 5. Frequency diagram to show the positioning of buttress roots around the base of the trunk of *Populus nigra* in relation to the frequency of wind direction. Scale bar represents a frequency of 5%. [drawn from data of Senn (1923)]

The above example illustrates dimorphism between separate roots, but dimorphism also exists within the same root, though at different stages of its development. For example, the pillar roots of *Ficus benghalensis* have different anatomies depending on whether they are in the aerial environment or in the soil. Further distinctions can be made between anatomies of various regions of the stilt roots of *Piper auritum* indicating trimorphism (Grieg and Mauseth, 1991). The three forms of the same root are a result of its varied response to different environments and/or root orientations: aerial, subterranean-vertical and subterranean-horizontal. The latter type of root is well supplied with phloem which may correlate with its ability to bear shoots. In this trimorphic root system, however, it is likely that the same genes are responsible for each of these three aspects of root development, though modifiers, genetic or environmental, are clearly responsible for the variations in anatomy.

Tabular Roots. Tabular, or buttress, roots are regular features of many tropical trees (Corner, 1978). Their peculiar morphology is the result of secondary thickening on the dorsal flank of roots which develop at the junction between the trunk and the tap root. There may also be a contribution from the cambium of the subtending trunk. Buttresses may, therefore, develop from roots of the "basal" category.

Buttresses have been noted on trees in windy sites and the frequency of their locations around the trunk has been correlated with the frequency of winds blowing from the various points of the compass. An example of this is given (Fig. 5) for *Populus nigra* studied by Senn (1923) in central Europe. (The buttresses of this tree are modest in size compared to those of tropical trees.) Because the buttresses were found predominantly on the windward side, they would appear to be traction structures tending to pull the tree back to the upright, rather than force-resistant structures that push it back. This correlation between buttress position and wind direction suggests that the latter somehow determines the former. Wind may not be the only factor responsible for buttress formation, however, because buttresses also develop in response to the distribution of mass within the canopy. All these observations suggest that the root buttress is a means of counteracting mechanical stresses sensed by the base of the trunk and that it forms as a result of some signal transmitted to the basal cambium of the roots. It may therefore be analogous to tension wood.

Distributor Subsystem

Air Roots. All roots initiate the internal distribution of aqueous nutrients to other regions of the plant, principally to the growing zones, though this may not be always obvious and may be only sporadic, viz. the internal roots mentioned above. The major nutrients usually considered are mineral ions, though equally important is air, the oxygen from which is required for respiration. This gas, as well CO_2 released from metabolism, require distribution into, out of and within the plant. One way this is accomplished, particularly in plants (helophytes) growing in anaerobic, waterlogged conditions, is via the intercellular spaces and the external lenticels of the roots. Sometimes, lenticels (which are rather disorderly clusters of cells) are replaced by more organized structures known as pneumatorhizae which are short, determinate lateral roots that have an opening to the atmosphere. Occasionally, air channels pervade the cortical tissue so extensively that the outer form of the root is conspicuously modified. This is observed in helophytes having a dimorphic root system comprised of water roots and air roots (Ellmore, 1981).

Air roots are part of the distributor subsystem and, in the example of *Ludwigia peploides* (Ellmore, 1981), are quite distinct from water roots. The former roots are short, grow upwards, and have an extensive cortical aerenchyma which is in continuity with the aerial environment above the water surface. The water roots, however, are long, grow downwards into the substratum, and have no aerenchyma. The same question arises about the genetical and environmental determinants of these dimorphic roots as was mentioned previously in the example of the clasping and feeding roots of *Philodendron melanochrysum*. As in this latter case, the two types of roots of *Ludwigia* arise at nodes at distinct locations in relation to the siting of the leaf bases. Moreover, in the related plant, *Jussiaea repens*, the type of root which emerges at a node can be influenced by application of different classes of plant growth regulators: (Samb and Kahlem, 1983) auxin and gibberellin promote water root formation, whereas cytokinin promotes the air root.

One of the best studied air-root systems is found in mangrove vegetation (Gill and Tomlinson, 1977). Shoot-borne roots of *Rhizophora mangle* form a succession of arches due to the remarkable ability of an aerial parent root axis to branch and produce a new root whose growth is oriented in the same direction. This possibly occurs as the result of the well known propensity for lateral roots to be induced on the convex surface of a curved root. Thus, successive branches arch out along radii from an initiating stem axis (Fig. 6A). Additional branch roots grow out in other directions, maybe because the most favored site, at the point of maximum convexity, is now occupied by the first branch root. From the initial branch point, the parent axis abruptly bends downwards and enters the aqueous environment. The root then develops an extensive intercellular air-space system which occupies about 50% of the root volume compared with a value of 6% for the aerial portion of root. The air spaces are in continuity for distances of at least 60 cm. In *Rhizophora* this aerated segment of root is concerned with gas exchange and is interposed between the aerial

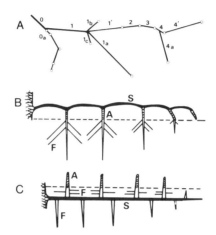

Figure 6. Forms of support roots and air roots of two mangrove species. A: plan view of the branching pattern of aerial support roots of *Rhizophora mangle*. 0, original aerial root axis. 1, 2, 3, 4 are the first branches from, respectively, the original axis, the first-order branch, the second-order branch and the third-order branch root. 0a, second branch from the original axis. 1a,b, c, secondary branches from the first-order branch; 4a, second branch from the third-order branch. 1' and 4' continue the first branch roots, but arise from the wound sites that affected the original growing tip. Open circles (o) indicate the insertion of a root into the substrate. B, C: diagrams showing the relationships between support roots (S), air roots (A) (the hatched portion being the region of gas exchange), and feeding and/or anchorage roots (F) in *Rhizophora* (B) and *Avicennea* (C). The level of the substate is indicated by the broken horizontal line. The support root is aerial in *Rhizophora* and submerged in *Avicennea*. [redrawn from Gill and Tomlinson (1977)]

supporting root and the more finely branched below-ground feeding and anchorage root system (Fig. 6B). In tidal parts of the mangrove swamp, air is forced by the rise and fall of the water level, in and out of this portion of the root via the lenticels. By contrast, in two other mangrove genera, *Avicennea* and *Sonneratia*, air roots grow upwards and break through the water surface having branched from a submerged horizontal root axis which also bears downward-growing feeding and anchorage roots (Fig. 6C). It is possible that the upward-growing air roots also have an effect in stabilizing the substratum of the swamp within which the trees grow.

Although in mangrove species the air roots are not formed in response to flooding but are an integral and normal part of the root system, this need not be generally true. For example, in *Rumex* spp., (Laan et al., 1989) aerenchymatous, shoot-borne roots are formed in response to waterlogging of the pole-borne root system. However, from this study it is not clear whether the new shoot-borne roots are intrinsically aerenchymatous or whether they become aerenchymatous only when they emerge into an already anaerobic environment. Different *Rumex* species showed different cellular arrangements and porosity of aerenchyma in the cortex which in turn correlated with the species" ability to withstand waterlogging. Similar intrinsic differences in aerenchyma volumes and forms have also been found in shoot-borne roots of various bog plants (Smirnoff and Crawford, 1983). These examples serve to illustrate the adaptive capacity of such roots with respect to the physiology of the whole plant as well as the diversity of the cellular responses within externally similar types of shoot-borne root.

Converter Subsystem

Photosynthetic Roots. The subsystem known as "converter" is exemplified at the plant organ level by the parenchymatous tissue of the leaf. Here, one form of energy (radiant light) is converted to another form (chemical) through the processes of photosynthesis and photorespiration. Although uncommon, some roots possess a parenchymatous cortex, also with photosynthetic capability through the possession of chloroplasts. In green, shoot-borne roots of *Bergia capensis*, for example, bubbles of oxygen can be seen issuing from cortical cells into an extensive aerenchyma (distributor subsystem) (D'Almeida, 1942). Clearly these roots are photosynthetic. However, nowhere is the photosynthetic capacity of roots more vividly demonstrated than in so-called shootless orchids such as *Taeniophyllum* and *Microcoelia*. Strictly speaking, these plants are not entirely shootless, but the shoot is certainly diminutive. The bulk of the plant consists of long, green, leaf-like roots attached to a minute stem. The constitution of the plant is probably $[S_1,R_1{}^+,{}_{S1}R_2]$ (Table 2).

Another curious group of plants in which the root has taken over from the shoot the latter's usual converter subsystem function, are members of the Podostemaceae, a family of plants found in swiftly running rivers. Here the root consists of a green thallus that grows firmly appressed to rocky surfaces. The thallus appears to be shoot-borne because it seems to arise at the base of the hypocotyl (i.e., it may be a basal root) soon after seed germination and takes the place of an underdeveloped (or non-existent, perhaps) pole-borne root. The status of the pole-borne shoot is unclear. Most, if not all, shoots form from buds on the root thallus. The constitution of such plants may thus be $[S_1,R_1-,{}_{S1}R_2,{}_{R2}S_2]$ (Table 2).

Storage Subsystem

Storage Roots. The modification of cylindrical shoot-borne roots into swollen tubers (usually they are basal roots arising from the hypocotyl) is well known and mentioned in most elementary botany texts in chapters dealing with energy stores. However, as Troll has reminded us in his monumental survey of plant forms (Weber, 1953) these storage forms take on various shapes, possibly as a result of different timings of cambial activity in the root. Swellings may occur at the base of the root, at the tip or even in intercalary positions. But the important point from the present perspective is that these root tubers are another means by which the plant removes the excess of photosynthetic carbon. The generally opportunistic life stratagem of the plant is to capitalize on this excess and optimize a storage subsystem which functions in conjunction with the reproducer subsystem. This is evident from the fact that root tubers (of *Dahlia* sp., *Ranunculus bulbosus*, for example) are all perennating organs with potent energy stores and with vigorous buds that emerge as root-borne shoots after a dormancy period. In *Raphanus sativus* the upper part of the tap (pole-borne) root normally swells to form a storage organ. If shoot cuttings are made, adventitious roots will form and develop as storage roots similar to the tap root (Ting and Wren, 1980). This indicates, at least in this species, that the root tuber phenotype might be genetically programmed and inherent to any root irrespective of whether it is pole-borne or shoot-borne and is not necessarily dependent on the response of an undifferentiated root to particular inductive conditions.

Motor Subsystem

Contractile Roots. It is often said in elementary texts that plants are sessile. But this is to forget the many bulbous plants (geophytes) which are capable of moving, sometimes as much as 30 cm during a year (Galil, 1980). Here, the roots which develop at the base of the bulb or corm are "contractile" and thus fulfil the role of "motor" subsystem. Root contraction is brought about by modification of cell growth in the inner cortex. In the case of *Hyacinthus orientalis*, microtubules below the plasma membrane reorient (Smith-Huerta and Jernstedt, 1989) and cause wall cellulose deposition also to be reoriented (Wilson and Honey, 1946); as a consequence, the cells shorten longitudinally and broaden radially. Many such cells acting in concert force the root to contract and, because the root is anchored at its tip, the bulb is pulled downwards into the soil. In *Oxalis hirta*, bulb movement and root contraction occur within a tube formed by the original root cortex with the endodermis lining the tube's inner surface (Davey, 1946). Contraction occurs in the vascular parenchyma accompanied by cellular collapse. Bulb movement may also be effected by the extension of the petiole, forcing the bulb down the cortical tube.

Extruder Subsystem

Ephemeral Roots. It has already been commented that roots are not necessarily long-lived organs and might even have quite short life-spans (see the chapter by Dickmann and Hendrick in this volume). Although no particular specialized morphological features draw attention to the ephemeral and fugaceous nature of such roots (and thus, through this feature, indicate their role as extruder subsystem), a fast root turnover rate may be enough to indicate that waste extrusion is their true significance. The annual shedding of shoot-borne roots of *Eriophorum angustifolium*, has already been mentioned (Shaver and Billings, 1975); the same is true of much of the fine root system of trees, though these roots are not necessarily on the shoot-borne system. In the water fern, *Azolla pinnata*, the shoot-borne roots are even more short lived; they cease growing after reaching a length of five cm and are shed from the shoot (frond) by means of an abscission zone after only a few days of growth. During their life they probably have a minimal impact on the nutrition of the plant through mineral capture. Essentially, they represent a carbonaceous waste product. Shedding of roots may have the additional advantage of eliminating cells or tissues that have accumulated mutations, such as unbalanced chromosome complements, due to stressful environmental conditions. New roots derived from dormant cells thus replace potentially defective roots. New roots are also again available to explore locations near the shoot into which minerals have been released from recently shed leaves and other deposits in the litter layer.

On a much larger scale, an extruder function may also be true of roots of the tree, *Metrosideros excelsa*. Here the roots never reach the ground and simply hang as festoons around the trunk earning them the name "broom roots"; whether or not they abscind is an open question. However, their red coloration might suggest that the root interior requires protection from harmful effects of sunlight and, hence, may signify that they fulfill some other subsystem function (metabolic conversions?).

PHYLOGENETIC ASPECTS OF SHOOT-BORNE ROOT DIVERSITY

Some attention was given earlier to the phylogeny of shoot-borne roots within the Gramineae, but this was largely from a theoretical point of view. It is of general interest, however, to discover whether any of the diverse types of shoot-borne roots relate to roles that they might have had in past evolutionary epochs. For example, certain types of root might have evolved relatively recently in response to changes in an environment that now supports the modern-day flora. In such an event, therefore, these types should be found in more advanced plant families and be absent from older families. Conversely, certain root types might be exhibited only by older families (evolutionarily speaking) and be absent in more recently evolved families.

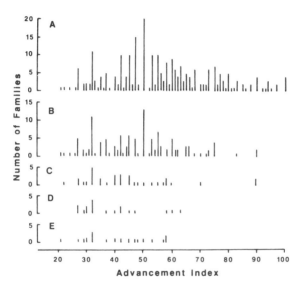

Figure 7. Relationship between the advancement index, and the presence of buttress roots, stilt roots and air roots in tree families represented in localities of Malaysia studied by Corner (1978). Frequencies in: A —all families; B —tree families. Frequencies in: C —tree families with buttress roots; D —tree families with stilt roots; E —tree families with air roots. [data in A and B from Sporne (1974); in C-E from Corner (1978)]

Sporne (1974) has developed an "advancement index" for families of dicotyledonous angiosperms to indicate their relative primitiveness or advancement. The index is calculated according to the presence or absence of certain anatomical and morphological characters deemed to be evidence of their evolutionary status. Accordingly, it can be used to judge whether the possession of shoot-borne roots of various types is an advanced or primitive character. In view of the presumed origin of shoot-borne roots early in plant evolution, it might be expected that they would be present in families with a low advancement index; they might also be retained in families with a higher index. To make a complete survey of the different types of shoot-borne roots in the dicotyledonous families would be time-consuming, but fortunately Corner (1978) has made a listing of the families of trees living in the swamp forests of Johore which bear shoot-borne roots and has further classified whether they are buttress, stilt or air roots.

The results of relating these root types to the advancement index of the families that possess them is shown in Fig. 7. The figure shows that the tree families surveyed tend to occupy the less advanced part of the range of advancement indices (Fig. 7C), but this might be a feature of trees in general (Fig. 7B), or of those tree families represented in this particular geographical region. Unfortunately, from Corner's data base (Corner, 1978) it is not possible to assess the range of advancement indices in the overall population which he studied, for the reason that he listed only tree species which exhibited shoot-borne roots (though he did list species that lacked buttress roots). However, assuming that these species constitute a sizeable fraction of the total number of tree species, then, as might be expected, all three types of root are associated with families whose advancement indices span the range associated with the tree families. Thus, for trees in this habitat at least, each type of root appears to be of ancient origin and retained into the present-day epoch.

CONCLUSION

The importance of shoot-borne roots cannot be over-emphasized. In some groups of plants (e.g., ferns and palms) they are the only members of the root system. Similarly, within important cereal groups they are the major type of roots and have probably been subject to unconscious selection from the earliest periods of domestication—for example, primitive maize possessed a pole-borne system in addition to the shoot-borne system, but in modern maize varieties the pole-borne system has largely been lost. Some major weed species also owe their success to the shoot-borne system. In Chile, for example, the rapid spread of bramble (*Rubus ulmifolius*) following its introduction from Europe and usage as a hedging plant, is mostly due to vigorous shoot-borne roots at the tips of exploratory stolons (Heslop-Harrison, 1959). Likewise, adventitious roots, formed on pre-existing roots after wounding, can make major contributions to natural root systems (Horsley, 1971); it has been estimated (Henderson et al., 1983), for example, that in *Picea sitchensis* about half of the major second-order lateral roots arise as a result of injury. Adventitious roots arising on pre-existing shoots are also the basis of the commercial exploitation of natural species for ornamental and other utilitarian purposes.

Shoot-borne roots present a diversity of forms which enable species to exist in particular environments (e.g., waterlogged swamps and aerial canopies of trees) that could not be entered if they were absent. Besides their wealth of forms, these roots also exhibit a diversity of relationships with the pole-borne roots, and with the shoot system. These interrelationships may in some manner determine the life-form and reproductive stratagem of a given species. It should now be possible to find some basis for an understanding of the significance of their diversity. The aim of this article has been to approach this diversity of form through the context of systems analysis in a way that might assist future research into the shoot-borne system.

REFERENCES

Aeschimann D., and Bocquet, G.,1980, Allorhizie et homorhizie, une reconsidération des définitions et de la terminologie, *Candollea* 35:19.

Barlow, P.W., 1986, Adventitious roots of whole plants: their forms, functions, and evolution, *in*: "New Root Formation in Plants and Cuttings," M.B. Jackson, ed., Martinus Nijhoff, Dordrecht.

Barlow, P.W., 1989, Meristems, metamers and modules and the development of shoot and root systems, *Bot. J. Linn. Soc.* 100:255.

Barlow, P.W., 1987, The hierarchical organization of plants and the transfer of information during their development, *Postepy Biol. Kom.* 14:63.

Blakely, L.M., Durham, M., Evans, T.A., and Blakely, R.M., 1982, Experimental studies on lateral root formation in radish seedling roots, I. General methods, developmental stages, an spontaneous formation of laterals, *Bot. Gaz.* 143:341.

Bonnett, H.T., Jr., and Torrey, J.G., 1966, Comparative anatomy of endogenous buds and lateral formation in *Convolvulus arvensis* roots cultured *in vitro*, *Amer. J. Bot.* 53:496.

Brown W.V., and Emery, W.P.H., 1957, Persistent nucleoli and grass systematics, *Amer. J. Bot.* 44:585.

Caldwell, M.M., 1979, Root structure: the considerable cost of belowground function, *in*: "Topics in

Plant Population Biology," O.T. Solbrig, S. Jain, G.B. Johnson, and P.H. Raven, eds, MacMillan Co., London.

Corner, E.J.H., 1978, Freshwater swamp-forest of South Johore and Singapore, *Gardens' Bull.* (suppl.) 1:1.

D'Almeida, J.F.R., 1942, A contribution to the study of the biology and physiology of Indian marsh and aquatic plants, Part II, *J. Bombay Nat. Hist. Soc.* 43:92.

Davey, A.J., 1946, On the seedling of *Oxalis hirta* L., *Ann. Bot.* 10:237.

Ellmore, G.E., 1981, Root dimorphism in *Ludwigia peploides* (Onagraceae): structure and gas content of mature roots, *Amer. J. Bot.* 68:557.

Esau, K., 1953, "Plant Anatomy," J. Wiley, New York.

Frangi, J.L., and Ponce, M.M., 1985, The root system of *Prestoes montana* and its ecological significance, *Principes* 29:13.

Galil, J., 1980, Kinetics of bulbous plants, *Endeavour* 5:15.

Gill, A.M., 1969, The ecology of an elfin forest in Puerto Rico, 6. Aërial roots. *J. Arn.. Arb.* 50:197.

Gill, A.M., and Tomlinson, P.B., 1977, Studies on the growth of red mangrove (*Rhizophora mangle* L.), 4. The adult root system, *Biotropica* 9:145.

Goller, W., 1977, Beiträge zur Anatomie adulter Gramineenwurzeln im Hinblick auf taxonomische Verwendbarkeit, *Beitr. Biol. Pfl.* 53:217.

Greig, N., and Mauseth, J.D.,1991, Structure and function of dimorphic prop roots in *Piper auritum* L., *Bull. Torr. Bot. Club* 118:176.

Groff, P.A., and Kaplan, D.R., 1988, The relation of root systems to shoot systems in vascular plants, *Bot. Rev.* 54:387.

Guédès, M., 1979, "Morphology of Seed Plants," J. Cramer, Vaduz.

Harper, J.L., Jones, M., and Sackville Hamilton, N.R., 1991, The evolution of roots and the problems of analysing their behaviour, *in*: "Plant Root Growth, An Ecological Perspective," D. Atkinson, ed., Blackwell Sci. Pubs., Oxford.

Henderson, R., Ford, E.D., Renshaw, E., and Deans, J.D., 1983, Morphology of the structural root system of Sitka spruce, 1. Analysis and quantitative description, *Forestry* 56:121.

Heslop-Harrison, Y., 1959, Natural and induced rooting of the stem apex of *Rubus*, *Ann. Bot.* 23:307.

Horsley, S.B., 1971, Root tip injury and development of the paper birch root system, *For. Sci.* 17:341.

Hoshikawa, K., 1969, Underground organs of the seedlings and the systematics of Gramineae, *Bot. Gaz.* 130:192.

Jablonka, E., Lachmann, M., and Lamb, M.J., 1992, Evidence, mechanisms and models for the inheritance of acquired characters, *J. Theor. Biol.* 158:245.

Jacob, F., 1977, Evolution and tinkering, *Science* 196:1161.

Laan, P., Berrevoets, M.J., Lythe, S., Armstrong, W., and Blom, C.W.P.M., 1989, Root morphology and aerenchyma formation as indicators of the flood-tolerance of *Rumex* species, *J. Ecol.* 77:693.

Lambers, H., 1980, The physiological significance of cyanide-resistant respiration in higher plants, *Plant Cell Environ.* 3:293.

Lamont, B., 1981a, Availability of water and inorganic nutrients in the persistent leaf bases of the grasstree *Kingia australis*, and uptake and translocation of labelled phosphate by the embedded aerial roots, *Physiol. Plant.* 52:181.

Lamont, B., 1981b, Morphometrics of the aerial roots of *Kingia australis* (Liliales), *Aust. J. Bot.* 29:81.

Lavender, E.A., Mackie-Dawson, L.A., and Atkinson, D., 1992, Genotypic variation in the development of structural roots in *Betula pendula*, *J. Exp. Bot.* 43 (suppl.):41.

Liu, M., Li, R.-J., and Liu, M.-Y., 1993, Adaptive responses of roots and root systems to seasonal changes, *Envir. Exp. Bot.* 33:175.

Lück J., and Lück, H.B., 1993, La notion de "distance morphogénétique," *in*: "Modèles et Transformations, La Biologie Théorique et P. Delattre," C. Bruter, ed., Polytechnica, Paris.

Lüttge, U., 1991, *Clusia*. Morphogenetische, physiologische und biochemische Strategien von Baumwürgen im tropischen Wald, *Naturwiss.* 78:49.

Luxová, M., 1986, The hydraulic safety zone at the base of barley roots, *Planta* 169:465.

McLean, R.C., and Ivimey-Cook, W.R., 1951, "Textbook of Theoretical Botany," vol. 1, Longmans Green and Co., London.

Miller, J.G., 1978, "Living Systems," McGraw Hill, New York.

Miller, J.G., and Miller, J.L., 1990, The nature of living systems, *Behav. Sci.* 35:157.

Nadkarni, N.M., 1981, Canopy roots: convergent evolution in rainforest nutrient cycles, *Science* 214:1023.

O'Brien, T.P., and Zee, S.-Y., 1971, Vascular transfer cells in the vegetative nodes of wheat, *Aust. J. Biol. Sci.* 24:207.

Piaget, J., 1971, "Structuralism," Routledge and Kegan Paul, London.

Putz, F.E., and Holbrook, N.M., 1989, Strangler fig rooting habits and nutrient relations in the llanos of Venezuela, *Amer. J. Bot.* 76:781.

Raechal, L.J., and Curtis, J.D., 1990, Root anatomy of the Bambusoideae (Poaceae), *Amer. J. Bot.* 77:475.

Rayner, R.J., 1983, New observations on *Sawdonia ornata* from Scotland, *Trans. Roy. Soc. Edin. Earth Sci.* 74:79.

Rayner, R.J., 1984, New finds of *Drepanophycus spinaeformis* Göppert from the Lower Devonian of Scotland, *Trans. Roy. Soc. Edin.Earth Sci.* 75:353.

Robinson, D., 1989, Phenotypic plasticity in roots and root systems: constraints, compensations and compromises, *in:* "Aspects of Applied Biology 22, Roots and the Soil Environment," Assoc. Appl. Biol., Wellesbourne.

Rouffa, A.S., 1978, On phenotypic expression, morphogenetic pattern, and synangium evolution in *Psilotum. Amer. J. Bot.* 65:692.

Row H.C., and Reeder, J.R., 1957, Root-hair development as evidence of relationships among genera of Gramineae, *Amer. J. Bot.* 44:596.

Salleo, S., Rosso, R., and Lo Gullo, M.A., 1982, Hydraulic architecture of *Vitis vinifera* L. and *Populus deltoides* Bartr. 1-year-old twigs: II—The nodal regions as "constriction zones" of the xylem system, *Giorn. Bot. Ital.* 116:29.

Samb, P.I., and Kahlem, G., 1983, Déterminism de l'organogénèse racinaire de *Jussiaea repens* L., *Z. Pflanzenphysiol.* 109:279.

Schatz, G.E., Williamson, G.B., Cogswell, C.M., and Stam, A.C., 1985, Stilt roots and growth of arboreal palms, *Biotropica* 17:206.

Scott, D.H., 1962, "Spermatophyta," Studies in Fossil Botany, vol. II, 3rd. ed., Hafner Pub. Co., New York.

Senn, G., 1923, Ueber die Ursachen der Brettwurzelbildung bei der Pyramiden-Pappel, *Verh. Naturforsch. Ges. Basel* 35:405.

Shaver, G.R., and Billings, W.D., 1975, Root production and root turnover in a wet tundra ecosystem, Barrow, Alaska, *Ecology* 56:401.

Smirnoff, N., and Crawford, R.M.M, 1983, Variation in the structure and response to flooding of root aerenchyma in some wetland plants, *Ann. Bot.* 51:237.

Smith-Huerta, N.L., and Jernstedt, J.A., 1989, Root contraction in hyacinth, III. Orientation of cortical microtubules visualized by immunofluorescence microscopy, *Protoplasma* 151:1.

Sobotik, M., and Kutschera-Mitter, L. 1991, Contribution to a key for determining taxa by anatomical features of roots, *in:* "Root Ecology and its Practical Application 2," L. Kutschera, E. Hübl, E. Lichtenegger, H. Persson, and M. Sobotik, eds, Verein für Wurzelforschung, Klagenfurt.

Sonnewald V., and Willmitzer, L., 1992, Molecular approaches to source-sink interactions, *Plant Physiol.* 99:1267.

Sporne, K.R., 1974, "The Morphology of the Angiosperms, The Structure and Evolution of Flowering Plants," Hutchinson.

Staff, I.A., and Waterhouse, J.T., 1981, The biology of arborescent monocotyledons, with special reference to Australian species, *in:* "The Biology of Australian Plants," J.S. Pate, and A.J. McComb, eds, Univ. West. Aust. Press, Nedlands.

Steinitz, B., Hagiladi, A., and Anav, D., 1992, Thigmomorphogenesis and its interaction with gravity in climbing plants of *Epipremum aureum, J. Plant Physiol.* 140:571.

Stewart, W.N., 1983, "Paleobotany and the Evolution of Plants," Cambridge Univ. Press, Cambridge.

Tillich, H.-J., 1977, Vergleichend-morphologische Untersuchungen zur Identität der Gramineen-Primärwurzel, *Flora* 166:415.

Ting, F.S., and Wren, M.J., 1980, Storage organ development in radish (*Raphanus sativus* L.), 1. A comparison of development in seedlings and rooted cuttings of two contrasting varieties, *Ann. Bot.* 46:267.

Ugolev, A.M., 1990, Concept of universal functional blocks and further development of studies on the biosphere, ecosystems, and biological adaptations, *J. Evol. Biochem. Physiol.* 26:331.

Waddington, C.H., 1957, "The Strategy of the Genes, A Discussion of Some Aspects of Theoretical Biology," G. Allen and Unwin, London.

Wallace, A.R., 1853, "Palm Trees of the Amazon and their Uses," van Voorst, London.

Wardlaw, I.F., 1965, The velocity and pattern of assimilate translocation in wheat plants during grain development, *Aust. J. Biol. Sci.* 18:269.

Warren Wilson, P.M., and Warren Wilson, J., 1977, Experiments on the rate of development of adventitious roots on *Sambucus nigra* cuttings, *Aust. J. Bot.* 25:367.

Weber, H., 1936, Vergleichend-morphologische Studien über die sproßbürtige Bewurzelung, *Nova Acta Leopold* NF 4:229.

Weber, H., 1953, "Die Bewurzelungsverhältnisse der Pflanzen," Verlag Herder, Freiburg.

Whipps, J.M., 1990, Carbon economy, *in*: "The Rhizosphere," J.M. Lynch, ed., J. Wiley & Sons, Chichester.

Wilson, K., and Honey, J.N., 1966, Root contraction in *Hyacinthus orientalis*, *Ann. Bot.* 30:47.

Zobel, R.W., 1986, Rhizogenetics (root genetics) of vegetable crops, *HortSci.* 21:956.

Zobel, R., 1989, Steady-state control and investigation of root system morphology, *in*: "Applications of Continuous and Steady-state Methods in Root Biology," J.G. Torrey, and L.J. Winship, eds, Kluwer Academic Pubs., Dordrecht.

COMMERCIAL IMPORTANCE OF

ADVENTITIOUS ROOTING TO AGRONOMY

John L. Kovar[1] and Rolf O. Kuchenbuch[2]

[1]Department of Agronomy
Louisiana State University
Baton Rouge, LA 70803-2110 USA

[2]Institut für Gemüse- und Zierpflanzenbau
Großbeeren/Erfurt e.V.
O-1722 Großbeeren, FRG

INTRODUCTION

Agronomic crops are those grown on a large scale, either for consumption by humans or livestock, or for production of raw materials. Those crops in which adventitious root formation is important can be divided into two broad categories—those that are vegetatively propagated and those for which final yield, whether fruit or dry matter, is influenced by the presence of adventitious roots. The number of commercially important species in the first category is small, with sugarcane (*Saccharum* spp. hybrid) and hybrid varieties of bermudagrass (*Cynodon dactylon* L. Pers.) being the most significant. The second category contains many species that play an important role, both directly and indirectly, in world food supply. This group includes cereal grains such as maize (*Zea mays* L.), rice (*Oryza sativa* L.), and wheat (*Triticum aestivum* L.), as well as forage species, such as perennial ryegrass (*Lolium perenne* L.) and red clover (*Trifolium pratense* L.). Agronomic species in which adventitious root formation is most important are, in general, members of the grass (*Gramineae*) family.

After a review of recent literature, it quickly becomes apparent that the formation of adventitious roots is the cornerstone of horticultural crop production (see the chapter by Davies and coworkers in this volume), due to the importance of adventitious rooting in vegetative propagation of desirable plants. Nearly all key agronomic crops, however, are propagated from seeds rather than cuttings, so that adventitious roots are a part of the functional root system. In light of this, a significant amount of research addressing root growth and function in agronomic species focuses on the root system as a whole, rather than on specific types of roots. Many times, it is assumed that the physiological functions of all root types are basically the same, in spite of differing origins. A variety of studies, however, have shown that this may not be the case.

Biology of Adventitious Root Formation, Edited by
T.D. Davis and B.E. Haissig, Plenum Press, New York, 1994

The purpose of this chapter will be to present a general overview of the role adventitious roots play in the production of agronomic crops. Our approach will first be to describe the growth and development of adventitious roots in a variety of species of agronomic importance, then to examine the ability of adventitious roots to sustain or enhance crop growth and vigor, and finally, to discuss adventitious rooting as an environmental response. It is not our intent to provide an exhaustive summary of the current state of knowledge regarding adventitious root formation in every species of agronomic importance. Turfgrass production, although commercially important, is not discussed in this chapter. In a strict sense, turfgrasses can be considered an ornamental, rather than an agronomic species.

ADVENTITIOUS ROOT GROWTH AND DEVELOPMENT

Adventitious root growth and formation are significantly different in monocotyledonous and dicotyledonous species. In monocotyledons, the root system consists of embryonic roots, which include the primary root and seminal roots that develop from embryonic nodes, and adventitious roots—often referred to as nodal or crown roots—that develop from primordia associated with lower stem nodes (Hoshikawa, 1969; Klepper et al., 1984). Lower stem nodes are known collectively as the "seedling crown" (Hayward, 1938). Depending on the species, both embryonic and adventitious roots can remain active for long periods (Klepper et al., 1984). In dicotyledons, the root system develops from the radicle, which elongates to form a taproot. Lateral roots then emerge in acropetal sequence at a relatively constant distance behind the growing tip (Charlton, 1991). The taproot and its laterals are considered seminal roots, whereas roots that develop either from true stem nodes or nodes of rhizomes and stolons are considered adventitious roots [Mitchell (1970); see the chapter by Barlow in this volume]. Due to these inherent differences, as well as those related to cultural practices, the relative importance of adventitious roots in the function of the entire root system is species-dependent. To more clearly illustrate these differences, a few examples will be given in the following discussion. Row-cropped species will be addressed first, followed by forage species.

Unless sugarcane, a monocotyledon, has been propagated from true seed, the plant has no primary roots, only adventitious roots. Pieces of stem tissue that have one or more vegetative buds are the planting material used by commercial growers. Root primordia are located at the base of each internode of the stem pieces, known as setts (Fig. 1). Under favorable conditions, sett roots and axillary buds begin growth about three days after planting (Clements, 1980). However, the shoot that arises soon develops its own root system (Fig. 1), which becomes part of the permanent root system. Sett roots, therefore, are not crucial to the survival of the shoot and disappear within two months of planting (Fageria et al., 1991). The appearance of roots from the shoot depends on sett placement at planting. If the sett is placed so that the most terminal bud is not nearest the soil surface, the resulting curved growth of the shoot hastens root development. This response likely is related to the downward translocation of auxin from the shoot meristem (Clements, 1980). The development of the shoot-root system occurs in two stages. The initial roots that develop from the shoot are thicker than sett roots and their growth rate is greater. These roots generally have few lateral branches and the angle of soil penetration is steep. Subsequent shoot roots that develop are finer and branch profusely. Glover (1967) reported that the average growth rate of these roots is 40 mm per day.

Maize, rice, wheat and other cereal grains also are monocotyledons, but unlike sugarcane, these species are annuals that are propagated from seeds. In maize, the radicle develops into the first seminal root, after which other seminal roots emerge from the embryo (Fig. 2). In the first four to six weeks after germination, the seminal roots support the shoot (Purseglove, 1985). After this time the importance of the seminal roots declines, and the seedling is supported by the developing nodal roots. These adventitious roots become the permanent root system of the plant (Purseglove, 1985). During the first three to four weeks of development, nodal roots may grow at a rate of five to six cm per day (Leonard and Martin, 1967).

The fibrous root system of rice develops in a similar fashion to that of maize (see the chapter by Morita and Abe in this volume.) After germination, the primary root emerges, followed by two or more seminal roots (De Datta, 1981). Adventitious roots then form from the basal nodes of the primary stem and tillers (Purseglove, 1985). After approximately 10 days, adventitious roots that eventually form the permanent root system rapidly replace the initial seminal roots (De Datta, 1981). Root development and shoot development of rice are highly correlated. Fageria et al. (1991) state that when leaf "n" is developing, roots emerge at node "n-3" of the same stem. The root system of upland rice tends to be more vigorous than that of lowland (flooded) rice. Adair et al. (1962) found that roots which developed at lower nodes of lowland rice spread laterally, but did not penetrate more than 15 cm into the soil.

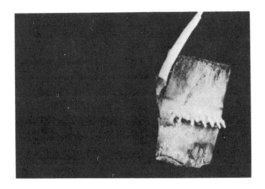

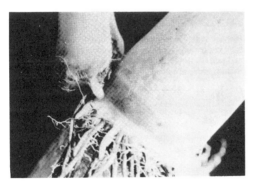

Figure 1. Adventitious root formation in sugarcane setts. Root development at germination (top). Close-up view of a node with a shoot and roots (bottom). [reproduced from Clements (1980) with permission]

Root system development in wheat, as well as other cereal grains such as barley (*Hordeum vulgare* L.), oats (*Avena sativa* L.) and rye (*Secale cereale* L.) is similar to that of maize and rice, since these species also are Gramineae (Fig. 3). After germination, the radicle, followed by two seminal roots, emerges and begins to elongate. Following emergence of the epicotyl, two to four additional seminal roots develop from another embryonic node known as the epiblast node (Cook and Veseth, 1991). Elongation of these seminal roots is governed by growing degree-days (phyllochrons). Nodal roots begin to develop from the bases of the second through sixth internodes at about the time of tiller

27

initiation. Development continues until upper nodes break the soil surface (Klepper et al., 1984). The precise number of nodes that form the crown depends on species, planting date and environmental conditions (Klepper et al., 1984). Adventitious or crown roots also develop on tillers when the three-leaf stage is reached (Cook and Veseth, 1991).

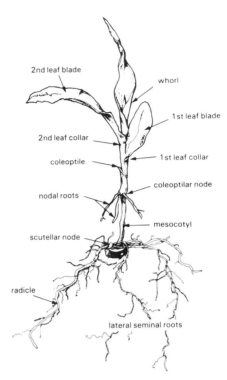

Figure 2. A maize seedling at the two-leaf stage, depicting the early development of seminal and nodal roots. [reproduced from Stevens et al. (1986) with permission]

The principle cultivated forages include both monocotyledonous and dicotyledonous species. Growth habits vary from annual to biennial to perennial. For the most part, the monocotyledons are members of Gramineae and the dicotyledons are members of the *Leguminosae*. The majority of forage species are propagated from seed. Bermudagrasses, especially hybrid cultivars, are a commercially important exception. More bermudagrass, which occupies more than 10 million hectares worldwide, is propagated by planting vegetative "sprigs" than by seeding (Burton and Hanna, 1985).

Adventitious roots develop in the forage grasses in a manner similar to that in other Gramineae. However, some perennial species, such as reed canarygrass (*Phalaris arundinacea* L.) and switchgrass (*Panicum virgatum* L.), produce horizontal underground stems, called rhizomes, from which adventitious roots develop (Metcalfe and Nelson, 1985). Other perennial grasses, such as buffalograss (*Buchloë dactyloides* Nutt.), produce creeping, aboveground stems, called stolons, from which adventitious roots also develop. Bermuda-grasses produce both rhizomes and stolons. Rhizomes and stolons have definite nodes, internodes and axillary buds (Metcalfe and Nelson, 1985). Adventitious roots develop at the nodes.

In most forage legume species, the primary root, from which the taproot and its lateral branches develop, supports the plant for its entire life. However, a few perennial species, such as white clover (*Trifolium repens* L.), produce stolons from which adventitious roots develop (Fig. 4). This species is taprooted as a seedling. The taproot, however, usually becomes diseased and dies within one or two years after planting (Westbrooks and Tesar, 1955). Prior to the death of the taproot, stolons are produced. The survival of this legume species thus becomes dependent on the development of adventitious roots.

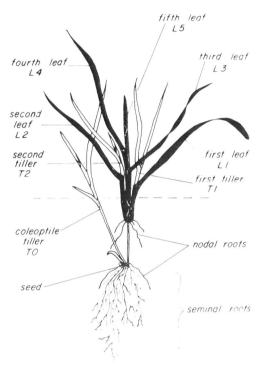

Figure 3. Diagram of a wheat seedling showing leaves, tillers and roots. [reproduced from Klepper et al. (1983) with permission]

ROLE OF ADVENTITIOUS ROOTS IN SUSTAINING CROP GROWTH

The differing roles of adventitious and seminal roots in sustaining or enhancing crop growth and vigor have not been much addressed in the literature. The weight, length or active surface area of the entire root system most often has been examined. Alternatively, research has focused on some physiological function of a single excised root. Part of the problem is methodological in that it is difficult to physically separate roots of each type without disrupting the integrity of the plant (see the chapter by Dickmann and Hendrick in this volume.) However, some research has been conducted on water and nutrient absorption by individual root types of plants grown under controlled conditions. In addition, several other studies have examined the effect of adventitious rooting on agronomic characteristics—e.g., lodging resistance, tiller formation and grain yield—of field-grown crops.

The absorption of water by the primary and adventitious roots of young maize plants was investigated by Navara (1987). Maize seedlings were grown for 30 days in lysimeters under controlled environmental conditions. Navara (1987) found that the primary roots supplied significantly more water to the shoot until the onset of growth of nodal roots at approximately 18 days after planting. For the entire 30-day period, nodal roots supplied

51.1% of the total water absorbed. However, when water uptake was calculated on a root dry-weight basis, water absorbed per gram of seminal root was two-fold that per gram of nodal root. Since seminal roots remain functional during the entire vegetative period of the maize plant (Weaver and Zink, 1945), this part of the root system may be important to the plant during periods of drought.

Figure 4. Diagram of a white clover stolon with developing adventitious roots. Stoloniferous branches emerge from axillary buds at nodes. [reproduced from Metcalfe and Nelson (1985) with permission]

The absorption of nutrients by seminal and adventitious roots also has been investigated. Russell and Sanderson (1967) investigated the uptake of phosphorus (P) by short segments of intact barley roots grown in solution culture. The activity of seminal axes, nodal axes and laterals was examined. Results indicated that lateral roots—both seminal and nodal—were responsible for 55% of the total P uptake of the 21- to 25-day-old barley seedlings. Nodal and seminal axes were responsible for 30% and 15%, respectively. In addition, Russell and Sanderson (1967) found that the P uptake rate per unit of root surface area was significantly greater for the nodal axes, compared with the seminal axes. This suggests that the nodal roots were somewhat more efficient than the seminals.

Similar results were obtained by Kuhlmann and Barraclough (1987) in a study that compared nitrogen (N) and potassium (K) uptake by the seminal and nodal roots of wheat. Plants in the tillering phase (36 days after planting) were grown in solution culture under greenhouse conditions for 20 days. Plants were harvested seven times during the period to compare N and K uptake over time. Kuhlmann and Barraclough (1987) found that seminal root growth over the 20-day period was linear, while nodal root growth was exponential. No change in relative N and K shoot composition occurred. When influx (uptake per unit root length) and specific absorption (uptake per unit root fresh weight) were calculated, differences in seminal and nodal roots were observed (Fig. 5). On the basis of influx, total uptake of N and K by nodal roots was approximately five-fold that of seminal roots at the beginning of the experiment and two-fold at the end. However, only small differences in specific absorption rates were found between root types (Fig. 5). Therefore, N and K uptake efficiencies of seminal and nodal roots were not significantly different. Uptake per unit root surface area was not calculated.

The presence of both the seminal and adventitious root systems appears necessary for maximum nutrient absorption and yield of wheat. Boatwright and Ferguson (1967) developed a technique to grow wheat with either a complete root system, only primary roots or only adventitious roots. The researchers found that, with adequate soil water and available P, grain yield was significantly decreased when either the seminal or adventitious roots were removed. When only seminal roots were present, grain yield was 57% of the maximum. When only adventitious roots were present, grain yield was 77% of the maximum. In addition, Boatwright and Ferguson (1967) found that plants with only adventitious roots translocated more P into the grain than plants with only primary roots. However, P translocation into the grain was significantly higher when both roots systems were present. Similarly, Kuhlmann and Barraclough (1987) found that N and K uptake were greatest when both adventitious and seminal roots were present. It can be argued that nutrient absorption and translocation by other Gramineae would be comparable, if both root types remain active during the growing season. However, it has been shown that not only species differences, but also cultivar differences can significantly affect adventitious root growth and nutrient uptake (Hockett, 1986).

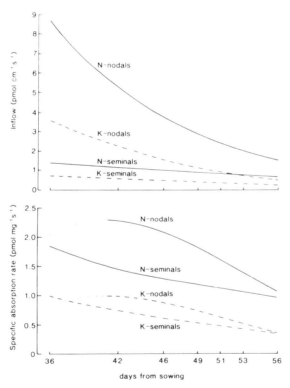

Figure 5. Influx (top) and specific absorption rate (bottom) for N and K by seminal and nodal roots of wheat. [reproduced from Kuhlmann and Barraclough (1987) with permission]

The role of adventitious roots in the resistance of stems to lodging is relevant to all Gramineae, yet for the most part has been ignored by researchers in recent years. Since secondary thickening of roots does not occur in monocotyledonous species, long fine seminal roots are not able to anchor the plant sufficiently. Thicker, slow-growing adventitious roots that develop from the stem nodes near or above the soil surface are required in order to transmit rotational forces into the soil (Plinthus, 1973). The "prop" or "brace" roots of maize are an obvious example of this. Recent research by Ennos (1991) addressed this characteristic in wheat. Results of a study examining the mechanics of anchorage in wheat plants indicated that lodging resistance would be improved either by longer, thicker coronal roots or an increase in the spreading angle of these roots away from the vertical.

With respect to dicotyledonous species, an area in which the role of adventitious rooting has been studied is that of forage persistence. Lack of persistence is a major limitation to the production of many forage legume species (Carlson et al., 1985). A number of studies have shown that the decline in persistence is related to the deterioration and eventual death of the taproot (Westbrooks and Tesar, 1955; Cressman, 1967). As stated previously, white clover survives by producing stolons (Metcalfe and Nelson, 1985). Therefore, most roots of this species present in established pastures are adventitious in origin. Recent research with red clover (Montpetit and Coulman, 1991) indicated that the persistence of this species, as expressed by spring vigor, is also related to the presence of adventitious roots. In that field study, the researchers found that only 5.3% of the plants sampled in the fourth year of the investigation had no adventitious roots. Monpetit and Coulman (1991) concluded that the presence of adventitious roots was the cause rather than consequence of superior spring vigor of red clover.

ADVENTITIOUS ROOTING AS AN ENVIRONMENTAL RESPONSE

The effect of the soil environment on growth and development of roots in agronomic species has been studied extensively. However, few investigations have specifically addressed the development of adventitious roots as a response to some environmental stress.

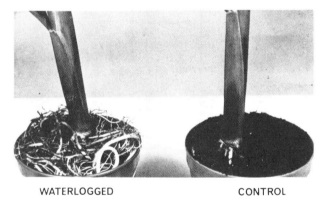

WATERLOGGED   CONTROL

Figure 6. Adventitious rooting of maize promoted by nine days of waterlogged conditions. [reproduced from Cannell and Jackson (1981) with permission]

The ability of certain agronomic species to tolerate anoxia (waterlogging) by the rapid production of adventitious roots is a response that has been studied (Kramer, 1951; Drew and Lynch, 1980; Hook, 1984). Adventitious roots develop on the stem, above the soil surface in some cases, under flooded or waterlogged conditions (Fig. 6). This response has been observed in a variety of agronomic species, including rice, maize and sunflower (*Helianthus annuus* L.). It is thought that these adventitious roots allow the plant to tolerate flooding or to recover more quickly and completely from flooding (Hook, 1984). The new roots, probably outgrowths of preformed primordia, are able to replace some of the functions of the original roots (Jackson, 1955). These roots adopt a horizontal rather than vertical orientation, so that a position near the soil-air interface is maintained (Jackson, 1955). The mechanism controlling this response is not well understood. In addition, Wample and Reid (1975) could find no apparent contribution of newly formed adventitious roots to the survival of sunflower plants.

The response of winter wheat to waterlogging at different soil temperatures was investigated by Trought and Drew (1982). As expected, growth of seminal roots was severely restricted by waterlogging at soil temperatures of 6°, 10°, 14° and 18° C. Nodal roots developed from the base of the shoots in response to waterlogging, however at all temperatures, root length and dry weight were less than in controls. Differences increased with increasing temperature, leading Trought and Drew (1982) to conclude that the greater tolerance of wheat to waterlogging at lower soil temperatures was due simply to a decrease in plant growth rate, rather than to a more efficient nodal root system. This response is significantly different from that commonly observed with maize, indicating that adventitious rooting as a response to anoxia is species-specific.

The relation of adventitious root development to plant tolerance of other environmental stresses also has been reported. Kannan (1981) found that the ability of iron-stressed sorghum (*Sorghum bicolor* L. Moench) to release acids and reductants into rhizosphere soil, in order to increase iron availability, was a function of adventitious roots only. In another study comparing the relative effects of soil salinity on sorghum growth and mineral partitioning, Boursier and Läuchli (1990) concluded that adventitious roots of salt-stressed sorghum represent a potential reservoir for the storage of sodium and chloride. The researchers suggest that cultivars with enhanced adventitious root growth would be better able to adapt to saline soil environments.

SUMMARY AND PERSPECTIVES FOR FUTURE RESEARCH

As is obvious from the preceding discussion, the formation of adventitious roots is of tremendous significance in the growth and development of agronomic species that are members of the family Gramineae. With the exception of sugarcane and hybrid bermuda-grass, the major species are propagated from seeds, rather than cuttings, so that the development of both seminal and adventitious roots play a role in the functional biology of the plants. Knowledge of the behavior and relative contribution of each of these root types to the total performance of the root system is limited in many cases. The distinct nature and unique behavior of each root type often is ignored. Perhaps some of the difficulty in relating results of experiments performed under controlled environmental conditions to plant response under field conditions would diminish, if the functional importance of each root type were considered. In other words, if the contribution of seminal roots is important only during early growth stages of the plant, care must be taken when relating results of a "pot experiment" with 25-day-old plants to plant response in the field over an entire season.

With respect to agronomic species, much remains to be learned about the role adventitious roots play in the production of fruit (grain) and dry matter that are harvested. Additional research is needed on the mechanisms that control partitioning of nutrients and assimilates during the plant life cycle, as well as the influence of environmental conditions on these control mechanisms. A better understanding of the effectiveness of adventitious roots, compared with seminal roots, in absorption of water and nutrients is necessary before genetic engineering can be used to "design" agronomic species with efficient root systems (see the chapter by Dickmann and Hendrick, by Friend and coworkers, and by Tepfer and coworkers in this volume).

REFERENCES

Adair, C. R., Miller, M. D., and Beachell, H. M., 1962, Rice improvement in the United States, *Adv. Agron.*, 14:61.
Boatwright, G. O., and Ferguson, H., 1967, Influence of primary and/or adventitious root systems on wheat production and nutrient uptake, *Agron. J.*, 59:299.
Boursier, P., and Läuchli, A., 1990, Growth responses and mineral nutrient relations of salt-stressed sorghum, *Crop Sci.*, 30:1226.

Burton, G.W., and Hanna, W.W., 1985, Bermudagrass, *in:* "Forages the Science of Grassland Agriculture," M.E. Heath, R.F. Barnes, and D.S. Metcalfe, eds., The Iowa State Univ. Press, Ames.

Cannell, R.Q., and Jackson, M.B., 1981, Alleviating aeration stresses, *in:* "Flooding and Plant Growth," T.T. Kozlowski, ed., Academic Press, Orlando.

Carlson, G.E., Gibson, P.B., and Baltensperger, D.D., 1985, White clover and other perennial clovers, *in:* "Forages the Science of Grassland Agriculture," M.E. Heath, R.F. Barnes, and D.S. Metcalfe, eds., The Iowa State Univ. Press, Ames.

Charlton, W.A., 1991, Lateral root initiation, *in:* "Plant Roots the Hidden Half," Y. Waisel, A. Eshel, and U. Kafkafi, eds., Marcel Dekker, New York.

Clements, H.F., 1980, "Sugarcane Crop Logging and Crop Control Principles and Practices," The Univ. Press of Hawaii, Honolulu.

Cook, R.J., and Veseth, R.J., 1991, "Wheat Health Management," APS Press, St. Paul.

Cressman, R.M., 1967, Internal breakdown and persistence of red clover, *Crop. Sci.,* 7:357.

DeDatta, S.K., 1981, "Principles and Practices of Rice Production," John Wiley & Sons, New York.

Drew, M.C., and Lynch, J.M., 1980, Soil anaerobiosis, micro-organisms and root function, *Annu. Rev. Phytopathol.,* 18:37.

Ennos, A.R., 1991, The mechanics of anchorage in wheat *Triticum aestivum* L., II. Anchorage of mature wheat against lodging, *J. Exp. Bot.,* 42:1607.

Fageria, N.K., Baligar, V.C., and Jones, C.A., 1991, "Growth and Mineral Nutrition of Field Crops," Marcel Dekker, New York.

Glover, J., 1967, The simultaneous growth of sugarcane roots and tops in relation to soil and climate, *Proc. South African Sugar Technol. Assoc.,* 41:143.

Hayward, H.E., 1938, "The Structure of Economic Plants," Macmillan, New York.

Hockett, E.A., 1986, Relationship of adventitious roots and agronomic characteristics in barley, *Can. J. Plant Sci.,* 66:257.

Hook, D.D., 1984, Adaptations to flooding with fresh water, *in:* "Flooding and Plant Growth," T.T. Kozlowski, ed., Academic Press, Orlando.

Hoshikawa, K., 1969, Underground organs of the seedlings and the systematics of Gramineae, *Bot. Gaz.* 130:192.

Jackson, W.T., 1955, The role of adventitious roots in recovery of shoots following flooding of the original root systems, *Amer. J. Bot.,* 42:816.

Kannan, S., 1981, The reduction of pH and recovery from chlorosis in Fe-stressed sorghum seedlings: the principal role of adventitious roots, *J. Plant Nutr.,* 4:73.

Klepper, B.L., Rickman, R.W., and Belford, R.K., 1983, Leaf and tiller indentification on wheat plants, *Crop. Sci.,* 23:1002.

Klepper, B.L., Belford, R.K., and Rickman, R.W., 1984, Root and shoot development in winter wheat, *Agron. J.* 76:117.

Kramer, P.J., 1951, Causes of injury to plants resulting from flooding of the soil, *Plant Physiol.,* 26:722.

Kuhlmann, H., and Barraclough, P.B., 1987, Comparison between the seminal and nodal root systems of winter wheat in their activity for N and K uptake, *Z. Pflanzenernaehr. Bodenkd.,* 150:24.

Leonard, W.H., and Martin, J.H., 1967, "Cereal Crops," Macmillan, New York.

Metcalfe, D.S., and Nelson, C.J., 1985, The botany of grasses and legumes, *in:* "Forages the Science of Grassland Agriculture," M.E. Heath, R.F. Barnes, and D.S. Metcalfe, eds., The Iowa State University Press, Ames.

Mitchell, R.L., 1970, "Crop Growth and Culture," The Iowa State University Press, Ames.

Montpetit, J.M., and Coulman, B.E., 1991, Relationship between spring vigor and the presence of adventitious roots in established stands of red clover (*Trifolium pratense* L.), *Can. J. Plant Sci.,* 71:749.

Navara, J., 1987, Participation of individual root types in water uptake by maize seedlings, *Biologia (Bratislava),* 42:17.

Plinthus, M.J., 1973, Lodging in wheat, barley, and oats: the phenomenon, its causes, and preventitive measures, *Adv. Agron.,* 25:210.

Purseglove, J.W., 1985, "Tropical Crops: Monocotyledons," Longman, New York.

Russell, R.S., and Sanderson, J., 1967, Nutrient uptake by different parts of the intact roots of plants, *J. Exp. Bot.,* 18:491.

Stevens, E.J., Stevens, S.J., Flowerday, A.D., Gardner, C.O., and Eskridge, K.M., 1986, Developmental morphology of dent corn and popcorn with respect to growth staging and crop growth models, *Agron. J.* 78:867.

Trought, M.C.T., and Drew, M.C., 1982, Effects of waterlogging on young wheat plants (*Triticum aestivum* L.) and on soil solutes at different soil tempertures, *Plant and Soil,* 69:311.

Wample, R.L., and Reid, D.M., 1975, Effect of aeration on the flood-induced formation of adventitious roots and other changes in sunflower (*Helianthus annuus* L.), *Planta*, 127:263.

Weaver, J.E., and Zink, E., 1945, Extent and longevity of the seminal roots of certain grasses, *Plant Physiol.*, 29:359.

Westbrooks, F.E., and Tesar, M.B., 1955, Tap root survival of Ladino clover, *Agron. J.*, 47:403.

COMMERCIAL APPLICATION

OF ADVENTITIOUS ROOTING TO FORESTRY

Gary A. Ritchie

Weyerhaeuser Company
G. R. Staebler Forest Resources Research Center
Centralia, WA 98531 USA

Planting cuttings of the fir along the roads; enjoying the cool air in the moonlight of the future... —Zhu Xi, Song Dynasty, China

INTRODUCTION

Zhu Xi's lyrical and evocative lines were penned over 800 years ago. The cuttings referred to are of the Chinese fir (*Cunninghamia lanceolata* [Lamb.] Hook.), China's major timber producing conifer. A scrutiny of ancient Chinese literature led Li (1992a) to conclude that this species has been propagated by cuttings in China for over 1,000 years. Interestingly, the propagation techniques used today in China have changed little from those of ancient times.

Succeeding the early Chinese, Japanese foresters were clonally propagating sugi (*Cryptomeria japonica* D. Don) on a commercial scale as far back as 500 years (Ohba, 1993). Instruction books describing detailed propagation methods for taking and rooting sugi cuttings were published in the 1600s and many cultivars in use today date back at least that far (Ohba, 1985). Recently the Japanese were planting nearly 30 million rooted sugi cuttings annually (Ritchie, 1991).

In the West, however, a vastly different picture unfolds. Here forestry is, and always has been, a seedling-based enterprise. In fact, up until the 1970s few forest trees, with the exception of poplars (*Populus* sp.) and willows (*Salix* sp.), had ever been vegetatively propagated on a commercial scale. Furthermore, vegetative propagation of conifers, which make up most of the West's timber supply, was relatively unheard of (Ritchie, 1991). However, this is rapidly changing. Vegetative propagation of forest planting stock through adventitious rooting, although eclipsed in the scientific and popular literature by more exotic developments, such as *in vitro* propagation, biotechnology and genetic engineering, is one of the most exciting emerging technologies in forestry. To wit, a recent symposium on the subject in Bordeaux, France, attracted over 200 delegates from 40 countries (Anon., 1992).

Biology of Adventitious Root Formation, Edited by
T.D. Davis and B.E. Haissig, Plenum Press, New York, 1994

As we will see, many aggressive, innovative and successful modern forestry production systems are being developed to exploit this essentially medieval technology. In this paper I will review the various incentives behind this technological revolution, describe a few of the production systems now in operation, and comment on their potential for advancement of modern, environmentally sound forestry practices.

SOURCES OF FOREST TREE SEED

In traditional seedling-based forestry systems two types of seed are used: wild seed and genetically improved seed. In North America, wild tree seed is often collected from natural or managed forests after logging, and marketed according to the geographic and elevation zone in which it was collected. European and Australian foresters make extensive use of the seed of exotic species. Notable among them are the Monterey pine, or radiata pine (*Pinus radiata* D. Don), indigenous to California and planted extensively in Australia, New Zealand and Chile. Sitka spruce (*Picea stichensis* [Bong.] Carr.), native to the northwestern coast of North America, now comprises over 50% of the trees planted throughout the British Isles (Rook, 1992). Before wide scale use is made of such exotics, they are extensively tested. Seed collections are made from across the range of the species, and seedling tests are planted in common garden experiments, called provenance trials, throughout the area to be reforested. The best provenances are then selected for commercial reforestation.

In contrast, genetically improved seed is normally produced in seed orchards, which fall into two categories. First are the more common "open pollinated" (OP), or wind pollinated, orchards. These are designed to facilitate production of large quantities of seed from selected, open pollinated trees (half-sib families). They are often established using grafts from phenotypically selected parent trees. When the grafts become sexually mature, crosses are made, often with mixes of orchard pollen being used to pollinate females, and seedling progeny are produced. These progeny are then performance tested across several field sites in statistically designed, replicated "progeny tests." Orchard parents yielding poor performing progeny are removed from the orchard (a process called roguing) leaving the parents of only superior progeny—i.e., the best genotypes. These are then cultivated and stimulated in various ways to produce seed for commercial use. Second generation orchards may be established using scions from superior phenotypes of superior families identified in the first generation progeny tests.

One of several drawbacks of OP orchards is that the selected females are pollinated by a pollen cloud, which may contain significant fractions of inferior pollen coming in from surrounding stands or from lower quality orchard trees. Hence, the seed produced may carry a lower level of genetic gain than those originally tested. Owing to this effect actual genetic gains from OP seedlings may range from only 50 to 80 percent of expected gains (Shelbourne, 1987).

This problem can be overcome by performing controlled crosses, i.e., pollinating flowers of a high-performing female with pollen from a high-performing male, then enclosing the flowers in some manner to prevent entrance of foreign pollen. Orchards designed to facilitate controlled pollination (CP) are now emerging as the most technologically advanced seed orchards. Although seed produced in this way can carry substantially higher genetic gain than that of OP seed, its production cost can be orders of magnitude greater than that of OP seed. For more detailed discussions of seed orchards see Faulkner (1975), Wright (1976) or Zobel and Talbert (1984).

SEED BASED PRODUCTION SYSTEMS

Once obtained, seed are sown either into containers in glasshouses (Tinus and McDonald, 1979) or outdoors into bareroot nurseries (Duryea and Landis, 1984) to produce planting stock. Seed may also be sown directly onto the forest floor, but this inefficient practice is rarely used in modern commercial forestry operations. Growing seedlings in glasshouses usually requires one year. In outdoor nurseries, one to four years may be needed depending upon the species, the geographic location and the type of stock desired. In very efficient systems the seed/plant ratio ranges from about 1.5 to 2.0. In other words, owing to losses at various production steps, more than one filled live seed is required to propagate one plantable seedling. In less efficient systems, this ratio may be as high as 10 or even more. This has important implications when expensive CP seed is being used.

SYSTEMS BASED ON ADVENTITIOUS ROOTING

Systems based on adventitious rooting are aimed at capturing benefits from either bulking or cloning (Talbert et al., 1993). Two key distinctions between bulking and cloning are: 1) the process of bulking makes relatively few copies of a large number of genotypes, and 2) in contrast, cloning makes large numbers of copies of relatively few genotypes. Clones are propagated, tracked, and deployed individually or in specific mixes, whereas bulked propagules are normally propagated, tracked and deployed by provenance or family, within which clones are intimately mixed and not of specific interest. The genetic gains associated with cloning can be substantially greater than those associated with bulking (Fig. 1). This is because bulking replicates at the family level while cloning can replicate the most elite individuals within a family.

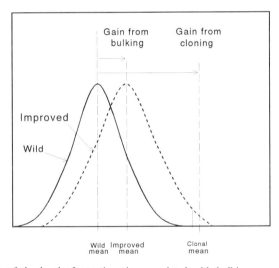

Figure 1. Illustration of the level of genetic gains associated with bulking versus cloning in forest tree improvement-propagation systems. When improved seed is bulked, the resulting propagules are expected to perform on average as the mean of the bulked population. When cloning is employed, however, it is possible to capture the performance gain of the most elite individuals within the population. For these elite individuals to be identified, however, all individuals must be rigorously tested.

Bulking Systems

Bulking has two distinct advantages over cloning. First, the elite families which produce the seed to be bulked have generally been evaluated previously in progeny tests. On average, rooted cutting performance can be expected to equal that of seedling performance, therefore there is no new expensive, time consuming testing requirement to overcome. This is a major consideration. Second, bulking is usually conducted with cutting donors that have been recently propagated from seed. Hence they are physiologically juvenile and, accordingly, relatively easily propagated.

Bulking systems have as their objective to decrease the seed/plant ratio, i.e., to produce more than one plant per seed, for seed which is both scarce and in demand. This is accomplished by producing stock plants, or cutting donors, from this valuable seed then harvesting branches that are rooted as cuttings. Following are some examples.

Bulking Wild Seed. The low value of wild seed rarely justifies its being bulked with cuttings. However, there are a couple of notable exceptions to this rule.

Alaskan yellow-cedar (*Chamaecyparis nootkatensis* [D.Don] Spach.) occurs along the northern Pacific coast of North America from the Cascade Mountains of Oregon and Washington, through British Columbia to coastal Alaska. The wood of this slow growing conifer is highly prized on the export market and, hence, natural stands of this species have been extensively harvested. Unfortunately, yellow-cedar is a recalcitrant producer of small, difficult-to-handle seeds which germinate only sporadically and require an extensive stratification period. Attempts to propagate from seed on a commercial scale have not been successful. Therefore, most logged yellow-cedar stands are regenerated with western hemlock (*Tsuga heterophylla* [Raf.] Sarg.), Douglas-fir *(Pseudotsuga menziesii* [Mirb.] Franco) or certain *Abies* species.

The high value of the wood of yellow-cedar, as well as pressures to conserve species diversity in its geographic region, have spawned a successful program aimed at producing planting stock from rooted cuttings. Cutting wood is collected from hedges, which are established from wild seed collections and maintained by frequent low shearing. Cuttings are rooted under mist in open greenhouses. From 1975 to 1986 over 50 percent of all yellow-cedar stock planted in British Columbia has been produced in this manner. Since 1987 an average of 650,000 have been planted annually, or about 90 percent of the yellow-cedar planting stock (Russell et al., 1990). Future production is planned from selected provenances and, eventually, families.

Another example of bulking wild seed comes from eastern Canada. Black spruce *(Picea mariana* [Mill.] B.S.P.) is an important timber species in this boreal region. A particular strain of black spruce, indigenous to Cape Breton Island, performs exceptionally well when planted in Nova Scotia. Unfortunately seed is in short supply. Hence, it is being bulked-up to commercial quantities using a rooted cutting system (Ritchie, 1991). Semi-hardwood cuttings are taken in early summer and set into containers in greenhouses and rooted under mist. In July they are moved outdoors for over-wintering under snow. They are shipped for field planting in the containers in which they were rooted.

Bulking Orchard Seed. As noted above, vegetative propagation of forest trees is most valuable when coupled with a breeding program where high performance OP or, especially, CP seed is produced. Most bulking programs have evolved in concert with such breeding programs and can produce large genetic gains (Burdon, 1982).

The principal conifer timber species in Scandinavia is Norway spruce *(Picea abies* L.). Much of the Norway spruce seed used in Scandinavia is produced in OP orchards. In the more northerly parts of the region many orchards have been slow and erratic seed producers, requiring 20 to 30 years to bear commercial quantities of improved seed. Hence, highly efficient systems have been developed for bulking many elite OP families throughout Scandinavia. [See Bentzer (1993) for a more extensive account of Norway spruce systems.]

The highest use of vegetative propagation for bulking, however, is when extremely valuable CP seed is available but in small quantities and at prohibitive cost. Several programs are focused on capturing that opportunity.

Sitka spruce is the most important timber producing conifer now being planted in

Great Britain and the Republic of Ireland. Although most planting of this species is with seedlings, an aggressive program of rooted cutting production has been developed and now contributes about six million plants annually for reforestation. Much of this bulking is of CP seed. Development of this program was spearheaded by the British Forestry Commission's Northern Research Station at Midlothian, Scotland (Mason, 1991). This system is somewhat unique in that a double production cycle is used. Seed is sown individually into large containers and grown in temperature controlled glasshouses for two years to produce large stock plants. Cuttings are harvested from these plants in winter and rooted under mist in rooting flats. They are then lined-out into a bareroot nursery in late summer and grown on for two additional seasons. Then another crop of cuttings is harvested from these nursery plants. This crop is rooted and grown out for field planting. By using this six-year, double-production cycle a bulking factor of between 500 and 1,000 plants per seed is achieved.

A recent survey of radiata pine seedling and cutting production in Australia (Duryea and Boomsma, 1992) revealed the following information. Eleven Australian nurseries produced 14.6 million radiata pine seedlings and 3.3 million rooted cuttings in 1992. Three of these grew only cuttings. Production is expected to increase from 18 percent to 31 percent over the next two years. Cuttings are produced generally because there is not enough CP seed to meet production needs. However, four of the 11 nurseries grow cuttings because of their superior form. Radiata pine rooted cuttings have smaller and fewer branches, thinner bark and less stem taper than seedlings. In a mill trial conducted in Australia, trees grown from cuttings yielded substantially more value than those from seedlings of the same age for these reasons (Spencer, 1987).

Figure 2. Eleven-month-old Douglas-fir stock plants growing in a greenhouse near Turner, Oregon (USA). The stock plants are cultured intensively and remain in free growth throughout the year. Each plant will yield about 50 cuttings. [photo by S. Altsuler, Weyerhaeuser Co.]

In New Zealand at least two companies are producing radiata pine from rooted cuttings. Carter Holt Harvey, Ltd. (formerly New Zealand Forest Products), set about five million cuttings in 1992, which is about 98 percent of their nursery production (W.J. Libby, personal communication). Tasman Forestry, Ltd., produces over a million rooted cuttings annually. It also produces commercial quantities of micropropagated radiata pine from CP seed, the only operation of its kind in the world.

Despite considerable interest and research by many private and governmental organizations in the United States, the only commercial conifer bulking program is Weyerhaeuser Company's Douglas-fir program in Washington state (Ritchie, 1992). Currently this program is producing about one million rooted cuttings annually using a stock plant system (Fig. 2). They are propagated from first generation CP families, which are selected for both high growth rate and desirable stem quality characteristics. In contrast, Canada has several commercial-scale rooted cutting programs. The largest of these are the black spruce (Picea mariana [Mill.] B.S.P.) programs of the Northern Clonal Forestry Centre, Ontario, and the St. Modeste Provincial Nursery in Québec. The Québec propagation system is quite unique and modeled after a micropropagation system. Rooting is accomplished in small acrylic trays in controlled environment rooms. Trays are arrayed on shelves and illuminated with fluorescent tubes. This system facilitates year-round harvest and rooting of succulent cuttings (Vallée, 1990).

Cloning Systems

Cloning, while a more powerful technique for exploiting genetic gain through capture of both additive and non-additive genetic variance, is burdened with its own unique set of biological and economic problems. The first is the testing problem. Cloning typically replicates an *untested individual* from a *tested family*. Hence the individual must be cloned and then tested on a range of field sites. Unlike many horticultural plants, which are selected for flower color, cold hardiness and other traits that are expressed early in the life of the plant, forest trees are generally selected for growth rate and stem quality. These traits often take many years to be expressed in the field. A considerable amount of research in forest genetics is aimed at shortening this testing period. While the testing process is occurring, each clone being evaluated must be maintained in a juvenile state in a clone bank. With trees, especially conifers, this is often very difficult or cumbersome to accomplish. Thus we encounter "Murphy's Law" of cloning:

By the time you know which one you want, it is too late to get it.

Fortunately, there are several commercially valuable tree species and genera which do not comply with "Murphy's Law." These are trees that, as mature individuals, produce juvenile tissue, often root or stump sprouts. When this occurs, it is possible to evaluate performance of individual trees late enough in their life to make proper selections, after which juvenile tissue can easily be obtained for propagation. The tree essentially maintains its own juvenile clone bank in the form of a collection of dormant stump buds while it is maturing. Not surprisingly, some of the most successful cloning programs in forestry have been built around such species.

Chinese Fir. As mentioned earlier, this species has been clonally propagated for centuries (Fig. 3). Information regarding its production is not generally available in the western literature, therefore, I will focus on it here in more detail than on other, better known species. What follows is summarized from Li et al. (1990) and Li (1992a,b).

Chinese fir is grown in 14 provinces of southern China and provides from 20 to 25 percent of that country's commercial timber. For centuries it has been propagated as cuttings from stump sprouts using the following procedures. After logging, the biggest stumps are selected and several vigorous sprouts from each stump are removed. These are inserted into the ground near the stump or elsewhere, where they root. Further silvicultural selection is accomplished through thinning out inferior individuals or those infected with insects or diseases. Thinning intensity can be high, as much as three trees removed for

Figure 3. The Chinese fir (*Cunninghamia lanceolata* [Lamb.] Hook.) is an important softwood timber species grown in China. Chinese fir produces prolific, juvenile stump sprouts and has been propagated clonally by adventitious rooting for perhaps 1,000 years. [photo courtesy of Prof. Li Minghe]

every four planted. By repeating this practice over and over, for centuries, Chinese foresters have apparently, and unwittingly, carried out a highly effective selection process in which the largest, healthiest and fastest growing genotypes have been selected and repropagated.

Assuming that the original plantations were established from seedlings, Li (1992a) estimates that more than 20 selection cycles have been carried out. Each is the equivalent of a 30-year clonal test! He goes on to argue that, because of this, the clonal plantations now in existence are much more uniform, faster growing and more pest resistant than seedling plantations. For example, trees in cutting plantations in Fujian Province are from 23 to 37 percent taller than equivalent seedling plantations. Yield from the most productive of these plantations is calculated at 1,170 m^3/ha at age 39 years, in comparison to average seedling plantation yield of 193.5 m^3/ha. Li stated:

> *Considering that vegetative propagation produces identical genotypes, that clonal selection leads to maximum genetic gain, and that a particular plantation may consist of only a few clones which came from the best trees among 1,000 or 10,000 individuals in the initial plantations, such high production would not be unexpected...*

In the 1950s, however, this all began to change. Large scale reforestation efforts undertaken at that time in China required massive quantities of cuttings. These were impossible to procure, especially in regions where Chinese fir had never been planted. In addition, the Chinese were, at that time, under strong influence of Russian genetic theories. These maintained that use of vegetative propagation would lead to erosion of the genetic quality of their forests. Accordingly, seedlings began rapidly to replace cuttings as preferred forest planting stock, a situation which persists to this day. To illustrate, prior to 1950 in Fujian Province, all plantations were established from cuttings. In 1978 only 25 percent were by cuttings, and cuttings are only rarely used today.

Professor Li and his colleagues have argued strenuously that this practice is destroying a valuable Chinese resource—its thousands of priceless, irreplaceable high yielding fir clones. Many Chinese tree breeders are falling in line with this reasoning and systems are being developed to perpetuate this clonal material and to bring new clonal material from breeding programs into production. About 60 million Chinese fir cuttings were rooted in 1991 (Li, personal communication).

Sugi. Japan also has had a long successful history of clonal forestry with sugi cultivars. Production in the late 1980s was estimated at over 30 million. About 27.7 million are grown by private land owners and another 3.7 million planted in National Forests (Ritchie, 1991). This amounted to about 25 percent of the total sugi planting in Japan.

There are estimated to be about 200 cultivars in use today. Thirty or more of these can be traced back across centuries of production. Clonal forestry in Japan is highly regionalized. The various Prefectures propagate cultivars which are particularly well adapted to their climatic and edaphic conditions. Most are planted in pure clonal stands, which tend to give higher yields than mixed stands. Some sugi is also being vegetatively propagated from elite orchard seed. Interestingly, the newer cultivars arising out of this material often do not root as readily as the old cultivars. This may reflect the fact that the old cultivars have been selected for their propagation characteristics as well as their volume yield and stem traits. Currently, however, use of rooted cuttings is on the decline in Japan, except for on the southern island of Kyushu. This arises apparently out of concern over narrowing the genetic base of production populations. Japanese foresters are now moving toward more conventional breeding programs for production of planting stock and are using cloning to exploit the most elite cultivars from these populations.

Spruces. Unfortunately, the vast majority of commercial tree species *do not* produce juvenile stump sprouts, or are otherwise difficult to propagate vegetatively. Therefore, cloning many highly valuable groups such as pines, spruces, larches and Douglas-fir requires an extraordinary effort to maintain thousands of clones in a juvenile state while they are, at the same time, being evaluated in extensive field tests. Below, I will outline one successful approach developed in Lower Saxony, Germany, to clone Norway spruce using the technique of serial propagation (taking cuttings from cuttings). This program began several decades ago and is internationally recognized as a pioneering effort.

A selection differential of one in 5,000 to 10,000 individuals is applied to seedlings based on their early height growth in a nursery. The selected individuals are vegetatively propagated via cuttings then again tested as clones in a nursery. About 30 percent of these clones are selected, serially propagated and field tested. This process repeats through five or more cycles until only about 200 clones from the original 5,000 to 10,000 remain in the production population. At each cycle, hedges are established from the selected clones so that they can be used to produce production numbers of cuttings from no more than 2,000 plants per clone in commercial propagation (Kleinschmit, 1992).

A particularly interesting feature of this program is that it combines bulking with cloning to select many clones from which relatively few copies are produced. Hence, they are able to impose a desired selection intensity, followed by multiplication of selected clones to a desired number of copies. In this manner, they have combined bulking and cloning giving a continuum with the intensities of selection and multiplication that allows them to extract benefits from both systems.

Table 1. Current estimated area, standing timer inventory, annual production and annual consumption of cutting-origin poplar and willow by country. [K=thousands; modified from Ball (1992)]

Country	Area (K ha)	Inventory (K m³)	Production (K m³)	Consumption (K m³)
Belgium	40	--	350	197
Bulgaria	232	--	250	--
Canada	--	2,945	--	--
China	140 (est.)	--	--	--
France	280	27,300	3,300	--
Germany	--	--	250-300 (est.)	220
Hungary	172	--	1,800	1,400
Italy	79	7,600	1,600	--
Morocco	0.45	--	4,140 tons	<10%
Netherlands	20	4,700	224	--
Pakistan	--	--	1,118	1,118
Portugal	3	--	9,000 tons	9,000 tons
Rep. Korea	19	--	29	29
Romania	214	--	797	797
Spain	93/yr	--	542	542
Switzerland	20	--	2,	--
Syria	12 planted in 1989	--	25,000 tons	--
Turkey	130	--	3,500	--
UK	11	--	4	--
USA	25	765,400	16,000	--
Yugoslavia	150	--	1,250	--

Poplars and Willows. Poplars and willows have been produced and cultivated as rooted cuttings throughout Europe, Mid-Asia, the near-East and other areas for centuries. Over 1,500,000 ha of plantations exist today. The stems of cuttings of both genera, however, contain preformed root initials, therefore, they do not exhibit true induced adventitious rooting (see chapters by Barlow, and Haissig and Davis in this volume). Nevertheless, their use is so important and widespread so as to deserve discussion in this review.

Ease of rooting, utility of the wood, facility of plantation establishment and adaptability to a variety of sites have led to the widespread use of willows and, especially, poplars as clonally produced timber species (Zsuffa, 1985). *Populus nigra* [L.] formed the basis of most early European poplar clones. However, introduction of the North American *P. deltoides* [March.] during the 17th century and its crossing with *P. nigra* yielded a stream of fast growing hybrids (*P. euroamericana*) which revolutionized poplar forestry. In North America today, a series of *P. deltoides* x *P. trichocarpa* hybrid clones is having a similar effect.

The International Poplar Commission, founded in 1947 under the Food and Agricultural Organization (FAO), compiles data on poplar and willow inventories, production and consumption for various countries. Current area, inventory, production and consumption estimates (Ball, 1992) are given in Table 1.

Both genera are extremely well suited to vegetative propagation. Clonal plantations can sometimes be established simply by inserting cuttings directly into the soil where they root readily. Plantations may take several forms, all of which fall broadly into two categories: row plantations and block plantations. Within these two types, plantation objectives are reflected in planting densities and silvicultural regimes.

Poplar plantations are almost always clonal. Interestingly, there are fewer than 100 registered poplar clones in use today around the world. Of these, only relatively few are in wide use. For example, in Hungary, only three clones represent 81 percent of the plantations (Zsuffa, 1985). Two clones account for 60 percent in the Netherlands and 70 percent in France. Fewer than 10 clones represent more than 50 percent of all poplar planting in the world. Two among these, *Robusta* and *I-214* have been extremely successful and are widely planted throughout the world.

Clonal poplar forestry is beginning to come into its own in the USA. Although the vast majority of poplar timber originates in natural aspen and cottonwood stands, there is appreciable interest in commercial planting of hybrid poplar clones. This amounts to roughly 20,000 to 30,000 ha annually (DeBell, 1992). Although the focus of activity had been in the southern and eastern USA, this is changing. James River Corp., Boise Cascade Corp. and Scott Paper Company have each established hybrid poplar plantations in Washington state, where they are growing clones developed by the University of Washington, Washington State University Poplar Research Program. These are used for pulp production.

It is particularly noteworthy that scientists at Iowa State University have genetically transformed hybrid poplars using *Agrobacterium tumefaciens*. Twenty such trees were planted into the field representing the first release of transformed woody plants in the USA. According to DeBell (1992): "...Requisite efforts to obtain permits, foster public awareness and understanding, and provide press coverage serve as models of high quality, responsible work in this area of biotechnology...." A major challenge in genetic engineering of many broadleaf tree species is identification and tissue-specific regulation of economically important genes or gene complexes. Gene transfer and plant regeneration methods are available. Application of these techniques to conifers, however, remains extremely difficult.

Eucalyptus. The most spectacular applications of vegetative propagation to modern forestry has been achieved in Brazil and in the Peoples' Republic of Congo with various eucalyptus species. These are presently the largest operational clonal forestry programs in the world (Zobel, 1993). Work in the Congo began in 1953 with species introductions, provenance testing and propagation system development (Leakey, 1987). The Industrial Afforestation Unit of the Congo (UAIC) was started in 1978, along with establishment of 650 ha of eucalypt plantations from superior genotypes. Annual planting since then has been about 3,000 ha and, according to Leakey (1987), about 1.2 million cuttings per year. The cuttings are planted at 5 x 5 m, their final spacing, in 50 ha monoclonal plots. Gains after six years from superior clones have been appreciable: 35 m³/ha vs 20-25 m³/ha for selected seedling provenances and 13 m³/ha for unselected seedlots. These clones are largely derived from the hybrids *Eucalyptus alba* Reinw. x *E. europhyla* and *E. tereticornis* Sm. x *E. saligna* Sm.

In 1967, closely following this successful program, the highly publicized Aracruz project began in Brazil. This massive undertaking was aimed at producing bleached wood pulp. In that project, efforts with *Eucalyptus grandis* W. Hill, *E. salign* Sm and *E. alba* Reinw. seedlings being inadequate to meet production goals, the company embarked on an ambitious, pioneering program to use rooted cuttings. This would enable them to make short term selections for several desirable traits at once and then clonally multiply the elite genotypes from stump sprouts.

Technical problems were nearly insurmountable, including: development of growth media, containers, controlled environments, fertilization regimes, rooting hormones, physiological conditioning of sprouts, establishment of field tests and clone banks, developing selection criteria and numerous sociological and human resource issues (Brandão, 1984).

The success of this project has become legendary. Not only have yields per hectare from the clonal forests more than doubled over yields from unimproved forests, but wood properties have been dramatically improved (Table 2). Additionally, uniformity of the trees, highly important from a conversion standpoint, has been greatly improved. Many clones have also been found to be resistant to serious pests such as fungal cankers and leaf eating insects. While these tabular values are dramatic, even greater gains have been made since these data were assembled in 1984 (Zobel, 1992).

Owing to the significance of their accomplishments, Drs. Leopold Brandão and Edgard Campinhos, Jr., Ms. Yara K. Ikemori and the late Mr. Ney N. Santos were presented with the prestigious Marcus Wallenberg Foundation Award in 1984. Many programs modeled after the Aracruz project have arisen in Brazil, Venezuela, South Africa, the Mediterranean and other regions (Zobel, 1992).

Table 2. Forestry productivity and pulpwood quality of *Eucalyptus grandis* clones from Aracruz Forestry Program. [modified from Brandão (1984)]

Statistic	Initial Forest at 7 Years	Improved Forest at 7 Years	Change (Quantity)	Change (%)
Yield (m^3/ha/yr)	33	70	+37	+112
Pulpwood density (kg/m^3)	460	575	+115	+25
Percent pulp	48	51	+3	+6
Pulp content (kg pulp/m^3)	238	293	+55	+23
Forest productivity (t pulp/ha/yr)	7.85	18.45	+10.6	+135

Tropical Hardwoods. Eucalyptus is not the only fast growing hardwood which is being successfully cloned. Gmelina (*Gmelina arborea* Roxb.), *Triplochiton scleroxylon,* K. Schum. and others are also being vegetatively propagated and planted in the tropics. Leakey (1987) presents a useful review of the use of these programs in Sabah, Nigeria, the Ivory Coast and Congo. A strategy for clonal forestry in the tropics is being developed with *T. scleroxylon* K. (Leakey, 1991).

FIELD PERFORMANCE OF PROPAGULES

> *...I myself will take a shoot from the very top of a cedar and plant it: I will break off a tender sprig from its topmost shoots and plant it on a high and lofty mountain ... it will produce branches and bear fruit and become a splendid cedar....*
> —Ezekiel 17:22

The key to field performance of adventitiously rooted trees relates to the *juvenility* of the donor plant: the more juvenile, the better the performance. Unbeknown to Ezekiel, but well understood by many modern tree propagators, the uppermost shoot of a tree, while chronologically young tissue, is the *least* juvenile part of the plant.

Cuttings from juvenile donor plants, or juvenile tissue, are essentially indistinguishable from seedlings when growing in the field (Frampton and Foster, 1993). However, as the donor tissue ages physiologically the performance of propagules from that donor plant will also behave in a more mature fashion. This has been demonstrated time and again across a wide range of forest tree species (cf. Greenwood et al., 1989). Loss of juvenility is normally manifest in slower growth, greater tendency for plagiotropism (branch-like growth habit) and differences in stem and foliage characteristics. These stem and foliage anomalies vary a great deal across species, but often result in stems from mature plants having less taper, fewer branches and smaller branch diameters than stems of seedling-origin trees. As a general rule, mature-like stems have greater value than juvenile stems, yet growth rates clearly favor the juvenile-origin material. Because the stem is the part of the tree having the greatest value in forestry, this phenomenon leads to some interesting compromises.

Australian and New Zealand foresters have developed methods to exploit these biological oddities in radiata pine cuttings. They harvest cuttings from donor plants which are around three years old, from seed, i.e., beyond the juvenile stage but not yet mature. They find that this combination gives them a juvenile-like rate of growth but with improved stem form (Menzies and Klomp, 1988).

DEPLOYMENT OF CLONAL MATERIAL

Many members of the public at large as well as various "environmentally oriented" groups are antagonistic to the use of clonal material in forestry whether it be poplars, eucalyptus or conifers. Most of their objections are biologically uninformed: soil poisoning, forming biological deserts, fostering "escapes" of mutated organisms into the environment, etc. However, the biological risks attended with growing large numbers of genetically uniform trees across extensive areas for long time periods are real and must be taken seriously.

The appropriate and responsible strategy for field deployment of clones in forestry is being vigorously debated. Deployment in pure monoclonal plantations is preferred with some genera and species (e.g., eucalyptus, poplars) because of higher productivity and greater uniformity, hence more efficient management. This technique also enables foresters to plant specific clones on specific sites to which they are well adapted. For example, late flushing clones may be planted in areas subjected to early frosts. In addition, when clones are planted in pure stands, their clonal characteristics, both positive and negative, can be easily observed and those clones exhibiting unfavorable characteristics can be removed from the production population. This is difficult or impossible to do with mixtures.

However, concerns over risks associated with narrowing the genetic base have led to the deployment of many types of trees, such as spruces, in clonal mixtures rather than pure blocks. Mixtures can be intimate, row-by-row, or mosaics of small clonal blocks. Although genetic diversity can be greatly reduced by planting only one clone, it can likewise be greatly broadened by planting mixtures of clones or clonal mosaics. Clonal mosaics seem to offer advantages of both mixtures and pure blocks. They yield high volume and uniformity, yet establish barriers to contain pests or pathogens.

The safe number of clones to use in clonal mixtures is also debated. Huehn (1987) contends that from 30 to 50 clones may be needed to achieve acceptable risk, while Libby (1982) argues that fewer than 10 clones may be safer than seedling populations. In Sweden and Germany, governmental regulations prescribe the number of clones and individuals within clones which may be deployed in production forestry. Unrealistic and uninformed regulation could severely limit the potential of vegetative propagation as a tool in reforestation (Zobel, 1992).

It is also important that vegetative propagation be backed up by a continuous breeding program, as indicated earlier. As clonal testing and roguing reduce the number of genotypes in a program these must be replaced with new and better genotypes through continuous breeding. Cloning by itself cannot create new and better genotypes, it can only make copies of existing genotypes. Therefore, advanced generation breeding must be carried out to insure a supply of new improved material for the future.

Finally, clonal testing is absolutely essential. To deploy commercial quantities of untested clonal material to the field is to court biological disaster. As cloning narrows the genetic base it also narrows the environmental tolerances of the plants. This can lead to serious maladaptation problems if clones are not first evaluated across the environments to which they will be planted. On the other hand, this close genotype x environment coupling can be exploited to capture unique adaptive and silvicultural benefits.

MEETING THE COST CHALLENGE

It is probably fair to state that in most commercial programs the techniques for achieving adventitious rooting and rooted cutting production are fairly well worked-out, albeit on an empirical level. Production of marketable quantities of well-adapted, field-fit, high performance planting stock has become relatively routine for many species. Earlier concerns expressed by foresters that rooted cuttings would not survive the rigors of real world reforestation sites, would be susceptible to wind throw or would exhibit prolonged plagiotropic growth, have not materialized. Remaining, however, is the considerable challenge of reducing production costs.

Although it is well recognized and accepted that planting expensive, elite genetic stock will pay off in the long run, foresters are continually faced with cost containment priorities in the short run. While every hectare planted with elite stock enjoys an immediate increase in Net Present Value (NPV), we are constantly reminded by field practitioners that: "You can't eat NPV."

It is also probably fair to suggest that if rooted cutting costs could be reduced to, perhaps, twice seedling cost, their wide acceptance would follow. However, achieving this level of cost reduction is not a trivial task by any means. Toward that goal, the following general principles are suggested:

- Achieve continuous improvement in every step of the production system so that yields are increased, losses are reduced, labor costs are minimized and production cycles are shortened.
- Focus on maximizing yields in the *later* production steps. As in every production system, the greatest costs are incurred by losses which occur late in the system, after a considerable investment has been made in every production unit.
- Cull early. Learn to identify culls as early as possible and remove them as soon as possible when the cost of doing so is least.
- Minimize labor inputs. Harvesting and setting cuttings is almost always done by hand. This is one of the most costly parts of the production cycle. Every effort should be made to increase the efficiency of these processes while automated systems are being developed.
- Share facilities. Certain steps, such as rooting, may occupy glasshouse space for only a few months of the year. If other crops can be grown, or if other cuttings can be rooted during down times, glasshouse fixed costs will be spread out over many more plants.
- Bypass production steps. Study the system to determine whether all production steps are needed. A classical case is direct nursery-rooting of cuttings which completely eliminates glasshouses and the high costs associated with them. This is not easy and may require some breakthrough thinking.
- Sell byproducts. In systems where stock plants are used only once and then discarded, it may be possible to market them for various specialty uses and thereby recapture some of the costs invested in them.
- Propagate competent material. By maintaining careful crop records in a database

it will be possible over time to identify and remove families or genotypes which are recalcitrant, hence more expensive, to propagate. This will increase overall system efficiency and lower costs.

• Deploy elite clonal material in mixtures with seedlings. While not reducing propagation costs *per se*, this strategy increases the land base across which a given amount of clonal material can be deployed. However, full employment of this strategy awaits considerable species- and site-specific research to verify its workability.

CONCLUSION

Forestry is and always has been a seed-based enterprise. The vast majority of forest planting stock is propagated directly from seed, a practice which will no doubt persist into the foreseeable future. However, with the development of production seed orchards around the world, particularly controlled pollinated orchards, and the imminent availability of elite, expensive seed, vegetative propagation of trees through adventitious rooting will play an increasingly important role in exploiting the high genetic potential of this material.

Vegetative propagation systems for timber trees can be divided into two types: 1) bulking systems, which make relatively few copies of a large number of genotypes, and 2) cloning systems, which make large numbers of copies of relatively few genotypes. Bulking systems capture only family level genetic gains but offer the distinct advantages of no testing requirement and juvenile starting material. In contrast, cloning offers the promise of substantially greater genetic gain but carries with it extensive costs associated with field testing and maintaining large banks of clones in a juvenile state for many years.

Certain tree species are remarkably simple to clone owing to the production of easily rooted juvenile stump or root sprouts by mature trees. Poplars and willows are examples. Other species, such as many valuable conifers, are difficult to clone and require elaborate juvenility maintenance, testing and propagation systems. Notable examples of some commercial bulking systems are those developed for use with Norway, Sitka and black spruce and radiata pine. Some very successful cloning systems are those in use with Chinese fir, sugi, hybrid poplars and various eucalyptus species. Some of these programs are literally centuries old.

Strategies for deployment of clones into production plantations must balance the benefits of high genetic gain and uniformity against the risks of clonal monoculture. Deployment in mosaics of small monoclonal blocks may offer a good compromise. If production and use of adventitiously rooted forest planting stock are to increase in the future, innovative approaches to production cost improvement must be developed and implemented.

ACKNOWLEDGMENTS

Many friends and colleagues from around the world have shared information with me as I prepared this review. I express particular gratitude to these individuals: Dr. David Boomsma, Dr. Dean DeBell, Dr. Jochen Kleinschmit, Prof. William Libby, Dr. Clements Lambeth, Dr. Roger Leakey, Dr. Mike Menzies, Prof. Kihachiro Ohba, Mr. R. Keith Orme, Dr. John Purse, Ms. Deborah Rogers, Dr. Tony Shelbourne, Mr. Robert Shula, Dr. Reinhard Stettler and Dr. Ken Wearstler. I am especially grateful to Professor Li Minghe, Central China Agricultural University, for generously sharing his most interesting information and photographs on Chinese fir propagation. To my knowledge, this has been heretofore unavailable in the West. Drs. John Frampton, Bruce Haissig and two anonymous reviewers provided many useful comments on the draft manuscript. I thank them for their efforts. Preparation of this manuscript was sponsored by Weyerhaeuser Company.

REFERENCES

Anon., 1992, Mass production technology for genetically improved fast growing forest tree species, *in*: Symp. Proc. AFOCEL, IUFRO, Sept 14-18, 1992, Bordeaux.

Ball, J., 1992, Synthesis of national reports on activities related to poplar and willow areas, production,consumption and the functioning of national poplar commissions, *in*: Int. Poplar Comm. 19th Session, Zaragoza, Sept. 23-25, 1992.

Bentzer, B., 1993, Strategies for clonal forestry with Norway spruce, *in*: "Clonal Forestry: Genetics, Biotechnology and Application," M.R. Ahuja, and W.J. Libby, eds., Springer-Verlag, New York. (*in press*)

Brandão, L.G., 1984, *in*: "The Marcus Wallenberg Foundation Symposium Proceedings: I. The New Eucalypt Forest," Lectures Given by the 1984 Marcus Wallenberg Prize Winners at the Symp. in Falun, Sept. 14, 1984, p. 3.

Burdon, R.D., 1982, The roles and optimal place of vegetative propagation in tree breeding strategies, *in*: "Proc. IUFRO Joint Meeting of Working Parties on Genetics about Breeding Strategies Including Multiclonal Varieties," Lower Saxony For. Res. Inst., FRG, p. 66.

DeBell, D.S., 1992, Activities related to poplar and willow cultivation, exploitation and utilization, *in*: Natl. Poplar Comm.of the U.S.A., period 1988-1991.

Duryea, M.L., and Landis, T.D. eds., 1984, "Forest Nursery Manual: Production of Bareroot Seedlings," Martinus Nijhoff/Dr. W. Junk Pub., The Hague, for the For. Res. Lab., Oregon State Univ., Corvallis.

Duryea, M.L., and Boomsma, D.B., 1992, "Producing Radiata Pine Cuttings in Australia." (unpublished manuscript)

Faulkner, R., ed., 1975, "Seed Orchards," British For. Comm. Bull.

Frampton, L.J., and Foster, S., 1993, Field performance of vegetative propagules, *in*: "Clonal Forestry: Genetics, Biotechnology and Application," M.R. Ahuja, and W.J. Libby, eds., Springer-Verlag, New York. (*in press*)

Greenwood, M,S., Hopper, C.A., and Hutchison, K.W., 1989, Maturation in larch: 1: Effect of age on shoot growth, foliar characteristics, and DNA methylation, *Plant Physiol.* 90:406.

Huehn, M., 1987, Clonal mixtures, juvenile mature correlations and necessary numbers of clones, *Silvae Genetica* 36:83.

Kleinschmit, J., 1992, Use of spruce cuttings in plantations, *in*: "Super Sitka for the 90s," D.A. Rook , ed., Brit. For. Comm. Bull. No. 103, p. 1.

Leakey, R.R.B., 1987, Clonal forestry in the tropics - a review of developments, strategies and opportunities. *Commonw. For. Rev.* 66:61.

Leakey, R.R.B., 1991, Towards a strategy for clonal forestry: some guidelines based on experience with tropical trees, *in*: "Tree Breeding and Improvement," J.E. Jackson, ed., Roy. For. Soc. of England, Wales and Northern Ireland, Tring, p. 27.

Li, M.H., 1992a, "Historical Development of Superior Clones of Chinese Fir in China," Dept. For., Central China Agric. Univ. (unpublished manuscript)

Li, M.H., 1992b, "Introduction to the Vegetative Propagation of Chinese Fir Through Photographs," Dept. For., Central China Agric. Univ. (unpublished manuscript)

Li, M.H., Yang, C., and Shen, B., 1990, Methods for mass production of improved stocks by rooting cuttings of Chinese fir, *Scientia Silvae Sinicae* 26:363.

Libby, W.J., 1982, What is a safe number of clones per plantation?, *in*: "Resistance to Diseases and Pests in Forest Trees," H.M. Heybroek, B.R. Stephan, and K. von Weissenberg, eds., Pudoc, Wageningen, p. 342.

Mason, W.L., 1991, Commercial development of vegetative propagation of genetically improved Sitka spruce (*Picea sitchensis* [Bong.] Carr.) in Great Britain, *in:* "The Efficiency of Stand Establishment Operations," M.I. Menzies, G. Parrott, and L.J. Whitehouse, eds., For. Res. Inst. Bull. No. 156, New Zealand Min. For.

Menzies, M.I., and Klomp, B.K., 1988, Effects of parent age on growth and form of cuttings, and comparison with seedlings, *in:*"Workshop on Growing Radiata Pine from Cuttings," New Zealand For. Res. Inst. Bull. No. 135, M.I. Menzies, J.P. Aimers, and L.J. Whitehouse eds., Min. For., Rotorua, p. 18.

Ohba, K., 1985, Cryptomeria in Japan, *in:* "Clonal Forestry: Its Impact on Tree Improvement and Our Future Forests," L. Zsuffa, R.M. Rauter, and C.W. Yeatman, eds., Proc. 19th Meeting Can. Tree Imp. Assoc., Toronto, Aug. 22-26, 1983, p. 145.

Ohba, K.,1993, Clonal forestry with sugi (*Cryptomeria japonica*), *in*: "Clonal Forestry: Genetics, Biotechnology and Application," M.R. Ahuja, and W.J. Libby, eds., Springer-Verlag, New York. (*in press*).

Ritchie, G.A., 1991, The commercial use of conifer rooted cuttings in forestry: a world overview, *New For.* 5: 247.

Ritchie, G.A., 1992, Commercial production of Douglas-fir rooted cuttings at Weyerhaeuser Company, *in:*"Mass Production Technology for Genetically Improved Fast Growing Forest Tree Species,"Symp. Proc. AFOCEL, IUFRO, Sept 14-18, 1992, Bordeaux, p. 363.

Rook, D.A., ed., 1992, "Super Sitka for the 90s," Brit. For. Comm. Bull. No. 103.

Russell, J.H., Grossnickle, S.C., Ferguson C., and Carson, D.W., 1990, Yellow-cedar stecklings: nursery production and field performance, FRDA Rep. No. 148, Brit. Columbia Min. For., Victoria.

Shelbourne, C.J.A., 1987, The role of cuttings in the genetic improvement of forest trees, *in:* "Symp. Sobre Silvicultura y Mjoramiento Genetico de Especies Forestales," Buenos Aires, April, 6-10, 1987, p. 36.

Spencer, P.J. 1987, Increased yields of high quality veneer and sawn timber from cuttings of radiata pine, *Aust. For.* 50:112.

Talbert, C.B., Ritchie, G.A., and Gupta, P., 1993, Conifer vegetative propagation: an overview from a commercialization perspective, *in:* "Clonal Forestry: I. Genetics and Biotechnology," M.R. Ahuja, and W.J. Libby, eds., Springer-Verlag, Berlin, p. 145. (*in press*)

Tinus, R.W., and McDonald, S.E., 1979, "How to Grow Tree Seedlings in Containers in Greenhouses," USDA. For. Serv. Gen. Tech. Rep. RM-60.

Vallée, J., 1990, Tree improvement and seedling production in Québec, presented at Northeastern Nurserymen's Conf. of the Northeastern State, Fed. and Prov. Assoc., Montréal, July 23-26, 1990.

Wright, J.W., 1976, "Introduction to Forest Genetics," Academic Press, Inc., New York.

Zobel, B., 1992, Vegetative propagation in production forestry, *J. For.* 90:29.

Zobel, B., and Talbert, J., 1984, "Applied Tree Improvement," John Wiley and Sons, Inc., New York.

Zobel, B., 1993, Clonal forestry in the Eucalypts, *in :*"Clonal Forestry: Genetics, Biotechnology and Application," M.R. Ahuja, and W.J. Libby , eds., Springer-Verlag, Berlin, p. 139. (*in press*)

Zsuffa, L., 1985, Concepts and experiences in clonal plantations of hardwoods, *in:* "Symposium on Clonal Forestry: Its Impact on Tree Improvement and Our Future Forests," L. Zsuffa, R.M. Rauter, and C.W. Yeatman, eds., Can.Tree Imp. Assoc., Toronto, Aug. 22-26,1993, p. 12.

COMMERCIAL IMPORTANCE OF

ADVENTITIOUS ROOTING TO HORTICULTURE

Fred T. Davies, Jr.[1], Tim D. Davis[2] and Dale E. Kester[3]

[1]Department of Horticultural Sciences
Texas A&M University
College Station, Texas 77843, USA

[2]Texas A&M Research & Extension Center
17360 Coit Road
Dallas, Texas 75252, USA

[3]Department of Pomology
University of California
Davis, California 95616, USA

INTRODUCTION

Vegetative or clonal reproduction is the most important propagation method used for the commercial production of many, if not most, horticultural crops (ornamentals, fruits, nuts and vegetables). One of the major advances in early agriculture was the discovery that important food crops such as figs, grapes and olives could be regenerated by inserting the base of their woody stems into the ground to induce the formation of adventitious roots and, hence, new plants. Because clonal reproduction may not occur naturally or at least easily in various plant species, the history of horticulture has evolved to a large extent around the development of new technologies for vegetative propagation. More elaborate propagation facilities and commercial asexual propagation techniques have continued to evolve to extend the process to more and more plants at lower costs per unit of production.

Vegetative propagation is generally more costly (per propagule) than sexual (seedling) propagation. Certainly, an advantage to reproduction through asexual propagation is the higher cash value placed on clonally regenerated cultivars (Hartmann et al., 1990). For instance, propagation of broad-leaved evergreens and leafy cuttings has required glasshouses and hot-beds and, later, mist and fog systems. For many species, the superiority of the clonally produced cultivars justified the higher cash value that was necessary to offset the added costs that invariably are associated with these more sophisticated systems.

Adventitious root formation is a prerequisite to successful clonal regeneration of

Biology of Adventitious Root Formation, Edited by
T.D. Davis and B.E. Haissig, Plenum Press, New York, 1994

53

propagules, with the possible exceptions of apomictic seed, and graftage and budding systems on seedling rootstock. With many woody plants, the low physiological capacity for adventitious root initiation led to the technology of grafting and budding onto rootstocks, which includes most fruit and nut crops. Although there are important reasons for utilizing specific rootstocks for individual cases, there are also important reasons to produce plants on their own roots (more economical production, avoidance of virus induced graft union disorders, ease of mass propagation, etc).

COMMERCIAL IMPORTANCE OF HORTICULTURAL CROPS

The horticultural industries comprise one of the largest combined commodity crops in agriculture. In fact, the USA export value of horticultural crops exceeds all other agriculture commodities, including forestry and agronomic crops (Anon., 1991). In the state of Texas, the combined horticultural commodities rank second behind beef commodities, exceeding cotton, dairy, poultry and all other agricultural commodities (C. Anderson, Texas A&M University, Department of Agricultural Economics; personal communication). The 1991 USA wholesale value of vegetable and citrus-fruit-nut crops were $5.0 and $9.7 billion, respectively, while ornamental crops (nursery crops, greenhouse, floriculture, foliage and bedding plants) cash receipts were $8.9 billion (Fig. 1). Total consumer expenditures for ornamental crops were $40 billion.

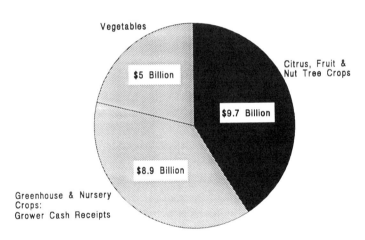

Figure 1. The 1991 wholesale value of USA production of greenhouse and nursery crops, vegetables, and fruit and nut crops. [National Agricultural Statistics Service, USDA; Steglin (1992)]

ADVENTITIOUS ROOT FORMATION IN HORTICULTURAL CROPS

Over 70% of the propagation systems used in the ornamental horticulture industries depend on successful rooting of cuttings (Fig. 2). The large, international floriculture industries (poinsettias, carnations, chrysanthemums, etc.) rely exclusively on asexual propagation by cuttings. Roses, which are the most commercially important international floriculture and landscape ornamental crop, are propagated with rootstock cuttings for graftage, or cultivars are "own-rooted" because they are propagated solely with cuttings (Halevy, 1986). The most common class of propagule used in the ornamental foliage plant industries is cuttings (Joiner, 1981).

In southern and western USA ornamental nurseries, seed propagation typically accounts for less than 10%; division from five to 15%; and graftage, spores and tissue culture liners less than 1% of propagation systems utilized, compared to 70 to 90% propagation via cuttings (Fig. 2). As a general rule, ornamental shrubs are asexually propagated by cuttings, whereas shade trees are propagated by seeds because clonal regeneration of cuttings is generally not cost-effective due to difficulty in rooting. Easy-to-root plant materials are much more widely propagated as tissue culture produced liners compared to recalcitrant species. Even though tissue culture can enhance a recalcitrant woody species ability to be clonally regenerated, frequently poor adventitious root formation limits their commercial production through tissue culture systems.

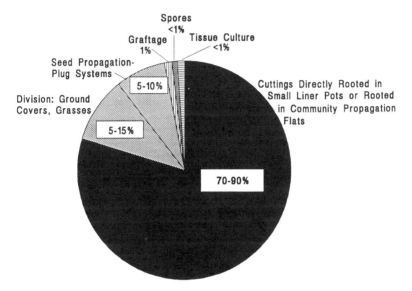

Figure 2. General range of commercial propagation systems utilized in containerized ornamental nurseries in the southern and western USA.

Various fruit and nut crops are propagated in part or solely by cuttings (Table 1). Some of the fruit crop genera listed are grafted or budded onto clonally rooted rootstock. Many small fruits and vines are commercially rooted from cuttings, such as grape, blueberry, cranberry, some kiwi, strawberry, blackberry, gooseberry and currant. Deciduous fruit and nut trees are generally more difficult to root, with some exceptions. In the genus *Prunus*, peach and plum can be routinely own-rooted whereas almond, apricot and cherry must be grafted onto rootstocks. Some cherry rootstocks, such as 'Colt,' can be clonally regenerated. Almond and peach can also be grafted onto hybrid peach-almond clonal rootstock. Most apple cultivars are now grafted onto clonal rootstock (either propagated by cuttings or mounded/stooled) for better size control (dwarfing/semidwarfing characteristics), earlier bearing and greater yield than graftage onto seedling rootstock. Filbert, mulberry, quince, fig and pomegranate are routinely propagated by cuttings. Selected evergreen tropical and subtropical species such as olive, pineapple (slips and suckers), some coffee and citrus (Persian lime), some date species, Feijoa, breadfruit, kola, litchi and avocado are routinely propagated by cuttings.

Table 1. Selected fruit and nut crops propagated in part or solely by cuttings.[1] [sources: Hartmann et al. (1990); Purseglove (1968, 1972)]

Genus	Common Name	Genus	Common Name
I Small Fruit & Vines			
Actinidia	Kiwifruit	*Grossularia*	Gooseberry
Vitus	Grape	*Ribes*	Currant
Vaccinium	Blueberry	*Fragaria*	Strawberry
Rubus	Blackberry		
II Deciduous Fruit & Nut Trees			
Corylus	Filbert	*Diospyros*	Persimmon
Morus	Mulberry	*Pyrus*	Pear
Prunus	Peach, plum, cherry, hybrids	*Zizyphus*	Jujube
Cydonia	Quince	*Punica*	Pomegranate
Ficus	Fig	*Malus*	Apple
III Evergreen Subtropical & Tropical Trees			
Anacardium	Cashew	*Theobroma*	Cocoa
Artocarpus	Breadfruit	*Coffea*	Coffee
Citrus	Persian lime	*Cola*	Kola
Litchi	Litchi	*Feijoa*	Feijoa
Olea	Olive	*Persea*	Avocado
Ananas	Pineapple	*Phoenix*	Date

[1]May include clonal rootstock used in graftage.

The challenge in the production of fruit and nut crops is to get individual cultivars to establish their own root system, and to extend the range of rootstocks that can be vegetatively propagated. With pecans, which are routinely propagated on highly heterozygous seedling rootstock and are sometimes susceptible to zinc deficiency under higher pH soils, the selection and clonal propagation of a rootstock with enhanced zinc uptake would be commercially desirable. Various deciduous and evergreen trees can be rooted by tissue culture methods, but this remains an expensive procedure compared to cutting propagation.

Table 2. Selected vegetable crops propagated by vegetative propagules. [sources: Purseglove (1968, 1972), Lorenz and Maynard (1988)]

Genus	Common Name	Vegetative Propagule
Allium	Onion, garlic chives	Bulbs, offsets division
Armoracia	Horseradish	Root cuttings, crown division
Asparagus	Asparagus	Crown division
Cichorium	Chicory	Root cuttings
Colocasia	Dasheen (taro)	Corms, tubers, suckers
Cynara	Artichoke	Root cuttings
Dioscorea	Yam	Tubers, bulbils
Helianthus	Jerusalem artichoke	Root cuttings
Ipomoea	Sweet potato	Stem cuttings, slips from roots
Mannihot	Cassava (manioc, tapioca)	Tubers, stem cuttings
Rheum	Rhubarb	Crown division
Solanum	Potato	Tuber

Certainly a large number of vegetable crops are sexually propagated. However, many genera are propagated by vegetative propagules requiring adventitious root formation, such as stem cuttings, bulbs, bulbils, offsets, divisions, root cuttings, corms, cormels, tubers, crowns, slips, etc. (Table 2). It is of interest that many of the world's vegetable crops with the greatest carbohydrate (starch) nutrient values are routinely asexually propagated, i.e., irish potato (*Solanum*), yam (*Dioscorea*), sweet potato (*Ipomoea*), cassava (*Mannihot*), taro or dasheen (*Colocasia*), etc.

Table 3. Horticulture propagation systems dependent on adventitious root formation.

I. Generally higher unit costs on a per propagule basis than seed, but more cost effective than propagules listed in II.
  - Stem cuttings (softwood, herbaceous, semi-hardwood, hardwood)
  - Leaf, Leaf-bud cuttings
  - Root cuttings
II. Generally highest unit costs on a per propagule basis compared with all other propagation systems.
  - Layering (tip, simple, trench, mound or stool, compound or serpentine, air)
  - Separation (bulbs, corms)
  - Division (rhizomes, offsets, tubers, tuberous roots, crowns)
  - Clonal rootstock (graftage/budding)
  - *In vitro* culture systems

LIMITATIONS OF ASEXUAL REPRODUCTION

Poor rooting continues to be a serious commercial limitation in the asexual propagation of many woody horticultural crops. Propagation systems commercially utilized to improve rooting success—such as mounding, stooling, layering, division, separation, graftage, budding, tissue culture, manipulation of stock plants through etiolation and banding—are costly and labor-intensive (Table 3) [Hartmann et al. (1990); see the chapter by Howard in this volume]. Labor costs contribute to 30% to 65% of ornamental crop production expenditures, and more than 80% of propagation costs. Hence, there is considerable financial incentive to streamline propagation techniques and improve rooting success. Direct-sticking or direct-rooting of cuttings in small liner-pots for rooting, as opposed to sticking into conventional flats or rooting trays, is an important technique for utilizing personnel and materials more efficiently. By direct-sticking, the additional production step of transplanting rooted cuttings and the potential for transplant shock due to disturbed root systems are avoided. However, recalcitrant species cannot be utilized in this more cost-effective system since rooting percentages must be 80% or greater to economically justify the additional propagation space required.

A current direction of the horticulture industries is to avoid the more costly graftage/budding systems and instead "own-root" desirable scions by cuttings without graftage. Examples include miniature roses and some of the newer Meidiland® landscape rose cultivars, and increased usage of "antique" roses that have been available for centuries and have greater disease and stress resistance than the newer, grafted hybrid-T roses (Henderson and Davies, 1990). Propagation of peaches by cuttings is done in the high-density meadow system where a large number of trees are required per hectare (Couvillon, 1985). Recalcitrant species, which are difficult to root, as well as species with weak, poorly developed and poorly adapted roots cannot be utilized in these systems; hence, they are more costly to produce. The increased time for reproducing recalcitrant species increases production costs. Increased times of plant residency mean that propagation space is not efficiently used, cuttings are subjected to greater environmental stresses during the longer propagation time, and more time is needed for producing plants (Davies, 1991).

Figure 3. Physiologically mature pinyon pine (*Pinus cembroides*) at the Texas A&M Research Center, Dallas, Texas. Tree is about six m tall and 15 years old.

The inability to induce adventitious root formation seriously limits our ability to propagate many potentially valuable horticultural crops, particularly mature woody species (see the chapter by Murray and coworkers and by Ritchie in this volume). As an example, there is a single extraordinary pinyon pine (*Pinus cembroides*) growing in Dallas which exhibits unusual characteristics for this species (Fig. 3). This tree is a "chance" seedling which came from a lot of seed collected from southwest Texas by Mr. Benny J. Simpson, a native plant specialist at Texas A&M University. Out of several hundred seedlings, this is the only plant that has survived the Dallas environment, which is generally too wet and has clay soils that are not conducive to the growth of pinyon pine and many other conifers. In addition to being hardy, the surviving plant exhibits an unusually rapid growth rate for this normally slow-growing species. In addition, the tree has an attractive, full canopy which contrasts with the more sparsely foliated nature of most pinyon pines. For these reasons, it is obvious to any landscape horticulturist that this plant would make an excellent ornamental tree or possibly even a "Christmas tree" if it could be propagated in large numbers. Unfortunately, the extraordinary qualities of this plant were recognized only after the tree began producing cones (i.e., after the tree began maturing). As is typical for mature conifers, cuttings from this tree have proven impossible to root. Furthermore, attempts to micropropagate the plant have failed because microshoots produced *in vitro* do not form adventitious roots. Thus, the tree remains a curiosity whose valuable characteristics cannot be cloned and exploited. There are many such examples in horticulture.

Successful rooting of recalcitrant species, which are currently uneconomical to root by cuttings, will lead to the development of new plant products and markets. The difficulty in rooting recalcitrant species and in developing unique clones opens up new opportunities for utilizing biotechnology, such as gene manipulation to increase tissue sensitivity to auxin and other limiting rooting factors, as well as to improve tissue culture systems [Debergh and Zimmerman (1991), Urbano (1992); see the chapter by Hamill and Chandler, by Palme and coworkers, by Tepfer and coworkers, and by Riemenschneider in this volume). Rooting and acclimation of tissue culture produced plants will need to be improved if biotechnology is to be incorporated into the propagation and production of horticultural crops.

CONCLUSION

Vegetative or clonal reproduction remains the most important propagation method used for the commercial production of many, if not most, horticultural crops. Because clonal reproduction does not naturally or at least easily occur in many plant species, the history of horticulture has evolved to a large extent around the development of new technologies for vegetative propagation. The utilization of auxin to stimulate rooting, and of intermittent mist and fog systems to enhance rooting of cuttings, has led to new horticultural products and industries—from chrysanthemum and poinsettia production, to woody ornamentals and fruit crops, to the rooting and acclimation of tissue culture produced liners. Unfortunately, poor rooting continues to be a serious limitation in the asexual propagation of many woody horticultural crops. Many costly and labor-intensive systems are employed to improve rooting success. Labor costs contribute to more than 80% of propagation expenditures; hence, there is considerable financial pressure to streamline propagation techniques and improve rooting success.

There is a tremendous need to extend the range of rooting to other selections, to increase the rooting percentages of individual selections and to make root systems physiologically stronger. The ability to successfully root recalcitrant species will lead to the development of new plant products and markets. The wide diversity of horticultural species and cultivars (an ornamental nursery may produce a product mix of 400-700 cultivars and species, with some nurseries producing up to 4000), as opposed to the few commercial forestry species used in a given region, has made it difficult to focus on just one species or cultivar. There is a need in horticulture to concentrate rooting efforts on a few select model systems (see the chapters by Ernst, and by Riemenschneider in this volume). The difficulty in rooting recalcitrant species and in developing unique clones opens opportunities for utilizing biotechnology. Rooting and acclimation of tissue culture-produced plants will need to be improved if biotechnology is to be commercially incorporated into the propagation and production of horticultural crops.

REFERENCES

Anon., 1991, National Agriculture Statistics Service, USDA/ARS, Washington.

Couvillon, G.A., 1985, Propagation and performance of inexpensive peach trees from cuttings for high density peach plantings, *Acta Hortic.* 173:271.

Davies, F.T., Jr., 1991, Back to the basics in propagation, *Proc. Int. Plant Prop. Soc.* 41:338.

Debergh, P.C., and Zimmerman, R.H., 1991, "Micropropagation: Technology and Application," Kluwer Academic Pubs., Dordrecht.

Halevy, A.H., 1986, Rose research - current situation and future need, *Acta Hortic.* 189:11.

Hartmann, H.T., Kester, D.E., and Davies, F.T., Jr., 1990, "Plant Propagation - Principles and Practices," 5th ed., Prentice Hall, Englewood Cliffs.

Henderson, J.C., and Davies, F.T., Jr., 1990, Drought acclimation and morphology of mycorrhizal *Rosa hybrida* L. cv. Ferdy, *New Phytol.* 115:503.

Joiner, J.N., 1981, "Foliage Plant Production," Prentice Hall, Englewood Cliffs.

Lorenz, O., and Maynard, D.E., 1988, "Knott's Handbook for Vegetable Growers," 3rd. ed., John Wiley & Sons, New York.

Purseglove, J.W., 1968, "Tropical Crops - Dicotyledons," John Wiley & Sons, New York.

Purseglove, J.W., 1972, "Tropical Crops - Monocotyledons," John Wiley & Sons, New York.

Steglin, F.E., 1992, Forecasting grower cash receipts, *Proc. South. Nurserymen's Res. Conf.* 37:387.

Urbano, C.C., 1992, Tissue culture: potential, problems and future, *Amer. Nurseryman* 175(5):30.

GENOMIC MANIPULATION OF PLANT MATERIALS

FOR ADVENTITIOUS ROOTING RESEARCH

Don E. Riemenschneider

USDA Forest Service
North Central Forest Experiment Station
Forestry Sciences Laboratory
Rhinelander, WI 54501 USA

INTRODUCTION

Knowledge of internal controls on adventitious root formation is necessary to understand both the fundamental developmental biology of rooting and to improve rooting for commercial purposes (Haissig et al., 1992). Yet, the array of physiological and biochemical processes involved in rooting is so large (Davis et al., 1988) that controls—i.e, factors that regulate rooting at will and without exception (Haissig et al., 1992)—have been difficult to identify. Such complexity suggests that studies designed to test single-factor hypotheses require more highly refined reductionist experimental approaches than are currently employed. One possible methodology involves contrasts between experimental plant genotypes or populations that do and do not consistently initiate roots when subjected to an experimental protocol [Haissig et al. (1992); see the chapter by Ernst in this volume]. Contrasting genotypes permit the comparison of putative single-factor controls in both the *on* (rooting) and *off* (non-rooting) condition without the need for traumatic experimental treatments that may have manifold side-effects. Unfortunately, suitable plant materials are not generally available, which limits genetic studies of rooting in many important species, especially woody plants.

Various genetic technologies have been proposed to overcome the lack of plant materials, including the application of molecular genetic (genetic engineering) and quantitative genetic (selective breeding) tools to achieve genomic modification. For example, three molecular approaches have been proposed (Haissig et al., 1992): 1) elucidation of the structure and function of known or supposed rooting-related genes, 2) identification of formerly unknown rooting-related genes from plant materials that differ phenotypically in rooting ability, and 3) identification of formerly unknown rooting-related genes from contrasting genotypes. Knowledge of molecular genetic effects on rooting would presumably lead to the genetic transformation of plants from the non-rooting to rooting genotype (or the reverse) after insertion of the proper control, whatever that might be.

Biology of Adventitious Root Formation, Edited by
T.D. Davis and B.E. Haissig, Plenum Press, New York, 1994

However, a major obstacle to both the molecular study and modification of rooting is the complexity of the relationship between gene-level knowledge and the phenotype, which involves *cis*-factors, *trans*-factors, protein-protein interactions and protein-DNA interactions (Haissig et al., 1992). Coupled with the foregoing are the difficulties associated with engineering recalcitrant species (Haissig et al., 1992). Thus, the choice of a molecular approach to the genomic modification of adventitious rooting should be based on the best possible theoretical guidelines to ensure both a high probability of practical success and the formulation of informational hypotheses. Yet, theoretical linkages between engineered single gene modifications and resultant effects on a discontinuous phenotype have not been widely discussed.

Quantitative genetics and its associated technology, selective breeding, can be used to achieve genomic modification in conjunction with, or in lieu of, genetic engineering. Genomic modification using selective breeding requires only that the selection and response criteria be under the control of segregating loci whose alleles differ in phenotypic effect. The selection criteria and response criteria may be the same trait or different traits, in which case a non-zero genetic intertrait correlation is required. Quantitative genetic control of adventitious rooting and other organogenic phenomena has been well established [Haissig and Riemenschneider (1988); see the chapter by Haissig and Davis in this volume]. However, the quantitative genetic properties of that control—i.e., additive or non-additive, the size and nature of genotype x environment interactions, etc.—are much less certain (Haissig and Riemenschneider, 1988). One reason for the lack of quantitative genetic knowledge of adventitious rooting may be seen from a listing of recent studies of root formation and development published in the crop improvement literature (Table 1). Clearly, genetic effects on root development has been a much studied subject. Yet, most investigations have focused on the growth and development of nonadventitious roots, on the formation of adventitious roots in species where such development is ontogenetically expected [i.e., maize and kleingrass (Table 1)], or some other aspect of root development apart from the formation of adventitious roots by isolated parts of plant shoots (i.e., cuttings). Thus, quantitative genetic knowledge of adventitious root formation—including correlations between different kinds of root development, and between adventitious rooting and shoot development—remains largely unavailable.

Table 1. Some recent studies of root formation and development from the crop improvement literature.

Species	Authors	Trait(s)
Alfalfa	Carter et al., 1982	Root system weight, length
	Salter et al.,1984	Root system weight, length, no. laterals, fibrousness
	Pederson et al., 1984	Root system weight
	Viands, 1988	Regeneration from pruned taproot
	Hansen & Viands, 1989	Selection for regeneration from pruned taproot
	White & Castillo, 1989	Reciprocal grafting study
Bean	Eissa et al., 1983	Root system length, weight
Cotton	Tischler et al., 1989	Adventitious rooting
Kleingrass	Jenison et al., 1981	Root system weight, spread
Maize	Arihara & Crosbie, 1982	Root system pulling resistance
Oat	Murphy et al., 1982	Root system volume, length, weight
Peanut	Ketring, 1984	Root system volume, weight
Tobacco	Crafts-Brandner et al., 1987 a,b	Reciprocal grafting study

Estimates of quantitative genetic parameters, needed to define efficient breeding strategies, can be readily estimated using standard biometrical genetic methods whenever the trait of interest is continuous and normally distributed on some scale (Mather, 1971). However, adventitious rooting is not always assessed as a continuous variable as, for example, when single observations (rooted vs non-rooted) are made due to the nature of the experimental material (small stock plants) or the logistical considerations inherent to selective breeding. Then, adventitious rooting is more properly interpreted as a threshold character; such characters have not been much discussed in the literature. Therefore, we should examine the theory of breeding for discontinuous threshold traits to identify efficient selective breeding guidelines.

Thus, the goal of this chapter is to explore the implications of a quantitative genetic model for genomic modification of adventitious rooting ability through genetic engineering and selective breeding. First, a simplistic quantitative model will be proposed to link quantitative genetic control with a discontinuous threshold phenotype. Second, metabolic control theory as developed by H. Kacser and others (Kacser and Burns, 1973; Cornish-Bowden, 1976; Kacser and Burns, 1973, 1979, 1981; Kacser, 1983; Porteous, 1983; Kacser and Beeby, 1984; Torres et al., 1986; Kacser and Porteous, 1987; Keightly and Kacser, 1987) will be used to explore the potential of genetic engineering to achieve genomic and phenotypic modification. Third, quantitative genetic theory applicable to selection for discontinuous, threshold traits (Robertson and Lerner, 1949; Dempster and Lerner, 1950; Gianola, 1979; Gianola and Norton, 1981) will used to set forth guidelines for efficient selective breeding. Last, the possibility of selection for rooting as an assimilated trait, a trait that is not expressed under normal environmental or developmental conditions, will be discussed.

Specific hypotheses based on the various evaluations of genomic modification techniques will be put forward where possible. The overall focus of this chapter is on adventitious rooting, but the principles should apply equally to any discontinuous trait that is based on a heritable underlying metric scale.

THE QUANTITATIVE GENETIC MODEL

The proposed model assumes that rooting is under quantitative genetic control as defined by Smith (1983) to mean that: "The population will respond smoothly to selection and single gene effects will not be detectable...," a condition that can be achieved by five or more segregating loci, each of equal effect (Smith, 1983). A quantitative genetic model is probably the most appropriate current paradigm for adventitious rooting because such models are the default outcome of two sequential existential hypotheses; one proven the other unproven. Quantitative models are thus characteristic of *intermediate* levels of genetic knowledge. The first existential hypothesis, that genetic effects on rooting and other organogenic phenomena exist, has been established nearly to the point of becoming axiomatic [Haissig and Riemenschneider (1988) and references therein]. In comparison, the second existential hypothesis, that genetic control can be mostly attributed to the qualitative inheritance of single genes with major effects, has not been widely demonstrated to be true in natural or artificially selected populations, apart from the effects of mutation (Zobel, 1975, 1986) or transformation (see the chapter by Tepfer and coworkers in this volume). Thus, quantitative genetic models provide a theory that agrees with what is known, but they are not dependent on that which remains unproven. Admittedly, major single-gene effects on adventitious rooting may exist, which would tend to diminish the future value of a quantitative genetic model as well as any model-based deductive reasoning. But, the disproof of such an existential hypothesis is beyond the scientific method (Bunge, 1967). Therefore, the quantitative genetic model is timely—i.e., in agreement with the known, but independent of the unknown or uncertain.

The model will further postulate that genetic effects on adventitious rooting are mediated through one or more biochemical pathways of undetermined complexity (Fig. 1A). Also, it is assumed that rooting is an inherently discontinuous trait with two states (roots vs no roots), with the threshold between the two states dependent on the flux through the quantitative genetically controlled biochemical pathways (Fig. 1B). The goal of genomic

63

modification for increased rooting ability is to alter the flux of the underlying biochemical pathway such that the rooting threshold is exceeded. The application of a threshold model on adventitious root formation is not novel. For example, a threshold model has been previously used in a mechanistic study of the effects of carbohydrate metabolism and translocation on root formation by leafy cuttings (Dick and Dewar, 1992). Clearly, a two-state threshold model embodies only the most simplistic view of adventitious root formation, the validity of which depends on the depth and method of assessment.

A. A hypothetical rooting pathway.

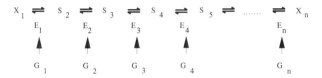

B. Relation between pathway and the rooting phenotype.

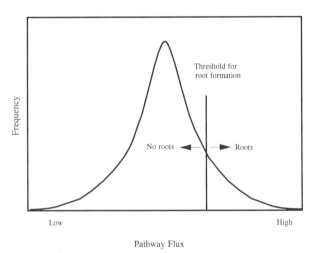

Figure 1. A quantitative genetic model for adventitious rooting. Genes (G_i) encode individual enzymes (E_i) that catalyze reactions in a biochemical pathway. Pathway flux constitutes a continuously (metrically) distributed underlying variable. Rooting results only when pathway flux exceeds some threshold, the exact position of which may vary among environments.

For example, the process of adventitious root formation may be divided into the phases of competency, root initiation and root development with any phase possessing two or more states that permit or preclude rooting (see the chapter by Mohnen, by Murray and coworkers, and by Haissig and Davis in this volume). Also, the state of *rooting* may be subdivided based on the method of assessment into quantitative (continuous) metrics such as root mass, root length, number of roots, time to root initiation, etc. But, as I will show in this chapter, certain special conditions may exist during the application of selective breeding to the genomic modification of rooting ability where the recognition of a simple threshold model is important. Thus, I encourage the reader to suspend disbelief in the simplistic so that the potential for genomic modification of adventitious rooting ability can be explored, first in relation to the relatively new theory of metabolic control and, second, in relation to the more aged theories of breeding for discontinuous characters and the possibility of breeding for rooting as a genetically assimilated trait.

THE QUANTITATIVE MODEL AND DIRECT GENETIC MODIFICATION

Evaluating the potential of direct genetic modification (genetic transformation) to effect a change in rooting ability requires that a predictive relation be established between the modification of a single gene (G_i) and the phenotypic expression of rooting (Fig. 1). That relation has been made possible by the application of metabolic control theory to quantitative genetics [reviewed by Dean et al. (1987)]. For example, the fundamental equations of steady state catalysis (Briggs and Haldane, 1925) have been used to derive a theory of metabolic control in which the flux of any biochemical pathway is derived as a function of the kinetics of individual reactions (Kacser and Burns, 1973). Metabolic control theory has been successfully applied to questions of molecular evolution (Kacser and Beeby, 1984), the molecular interpretation of dominance (Kacser and Burns, 1981) and the epistatic interactions between intra- and inter-locus mutations (Szathmary, 1993). Briefly, the flux through a biochemical pathway, the proposed underlying scale of Figure 1, is represented mathematically as:

$$F = \frac{(X_1 - X_n)K_{1n}}{\dfrac{M_1}{V_1} + \dfrac{M_2 K_{12}}{V_2} + \dfrac{M_3 K_{13}}{V_3} + ... + \dfrac{M_n K_{1n}}{V_n}} \tag{1}$$

where F, X_i and K_{1n} are pathway flux, substrate and product concentrations, and equilibrium constants, respectively. M_i and V_i are the Michaelis constants, K_m (a measure of the affinity of an enzyme for its substrate) and V_{max} (a measure of the velocity constant of the breakdown of the enzyme-substrate complex). Reaction rates are inversely proportional to K_m and directly proportional to V_{max}. Equation (1) can be simplified (Kacser and Burns, 1981) to:

$$F = \frac{C}{\dfrac{1}{E_1} + \dfrac{1}{E_2} + \dfrac{1}{E_3} + ... + \dfrac{1}{E_i} + ... + \dfrac{1}{E_n}} \tag{2}$$

where C is a constant and E_i are composite functions of K_{1n}, M_i and V_i, representing the activity of the ith enzyme. E_i can be decreased by alteration of coding gene sequences (mutation) or transformation with so-called "antisense" constructs. E_i can be increased by genetic transformation with multiple gene copies (gene amplification) or transformation with constitutive or inducible promoter sequences. The change in pathway flux in relation to the change in the activity of any one enzyme ($dF/dE_i = Z_i$, the control coefficient of the ith enzyme) is of the form:

$$Z_i = \frac{\dfrac{1}{E_i}}{\dfrac{1}{E_1} + \dfrac{1}{E_2} + \dfrac{1}{E_3} + ... + \dfrac{1}{E_i} + ... + \dfrac{1}{E_n}} \tag{3}$$

Kacser and Burns (1981) used equations (1) and (3) to explore pathway flux as a function of enzyme number and activity (Fig. 2). The function, in general, became less linear as the number of enzymes in the pathway increased. Thus, the effect of changes in the activity of one enzyme on total pathway flux ($dF/dE_i = Z_i$, the control coefficient) would be small for most deviations from wild-type activity, even for modestly complex systems. An evaluation of dominance phenomena suggested that non-linearity was common and that control coefficients (Z_i) tended to be small (Kacser and Burns, 1981).

The foregoing theory can also be used to evaluate the possibility of modifying rooting by genetic engineering strategies such as gene amplification or constitutive-promoter fusions. I estimated the effect on pathway flux of increasing the activity of a single enzyme, up to 10-fold in excess of the wild-type activity, by using equation (2). Results indicated that the effect of increased activity of a single enzyme depended on the number of enzymes in the pathway (Fig. 3). Thus, pathway flux was linear with E_i for a one-enzyme system, but became progressively smaller and less linear as the number of enzymes (n) was increased (Fig. 3).

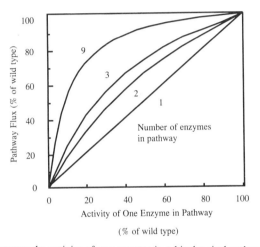

Figure 2. The relation between the activity of one enzyme in a biochemical pathway and the pathway flux. Flux is less linear with single enzyme changes as the number of enzymes in the pathway increases. [recalculated from Kacser and Burns (1981)]

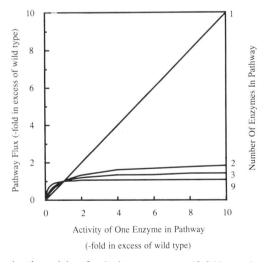

Figure 3. Effect of increasing the activity of a single enzyme up to 10-fold over the activity of the wild-type. Pathway flux is not significantly increased when the number of the enzymes in the pathway is large.

In general, assuming that all E_i are initially equal in the wild-type (all $E_{i>1} = 1.0$ for this example), pathway flux is scaled to unity ($C=n$) and the activity of E_1 is increased by hypothetical genetic engineering of G_1, then the pathway flux can be written as:

$$F = \frac{n}{\dfrac{1}{E_1} + (n-1)} \tag{4}$$

which has a finite limit for n > 1 of:

$$\lim_{E_1 \to \infty} F = \frac{n}{(n-1)} \tag{5}$$

Thus, if rooting involves the expression of at least 100 genes, a conservative number based on Zobel (1986), and if all encoded enzymes have equal control coefficients—and there seems no compelling reason to reject those hypotheses *a priori*—then the maximum achievable flux of such a pathway modified by genetic engineering of a component enzyme would be only 101% of the wild-type. Kacser and Burns (1981) state that: "...There is an asymmetry, in that possible mutants with higher activity than that of the wild-type (normal) will almost always evade detection because their phenotypic effect is very small, and that mutations to increased fluxes that can be detected are unlikely to occur...." The foregoing conclusion applies whether increased fluxes are due to random mutation or to directed genetic modification. Therefore, the probability that traits under quantitative genetic control can be modified by the genetic engineering of a single component seems small, except under very special circumstances.

The demonstrable successes of genetic engineering for agronomic crop improvement suggest what the aforementioned circumstances might be. First, metabolic control theory predicts that genetic engineering of single genes should be successful when the application alters the activity of an encoded enzyme in the range between metabolic blockage and the Mendelian threshold (Fig. 4). The Mendelian threshold (Kacser and Burns, 1981) is the point below which an enzyme has a high control coefficient (dF/dE is high) and above which an enzyme has a low control coefficient (dF/dE low, the plateau region of the flux-activity curve). Activity changes between complete metabolic blockage and the Mendelian threshold are most likely to result in observable phenotypic changes (Kacser and Burns, 1981). For example, the two most prominent successes of genetic engineering are herbicide tolerance and the modification of post-harvest physiology using antisense engineering. Genetically engineered herbicide tolerance (Fillatti et al., 1987) increases target enzyme activity from the point of chemically induced metabolic blockage to some point beyond the Mendelian threshold. Conversely, antisense engineering (Oeller et al., 1991) accomplishes the reverse, movement of enzyme activity from the wild-type to a point sufficiently near metabolic blockage to disable a biosynthetic pathway. Genetic engineering has not been widely applied to the plateau region of the relation between F and E_i. In addition, the one adventitious rooting-specific mutation identified in a mutable tomato line [designated *ro* (Zobel, 1975, 1986)] inhibited rooting. No mutants with increased adventitious rooting were reported (Zobel, 1975, 1986). Furthermore, research on the effects of *Agrobacterium rhizogenes* transformation has demonstrated that one important mediator of altered rooting phenotype may be *decreased* polyamine biosynthesis (see the chapter by Tepfer and coworkers in this volume). Thus, single gene genomic modification seems more effective in *decreasing* pathway flux, thereby inhibiting phenotypic expression, than in *increasing* flux and promoting expression. Therefore, I will suggest several hypotheses, based on metabolic control theory, concerning the application of genetic engineering for the modification of rooting ability: namely that, genetic engineering will most probably modify rooting ability when...

- the objective is to increase the flux of essentially single-enzyme pathways (i.e. qualitatively inherited loci), or
- the objective is to decrease pathway flux, for example to eliminate inhibitor expression or to modify genotypes from the rooting to non-rooting state, and the target modification lies between metabolic blockage and the Mendelian threshold (Fig. 4), or
- knowledge is sufficient to target a regulatory locus such that one modification affects expression of many loci and thus the activity of many enzymes simultaneously.

In addition, engineering approaches are suggested when breeding approaches would result in the loss of cultivar identity, as is the case with many vegetatively propagated horticultural crops. But, genetic engineering would not be expected to be successful...

- when the objective is to increase pathway flux by the amplification or "up-regulation" of genes encoding enzymes of average or unknown sensitivity coefficients.

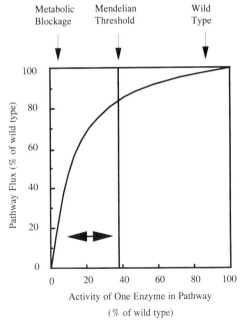

Figure 4. The effect of genetically engineered changes in the activity of a single enzyme on total pathway flux, and thus phenotype, is greatest when the change occurs between metabolic blockage and the Mendelian threshold (double arrow) (Kacser and Burns, 1981). The Mendelian threshold is that point below which pathway flux must decrease before a phenotypic effect is observed. Good examples are genetically engineered herbicide tolerance (blockage to threshold) and alteration of post-harvest physiology via antisense engineering (threshold to blockage).

THE QUANTITATIVE MODEL AND SELECTIVE BREEDING

Application of selective breeding requires that at least some genes controlling the trait of interest segregate multiple forms (alleles), which encode enzymes with varying catalytic properties, and that a correlation exists between allelic frequency and phenotype. Selection alters the frequency at which different alleles occur in the population, which results in a corresponding change in mean phenotype. The rate of change in allelic frequencies depends on several factors including selection intensity and heritability (h^2, the proportion of variation in the phenotype that is attributable to genetic variation). Heritability can be estimated on the basis of the underlying metric scale or the discontinuous scale. Heritability on the underlying scale is important for two reasons. First, the amount of genetic variation in a population determines the range of genotypic values along the underlying scale and, thus, the probability that some individuals will exceed any fixed threshold. Second, non-genetic (environmental) variance determines the range of values one genotype may possess along the underlying scale and, thus, both the frequency with which the genotype will form roots and the precision with which the genotypic (breeding) value may be estimated. Heritability on the discontinuous scale, as will be discussed in the following, is a function of heritability on the underlying scale and the location of the threshold. Selective breeding is in general a more powerful tool than genetic engineering, when knowledge of the molecular and biochemical catalytic properties of the phenotype are poorly developed, and genetic engineering targets are unknown.

Rooting as a Discontinuous Trait

Standard biometrical plant breeding methods are properly applied to adventitious rooting when rooting is evaluated as a continuously distributed metric variable. A mountain of plant breeding literature, not discussed here, provides suitable guidelines for the genomic modification of such traits. Yet, rooting may not always be assessed or assessable as a continuous trait. For example, selection in the positive direction (increased rooting ability), based on a continuous variable, requires that the breeder distinguish somehow between individuals that form roots. This may be accomplished in at least two ways: first, by the use of one or more auxiliary variables such as number of roots, root mass, root length, time to root formation, etc; and, second, by scoring the fundamental measure of rooting (roots vs no roots) on several cuttings from each individual. Mean percent rooting would approach normality at high sample sizes. However, the large populations needed in breeding programs to support high selection intensities, coupled with the need for non-destructive evaluation, suggest that auxiliary variables may not always be readily assessed. In addition, multiple observations per individual may not always be possible when the experimental population consists of small plants with single stems. Furthermore, selection in the negative direction (decreased rooting ability) based on a continuous variate may be even more problematic because the breeder is obliged to discriminate, based on rooting ability, among individuals that do not form roots. That seems on the surface to be a difficult task.

Based on the foregoing arguments, a discussion of threshold breeding methods for rooting ability seems warranted. In the following, I will discuss some aspects of the theory of breeding for discontinuous characters, with the objective of providing experimental guidelines to maximize response to selection. I will discuss breeding in the positive "direction" (increased rooting ability), for convenience and to appeal to those who are interested in commercial plant propagation (see the chapter by Davies and coworkers, by Kovar and Kuchenbuch, and by Ritchie in this volume)—but the arguments equally apply to the opposite direction of selection.

Various theoretical frameworks with which to evaluate selection for discontinuous characters have been provided for phenotypes with two classes (Robertson and Lerner,

1949; Dempster and Lerner, 1950), more than two classes (Gianola, 1979) and more than two classes with scaling (Gianola and Norton, 1981). Also, the heritability of discontinuous traits in relation to an underlying scale has been established, mostly through research on animal skeletal traits (Gruenberg, 1951; Self and Leamy, 1978; Hagen and Blouw, 1983). The work of Dempster and Lerner (1950) provides the most convenient framework for my presentation here because the formulations are algebraically convenient and they provide the starting point for multiclass evaluations (Gianola, 1979; Gianola and Norton, 1981). According to Dempster and Lerner (1950), the formulation is applicable when there is an underlying normally distributed variate that is subject to both genetic and environmental effects, when the character is present in only those individuals in which the underlying variate exceeds some threshold value, and when effects are additive. Thus, assuming that a population varies in some underlying scale—e.g., that a proportion (p) of the individuals that have been selected as parents form roots—the narrow sense heritability of rooting in the discontinuous (p) scale is:

$$h_{pa}^2 = \frac{z^2 h_x^2}{p(1-p)} \tag{6}$$

where z^2 is the ordinate of the threshold on the underlying normal scale, h_x^2 is the heritability of the underlying character and p is the proportion rooted. And, predicted gain in the underlying scale is:

$$G = \frac{h_x^2 z}{p} \tag{7}$$

where h, z and p are as before. Gain in the p scale is estimated by determining the position of the threshold in the next generation, which amounts to moving the threshold of the parental generation to the left a distance equal to the gain. The area to the right of the new threshold is the frequency of rooting in the next generation (Dempster and Lerner, 1950). I recalculated gain in the p scale according to equation (7) for heritability values on the underlying scale of 0.1, 0.3, 0.5, 0.7 and 0.9 (Fig. 5).

Predicted response to selection is greatest at rooting incidence of about 0.1 to 0.2, depending on heritability of the underlying trait (Fig. 5). Gain is low at very low frequencies because heritability (equation 5) is low. Gain is also low at high rooting frequencies because selection intensities become small—i.e., when most individuals possess the character under selection there is little difference between the mean of the selected population and the mean of the overall parental population. An important special case of selection for a threshold character occurs when trait frequency is equal to the proportion of individuals selected to serve as parents of the next generation. In that case, selection for the threshold character is exactly as efficient as if selection were based on the underlying trait (Dempster and Lerner, 1950).

The above discussion applies to mass selection where only a single observation (roots vs no roots) is made on each individual. Thus, special considerations associated with threshold trait selection would not apply to selection among families where the mean genotypic value would be a metric trait (although not necessarily normally distributed). Among-family selection is heavily weighted in advanced-generation breeding populations where genetic variation tends to be redistributed among families as opposed to within families. In comparison, more genetic variance occurs between individuals within families in rudimentary pedigrees. For example, ¾ of additive genetic variation occurs within half-sib, outcrossed F_1 families and ½ occurs within full-sib, outcrossed families. Thus, significant opportunity exists in early generations for genetic advance by selection among individuals within families, perhaps as much as 60% (Table 2).

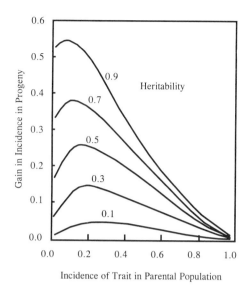

Incidence of Trait in Parental Population

Figure 5. Response to selection (gain) is greatest for a discontinuous variate with two phenotypic classes, i.e., rooting and non-rooting, when the incidence of the favorable condition (rooting) is in the range 0.15-0.25, regardless of the heritability of the underlying continuous trait.

Table 2. Gain from selection among and within families assuming various additive genetic variances and a constant environmental variance of 10 (arbitrary scale). Gain computations assume 20 individuals per family and a selection intensity of 10% among and within families in all cases.

| | | | | | Genetic Gain from Selection | | | |
| | | Half-Sib Families | | | | Full-sib families | | |
Additive genetic variance	h^2	Among	Within	Within (%)		Among	Within	Within (%)
100	0.91	8.7	14.2	62.2		12.3	11.3	47.8
80	0.89	7.7	12.5	61.9		11.0	9.9	47.4
60	0.86	6.7	10.6	61.4		9.5	8.3	46.6
40	0.80	5.4	8.3	60.6		7.7	6.4	45.3
20	0.67	3.7	5.2	58.5		5.4	3.9	42.0
10	0.50	2.5	3.1	55.4		3.7	2.3	37.7
5	0.33	1.7	1.8	51.7		2.6	1.2	32.9
1	0.09	0.5	0.4	44.2		0.9	0.3	23.6

That advance could be lost if selection is practiced on the discontinuous scale and if individuals within selected families root at frequencies near 1.0 or 0.0 (Fig. 5), which would be expected when among-family selection is practiced for either high or low rooting ability. Simple half- and full-sib family structures are most typical of breeding programs for forestry and horticultural crops, and less typical for agronomic crops. Thus, consideration of the threshold-nature of the rooting phenotype would be more important in the former than in the latter. Loss of within-family response to selection could be prevented by...

  • modifying the environment such that rooting frequency within the best-rooting families is reduced (selected families, positive selection) to near optimum frequencies (Fig. 5), or
  • using correlated variables such as number of roots, root weight, time to rooting, etc., that are likely to be metrically distributed.

Rooting and Genetic Assimilation

The current and past (Haissig and Riemenschneider, 1988) discussions of selective breeding for adventitious rooting have assumed that rooting is phenotypically expressed by at least some individuals in the population. Yet, that assumption is untrue in cases where the needs for genomic modification of rooting ability are most pressing. For example, it is often difficult or impossible to induce cuttings from mature woody plants to form roots regardless of the experimental treatment or environmental conditions [Hackett (1988); see the chapter by Haissig and Davis, and by Murray and coworkers in this volume]. Thus, there exist conditions where the desired phenotype occurs at a frequency of 0.0 [all individuals lie to the left of the threshold (Fig. 1)], genetic variation on the discontinuous threshold scale is nil and there is no direct breeding approach to genomic modification. The genetic amelioration of such conditions requires the appearance of a heritable trait where that trait did not previously exist: by definition, "genetic assimilation," also termed the inheritance of an acquired character (Waddington, 1942, 1953a,b).

That such amelioration could be accomplished was demonstrated by Waddington (1953a), whose interest in the modification of a discontinuous trait arose from a need to explain such change in an evolutionary context. Waddington's (1942) research began with the simple thesis that:

...Developmental reactions, as they occur in organisms submitted to natural selection, are in general canalized. That is to say, they are adjusted so as to bring about one definite end-result regardless of minor variations in conditions during the course of the reaction...

As evidence for this thesis, Waddington (1942) noted that, whereas tissue development and differentiation might be directed into one of many possible paths, it was difficult to cause the tissue to differentiate permanently into an intermediate state. Waddington (1942) also noted that the "constancy" of the wild-type required buffering of the genotype against environmental variations. Waddington's goal was to demonstrate experimentally that a novel developmental characteristic could arise in a population and become canalized. That goal was achieved a decade later (Waddington, 1953a,b) with the classic demonstration that genetic selection could result in the appearance of a formerly absent wing-vein character in population of flies. The wing character originally appeared only at one temperature in an experiment where 14 generations of selection for the trait were conducted. But, the trait also appeared spontaneously at a different temperature, at which the trait was formerly absent (Waddington, 1953a). Thus: "...Selection for the ability to respond to an environmental stimulus has built a strain in which the abnormal phenotype comes to be produced in the absence of any abnormal environment...." (Waddington, 1953a).

Waddington's result was clarified by Stern (1958) in terms that are more consistent with the current discussion:

If, in a population, genes are present which produce a certain phenotype in both environments A and B, and if selection for these genotypes can be accomplished initially in B only, then, by definition, selection in B will accomplish production of the trait in both A and B ... We may speak of them as subthreshold differences in A and supra-threshold differences in B....

Stern's (1958) reference to a threshold defined by some underlying metric scale thus resembles the model presented above (Fig. 1). And, it is reasonable to hypothesize that...

• selection can assimilate adventitious rooting ability into a population where rooting did not previously occur, if selection is conducted under conditions, or at a stage of development, where rooting can be observed.

A test of the assimilation hypothesis can be based on an appropriate selection experiment, with recognition of an underlying scale. Consider, for example, the genetic assimilation of a novel phenotype in relation to indirect selection where the selection and response criteria are observed under different environmental or developmental circumstances (Falconer, 1960). The expected correlated response to selection in an environment where rooting is not normally observed when a population is selected for rooting ability under conditions where rooting occurs is:

$$G = i h_R h_{NR} r_A \sigma_{P_{NR}} \tag{8}$$

where G is genetic gain, i is the selection intensity, r_A is the genetic correlation, h_R and h_{NR} are heritabilities for rooting in the two environments and $\sigma_{P_{NR}}$ is the phenotypic standard deviation in the non-rooting environment. Equation (8) emphasizes the importance of the underlying metric scale because on the outward scale (% rooted cuttings) h_{NR}, r_A, $\sigma_{P_{NR}}$ and response to selection (G) are all nil. The hypothesis that all components of equation (8) are non-zero in some underlying scale, and thus that rooting can be an assimilated trait, could be tested by a selection experiment where the non-rooting conditions are slightly subthreshold, such that one or a few generations of selection would cause the appearance of some suprathreshold genotypes. The conditions for such an experiment could be achieved using a population of perennial woody plants where selection is imposed at a young age and response is observed in select individuals during subsequent aging. An ideal experimental population would rapidly mature in loss of rooting ability (to quickly observe a supra-threshold response) and concomitantly hasten the onset of sexual reproduction (to produce advanced generations). The observation of rooting in select individuals at a later age than is observed in unselected controls would be strong evidence that the hypothesis of genetic assimilation is applicable to rooting. A suitable north temperate conifer species for such a test would be jack pine (*Pinus banksiana* Lamb.)

CONCLUSIONS

I began this chapter with the premise that genomic modification using the technologies of genetic engineering and selective breeding can produce plant materials suitable for fundamental studies of adventitious rooting. Then, I used a quantitative genetic model as the basis for discussing applicable metabolic control and quantitative genetic theories to provide guidelines for genomic modification. Application of metabolic control theory to an evaluation of genetic engineering suggests that engineering technologies have limits when the objective is to increase the flux of rooting-related biochemical pathways. However, genetic engineering might be an effective strategy when rooting can be modified by reducing pathway flux or when the activity of many pathway enzymes can be altered simultaneously via the engineering of regulatory loci. The potential power of metabolic control theory to evaluate genetic engineering strategies suggests that the theory be critically examined to test whether it is truly applicable to the processes involved in adventitious rooting. For example, the application of metabolic control theory to the formation of adventitious roots presupposes that steady state kinetics are applicable to rooting-related biochemical processes. Yet, the practice of cuttage is known to produce marked changes in tissue- and organ-related distribution of substrate and product pools [Davis et al. (1988) and references therein]. Recent work (Dick and Dewar, 1992) suggests that both static kinetics and threshold concepts may apply. Metabolic control theory also points to the need for knowledge of more than the characteristics of a target gene/enzyme. For example, it is doubtful that genetic engineering can be used to modify a trait unless the target enzyme has a sufficiently high control coefficient [Z_i (Kacser and Burns, 1981)] to effect change in the non-linear range of the flux-activity curve (Fig. 4). The flux control coefficient of a target enzyme cannot be estimated apart from other reactions in the pathway (equation 3). Thus, it is difficult to envision how a target enzyme could be identified without substantial knowledge of the pathway(s) (Fig. 1) to which it contributes. The strong appeal of metabolic control theory may be its ability to spawn an array of deductive hypotheses, which is characteristic of hypothetico-deductive systems (Bunge, 1967). Research based on hypothetico-deductive systems may be especially powerful when linked to the methods of multiple hypotheses (Chamberlin, 1897) and strong inference (Platt, 1964), and thus much needed in future studies of adventitious rooting [Haissig (1988); see the chapter by Haissig and Davis in this volume].

In contrast to genetic engineering, the technology of selective breeding is applicable to any population where the selection and response criteria are quantitatively inherited (Haissig and Riemenschneider, 1988) because: "...All variation is always open to such attack..." (Mather, 1971). In addition, breeding guidelines exist when rooting is observed as a discontinuous trait. Furthermore, selective breeding may be applicable when rooting is only observable in certain environments or during certain stages of development. The hypothesis that rooting can be a genetically assimilated trait, if true, suggests that selective breeding can introduce novel developmental processes into populations, an objective generally assigned to genetic engineering approaches in contemporary plant improvement.

While selective breeding and genetic engineering can be readily contrasted as technologies, the two strategies are congruent in their need for increased knowledge of continuous traits that underlie the discontinuous rooting phenotype. For example, it has already been noted that the application of genetic engineering requires substantial knowledge of underlying pathways. In addition, knowledge of an underlying scale would allow the use of continuous (metrical) selection criteria, which overcome inefficiencies associated with selection for a threshold trait (Fig. 5). Also, knowledge of an underlying scale would permit estimation of the correlated effects of selection for rooting on developmental stages where rooting is not normally observed (i.e., progress towards the genetic assimilation of an acquired trait). A suitable underlying scale for selective breeding could be based on biometrical association without knowledge of functional cause-effect relationships [i.e. information as opposed to knowledge (see the chapter by Rauscher in this volume)]. Thus, the continuous trait could be any morphological, anatomical, physiological or biochemical process that is related to rooting based on non-zero estimates of intertrait genetic correlations. The irrelevance of cause-and-effect knowledge to selective breeding was noted by Mather (1971): "In genetics it must be very rare for the metric we use in measuring the expression of a character to be based on biological considerations or to have any deep-seated biological significance...." Finally, there is a critical need for knowledge of correlated effects of genomic modification of adventitious rooting ability. For example, 33% of the mutations that affected root development in a mutable line of tomato also affected shoot characteristics (Zobel, 1975). Also, Kacser and Burns (1981) noted, based on metabolic control theory, that:

> ...There is universal pleiotropy. All characters should be viewed as quantitative since, in principle, variation anywhere in the genome affects every character....

Thus, correlated effects should be expected whether genomic modification of rooting ability is achieved by mutation, genetic engineering or selective breeding for naturally variant alleles. The methods of quantitative genetic analysis seems particularly well-suited to the study of such effects.

REFERENCES

Arihara, J., and Crosbie, T.M., 1982, Relationships among seedling and mature root system traits of maize, *Crop Sci.* 22:1197.

Briggs, G.E., and Haldane, J.B.S., 1925, A note on the kinetics of enzyme action, *Biochem. J.* 19:338.

Bunge, M., 1967, "Scientific Research I, The Search for System," Springer-Verlag, New York.

Carter, P.R., Sheaffer, C.C., and Voorhees, W.B., 1982, Root growth, herbage yield, and plant water status of alfalfa cultivars, *Crop Sci.* 22:425.

Chamberlin, T.C., 1897, Studies for students, The method of multiple working hypotheses, *J. Geol.* 5(8):837; reprinted as "Multiple Hypotheses, A Method for Research, Teaching, and Creative Thinking, Inst. for Humane Studies, Inc., Stanford.

Cornish-Bowden, A., 1976, The effect of natural selection on enzymic catalysis, *J. Mol. Biol.* 101:1.

Crafts-Brandner, S.J., Leggett, J.E., Sutton, T.G., and Sims, J.L., 1987a, Effect of root system genotype and nitrogen fertility on physiological differences between burley and flue-cured tobacco, I. Single leaf measurements, *Crop Sci.* 27:535.

Crafts-Brandner, S.J., Sutton, T.G., and Sims, J.L., 1987b, Root system genotype and nitrogen fertility effects on physiological differences between burley and flue-cured tobacco, II. Whole plant, *Crop Sci.* 27:1219.

Davis, T.D., Haissig, B.E., and Sankhla, N., eds., 1988, "Adventitious Root Formation in Cuttings," Adv. in Plant Sci. Ser., vol.2, Dioscorides Press, Portland.

Dean, A.M., Dykhuizen, D.E., and Hartl, D.L., 1987, Theories of metabolic control in quantitative genetics, *in:* "Proc. 2nd Int. Conf. on Quant. Genet.," B.S. Weir, M.M. Goodman, E.J. Eisen, G. Namkoong, eds., Sinauer Assoc., Inc. Sunderland, p. 536.

Dempster, E.R., and Lerner, I.M., 1950, Heritability of threshold characters, *Genetics* 35:212.

Dick, J. McP., and Dewar, R.C., 1992, A mechanistic model of carbohydrate dynamics during adventitious root development in leafy cuttings, *Ann. of Bot.* 70:371.

Eissa, A.M., Jenkins, J.N., and Vaughan, C.E., 1983, Inheritance of seedling root length and relative root weight in cotton, *Crop Sci.* 23:1107.

Falconer, D.S., 1960, "Introduction to Quantitative Genetics," The Ronald Press, New York.

Fillatti, J.J., Sellmer, J., McCown, B., Haissig, B., and Comai, L., 1987, *Agrobacterium* mediated transformation and regeneration of *Populus, Mol. Gen. Genet.* 206:192.

Gianola, D., 1979, Heritability of polychotomous characters, *Genetics* 93:1051.

Gianola, D., and Norton, H.W., 1981, Scaling threshold characters, *Genetics* 99:357.

Gruenberg, H., 1951, Genetical studies on the skeleton of the mouse, IV. Quasi-continuous variations, *J. Genet.* 51:95.

Hackett, W.P., 1988, Donor plant maturation and adventitious root formation, *in:* "Adventitious Root Formation in Cuttings," T.D. Davis, B.E. Haissig, and N. Sankhla, eds., Adv. in Plant Sci. Ser., vol. 2, Dioscorides Press, Portland, p. 11.

Hagen, D.W., and Blouw, D.M., 1983, Heritability of dorsal spines in the fourspine stickleback (*Apeltes quadracus*), *Heredity* 50:275.

Haissig, B.E., 1988, Future directions in adventitious rooting research, *in:* "Adventitious Root Formation in Cuttings," T.D. Davis, B.E. Haissig, and N. Sankhla, eds., Adv. in Plant Sci. Ser., vol.2, Dioscorides Press, Portland, p. 303.

Haissig, B.E., Davis, T.D., and Riemenschneider, D.E., 1992, Researching the controls of adventitious rooting, *Physiol. Plant,* 84:310.

Haissig, B.E., and Riemenschneider, D.E., 1988, Genetic effects on adventitious rooting, *in:* "Adventitious Root Formation in Cuttings," T.D. Davis, B.E. Haissig, and N. Sankhla, eds., Adv. in Plant Sci. Ser., vol.2, Dioscorides Press, Portland, p. 47.

Hansen, J.L., and Viands, D.R., 1989, Response from phenotypic recurrent selection for root regeneration after taproot severing in alfalfa, *Crop Sci.* 29:1177.

Jenison, J.R., Shank, D.B., and Penny, L.H., 1981, Root characteristics of 44 maize inbreds evaluated in four environments, *Crop Sci.* 21:233.

Kacser, H., 1983, The control of enzyme systems *in vivo*: Elasticity analysis of the steady state, *Biochem. Soc. Trans.* 11:35.

Kacser, H., and Beeby, R., 1984, Evolution of catalytic proteins or on the origin of enzyme species by means of natural selection, *J. Mol. Evol.* 20:38.

Kacser, H., and Burns, J.A., 1973, The control of flux, *Symp. Soc. Exp. Biol.* 27:65.

Kacser, H., and Burns, J.A., 1979, Molecular democracy: Who shares the controls?, *Biochem. Rev.* 7:1149.

Kacser, H., and Burns, J.A., 1981, The molecular basis of dominance, *Genetics* 97:639.

Kacser, H., and Porteous, J.W., 1987, Control of metabolism: what do we have to measure?, *Trends Biochem. Sci.* 12:5.

Keightly, P.D., and Kacser, H., 1987, Dominance, pleiotropy and metabolic structure, *Genetics* 117:319.

Ketring, D.L., 1984, Root diversity among peanut genotypes, *Crop Sci.* 24:229.

Mather, K., 1971, On biometrical genetics, *Heredity* 26:349.

Murphy, C.F., Long, R.C., and Nelson, L.A., 1982, Variability of seedling growth characteristics among oat genotypes, *Crop Sci.* 22:1005.

Oeller, P.W., Min-Wong, L., Talor, L.P., Pike, D.A., and Theologis, A., 1991, Reversible inhibition of tomato fruit senescence by antisense RNA, *Science* 254:437.

Pederson, G.A., Kendall, W.A., and Hill, R.R., Jr., 1984, Effect of divergent selection for root weight on genetic variation for root and shoot characters in alfalfa, *Crop Sci.* 24:570.

Platt, J.R., 1964, Strong inference, *Science* 146:347.

Porteous, J.W., 1983, Catalysis and modulation in metabolic systems, *Trans. Biochem. Soc.* 11:29.

Robertson, A., and Lerner, I.M., 1949, The heritability of all-or-none traits: Viability of poultry, *Genetics* 34:395.

Salter, R., Melton, B., Wilson, M., and Currier, C., 1984, Selection in alfalfa for forage yield with three moisture levels in drought boxes, *Crop Sci.* 24:345.

Self, S.G., and Leamy, L., 1978, Heritability of quasi-continuous skeletal traits in a randombred population of house mice, *Genetics* 88:109.

Smith, J.M., 1983, The genetics of stasis and punctuation, *Ann. Rev. Genet.* 17:11.

Stern, C., 1958, Selection for subthreshold differences and the origin of pseudoexogenous adaptations, *Amer. Nat.* 92:313.

Szathmary, E., 1993, Do deleterious mutations act synergistically? Metabolic control theory provides a partial answer, *Genetics* 133:127.

Tischler, C.R., Voigt, P.W., and Holt, E.C., 1989, Adventitious root initiation in kleingrass in relation to seedling size and age, *Crop Sci.* 29:180.

Torres, N.V., Mateo, F., Melendez-Hevia, E., and Kacser, H., 1986, Kinetics of metabolic pathways. A system *in vitro* to study the control of flux, *Biochem. J.* 234:169.

Viands, D.R., 1988, Variability and selection for characters associated with root regeneration capability in alfalfa, *Crop Sci.* 28:232.

Waddington, C.H., 1942, Canalization of development and the inheritance of acquired characters, *Nature* 150:563.

Waddington, C.H., 1953a, Genetic assimilation of an acquired character, *Evolution* 7:118.

Waddington, C.H., 1953b, The "Baldwin effect," "genetic assimilation" and "homeostasis," *Evolution* 7:386.

White, J.W., and Castillo, J.A., 1989, Relative effect of root and shoot genotypes on yield of common bean under drought stress, *Crop Sci.* 29:360.

Zobel, R.W., 1975, The genetics of root development, *in*: "The Development and Function of Roots," J.G. Torrey, and D.T. Clarkson, eds., Academic Press, New York, p. 261.

Zobel, R.W., 1986, Rhizogenetics (root genetics) of vegetable crops, *HortSci.* 21:956.

MODEL SYSTEMS FOR STUDYING

ADVENTITIOUS ROOT FORMATION

Stephen G. Ernst

101 Plant Industry
University of Nebraska
Lincoln, NE 68583-0814 USA

INTRODUCTION

Organogenesis involves a series of complex events, and the molecular mechanisms controlling organogenesis are not well understood in plant or animal systems. Apical meristems exert the greatest influence on organogenesis in plant development relative to other meristematic tissues, and are, therefore, important foci in studying plant development. The root apical meristem (primary, lateral and adventitious) is simpler in cellular composition compared to the shoot apical meristem, and therefore may be more amenable to genetic and developmental characterization relative to shoot apical meristems. However, the basic biology of *de novo* (adventitious) root apical meristem formation is still poorly understood, and will require a concerted effort in the scientific community to dissect the developmental events involved in the formation of this meristem (see the chapter by Barlow and by Haissig and Davis in this volume).

Limb formation in vertebrates is one of the better understood organogenic events, in part due to the focus on one or two key model systems—wing development in the chick embryo and limb regeneration in salamander. Wing development in the chick embryo has been standardized using one of several interchangeable measures of embryo development [e.g., days post fertilization, Hamburger-Hamilton stages as described by Hamburger and Hamilton (1951)], which makes it easier for researchers in different laboratories to interpret, chronologically and developmentally, the biochemical and morphological data associated with limb formation. For example, the chick limb bud at stage 20 (three days post-fertilization) is comprised of an ectodermal sheath enclosing mesenchymal tissue that at the histological level appears to be undifferentiated. However, using chick/quail chimeras, Christ et al. (1977) demonstrated that the mesenchymal cells have different origins and will differentiate into different tissues—the cells derived from the somites give rise to the striated muscles; the cells derived from the somatopleural cells form the skeleton and other soft tissues. Mesenchymal tissue in the posterior portion of the limb bud, known as the zone of polarizing activity (ZPA), is also histologically indistinguishable from the surrounding tissue, yet when transplanted to an anterior site in the limb bud it results in a duplication of the host wing (mirror-image) (Eichele, 1989). It was also determined that the steroid hormone retinoic acid (RA), when placed at the same anterior site, results in the same duplication as the transplanted ZPA (Tickle et al., 1985), and concentration gradients of RA across the limb bud are generated (Eichele and Thaller, 1987). However, there are some differences at the molecular level when comparing the effects of the ZPA and exogenously applied RA (Noji et al., 1991), and further studies are needed to understand the biochemical effects of the ZPA in controlling limb development.

This example of chick wing development is cited not so much to demonstrate biological parallels of organogenesis in animals versus plants, but to provide evidence of the gains that can be achieved by the scientific community focusing their efforts and resources on one or a few model systems. Many other examples from animals could be cited to document the gains possible when focusing on one or a few key model systems. The basic tenet of this paper is quite simple: it would behoove scientists working in adventitious rooting research to focus on one or a few model systems. This will result in a net increase in the critical mass of scientists and resources available to work on this problem, and make results from different laboratories more comparable. Once the genetic and physiological mechanisms involved in adventitious root formation are understood, these mechanisms can be manipulated to effect adequate rooting in commercially desirable species. Adventitious meristem formation is a very complex process, and the current dilution of efforts across several species slows progress in understanding the underlying mechanisms associated with the formation of apical meristems, whether embryonic or adventitious.

The major premise of this paper is that if the basic biology of root forma-tion—including formation of primary (embryonic), lateral and adventitious roots *in vivo* and *in vitro*—were better understood, then we would be in a position to more easily manipulate adventitious root formation in commercially important species and genotypes that are currently recalcitrant to efficient artificial induction of adventitious roots.

It my view there are two critical areas of basic plant biology that are relatively poorly understood and, therefore, limit our ability to manipulate plant organs, tissues or cells to desired endpoints, including adventitious rooting. First and foremost is a lack of understanding of the modes of action, and specifically signal transduction, of identified (or of as yet unidentified) plant hormones or growth factors. The second area is the lack of development of coordinated organ, tissue and cellular systems for studying differentiation, development and homeostasis. Development of one area is obviously dependent on advances in the other. For example, in the 1930s the fields of plant and animal hormone biology were relatively equal, especially in regard to identification and characterization of the respective hormones. However, for the last several decades advances in animal biology have far outpaced those in plant biology. While this is due in part to the greater critical mass of scientists and the greater amounts of research monies involved in animal versus plant research, these advances are also due to the availability of good animal cell and tissue models for studying causal mechanisms (e.g., signal transduction components) in specific stimulus-response systems. The animal developmental, growth factor and signal transduction literature is dominated by cellular models, and it would be advantageous for plant biologists to develop and use similar homogeneous cellular model systems. Adventitious root formation is obviously a very complex multicellular event, and a mixture of cellular, tissue and organ systems will be required to identify and characterize the many genetic and physiological components of this developmental process.

The objective of this paper is to identify possible model systems that would be useful for enhancing the research infrastructure and relevant knowledge base associated with the basic biology of adventitious root formation. First, an ideotypic model system at the whole plant and organ level will be considered, followed by a brief justification as to why the use of cellular model systems is essential as part of the long-term effort to better understand the biology of adventitious root formation.

SPECIES MODELS: A MACRO STRATEGY

The selection and use of model species or systems is totally dependent on the goals of a research program, such as whether it is basic or applied, or whole plant or cellular, in focus. In this paper, the focus will first be on the construction of an ideotypic model system for the study of adventitious rooting. This ideotype will then be compared to species or model systems previously or currently used to study general root morphogenesis and adventitious rooting. Only species or systems of greater utility for studying some of the underlying biological mechanisms associated with adventitious rooting will be included here. The concept of homogeneous cell lines will then be considered, to document how research at the cellular level can be used to better understand molecular processes in a complex developing organ such as an adventitious apical meristem.

Table 1. Characteristics of an ideotypic model system for studying causal mechanisms in adventitious rooting.

Genetic Characteristics
1. Well characterized genome.
   a. Physical and genetic linkage maps available.
   b. Small genome size with few linkage groups (chromosomes).
   c. Large depository of characterized and mapped mutants (biochemical and phenotypic) for traits of interest, and new mutants easy to make and screen.
   d. Large recombinant genomic library available for screening (e.g., in YACs).
   e. Little non-coding DNA.
2. Self- and cross-fertile, and easy to make crosses (easy to emasculate, small size).
3. High seed number per cross (i.e., per fruit) and per plant.
4. Short generation intervals (weeks to a few months).
5. Genetic variability for traits of interest.
6. Gene tagging system available (transposons, T-DNA, etc.).
7. Homologous recombination systems available for site-directed mutagenesis.
8. Genetic complementation available and relatively easy using traditional breeding or with genomic libraries.
9. Well developed cell and tissue transformation systems.

Developmental Characteristics
1. Rapid inducibility of adventitious root meristems.
2. Methods for differentially altering competence, induction and determination of adventitious root formation established.
3. Highly defined and precise chronology established for competence, induction and determination.
4. Able to go from whole plant to *in vitro* organ, tissue or cellular systems, and back to whole plant.
5. Established molecular and anatomical markers available for monitoring adventitious root meristem differentiation and development, and standardized at measurable developmental increments.
6. Established homogeneous cell lines that exhibit *in vivo* characteristics and that can be differentially induced to differentiate *de novo* root meristems and roots.
7. Physiological basis of genetically characterized mutants established.

General Characteristic
   Results transferable to other plant species or systems.

Ideotypic Model System

Since first formalized by Donald (1968), the concept of ideotype has been widely used in crop breeding, with both positive and negative results (e.g., Kelly and Adams, 1987; Rasmusson, 1987). The concept of an "ideal type" will be used here to help construct a "wish list" of characteristics that would be desirable to have in a model system for studying causal mechanisms associated with adventitious rooting.

Some of the characteristics that may be desirable in an ideotypic model system for the study of causal mechanisms of adventitious rooting are listed in Table 1. The characteristics listed in Table 1 are directed first at the species or cultivar level, with integration of whole plant and organ, tissue or cellular systems within that context. The list in Table 1 has two major subdivisions: genetic and developmental characteristics desirable in a model system. The power of a genetic approach when studying a given developmental event is well documented, e.g., *Drosophila* and *Caenorhabditis elegans* (see the chapter by Riemenschneider in this volume). The rapid rise of *Arabidopsis thaliana* as a model system in plants is in large part due to the relative ease in genetically characterizing developmental events using artificially induced mutants. The utility of many of the genetic characteristics listed in Table 1 has been amply documented and justified in the *Arabidopsis* literature and, therefore, will not be repeated here [see, e.g., reviews by Meyerowitz and Pruitt (1985) and in Koncz et al. (1992a)].

An argument can be made that such a genetically well-characterized system is not needed for studying the biology of adventitious rooting. However, it is my opinion that because adventitious root formation is such a complex process, mutation analysis will provide the best system for elucidation of the many steps involved in alteration of cellular competence (predisposition), induction and development of the *de novo* meristem. The many physiological and biochemical pathways involved need to be dissected; and, pathways, enzymes, structural elements, growth factors, etc. with major influences need to be determined. A simple comparison between recalcitrant and responsive genotypes will not allow for such dissection. Rather, I argue that a focused effort by many scientists, using one or a few model systems that utilize an array of tagged or characterizable mutants, has a much better chance of success, with various research groups focusing on key elements of *de novo* meristem formation at the genetic and biochemical level based on standardized developmental times for that model system.

While the list of genetic characteristics in Table 1 is more general, the list of developmental characteristics should be more specific to the phenomenon of interest, i.e., adventitious root formation. It is probably more realistic to be general at this stage, and to describe a model dealing with *de novo* meristem formation, regardless of whether it is shoot, flower or root. However, root apices are much simpler in cellular composition than shoot or floral apices (Steeves and Sussex, 1989). Therefore, the inclusion of each of these characteristics will be justified in regard to how a given characteristic will aid in understanding the developmental biology of adventitious root formation. This list was written with the assumption that primary and lateral root formation may also be useful models for understanding adventitious root formation. For example, Mayer et al. (1991) used saturation mutagenesis to document how disruption of the basic body pattern in the proembryo (mutations affecting apical-basal and radial organization, and cell shape) can affect morphology of the seedling, suggesting that positional effects may be important in both plant and animal embryogenesis, including the formation of the primary root apical meristem.

Considering the developmental characteristics listed in Table 1, several items are included for ease of experimental manipulation and monitoring. Rapid inducibility (hours to a few days) is included for convenience in monitoring response at the morphological and biochemical level. Specific mechanisms for altering competence, induction and determination events, and a defined and precise chronology for such, will allow for rapid comparison of experiments conducted in different laboratories. A standardized set of anatomical and molecular markers for monitoring the various developmental events will also aid in "phenocritical" comparisons. This would be similar to the cluster designation (CD) system used for characterization of hemopoietic cells, and the leaf plastochron index of the plant shoot apex. Medford (1992) provides a summary of the current reports for gene expression in shoot apices and apical meristems. A goal would be a system whereby one could map gene expression in the various cells and cell types that making up a *de novo* meristem at various developmental times.

For *de novo* root meristem formation, the molecular events associated with changes in competence and induction are of key importance in transfer of this technology to commercially important plants. An understanding of competence, induction and determination events will aid in characterization of mutants, and the mutants will be useful in separating biochemical events within a given competence, induction and determination framework.

It would be ideal if homogeneous cell lines were available that exhibited the full range of *in vivo* character and potential of each of the cell types that make up the developing *de novo* meristem. While the cell lines would not, in isolation, be expected to develop into a *de novo* meristem, they would be valuable in determining how a specific cell type at a given stage in the developmental process would respond in isolation to a specific manipulative event. Perhaps different cell lines could then be cultured in combination to monitor interactive effects, whether due to joint manipulation or to response of one cell type when another cell type is manipulated. To achieve the full capability of such a cell system, it would be ideal to be able to go from the whole plant down to the organ, tissue or cell level *in vitro*, and back to the whole plant, all the while maintaining as much *in vivo* character as possible. Later in this chapter I will present a more detailed discussion of homogeneous plant cell lines.

Characterization of the physiological basis of selected mutants would aid in dissection of the many biochemical events involved in formation of adventitious meristems. Such characterization would be required for determination of key developmental events, especially in controlling competence for *de novo* root meristem induction.

Listed in Table 1 is a third category of General Characteristics, with the goal of having a system that is readily transferable to other plant species or systems. Technology transfer is currently not well supported in the USA, and demonstration of transferability of this knowledge base to other organisms should be given high priority. Ideally, this is where privately funded or joint public-private ventures can be most useful.

Model Systems Currently Used to Study Adventitious Rooting

There are a plethora of possible model species and systems for adventitious root formation in the literature [e.g., Jarvis (1986), O'Toole and Bland (1987), Zobel (1991; see also the chapter by Haissig and Davis in this volume)]. The species and systems outlined here include only those that have some kind of genetic analysis component; e.g., linkage maps, identification and analysis of mutants, quantitative genetics of adventitious root formation, etc. There are many other good model systems that are not well characterized genetically (e.g., onion, pea, mung bean, sunflower), but which may be very useful on a physiological level or if genetic analyses were more complete. The species and systems outlined here include mutants associated not only with adventitious root formation, but also those with mutations in primary and lateral root growth and development. No woody plant systems are included because of their usually long generation intervals and the relative lack of genetic and physiological characterization. Also, I do not intend to imply that only one model species should be used exclusively. However, if there was one predominant model species that incorporated both whole plant and cellular systems, the research community might be more integrated and focused in dissection of the causal mechanisms of adventitious root formation (e.g., by sharing of mutants, molecular markers for specific developmental events, cell lines, etc.).

Schiefelbein and Benfey (1991) summarized a wide array of genetically characterized root mutants. Of the eight species listed, the mutants in *Arabidopsis* and tomato

(*Lycopersicon esculentum*) were the most extensive and best characterized. In addition, these two species have excellent genetic and physiological infrastructure. *Arabidopsis* has a more limited history in regard to the study of adventitious root formation, but potential exists using bolt (stem) or leaf explants. However, *Arabidopsis* may be the species of choice for studies of lateral or primary root formation and development; the resulting information could have wide applicability to adventitious root formation [e.g., see root mutants in Schiefelbein and Benfey (1991) and Aeschbacher and Benfey (1992)]. Maize has excellent genetic characterization, including prospects for studying root mutants due to transposon mutagenesis, but its utility in studying adventitious root formation may be limited. Tobacco also has several characterized mutants, but future work in tobacco should probably be focused in one or a few species, and primarily those with a somatic chromosome number of 24 (e.g., vs 48 in the amphiploid *Nicotiana tabacum*) to minimize the effects of gene duplication.

For studying adventitious root formation, *Lycopersicon* spp. (primarily *Lycopersicon esculentum*; tomato) currently best resembles the ideotype outlined in Table 1. Tomato generally forms adventitious roots readily from several explant types such as stems and leaves (Koblitz, 1991). There are also many characterized mutations, including mutations in adventitious root formation on the stem (Zobel, 1991). Tomato is well characterized genetically, including genetic and physical linkage maps (Bernatzky and Tanksley, 1986; Helentjaris et al., 1986; Mutschler et al., 1987; DeVerna and Paterson, 1991), a wide genetic base including use of wild relatives (DeVerna and Paterson, 1991), small genome (2C DNA content of 1.48 pg and 10^9 bases per haploid genome) with few chromosomes (12 haploid) (Kalloo, 1991), a large number of single gene mutants [e.g., Rick (1984)], short generation time with 5,000 to 25,000 seeds per plant and self-fertile (Kalloo, 1991), several gene transfer methods (Hille et al., 1991), and a relatively low level of gene duplication [10%; Zobel (1991)]. The utility of transposons from corn and snapdragon are also being investigated. Tomato is also relatively well characterized physiologically, although primarily in fruit characteristics, environmental stress and pest resistance. More research directed towards hormone physiology and root differentiation and development would be essential. Much research has been directed toward establishment of cell, tissue and organ culture systems of tomato (Koblitz 1991), including somaclonal variation (Sree Ramulu, 1991). However, investigation of altering cellular competence, induction and determination of adventitious rooting must be initiated, including setting up a chronologically repeatable system. In addition, the genetic variability of the competence/induction response for adventitious rooting should be investigated. Then some initial anatomical and molecular markers for root formation can be identified. Also, the physiological basis of currently available adventitious root formation mutants should be initiated.

Arabidopsis is superior to tomato in terms of molecular genetic potential, especially in regard to the effort required to make and characterize mutants via classical and T-DNA insertion methods [see, e.g., reviews by Meyerowitz and Pruitt (1985) and in Koncz et al. (1992a)]. Separation of competence and induction has been demonstrated for shoot regeneration from root explant tissue (Valvekens et al., 1988), and this could easily be extended to *de novo* root formation on the root explants. Extension to bolt (stem) or leaf explants will require more work. However, there is good genotypic variation in response to culture *in vitro* (Koncz et al., 1992b), and this could be used to advantage in determining competence response and in identifying molecular markers associated with root regeneration.

There are many other systems with good physiological characterization specifically for adventitious rooting [e.g., the sunflower hypocotyl system; see Liu et al. (1990)], but which lack good genetic characterization. Whether the genetic characterization can be developed with sufficient speed to make them competitive with tomato or *Arabidopsis* is an obvious point for consideration.

CELL LINE MODELS: A REDUCTIONIST STRATEGY

General Comments

Relative to animal cell systems, plant cell systems are very crude and heterogeneous. Most *in vitro* plant culture systems use relatively intact tissue—e.g., stem, leaf, root or floral sections, which are highly heterogeneous in regard to cell type. Most of the plant cell lines that have been developed are cell suspension or callus lines initiated by supplying high levels (i.e., well above *in vivo* levels) of exogenously supplied growth regulators to stimulate cell proliferation. However, these "cell lines" are actually quite heterogeneous in cell composition, as determined by monitoring cellular metabolism and response of the cells to exogenously supplied factors [changes in nutrients, growth regulators, etc.; e.g., King et al. (1973)]. In addition, these cell cultures are quite different from the *in vivo* cells from which they originated, and may not be appropriate model systems to study the majority of developmental and physiological features of specific *in vivo* cell types. The plant cell system that best approaches an animal cell system in homogeneity of cell type and state is the suspension cultures of *Zinnia elegans* derived from leaf mesophyll cells and used for induction of xylogenesis (Fukuda and Komamine, 1980). However, the cellular heterogeneity of this system is still more similar to a primary culture of animal cells than the homogeneous cell systems available for animals (including immortalized and non-immortalized cell systems). In other words, the homogeneous cell lines envisioned here are much different than any cell line currently available in plants. The biochemistry, physiology and development of the cell lines would be much more similar to cells *in vivo*. In addition, they would be homogeneous and could be synchronized for a given developmental pathway for ease of molecular characterization of causal factors.

To study the specific signalling factors associated with growth, differentiation, cell-to-cell communication, etc. in plants with minimal "noise," the establishment of homogeneous cell systems from a wide variety of cell types is essential. To accomplish this, we first need to determine methods for sorting and growing relatively intact cells [i.e., cells with the majority of the cell wall still intact to maintain extracellular matrix (ECM) integrity], and put them in an *in vitro* environment which will allow them to maintain as much *in vivo* character as possible. These cell lines can then be used to molecularly and temporally dissect the events associated with the formation of an adventitious meristem. In most cases, *de novo* meristem formation would not occur in the cell line. Rather, the cell line would be used to study the effects of specific factors on cellular differentiation, including cell types that make up a *de novo* meristem at various stages of development. For example, it is generally recognized that lateral root formation occurs in the pericycle adjacent to vascular poles (the type and proximity of the pole depending on the species) (Charlton, 1991). Perhaps a homogeneous culture of pericycle cells could be developed that retains enough *in vivo* character that some of the causal factors associated with competence changes in lateral root induction could be studied in isolation. The effects of growth factors, artificial vascular poles, cell-to-cell communication via diffusion, plasmodesmata, or ECM interactions, cellular sensitivity to growth factors, possible second messengers and a variety of other factors could be considered.

Requirements for Establishing Homogeneous Plant Cell Lines

There are at least three major elements that must be addressed in the development of homogeneous cell lines in plants: the effects of the cell wall, the selection of specific cell types and the *in vitro* environment. The rigid cell wall is obviously a deterrent to the separation of specific cell types, yet cell wall components may also be determinants of cellular differentiation and homeostasis (Ryan and Farmer, 1991). Very little is known regarding how much *in vivo* character is maintained in "protoplasted" cells, including the newly regenerated cell wall, in part because they are usually cultured in a medium and environment that would not allow them to express those characteristics. Other methods may also be necessary for separation of cells, e.g., partial enzymatic degradation followed by sonication or other physical separation methods. Biochemical and immunocytological tools are available to study the effects of "protoplasting" cells relative to maintenance of *in vivo* character. However, cell type-specific molecular markers are currently insufficient for such.

Once the plant tissue has been dissociated into individual plant cells that maintain sufficient *in vivo* character, the cell type or types one wishes to work with must be selected and separated for the establishment of the primary culture. Typical mechanisms used in animal cell culture are dilution, biochemical selection usii g physiological determinants or antibodies for cell sorting (Freshney, 1987). Dilution via repeated subculturing may be the easiest option if proper *in vitro* environment can be maintained for an extended time period, and if apoptosis (programmed cell death) is not a factor within the time frame needed for establishment and testing. The use of biochemical selection will require much more information regarding the physiology of specific cell types, including metabolism, growth factor requirements for homeostasis or development, etc. Cellular selection will most generally require the use of cell type-specific molecular markers and, therefore, these markers must first be identified. Very few markers are currently available, but studies of gene expression at the cellular and tissue level are rapidly adding to that data base (see the chapter by Mohnen, and by Hand in this volume.) However, this is where a common model species and system(s) becomes most important. For example, if several groups of scientists were working with a homogeneous cell line of pericycle cells from tomato or *Arabidopsis* to study competence for lateral root meristem induction *in vitro*, and common methods of developmental synchronization were applied, a data base of cellular markers could be developed and used to standardize chronologically and developmentally the development of the meristem, including important intermediate states that are not anatomically apparent. This is analogous to the cluster designation (CD) factors identified for hematopoietic cell development in animals or to the use of 2D gel protein data bases [see, e.g., Bauw et al. (1989)]. Immobilized antibodies would also be a very efficient means of selecting and separating specific cell types, but will depend on the developmental specificity of cell wall antigens [e.g., Hong et al. (1989) and de Oliveira et al. (1990)], and whether the cell wall must be maintained to retain *in vivo* character (i.e., whether plasma membrane antigens would be available to immobilized antibodies).

It is the *in vitro* environment that will probably require the greatest amount of research and innovation in the development of homogeneous cell lines in plants. The factors that need be considered for the *in vitro* environment are too numerous to be detailed here. However, in development of the *in vitro* environment, one must be constantly cognizant of the *in vivo* environment from which the cells were derived. For example, plant cells are constantly bathed in both passively (i.e., diffused) and actively maintained bipolar or unipolar streams of nutrients, growth factors, ionic constituents (including maintenance of pH) and intercellular and transmembrane potentials (electrical and chemical); plant cells are under constant pressure from adjacent cells, and the cell walls of those adjacent cells may also exert biochemical and/or physical developmental influences. For establishment of cell suspensions the cells are typically removed from the above-described *in vivo* environment and placed in an *in vitro* environment that has abnormal concentrations, mixtures and polarities or flows of growth factors, and adjacent cell types. Additionally, associated physical barriers are not maintained, and intercellular and transmembrane potentials are ignored, and there may be other abnormalities associated with a host of non-identified factors. For example, pressure and adjacent cell type have been implicated in possibly altering development in plant tissue cultures (Brown and Sax, 1962; Tran Thanh Van, 1980). Perhaps some type of confinement system with added specific ECM components with specified flow rates of specific growth factors will be required to maintain certain homogeneous cell lines, similar to the artificial capillary systems used for endothelial cell research. In addition, the current generation of plant cell culture media are probably not adequate for the maintenance of such homogeneous cell lines. For example, it was recently demonstrated that the iron-EDTA composition used in most popular plant tissue culture media is readily converted by visible light into formaldehyde, with a concomitant precipitation of the iron, resulting in poorer *in vitro* growth (Hangarter and Stasinopoulos, 1991). Also, if metabolism of specific cell types were better understood, more complete and specialized media could be developed.

SUMMARY

In summary, one means of enhancing the research infrastructure for investigation of causal factors of adventitious rooting may be the identification and development of ideotypic model species, and the development of homogeneous cellular model systems. These models would allow for more coordinated and integrative approaches in the identification and characterization of causal factors, and thereby enhance the pace of our understanding of the basic biology of adventitious root formation.

Arabidopsis is superior to tomato in terms of molecular genetic potential, including efforts required to genetically and molecularly characterize mutants. Study of adventitious root formation would be relatively new to the *Arabidopsis* model system, but the current and future data base for primary and lateral root formation in *Arabidopsis* should complement research in adventitious root formation. Concomitant development of homogeneous cell lines will be more difficult, but is necessary for determination of many of the specific factors involved in formation of a *de novo* root meristem.

ACKNOWLEDGMENTS

I thank Drs. Sandy L. Smith and Carol M. Schumann, and three anonymous reviewers for critical comments on earlier versions of this manuscript.

REFERENCES

Aeschbacher, R.A., and Benfey, P.N., 1992, Genes that regulate plant development, *Plant Sci.* 83:115.

Bauw, G., van Damme, J., Puype, M., Vandekerckhove, J., Gesser, B., Ratz, G.P., Lauridsen, J.B., and Celis, J.E., 1989, Protein-electroblotting and -microsequencing strategies in generating protein data bases from two-dimensional gels, *Proc. Natl. Acad. Sci. USA* 86:7701.

Bernatzky, R., and Tanksley, S.D., 1986, Toward a saturated linkage map in tomato based on isozymes and random cDNA sequences, *Genetics* 112:887.

Brown, C.L., and Sax, K., 1962, The influence of pressure on the differentiation of secondary tissues, *Amer. J. Bot.* 49:683.

Charlton, W.A., 1991, Lateral root initiation, *in*: "Plant Roots, The Hidden Half," Y. Waisel, A. Esheland, and U. Kafkafi, eds., Marcel Dekker, Inc., New York, p. 103.

Christ, B., Jacob, H.J., and Jacob, M., 1977, Experimental analysis of the origin of the wing musculature in avian embryos, *Anat. Embryol.* 150:171.

de Oliveira, D.E., Seurinck, J., Inze, D., van Montagu, M., and Botterman, J., 1990, Differential expression of five *Arabidopsis* genes encoding glycine-rich proteins, *Plant Cell* 2:427.

DeVerna, J.W., and Paterson, A.H., 1991, Genetics of *Lycopersicon*, *in*: "Genetic Improvement of Tomato," G. Kalloo, ed., Springer-Verlag, New York, p. 21.

Donald, C.M., 1968, The breeding of crop ideotypes, *Euphytica* 17:385.

Eichele, G., 1989, Retinoids and vertebrate limb pattern formation, *Trends Genet.* 5:246.

Eichele, G., and Thaller, C., 1987, Characterization of concentration gradients of a morphogenetically active retinoid in the chick limb bud, *J. Cell Biol.* 105:1917.

Freshney, R.I., 1987, "Culture of Animal Cells," Wiley-Liss, Inc., New York.

Fukuda, H., and Komamine, A., 1980, Establishment of an experimental system for the study of tracheary element differentiation from single cells isolated from the mesophyll of *Zinnia elegans*, *Plant Physiol.* 65:57.

Hamburger, V., and Hamilton, H.L., 1951, A series of normal stages in the development of the chick embryo, *J. Morph.* 88:49.

Hangarter, R.P., and Stasinopoulos, T.C., 1991, Effect of Fe-catalyzed photooxidation of EDTA on root growth in plant culture media, *Plant Physiol.* 96:843.

Helentjaris, T., Slocum, M., Wright, S., Schaefer, A., and Nienhuis, J., 1986, Construction of genetic linkage maps in maize and tomato using restriction fragment length polymorphisms, *Theor. Appl. Genet.* 72:761.

Hille, J., Zabel, P., and Kornneef, M., 1991, Genetic transformation of tomato and prospects for gene transfer, *in*: "Genetic Improvement of Tomato," G. Kalloo, ed., Springer-Verlag, New York, p. 283.

Hong, J.C., Nagao, R.T., and Key, J.L., 1989, Developmentally regulated expression of soybean proline-rich cell wall protein genes, *Plant Cell* 1:937.

Jarvis, B.C., 1986, Endogenous control of adventitious rooting in non-woody cuttings, *in*: "New root Formation in Plants and Cuttings," M.B. Jackson, ed., Martinus Nijhoff Pubs., Dordrecht, p. 191.

Kalloo, G., ed., 1991, "Genetic Improvement of Tomato," Springer-Verlag, New York.

Kelly, J.D., and Adams, M.W., 1987, Phenotypic recurrent selection in ideotype breeding in pinto beans, *Euphytica* 36:69.

King, P.J., Mansfield, K.J., and Street, H.E., 1973, Control of growth and cell division in plant cell suspension cultures, *Can. J. Bot.* 51:1807.

Koblitz, H., 1991, Cell, tissue and organ culture in *Lycopersicon*, *in*: "Genetic Improvement of Tomato," G. Kalloo, ed., Springer-Verlag, New York, p. 231.

Koncz, C., Chua, N.-H., and Schell, J., eds., 1992a, "Methods in Arabidopsis Research," C. Koncz, N.-H. Chua, and J. Schell, eds., World Sci., River Edge, p. 482.

Koncz, C., Schell, J., and Redei, G.P., 1992b, T-DNA transformation and insertion mutagenesis, *in*: "Methods in Arabidopsis Research," C. Koncz, N.-H. Chua, and J. Schell, eds., World Sci., River Edge, p. 224.

Liu, J.-H., Mukherjee, I., and Reid, D.M., 1990, Adventitious rooting in hypocotyls of sunflower (*Helianthus annus*) seedlings, III. The role of ethylene, *Physiol. Plant.* 78:268.

Mayer, U., Torres Ruiz, R.A., Berleth, T., Misera, S., and Jurgens, G., 1991, Mutations affecting body organization in the *Arabidopsis* embryo, *Nature* 353:402.

Medford, J.I., 1992, Vegetative apical meristems, *Plant Cell* 4:1029.

Meyerowitz, E.M., and Pruitt. R.E., 1985, *Arabidopsis thaliana* and plant molecular genetics, *Science* 229:1214.

Mutschler, M., Tanksley, S.D., and Rick, C.M., 1987, 1987 linkage maps of the tomato (*Lycopersicon esculentum*), *Rep. Tomato Genet. Coop.* 37:5.

Noji, S., Nohno, T., Koyama, E., Muto, K., Ohyama, K., Aoki, Y., Tamura, K., Ohsugi, K., Ide, H., Taniguchi, S., and Saito, T., 1991, Retinoic acid induces polarizing activity but is unlikely to be a morphogen in the chick limb bud, *Nature* 350:83.

O'Toole, J.C., and Bland, W.L., 1987, Genotypic variation in crop plant root systems, *Adv. Agron.* 41:91.

Rasmusson, D.C., 1987, An evaluation of ideotype breeding, *Crop Sci.* 27:1140.

Rick, C.M., 1984, Stock list, *Rep. Tomato Genet. Coop.* 34:22.

Ryan, C.A., and Farmer, E.E., 1991, Oligosaccharide signals in plants: a current assessment, *Annu. Rev. Plant Physiol. Mol. Biol.* 42:651.

Schiefelbein, J.W., and Benfey, P.N., 1991, The development of plant roots: new approaches to under-ground problems, *Plant Cell* 3:1147.

Sree Ramulu, K., 1991, Genetic variation in *in vitro* cultures and regenerated plants in tomato and its implications, *in*: "Genetic Improvement of Tomato," G. Kalloo, ed., Springer-Verlag, New York, p. 259.

Steeves, T.A., and Sussex, I.M., 1989, "Patterns in Plant Development," Cambridge Univ. Press, New York.

Tickle, C., Lee, J., and Eichele, G., 1985, A quantitative analysis of the effect of all-*trans*-retinoic acid on the pattern of chick wing development, *Dev. Biol.* 109:82.

Tran Thanh Van, K., 1980, Control of morphogenesis by inherent and exogenously applied factors in thin cell layers, *Int. Rev. Cytol.* 11A:175.

Valvekens, D., van Montagu, M., and van Lijsebettens, M., 1988, *Agrobacterium tumefaciens*-mediated transformation of *Arabidopsis thaliana* root explants by using kanamycin selection, *Proc. Natl. Acad. Sci. USA* 85:5536.

Zobel, R.W., 1991, Genetic control of root systems, *in*: "Plant Roots, The Hidden Half," Y. Waisel, A. Eshel, and U. Kafkafi, eds., Marcel Dekker, Inc., New York, p. 27.

NOVEL EXPERIMENTAL SYSTEMS FOR

DETERMINING CELLULAR COMPETENCE AND DETERMINATION

Debra Mohnen

Complex Carbohydrate Research Center and
Department of Biochemistry
University of Georgia
220 Riverbend Road
Athens, GA 30605 USA

INTRODUCTION

Plants are aggressive in their development. They can overcome severe injuries, such as removal of a limb or decapitation, by forming new centers of cell division called meristems that develop into roots, vegetative shoots or flowers. A major, albeit elusive, goal in plant developmental biology is to understand the molecular basis for the plasticity that allows differentiated cells to embark upon new developmental programs.

Organs that arise from sites in the plant other than their normal sites in the embryo or the primary root are defined as adventitious (Esau, 1977; Lovell and White, 1986; see the chapters by Barlow, and by Haissig and Davis in this volume). Thus roots that arise on stems or leaves, or on explants from stem or leaf tissues, are termed adventitious roots. Adventitious is derived from the Latin *adventicious*, meaning foreign. The formation of organs within tissues or cell clusters in which they are normally foreign requires that a cell or group of cells embark upon a new developmental program. Several questions immediately emerge from this realization. Do all cells/tissues have the potential for adventitious organ formation? If so, do all cells/tissues require the same inducers in order to form adventitious organs? What is the molecular mechanism that controls adventitious organ formation? How many different signals are needed for induction of a specific type of organ (i.e., root, vegetative shoot or flower)? It is clear that a multiplicity of experimental approaches, including developmental, anatomical and molecular, will be needed to answer these questions. One experimental approach, transfer experiments using tissue explants to study the developmental stages of root organogenesis, will be discussed in this paper.

The specific goals of this paper are threefold. One is to discuss two early developmental steps required for adventitious root formation: competence and determination (see chapter by Murray and coworkers in this volume). This will include a review of

Biology of Adventitious Root Formation, Edited by
T.D. Davis and B.E. Haissig, Plenum Press, New York, 1994

several *in vitro* experimental systems that have been used to study competence and determination of cultured tissue explants for root formation. The second goal is to introduce those unfamiliar with thin cell-layer explants (TCLs) to the potential advantages of these tissue explants for studies of root organogenesis. The *in vitro* organogenesis of tobacco TCLs is a useful experimental system for studies of plant developmental plasticity since the same tissue-origin TCL can be induced to form vegetative shoots, flowers or roots, by culture on media that differ only in phytohormone content (Mohnen et al., 1990). The timing of competence and determination of TCLs for root formation will be compared with other *in vitro* systems. Finally, possible future directions to take in order to answer unresolved questions regarding competence and determination of tissues for adventitious root formation will be discussed.

DEFINITIONS OF COMPETENCE AND DETERMINATION

Competence and determination are operationally defined, related terms which describe the state of reactivity of, and response of, cells, tissues or organs to inducing factors (Meins, 1986). Both terms have been adapted to plants from animal studies. The term competence was first used in animal studies to represent the transient period of time during which embryo cells could become determined for a specific developmental progression (Waddington, 1934). Competence in plants is broadly defined as a state of reactivity of cells to respond to specific stimuli (Meins, 1986). In this paper competence for root formation will be defined as the ability of cells within tissues to respond to specific root-inducing stimuli by the formation of roots. Once competent cells/tissues have been exposed to an inducer the tissues may become determined for root formation. Determination is defined as the commitment of cells to a specific developmental fate (Meins and Binns, 1979). Thus, once a cell, or group of cells, has received a signal for root formation, they will remain committed to root formation even upon the removal of the signal, i.e., they will continue root organogenesis in the absence of the root-inducing factor. Christianson and Warnick (1985) have outlined a minimum of three distinct phases that are required for root organogenesis of tissue explants (Fig. 1).

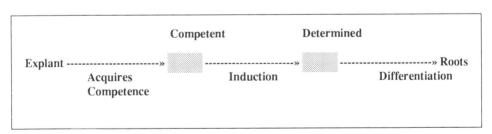

Figure 1. Phases in the organogenic process. [adapted from Christianson and Warnick (1985)]

The state of determination can only be ascertained by experimental manipulation of cells, tissues or organs (Meins and Binns, 1979). The experimental manipulation most often used for studies of determination for organogenesis has entailed transfer of tissue explants from media containing a root-inducing factor, so-called root-inducing medium (RIM) to a medium without the factor (e.g., basal medium) and counting the number of roots after a fixed period of time. The root-inducing factor(s) is usually auxin or auxin in combination with relatively low levels of cytokinin. There is evidence that the phytohormones auxin and cytokinin are necessary for competence and determination of tissue explants to form roots. For example, it has been suggested that in *Convolvulus*, auxin is required for both the attainment of competence for primordia formation and for the determination to form roots (Bonnett and Torrey, 1965; Lyndon, 1990).

Tissues are said to have become determined for root formation at that point in time when, after being removed from the root-inducing factor, they continue with root formation. Fig. 2 illustrates various expected results in a hypothetical experiment designed to test for the period of competence and determination of a tissue for root formation. In Fig. 2A an explant is cultured for a period of time on RIM, transferred to basal medium (BM) which has no root-inducing factor and, after a fixed period of time scored for root formation. Since no roots formed with this treatment it is concluded that the time period on RIM was insufficient for the tissue to become determined for root formation. Fig. 2B, however, depicts a tissue that was cultured on RIM for a period of time just sufficient to allow cells within the tissue to become determined for root formation. Thus, Fig. 2B illustrates the point of determination of the tissue explant for root formation.

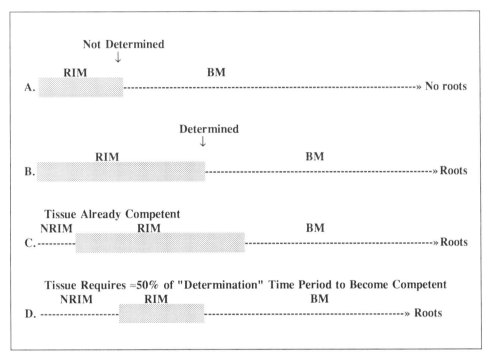

Figure 2. Diagram demonstrating how transfer experiments can be used to ascertain the time of determination and competence for root formation. (A) Culture of an explant for insufficient time on RIM to allow the explant to become determined for root formation. (B) Culture of an explant on RIM for sufficient time for the explant to become determined for root formation. (C) Example of an explant that is competent for root formation at the beginning of the culture period. Preculture on NRIM does not shorten the length of culture required for determination (cf. Fig. 2B), thus indicating that the entire time on RIM is necessary for attainment of determination. (D) Example of an explant that is not competent for root formation at the beginning of the culture period. Preculture on NRIM reduces the length of time required on RIM (cf. Fig. 2B) in order to induce root formation. Dashed rectangles (▨) represent culture on root-inducing medium (RIM); horizontal lines (----) denote culture on a non-root-inducing medium (NRIM, medium without the root-inducing factor) or on a basal medium (BM, medium without phytohormones).

Elucidation of the time required for determination leads to several questions: Was the tissue explant competent to respond to the root-inducing factor at the beginning of the culture period?—i.e., was the tissue explant competent for root formation when it was removed from the plant? Alternatively, is the entire time period depicted in Fig. 2B required for determination of the explant for root formation? or is a portion of the time required for the tissue to become competent to respond to the root-inducing factor? These questions can be experimentally addressed by culture of the explant on a non-root-inducing (NRIM) medium prior to culture on a root-inducing medium as shown in Fig. 2C and 2D. The logic is that by culturing the explant under conditions similar to root-inducing conditions, except that the root-inducing factor is absent (i.e., culture on NRIM), the tissue may be able to undergo any physiological responses necessary for competence. The assumption made is that the root-inducing factor(s) is not also necessary for induction of competence. Fig. 2C represents a tissue which requires the full period of culture on RIM (as ascertained in the experiment shown in Fig. 2B) in order to form roots. The conclusion here is that, since preculture on NRIM did not shorten the period of culture on RIM required for determination, the tissue was competent to react to the root-inducing factor at the start of the experiment. Fig. 2D depicts a tissue which is not competent to respond to the root-inducing factor at the start of the experiment. Prior culture on NRIM shortened the length of time required for the explant to be cultured on RIM in order to form roots (compare Figs. 2B and 2D). This is interpreted to mean that about 50% of the determination period depicted in Fig. 2B is required for the tissue to become competent to respond to a root-inducing factor and 50% of the time is required for the tissue to react to the root-inducing factor by becoming committed to root formation. The molecular basis for competence is not known, although one possibility is that competence reflects the expression of receptors for the root-inducing factor (Wareing and Graham, 1984).

Table 1. Examples of *in vitro* systems for studying adventitious root formation.

Tissue	Plant	Reference
Shoots from shoot cultures	*Malus pumila* Mill.	James, 1983
Microcuttings from shoot cultures	*Malus domestica*	Mullins, 1985; Sriskandarajah et al., 1982
De-bladed petioles	*Hedera helix*	Geneve et al., 1988
Leaf cuttings	*Hedera helix*	Geneve et al., 1991
Callus	*Medicago sativa*	Walker et al., 1979
Leaf explant	*Nicotiana tabacum*	Attfield & Evans, 1991b
Leaf explant	*Convolvulus arvensis*	Christianson & Warnick, 1985; Warnick, 1992
Thin cell-layer explants (TLCs)	*Nicotiana tabacum*	Mohnen; Mohnen, Eberhard, Albersheim & Darvill; unpublished

SYSTEMS FOR STUDYING COMPETENCE AND DETERMINATION FOR ADVENTITIOUS ROOT FORMATION

The observations by Skoog and Miller (1957) that root and shoot organogenesis in tobacco tissue cultures can be controlled by manipulation of the concentration of auxin and cytokinin in the medium led to an appreciation that in many plant species organogenesis of tissue explants can be controlled by auxin, or auxin and cytokinin. With this apparent ease of manipulating *in vitro* organogenesis, it is surprising that so few studies on the competence and determination of tissues for root formation have been reported (McDaniel, 1984). Table 1 lists examples of non-whole-plant (*in vitro*) studies, of competence and/or determination for root formation. Three general types of excised tissues have been used to study competence and determination for root formation: organ cultures such as shoot cultures and leaf cuttings, callus cultures and tissue cultures such as leaf explants and thin cell-layer explants. Several reports of the early developmental events associated with root formation in *in vitro* cultured tissues will be discussed below in an effort to emphasize unifying principles. The tissues to be discussed include apple shoot cultures, English Ivy leaf cuttings, alfalfa callus, field bindweed leaf explants, and tobacco leaf explants and thin cell-layer explants.

Apple Shoot Cultures

Excised shoots of apple cultures have been used to study factors that regulate the formation of roots and to elucidate the length of exposure to auxin required for determination for rooting. Excised shoots of the apple rootstock M.9 can be induced to root by *in vitro* culturing in the presence of 10 to 15 µmol L^{-1} indolebutyric acid (IBA) (James and Thurbon, 1979; James, 1983). Excised shoots exposed to IBA-containing media for as little as 12 hours, followed by transfer to hormone-free media, formed roots (James, 1983). Thus, the excised shoots were determined for root formation in response to IBA in ≤ 12 hours, although optimal rooting required four days. The 12-hour period was not further subdivided to establish whether there was a requirement for the organs to become competent to respond to IBA, however, the fact that different rootstocks differ in their rooting potential in response to auxin (i.e., high-root potential rootstock M.26 versus low rooting potential rootstock M.9) suggests that different rootstocks may differ in their competence for rooting (James and Thurbon, 1981). James (1983) compared the optimal rooting response of different types of organ culture in a range of plants, and concluded that the length of exposure to auxin required to yield optimal rooting (i.e., the time required for determination for rooting) is dependent on the type and concentration of auxin used and on the species and variety of plant.

Studies with microcuttings of apple (leafy explants 50 mm in length and containing three to four nodes) also support the premise that changes in the competence to respond to auxin can play an important role in regulating the ease with which cuttings can be rooted (Sriskandarajah et al., 1982). Proliferating apple shoot cultures were produced by growth of isolated buds of difficult-to-root apple cultures Jonathan and Delicious on medium containing 10 µmol L^{-1} benzyladenine. Repeated subculture of the shoot cultures in continuous light at 26° C yielded a progressive increase in the ability of microcuttings from the shoots to produce adventitious roots when the microcuttings were cultured on a medium containing 10 µmol L^{-1} naphthaleneacetic acid (NAA). For example, freshly cut microcuttings did not form roots, while after the ninth subculture 95% of Jonathan microcuttings formed roots when cultured on NAA-containing medium (Sriskandarajah et al., 1982). These results suggest that within the same genotype the competence of cuttings for root formation can be altered by preculture of the tissue under defined conditions. This change in competence was maintained in small tissue explants such as leaf lamina, petiole explants and internode explants, i.e., explants from microcuttings with the potential to form roots were able to root, whereas explants from microcuttings that did not have the ability to root were not able to root (Mullins, 1985). Thus, competence and determination studies using tissue explants can yield information that reflects the rooting potential of more complex cultures such as shoot cultures.

Hedera Helix Leaf Cuttings and Debladed Petioles

Hackett and colleagues have studied adventitious root initiation in juvenile- and mature-phase debladed petioles and leaf cuttings of *Hedera helix* (English ivy). One advantage of this system is that the effect of ontogenetic age of the stock plant on the rooting potential of cuttings can be studied. Geneve et al. (1988) incubated sterilized debladed petioles from juvenile- and mature-phase *Hedera helix* in a medium containing 100 µmol L^{-1} NAA. Using this *in vitro* system they demonstrated that culture of juvenile petioles on NAA containing medium for as little as one day allowed some root formation, while longer exposures up to 18 days yielded greater numbers of roots to form. Thus, juvenile petioles require as little as one day on a root-inducing medium to become determined for root formation. Since root meristems were first formed after ca. 12 days of culture it is clear that determination for root formation preceded meristem formation. Root meristems in juvenile petioles originated in cortical parenchyma cells adjacent to vascular bundles. In contrast, the few roots which formed in mature petioles developed from callus that developed from cortical cell division. The rooting response of debladed petioles was shown to be similar to that of *Hedera helix* stem cuttings and excised shoot apices (Geneve et al., 1988), thus demonstrating that relatively simple tissues can be used to study the mechanism of adventitious root formation.

Reciprocal grafts of leaf petioles and lamina of juvenile and mature *Hedera helix* have recently been used to demonstrate that a factor(s) from juvenile lamina appears to modify the competence of mature petioles for root formation (Geneve et al., 1991). When juvenile leaf cutting scions (lamina and attached petiole) were grafted onto mature petioles (stock), the mature petiole was induced to initiate more roots more rapidly than either a mature lamina grafted onto a mature petiole or than a mature leaf cutting.

Callus

A rapidly-growing callus culture of alfalfa, *Medicago sativa* L. cv. Regen S, derived from immature ovaries was used to study competence and determination for root organogenesis (Walker et al., 1979). Alfalfa callus does not form apical meristems if the callus is kept on a hormone-containing medium; rather, the callus must be transferred to a hormone free medium subsequent to exposure to an inductive medium in order for meristems to form. This is different than the classical work of Skoog and Miller (1957) in which culture of tobacco callus on medium with relatively high auxin to cytokinin concentrations led to root formation. When the alfalfa callus was subcultured on medium containing 25 mol L^{-1} α-NAA and 10 µmol L^{-1} kinetin (non-inductive medium), or when the callus was subsequently transferred to a hormone-free regeneration medium, no roots formed. However, transfer of the callus to a root-inducing medium containing 5 µmol L^{-1} 2,4-dichlorophenoxyacetic acid (2,4-D) and 50 µmol L^{-1} kinetin (inductive medium) and subsequent transfer to a hormone-free regeneration medium resulted in root formation. Maximum root formation occured in alfalfa callus that had been cultured for four days on the root-inducing medium. The authors concluded that 2,4-D is an inducer of root formation in alfalfa callus cultures that are already competent to respond. It should be noted, however, that without the required transfer experiments it cannot be excluded that part of the four day period required to obtain maximum root formation was actually required for the tissue to become competent to respond to 2,4-D. Nevertheless, since maximum root formation occurred within four days, it appears that the alfalfa callus cultures were determined for root formation within four days of exposure to the root-inducing medium. Interestingly, small pieces of callus, <150 µm in diameter (i.e., less than six cells), were unable to produce roots. If small calli were allowed to grow to a larger size they regained the ability to form roots. Thus, it appears that a minimum size is required for alfalfa callus to become competent for induction of root formation.

Leaf Explants

The significance of results from studies with callus cultures for problems dealing with adventitious roots that form directly on the plant is not obvious. Several studies of competence and determination for root formation have been done in which organogenesis occurs more-or-less directly on tissue explants removed from the plant. Christianson and Warnick have done a series of extensive studies on the competence and determination of *Convolvulus arvensis* L. (field bindweed) leaf explants for root formation (Christianson and Warnick, 1985; Warnick, 1992) and shoot formation (Christianson and Warnick, 1983, 1984, 1985, 1988). The results of Warnick (1992) dealing with root formation will be discussed here. When leaf explants ($\approx$3 x 4 mm) from *Convolvulus arvensis* are placed on agar-solidified medium containing 59 µmol L^{-1} IBA, and cultured for three weeks, roots form from callus produced along the cut edges of the explant. The precise number of roots per explant is genotype specific with the mean number of roots per explant for a given genotype ranging from zero to greater than seven. Root primordia form within seven to 13 days of culture. The length of time in culture required for *Convolvulus* leaf explants to become determined for root formation was assayed by transferring explants from RIM to a basal medium (no hormones) and counting the number of roots after 30 days of total culture. The length of time required for determination was genotype specific and ranged from three to 10 days. No roots or root primordia were observed at or before the time of determination. It was noted that transfer of explants from RIM to BM after a culture period sufficient for determination yielded longer and more numerous roots.

In order to ascertain when the *Convolvulus* leaf explants became competent to respond to RIM, the explants were pre-cultured on a non-root-inducing medium (NRIM). Attempts to use a hormone-free medium (BM) for competence studies were unsuccessful since pre-culture on BM resulted in either death of the explants or a marked reduction in the number of roots that formed. Therefore, for *Convolvulus*, it was necessary to use a NRIM which contained some phytohormone. Leaf explants were pre-cultured on either a callus-inducing medium (CIM) containing 17 µmol L^{-1} indoleacetic acid (IAA) and 1.4 µmol L^{-1} kinetin, or on a shoot-inducing medium (SIM) containing 0.29 µmol L^{-1} IAA and 34.4 µmol L^{-1} 2-isopentenyl adenine. The amount of time that SIM or CIM could substitute for culture on RIM and still allow root formation was taken as the period of time required for the explants to become competent for root formation in response to IBA. Results of such studies showed that for all genotypes tested, *Convolvulus* leaf explants require about three days to become competent for root formation.

Christianson and Warnick (1985), by comparing various genotypes of *Convolvulus* for competence and determination for root formation, concluded that lack of root formation in some genotypes was due to a lack of the explants to acquire competence for root induction. Based on this information, a scheme for sequential exposure of explants to standard culture media was devised that promoted competence for root formation with subsequent root organogenesis. Thus, studies of competence and determination yielded information which allowed development of a culture scheme that promoted root formation in otherwise recalcitrant genotypes. An important contribution of this study is that it led to the awareness that the acquisition of competence for root induction may require different hormonal signals than the actual induction of root formation.

The stages of competence and determination of tobacco leaf explants for root formation have recently been studied (Attfield and Evans, 1991b). Leaf explants (1.0 cm^2) from *Nicotiana tabacum* cv. Xanthi cultured on liquid RIM medium containing 5 µmol L^{-1} IBA produce roots directly from the bundle-sheath and vein parenchyma cells (Attfield and Evans, 1991a). Root primordia appear within approximately six days of culture (Attfield and Evans, 1991a). Transfer of leaf explants from RIM to BM revealed that explants become determined for root formation after only one day of culture, although eight days on RIM were required to yield the maximum number of roots per explant (Attfield and Evans, 1991b). The one day required for determination was not further subdivided to ascertain whether part of that period was required for attainment of competence. However, culture of the explants on BM for various amounts of time and subsequent transfer to RIM led to a decline in the number of roots formed such that after 10 days on BM very few roots formed. Thus, it appears that the competence of tobacco leaf explants for root formation was reduced by prior culture on BM.

USE OF TOBACCO THIN CELL-LAYER EXPLANTS (TCLs) FOR STUDIES OF ROOT FORMATION

Thin strips of tissue cut from the outer layers of tobacco floral branches are able to undergo root, vegetative shoot and flower organogenesis (Tran Thanh Van, 1973; Tran Thanh Van et al., 1974). Such tissues are called thin cell-layer explants (TCLs). Studies of the organogenesis potential of TCLs from the floral branches, petioles or stems of a number of species have demonstrated that the type of organogenesis is dependent upon the species, the precise tissue being used, and the hormone content of the medium [Monacelli et al., (1984), Klimaszewska and Keller (1985), Charest et al. (1988), Detrez et al. (1988), Mulin and Tran Thanh Van, (1989), Pelissier et al. (1990); reviewed in Compton and Veilleux (1992)]. For example, TCLs from the lower stem of tobacco produce vegetative shoots while TCLs from the flower branch of the same plant produce flowers (Tran Thanh Van, 1973). Interestingly, in at least one case the organogenic potential of TCLs for embryogenesis has also been documented (Pelissier et al., 1990). Thus, TCLs can be used to study the competence of various tissues for different types of organogenesis. TCLs are particularly useful for organogenesis studies since the organogenesis of roots, vegetative shoots and flowers can occur directly, i.e., without intervening callus formation. Tobacco TCLs, which are $\approx$ 1 x 10 mm in area and four to ten cell layers thick are composed of one cell layer of epidermis, two to four cell layers of chlorenchyma and three to six cell layers of parenchyma and collenchyma (Mohnen et al., 1990). Since the starting tissue contains no meristems or centers of cell division, organogenesis in TCLs is truly *de novo* organ formation.

TCLs have been used to study adventitious root formation (Tran Thanh Van et al., 1974, 1985; Tran Thanh Van and Chlyah, 1976; Thorpe et al., 1978; Tran Thanh Van and Cousson, 1980; Tran Than Van 1980a,b, 1981; Nassogne et al., 1985; Le Guyader, 1987; Cousson et al., 1989; Eberhard et al., 1989; Mohnen et al., 1990; Altamura et al., 1991; Altamura and Capitani, 1992). TCLs from stem, floral branch or pedicel tissue cultured on rooting inducing medium (containing 5 to 20 μmol L^{-1} IBA and 0.1 to 0.5 μmol L^{-1} kinetin) form two to 12 roots per TCL after 24 to 30 days in culture (Tran Thanh Van et al., 1974; Torrigiani et al., 1989; Mohnen et al., 1990). Roots generally arise from the subepidermal layer or from layers slightly deeper within the tissue (Tran Thanh Van et al., 1974; Altamura et al., 1991; unpublished results of Gruber, Liljebjelke, and Mohnen).

The competence and determination of tobacco (*Nicotiana tabacum* cv. Samsun) floral branch-derived TCLs (Mohnen et al., 1990) for root formation were ascertained. TCLs were cultured for zero to 16 days on RIM (15 μmol L^{-1} IBA, 0.5 μmol L^{-1} kinetin) and subsequently transferred to basal medium for a total culture period of 25 days. In three out of five experiments, TCLs required a minimum of three days on RIM before roots formed (unpublished results of Mohnen; and Mohnen, Eberhard, Albersheim, and Darvill). These results suggest that TCLs require three days of exposure to RIM to become determined for root formation. In two experiments, however, roots formed in explants cultured for only one day on RIM. Thus, there appears to be some variation in the amount of time required for determination of TCLs for root formation, depending upon the physiological state of the explant. Maximal root formation required four or more days of culture.

In order to determine whether tobacco TCLs are initially competent to respond to RIM, or whether they require a period of time to become competent, TCLs were first pre-cultured on BM for three days before being transferred to RIM for zero to 16 days and subsequently transferred to BM for a total of 25 days (Mohnen, unpublished results). TCLs precultured on BM formed roots after only two days of culture on RIM. Thus, the three days of culture required for TCLs to become determined for root formation can be subdivided into one day on RIM for development of competence and two days on RIM for determination for root formation. Since root meristems form by the 14th day of culture (Gruber, Liljebjelke and Mohnen, unpublished results) it is clear that in tobacco TCLs, as in *Convolvulus* and tobacco leaf explants (Christianson and Warnick, 1985; Attfield and Evans, 1991a,b; Warnick, 1992), determination for root formation occurs well before meristems are formed.

CONCLUSION

Several conclusions can be made from a comparison of the period of competence and determination of organ cultures, callus, leaf explants and TCLs for root formation (Table 2). Determination of explants for root formation is a lengthy process, usually requiring days rather than hours. This is sufficient time to allow cell division to occur. However, a detailed study of histological changes during the period required for determination remains to be done in order to ascertain whether there is a correlation between cell division and determination. A period of culture was required for at least some of the explants to become competent to respond to root-inducing factors (e.g., auxin; see chapter by Blakesley in this volume). Thus, it appears that tissues taken directly from the plant may not be competent to respond to root-inducing factors and that, as discussed by Christianson and Warnick (1985), maximization of rooting potential may require attention to the initial attainment of competence for root-inducing factors. Finally, in all cases tested, determination for root formation preceded formation of a root primordium. Thus determination, if it is associated with an anatomical change in the tissue as opposed to a purely physiological change within cells, would have to be associated with a pre-meristem event.

The state of competence for root formation becomes important in those situations where a potential root initiation site is not preexisting in a tissue, but must be created (Lovell and White, 1986). In theory, development of competence for root formation could be required whenever adventitious rooting occurs in the absence of a preformed root primordium. The fact that some plant tissues, both stem cuttings and explants, appear to differ in their competence to form adventitious roots, suggests that the progression of cells into a state competent for root formation may not occur in plants that are difficult to root. For example, stem cuttings from juvenile- and mature-phase *Hedera helix* differ in their potential to form adventitious roots with mature-phase cuttings being difficult to root (Geneve et al., 1988). This difference is thought to be a characteristic of cells at the site of root initiation rather than related to the translocation of promoters or inhibitors.

The competence and determination experiments with *in vitro*-cultured explants discussed above illustrate that relatively straightforward and simple experiments can elucidate whether a responsive tissue is likely to be competent to respond to a given root-inducing factor *in planta*, or whether the tissue first requires a developmental change in order to become competent to respond to root-inducing factors. The prerequisites for such studies are that: 1) the conditions are known which allow adventitious root formation in the plant or explant of interest, and 2) the rooting response occurs *in vitro*. It is clear from the above-mentioned studies that both relatively complicated explants such as leaf or shoot cuttings, as well as less complex tissue explants such as leaf and thin cell-layer explants can be used to study competence and determination for rooting. The choice of material will depend upon the questions being asked. Leaf and shoot cuttings may give results directly applicable to a specific horticultural or agricultural problem. However, the relatively simpler tissues such as leaf or stem explants, which contain many fewer cells and cell types, may afford an experimental system more amenable to direct analysis of the mechanism(s) of adventitious root formation, compared to conventional cuttings.

Table 2. Competence and determination of explants for root formation (time in days).

Tissue	Competence	Determination	Primordium	Reference
Excised shoots (*Malus pumila*)	ND[1]	0.5	----	James, 1983
Debladed petioles (*Hedera helix*)	ND	1	≈12	Geneve et al., 1988
Callus (*Medicago sativa*)	ND	4	----	Walker et al., 1979
Leaf explant (*Convolvulus*)	3	3-10	7-13	Christianson & Warnick, 1985; Warnick, 1992
Leaf explant (*Nicotiana tabacum*)	ND	1	≈6	Attfield & Evans, 1991a,b
TCL[2] (*Nicotiana tabacum*)	1	3	≈14	Mohnen; Mohnen, Eberhard, Albersheim & Darvill, upublished
	ND	≤1		

[1]The period of competence within the time-frame of determination was not ascertained.
[2]*Nicotiana tabacum* TCLs became determined for root formation after either about one day or three days, depending upon the experiment.

Many questions remain regarding competence and determination of cells and/or tissues for root formation. How do developmental changes coincide with anatomical changes? What are the molecular events that lead to the states of competence and determination? Do results with *in vitro*-cultured tissues reflect developmental states *in planta*? It is clear that in order to answer these questions one experimental system should be studied in great detail at the developmental, anatomical, molecular and genetic levels. There is a danger in using a model system, however, in that the regulatory mechanisms for plasticity of organ formation may have many species-specific nuances. Nevertheless, only when molecular events can be matched to developmental stages, will a systematic approach to overcoming recalcitrance of rooting in some plants and plant tissues be possible.

> *...There is substantial variation between species, and even between cuttings from young and old plants of the same species, in events leading to the creation or locating of a potential root initiation site. We believe that it is these processes that need our detailed attention in the future because they may hold the key to our understanding of the physiological basis for adventitious root production with profound implications for applied research in forestry, horticulture and agriculture....* (Lovell and White, 1986)

Lovell's and White's above-quoted statement regarding the necessity for more studies of the early anatomical changes associated with root formation is equally valid for the requirement of more studies of the early developmental events that occur during root formation. Developmental studies by themselves, however, only yield information about the timing of and numbers of developmental steps required for root formation. Developmental studies are most beneficial, therefore, when they are accompanied by anatomical, biochemical/molecular and genetic studies using the same plant system (see chapters by Ernst, by Hand, and by Riemenschneider in this volume). Such an all inclusive research plan requires a long term commitment of research time and funds.

ACKNOWLEDGMENTS

I thank Karen Howard and Stefan Eberhard for help in preparing the manuscript, and Alan Darvill and Peter Albersheim for sharing unpublished results. Unpublished work presented was funded in part by the Department of Energy-funded Center for Plant and Microbial Complex Carbohydrates (DE-FG05-93ER20097).

REFERENCES

Altamura, M.M., and Capitani, F., 1992, The role of hormones on morphogenesis of thin layer explants from normal and transgenic tobacco plants, *Physiol. Plant.* 84:555.

Altamura, M.M., Capitani, F., Serafini-Fracassini, D., Torrigiani, P., and Falasca, G., 1991, Root histogenesis from tobacco thin cell layers, *Protoplasma* 161:31.

Attfield, E.M., and Evans, P.K., 1991a, Developmental pattern of root and shoot organogenesis in cultured leaf explants of *Nicotiana tabacum* cv. *Xanthi* nc, *J. Exp. Bot.* 42:51.

Attfield, E.M., and Evans, P.K., 1991b, Stages in the initiation of root and shoot organogenesis in cultured leaf explants of *Nicotiana tabacum* cv. *Xanthi* nc, *J. Exp. Bot.* 42:59.

Bonnett, H.T., Jr., and Torrey, J.G., 1965, Chemical control of organ formation in root segments of *Convolvulus* cultured *in vitro*, *Plant Physiol.* 40:1228.

Charest, P.J., Holbrook, L.A., Gabard, J., Iyer, V.N., and Miki, B.L., 1988, *Agrobacterium*-mediated transformation of thin cell layer explants from *Brassica napus* L., *Theor. Appl. Genet.* 75:438.

Christianson, M.L., and Warnick, D.A., 1983, Competence and determination in the process of *in vitro* shoot organogenesis, *Dev. Biol.* 95:288.

Christianson, M.L., and Warnick, D.A., 1984, Phenocritical times in the process of *in vitro* shoot organogenesis, *Dev. Biol.* 101:382.

Christianson, M.L., and Warnick, D.A., 1985, Temporal requirement for phytohormone balance in the control of organogenesis *in vitro*, *Dev. Biol.* 112:494.

Christianson, M.L., and Warnick, D.A., 1988, Organogenesis *in vitro* as a developmental process, *Hortic. Sci.* 23:515.

Compton, M.E., and Veilleux, R.E., 1992, Thin cell layer morphogenesis, *in*: "Horticultural Reviews," J. Janick, ed., John Wiley & Sons, New York.

Cousson, A., Toubart, P., and Tran Thanh Van, K., 1989, Control of morphogenetic pathways in thin cell layers of tobacco by pH, *Can. J. Bot.* 67:650.

Detrez, C., Tetu, T., Sangwan, R.S., and Sangwan-Norreel, B.S., 1988, Direct organogenesis from petiole and thin cell layer explants in sugar beet cultured *in vitro*, *J. Exp. Bot.* 39:917.

Eberhard, S., Doubrava, N., Marfà, V., Mohnen, D., Southwick, A., Darvill, A.G., and Albersheim, P., 1989, Pectic cell wall fragments regulate tobacco thin-cell-layer explant morphogenesis, *Plant Cell* 1:747.

Esau, K., 1977, "Anatomy of Seed Plants," John Wiley and Sons, New York.

Geneve, R.L., Hackett, W.P., and Swanson, B.T., 1988, Adventitious root initiation in de-bladed petioles from the juvenile and mature phases of English ivy, *J. Amer. Soc. Hortic. Sci.* 113:630.

Geneve, R.L., Makhtari, M., and Hackett, W.P., 1991, Adventitious root initiation in reciprocally grafted leaf cuttings from the juvenile and mature phase of *Hedera helix* L., *J. Exp. Bot.* 42:65.

James, D.J., 1983, Adventitious root formation 'in vitro' in apple rootstocks (*Malus pumila*), I. Factors affecting the length of the auxin-sensitive phase in M.9, *Physiol. Plant*, 57:149.

James, D.J., and Thurbon, I.J., 1979, Rapid *in vitro* rooting of the apple rootstock M.9, *J. Hortic. Sci.* 54:309.

James, D.J., and Thurbon, I.J., 1981, Phenolic compounds and other factors controlling rhizogenesis *in vitro* in the apple rootstocks M.9 and M.26, *Z. Pflanzenphysiol. Bd.105. S.*, 11.

Klimaszewska, K., and Keller, W.A., 1985, High frequency plant regeneration from thin cell layer explants of *Brassica napus*, *Plant Cell Tissue Organ Cult.* 4:183.

Le Guyader, H., 1987, Ethylene et differenciation de racines par des cultures de couches cellulaires minces de tabac, *C.R. Acad. Sci. Paris* 305:591.

Lovell, P.H, and White, J., 1986, Anatomical changes during adventitious root formation, *in*: "New Root Formation in Plants and Cuttings," M.B. Jackson, ed., Martinus Nijhoff Pubs., Boston.

Lyndon, R.F., 1990, "Plant Development the Cellular Basis," Unwin Hyman Ltd., London.

McDaniel, C.N., 1984, Competence, determination, and induction in plant development, *in*: "Pattern Formation, A Primer in Developmental Biology", G.M. Malacinski and S.V. Bryant, eds., Macmillan Pub. Co., New York.

Meins, F., 1986, Determination and morphogenetic competence in plant tissue culture, *in*: "Plant Cell Culture Technology," M.M. Yeoman, ed., Blackwell Sci. Pub., Boston.

Meins, F., and Binns, A., 1979, Cell determination in plant development, *BioScience* 29:221.

Mohnen, D., Eberhard, S., Marfà, V., Doubrava, N., Toubart, P., Gollin, D.J., Gruber, T.A., Nuri, W., Albersheim, P., and Davill, A., 1990, The control of root, vegetative shoot and flower morphogenesis in tobacco thin cell-layer explants (TCLs), *Development* 108:191.

Monacelli, B., Pasqua, G., Altamura, M.M., and Mazzolani, G., 1984, Photocontrol of the morphogenesis *in vitro* in a short-day tobacco, *Ann. di Bot.* XLII:1.

Mulin, M., and Tran Thanh Van, K., 1989, Obtention of *in vitro* flowers from thin epidermal cell layers of *Petunia hybrida* (Hort.), *Plant Sci.* 62:113.

Mullins, M.G., 1985, Regulation of adventitious root formation in microcuttings, *Acta Hortic.* 166:53.

Nassogne, C., Havelange, A., Marcotte, J.-L., Bernier, G., and Tran Thanh Van, K., 1985, Changes in mitotic activity and cell shape during flower and root neoformation in thin cell layers (TCL) from tobacco inflorescences, *Soc. Physiol. Vég.* 93:14.

Pelissier, B., Bouchefra, O., Pepin, R., and Freyssinet, G., 1990, Production of isolated somatic embryos from sunflower thin cell layers, *Plant Cell Rep.* 9:47.

Skoog, F., and Miller, C.O., 1957, Chemical regulation of growth and organ formation in plant tissues cultured *in vitro*, *Symp. Soc. Exp. Biol.* XI:118.

Sriskandarajah, S., Mullins, M.G., and Nair, Y., 1982, Induction of adventitious rooting in vitro in difficult-to-propagate cultivars of apple, *Plant Sci. Lett.* 24:1.

Thorpe, T.A., Tran Thanh Van, M., and Gaspar, T., 1978, Isoperoxidases in epidermal layers of tobacco and changes during organ formation *in vitro*, *Physiol. Plant.* 44:388.

Torrigiani, P., Altamura, M.M., Capitani, F., Serafini-Fracassini, D., and Bagni, N., 1989, *De novo* root formation in thin cell layers of tobacco: changes in free and bound polyamines, *Physiol. Plant.* 77:294.

Tran Thanh Van, M., 1973, Direct flower neoformation from superficial tissue of small explants of *Nicotiana tabacum* L., *Planta* 115:87.

Tran Thanh Van, K., 1980a, Control of morphogenesis by inherent and exogenously applied factors in thin cell layers, *Int. Rev. Cytol.* 11a:175.

Tran Thanh Van, K., 1980b, Control of morphogenesis or what shapes a group of cells, *in*: "Plant Tissue Culture," A. Fiechter, ed., Springer-Verlag, Berlin.

Tran Thanh Van, K., 1981, Control of morphogenesis in *in vitro* cultures, *Annu. Rev. Plant Physiol.* 32:291.

Tran Thanh Van, K., and Chlyah, A., 1976, Différenciation de boutons flouraux, de bourgeons végétatifs, de racines et de cal à partir de l'assise sous-épidermique des ramifications florales de *Nicotiana tabacum* Wisc.38, Etude infrastructurale, *Can. J. Bot.* 54:1979.

Tran Thanh Van, K., and Cousson, A., 1980, Microenvironment-genome interactions in *de novo* morphogenetic differentiation on thin cell layers, *in*: "NSF-CNRS Colloque Proc.," Praeger, New York.

Tran Thanh Van, M., Dien, N.T., and Chlyah, A., 1974, Regulation of organogenesis in small explants of superficial tissue of *Nicotiana tabacum* L., *Planta* 119:149.

Tran Thanh Van, K., Toubart, P., Cousson, A., Darvill, A.G., Gollin, D.J., Chelf, P., and Albersheim, P., 1985, Manipulation of the morphogenetic pathways of tobacco explants by oligosaccharins, *Nature* 314:615.

Waddington, C.H., 1934, Experiments on embryonic induction, Part I. The competence of the extra-embryonic ectoderm in the chick, *J. Exp. Biol.* 11:211.

Walker, K.A., Wendeln, M.L., and Jaworski, E.G., 1979, Organogenesis in callus tissue of *Medicago sativa*, The temporal separation of induction processes from differentiation processes, *Plant Sci. Lett.* 16:23.

Wareing, P.F., and Graham, C.F., 1984, Determination and pluripotentiality, *in*: "Developmental Control in Animals and Plants," C.F. Graham, and P.F. Wareing, eds., Blackwell Sci. Pubs., Boston.

Warnick, D.A., 1992, "Developmental Biology of Rhizogenesis *in vitro* in *Convolvulus arvensis*," M.S. thesis, San Jose State Univ., Dept. Biol. Sci., San Jose.

DIFFERENTIAL COMPETENCE FOR ADVENTITIOUS ROOT FORMATION IN HISTOLOGICALLY SIMILAR CELL TYPES

John R. Murray, M. Concepcion Sanchez, Alan G. Smith
and Wesley P. Hackett

Department of Horticultural Science
University of Minnesota
St. Paul, MN 55108 USA

INTRODUCTION

Shoot tissue formed during the protracted juvenile phase of woody perennials lacks the ability to flower. With the transition to the mature phase, shoot apices or axillary meristems of newly formed tissue gain the ability to flower and this ability is maintained in subsequently formed shoot tissue (Zimmerman, 1973, 1976). In addition to this phase-dependent difference in ability to form flowers, other persistent phenotypic differences exist between shoot tissue of the basal (i.e., juvenile) and apical (i.e., mature) portions of a plant (Hackett, 1985; Poethig, 1990). Due to the protracted nature of both the juvenile and mature phases of woody plants, phase-dependent phenotypic characters are stably expressed through a large number of cell divisions over years of growth within a phase. It is presumed that phase-dependent characters do not result from a genetic change, but result from an epigenetic difference in the capacity to express genes that permit or prevent expression of a phenotype.

The capacity of a tissue to respond in a specific way to inductive stimuli or signals can operationally be defined as competence (McDaniel, 1982). Conceptually, this definition has been useful in developing a framework for the analysis of plant developmental processes, however, little is known about the mechanistic basis of competence. The concept of competence is not new to studies of plant developmental processes, but other synonymous terms, such as capability, capacity, potential or even predisposition, have been used more often (see the chapter by Mohnen in this volume.)

In most woody perennial species there is a decline or loss of competence for the formation of adventitious roots in mature-phase shoot tissue, as compared to juvenile-phase tissue (Gardner, 1929). This mature-phase "difficult-to-root" characteristic presents a significant challenge to the clonal propagation of individuals that are selected based on mature-phase growth habit, flower appearance or fruit quality (Howard et al., 1988; see the chapter by Howard in this volume). Although the reduced competence for adventitious root formation in mature phase tissue is stably maintained, it is possible to alter this competence by several experimental approaches (Hess, 1969).

Biology of Adventitious Root Formation, Edited by
T.D. Davis and B.E. Haissig, Plenum Press, New York, 1994

The growth of shoots in darkness (etiolation) or the exclusion of light from tissue that was initially light-grown (blanching) (Harrison-Murray, 1982), are experimental approaches that can have a substantial effect on competence for rooting that persists for weeks to months subsequent to the application of the treatment. There is a general applicability of light exclusion for the improvement of adventitious root formation across woody perennial species (Maynard and Bassuk, 1988). We will discuss a limited number of examples to demonstrate the differential competence of cells to form adventitious roots as a function of light treatment and the persistent nature of the effect of light exclusion on rooting. This persistence, which is maintained in the treated tissue as well as newly formed tissue, suggests that transient effects on the levels of auxin or other metabolites are not the basis of the change in competence for root formation. If the difficult-to-root character of mature-phase tissue is epigenetically regulated, a persistent gain of competence for rooting due to light exclusion is likely to result from a change in the capacity to express genes. This hypothesis raises a series of questions. What rooting-related genes are differentially competent to be expressed in juvenile- and mature-phase tissues? What is the basis of the difference in competence to express these genes?—is the epigenetic effect imposed on the structural genes that encode products that are involved in the process? or is it imposed on regulatory genes?

To demonstrate an approach to address these questions, we will discuss work conducted with English ivy (*Hedera helix* L.). That section will assess the histology of juvenile- and mature-phase tissues, the phase-dependent difference in rooting response, and will describe initial experiments conducted to isolate rooting-related genes. Additionally, we will discuss the isolation and characterization of an anthocyanin biosynthetic gene which is competent to be expressed in juvenile- but not mature-phase ivy. Although it is likely this gene is not related to adventitious root formation, it will be discussed to present an approach to determining the basis of the lack of competence for expression of a phase-dependent characteristic. An understanding of the basis of competence, or lack of competence, for one developmental process might provide an understanding of the epigenetic control of other phase-dependent processes, such as rooting.

CHARACTERISTICS OF ETIOLATION-INDUCED CHANGE IN ROOTING COMPETENCE

A general or localized exclusion of light from shoot tissue during the early stages of growth from apical or axillary buds can influence the competence for adventitious root formation in cuttings taken from the treated shoots (Gardner, 1936; Frolich, 1961; Stoutemyer, 1961; Herman and Hess, 1963; Harrison-Murray, 1982; Maynard and Bassuk, 1988). Gardner (1936), working with the difficult-to-root apple variety McIntosh, used black tape wrapped around young shoot tissue to exclude light. Subsequently, the cuttings were taken with the basal cut in the blanched area of the shoot, then placed in the propagation bench without the application of exogenous auxin. The effect of this blanching treatment on increased rooting potential was greatest when the shoot was wrapped with tape as near the shoot tip as possible, with up to 70% of the cuttings rooting. In comparison, when the expanding shoots were taped at greater distance basipetal to the shoot tip, the formation of adventitious roots by cuttings taken from these shoots was progressively reduced with the greater distance. These data indicate that the exclusion of light was most effective in changing or maintaining competence for rooting in the less differentiated, acropetal tissue of the growing shoot tip. Gardner also combined complete exclusion of light during the initial 7.5 cm of growth of McIntosh shoots in the spring, with the taping of the basipetal portion of the shoots at the time of transfer to daylight conditions. This approach allowed testing the rooting potential of tissue at the base of the cutting that had not undergone light-grown greening processes either prior to or after dark treatments. When these shoots were used as hardwood cuttings the following March, 97% of the cuttings rooted, an apparent increase in rooting as compared to the response to blanching described above. These results indicate that when the etiolation treatment is maintained over many months, competence for root formation is maintained.

Howard and coworkers have extensively studied the influence of dark treatments on adventitious root formation in softwood cuttings of the apple rootstock variety M.9. The average treatment response over four years of studies demonstrated that there was a seven-fold increase in rooting percentage for auxin treated cuttings that were etiolated initially, but acclimated in light for two weeks (78%), as compared to auxin treated, light-grown controls (11%) (Harrison-Murray, 1982). With the continued exclusion of light imposed by taping the basal portion of these etiolated shoots during the two week light acclimation period, there was an additional, small increase in the average rooting to 85%. When tape was applied to blanch non-etiolated shoots (ca. 4-cm-long), there was only a three-fold increase in rooting of the blanched softwood cuttings as compared to non-etiolated controls (Howard, 1981). These data indicate that for M.9, blanching light-grown shoot tissue is significantly less effective than etiolation, with or without taping, in changing or maintaining competence. However, the initial growth of shoots under heavy shading has been shown to be as effective as complete darkness in influencing the rooting potential of M.9 cuttings (Howard, 1984, 1985).

To test the effect of the time of imposition of dark treatment, relative to the stage of shoot growth, dark treatments were applied to shoots that had already been growing for three weeks or applied to newly breaking buds (Howard, 1982). After the dark treatments were imposed, there was a gradual change in the newly formed shoot tissue to the typical, etiolated phenotype. Two weeks after the termination of the dark treatment, distal cuttings from the dark-grown part of shoots rooted at a level comparable to basal cuttings that were etiolated beginning at bud-break (75% vs 73%, respectively). Thus, we can conclude that chronologically-young shoot tissue which is undergoing growth and differentiation is sensitive to dark treatment, whether given early or late in the growth from buds.

To assess the persistence of the effect of light exclusion on the maintenance of competence, etiolated shoots were acclimated in light for different periods of time, up to over three weeks, prior to taking cuttings (Howard, 1980). There was no effect of the length of light acclimation on rooting; all cuttings rooted at a high percentage. It is interesting to note that these shoots became green after a few days of exposure to light, indicating that the effect of etiolation on competence for rooting persisted subsequent to the expression of processes induced by light. In a later study (Howard, 1982), hardwood cuttings taken from shoots that were etiolated and subsequently allowed to green nine months earlier, rooted at a rate of 53% compared to 12% for the non-etiolated controls. This level of rooting for the etiolated hardwood cuttings was significantly less than the response obtained with softwood cuttings taken two weeks after etiolation the previous summer; however, it does demonstrate that an intermediate level of competence persisted nine months after the etiolation treatment. Furthermore, when cuttings were taken from shoot tissue, which extended after the termination of the etiolation treatment, they rooted at a rate of 41% compared with 12% of equivalent, distal cuttings of non-etiolated shoots (Howard, 1981). The newly formed leaves that developed in the weeks immediately following the onset of light treatment showed signs of juvenile morphology. These observations suggest that the influence of etiolation persisted in tissue which was present during the etiolation treatment, as well as in tissue formed after exposure to light. However, this subsequently formed tissue was derived from activity of the apical meristem that had been exposed to the etiolation treatment.

One important factor in determining the degree of persistence of the effect of etiolation is the duration of the dark treatment. For instance, when cuttings were removed from Hass avocado shoots grown in the dark to a length of 7.5 cm, they had a high proportion (92%) of adventitious root formation (Frolich, 1961). However there was a rapid decline in the rooting of cuttings of these initially etiolated shoots as a result of each day of light exposure, with only 33% of the cuttings forming roots by day seven of light treatment. This leads to the conclusion that the influence of etiolation is short lived in avocado. However, when the basal portions of the initially etiolated shoots were maintained in darkness for up to five additional weeks, followed by exposure to light for seven days prior to taking cuttings, there was an increase in the rooting of cuttings. The cuttings that received four weeks of additional darkness rooted at a rate of 88%, even though they were exposed to light for seven days. Thus, competence for rooting was maintained at a high level in the light when the duration of the dark treatment was sufficiently long.

In total, these experiments with apple and avocado indicate that the effect of etiolation is not ephemeral, but persists up to many months subsequent to the exposure to light, as well as during the further growth and differentiation of the shoot tissue derived from apical meristems exposed to the dark treatment.

Two important questions arise from these observations about the influence of light exclusion on competence for adventitious root formation. First, is there an effect of light exclusion in other species? and, if so, is it a general response that permits the manipulation of rooting potential in mature-phase tissues of a number of species? Maynard and Bassuk (1987) tested the effect of light exclusion on the rooting of softwood cuttings of 21 difficult-to-root taxa. Light exclusion significantly improved adventitious root formation of cuttings in 18 of the 21 taxa. Furthermore, in a review of this subject, they list at least 42 taxa that have been reported to have increased rooting potential in response to full or partial light exclusion (Maynard and Bassuk, 1988). Therefore, it appears that the influence of light exclusion on competence for adventitious root formation is a general response.

Second, what is the basis of the difference in rooting competence between dark-treated and light-grown control tissue of difficult-to-root species? A considerable amount of research has focused on the effect of light exclusion on physiological factors, such as the level of endogenous auxin and rooting cofactors (Herman and Hess, 1963; Kawase, 1965; Kawase and Matsui, 1980; Maynard and Bassuk, 1988). However, the persistent effect of etiolation with subsequent exposure to light on the rooting of M.9 hardwood cuttings nine months later (Howard, 1982), suggests there is a fundamental change that is not related to the level of endogenous auxin or rooting cofactors, or other transient metabolic effects. Furthermore, the maintenance of competence in tissue formed by meristematic activity after the exclusion of light, suggests there is either a transmissible factor produced in the initially etiolated tissue that influences rooting in later-formed tissue or, alternatively, a persistent effect on the apical meristem and cells derived from the meristem (Howard, 1981). There is little evidence (Kawase and Matsui, 1980; Howard, 1983) or no evidence (Frolich, 1961; Doss et al., 1980) for a factor that is transmissible over a significant distance from the region exposed to dark treatment.

What influence, then, could etiolation have on the apical meristem that persists with cell divisions, growth and differentiation in light? Perhaps the exclusion of light from the apical meristem of chronologically young, undifferentiated shoot tissue results in a persistent change (but not a change in the nucleotide sequence) in the ability of cells to express genes that are required for rooting or that prevent rooting. For instance, it can be hypothesized that etiolation might result in a change in the chromatin which permits or prevents the expression of key regulatory or structural genes, and that this altered chromatin is mitotically heritable in newly formed tissue. To test this hypothesis or alternative hypotheses, phase-dependent genes related to rooting or other phase characteristics first must be isolated and characterized with respect to the level at which they are regulated (i.e., transcriptionally or post-transcriptionally), and then tested in light exclusion experiments.

Finally, one obvious effect of etiolation or early blanching treatment is on the histology of the treated stem tissue. The blanching of *Phaseolus vulgaris* hypocotyls and *Hibiscus rosasinensis* stem tissue resulted in decreased cell wall thickness (i.e., reduced deposition of cellulose and hemicellulose) and decreased lignification of phloem and pericyclic fibers, relative to non-blanched controls (Herman and Hess, 1963). Additionally, a greater proportion of cells remained parenchymatous. Similar observations have been made in other species in regard to reduced sclerification (fibers or sclereids) in response to light exclusion (Stoutemyer, 1961; Doud and Carlson, 1977). It is interesting to note that Stoutemyer (1937) states in a review of adventitious root formation in *Malus* spp. that "...Practically the only distinction which can be pointed out is that the mature phase stem of Virginia Crab contains more pericyclic fibers than are evident in stems of the juvenile form....". Similar observations have been made in juvenile- and mature-phase stem tissues of other *Malus* spp. (Beakbane, 1961). Therefore, light exclusion with mature-phase stems of apple (Doud and Carlson, 1977) results in stem tissue histology that is at least initially similar to the juvenile-phase characteristic of reduced fiber formation. However, we are not aware of any studies that tested the persistent effect of light exclusion on the histology of stem tissue subsequent to prolonged light treatment. If this altered histology *per se* is the primary determinant of altered competence for rooting, it would be expected that the characteristics of the cell types would persist after exposure to light. In lieu of altered histology, differential competence for the expression of rooting-related genes might persist after exposure to light. The following discussion provides insight into the possibility of differential gene expression of rooting-related genes in the juvenile- and mature-phase of English ivy.

COMPARATIVE INVESTIGATIONS WITH IVY

To initiate an analysis of the basis of phase-dependent characters, we have used both phases of ivy to compare tissue that is competent and incompetent for adventitious root formation as well as anthocyanin biosynthesis. This comparative approach permits determining whether or not the tissues from both phases are capable of perceiving the inductive treatments, and if so, in what way they diverge in their response.

The phenotype of juvenile-phase shoot tissue of ivy is markedly different from the phenotype of mature-phase shoot tissue (Stein and Fosket, 1969; Rogler and Hackett, 1975). The phenotypic characters that differ between juvenile and mature shoots include differences in growth habit, leaf shape, phyllotaxy, photosynthetic capacity, stem pigmentation and the presence of adventitious roots. In contrast to this marked difference in phenotypes, the histology of the cell types of the shoot tissue is surprisingly similar between juvenile- and mature-phase ivy (Geneve et al., 1988). This similar histology of cell types suggests that the regulation of the primary differentiation of cell types during histogenesis is similar or common in each phase. However, these analogous cell types of each phase can have differential competence to express developmental processes.

It has been a premise of our research that the difference between phenotypes, or the difference in competence to express developmental processes in juvenile- and mature-phase tissues, results from a difference in the expression of genes that function in the differentiation of phase-dependent characters. The products of these genes could have promotive or preventive effects on the differentiation of these characters. However, it is unlikely that all phase-dependent genes are controlled by a single regulatory mechanism due to potential differences in the timing and localization of their expression, and in the stimuli required for their induction of expression in competent tissue. It can be postulated that genes are differentially expressed, in either a quantitative or qualitative manner, depending upon the phase. Within histologically similar cell types of vegetative shoot tissue it is likely there are genes that are actively expressed or competent of being expressed in one phase of maturation, but are maintained in an inactive state and are incompetent of being expressed during the other phase. A gene that encodes for an enzyme of anthocyanin biosynthesis displays this pattern of qualitative difference in expression in ivy. We might also expect that there are genes that are competent of being expressed in both juvenile- and mature-phase tissue, but that a phase-dependent difference in the level of expression results in a different phenotype. In excised petioles of ivy, a gene that encodes a proline-rich protein is expressed at higher levels in mature- than juvenile-phase petioles.

We have taken two approaches to isolate the genes mentioned above that are differentially expressed with respect to phase. First, genes from other species that encode products of known function have been used to assess their level of expression in both phases of a single clone of ivy. Second, a cDNA library, constructed from RNA expressed in juvenile-phase tissue, was screened by differential hybridization using labeled probes prepared from RNAs from juvenile- and mature-phase tissue.

ROOT INITIATION RESPONSE IN EXCISED PETIOLES

The loss or reduction in competence to form adventitious roots in cuttings of the mature-phase, as compared to juvenile-phase ivy (Hess, 1962; Girouard, 1967a,b), is typical of the phase-dependent difference in competence for root initiation in many species (Gardener, 1929; Schreiber and Kawase, 1975; Davies et al., 1982). Geneve (1985) developed an *in vitro* system using de-bladed petioles to study adventitious root initiation in a single clone of ivy. For both juvenile- and mature-phase petioles treated *in vitro* without auxin, there is no observable cell division (Geneve et al., 1988). Juvenile-phase petioles initiate adventitious roots in response to auxin treatment, whereas mature-phase petioles produce only callus and do not initiate roots. In both juvenile- and mature-phase petioles treated with radiolabeled auxin, the distribution and number of counts within the petioles are similar. These results indicate there is a low endogenous level of auxin in the detached, de-bladed petioles, that both phases are responsive to similar levels of auxin by undergoing cell divisions, but only juvenile-phase petioles are competent to form adventitious roots. Woo (1992) found that petioles from partially expanded (chronologically young), mature-phase leaves will initiate a few adventitious roots per petiole in 30-50% of the petioles that are treated with auxin. No roots are initiated in these petioles without auxin. This observation suggests that the lack of competence for adventitious root formation in mature-phase shoot tissue is not absolute, and that competence may vary with the degree of differentiation, developmental stage or chronological age within the mature phase. This response of chronologically young, mature-phase petioles provides a useful experimental tissue that is intermediate in its level of competence for rooting. The intermediate competence of chronologically young, mature-phase petioles is somewhat analogous to the effect of etiolation described above, in that etiolation of chronologically young, acropetal tissue of the growing shoot tip resulted in the acquisition or maintenance of a higher level of competence for rooting than differentiated, basal shoot tissue. Additionally, mature-phase petioles from partially expanded leaves have few phloem fibers, whereas there are bands of phloem fibers in the petioles by the time the mature-phase leaves are fully expanded.

Microscopic observations indicate that the histology of juvenile- and mature-phase ivy petioles is similar prior to treatment, except for the greater numbers of phloem fibers in mature-phase petioles (Geneve et al., 1988). At day six of auxin treatment, in both phases, cell divisions are first observed in epithelial cells of ducts that are adjacent, but exterior to, the vascular bundles of the petioles. Observations at day nine of treatment show localized cell divisions of phloem parenchyma and inner cortical parenchyma adjacent to the vascular bundles in juvenile-phase petioles. These localized cell divisions lead to the formation of well-defined root primordia by day 12 of treatment in juvenile petioles. In contrast to the juvenile petioles, the early divisions of the epithelial cells of the ducts is followed by the onset of cell divisions in the phloem and throughout the cortical parenchyma of mature-phase petioles. This lack of localized cell divisions and apparently random orientation of the planes of divisions within the cortex result in a lack of formation of organized meristems or root initials in the mature-phase petioles. Juvenile-phase petioles also have some randomly oriented cell divisions throughout the cortical parenchyma. Thus, mature-phase petioles are responsive to auxin, but an apparent limitation in the control of the location and orientation of the divisions of cortical parenchyma prevents morphogenesis of an organized root primordia.

In order to isolate clones of genes that are differentially expressed just prior to the onset of cell division in rooting-competent (juvenile) and -incompetent (mature) petioles, a cDNA library constructed from RNA isolated from juvenile-phase petioles five days after auxin treatment was screened with probes prepared from RNAs from both juvenile- and mature-phase petioles (Woo, 1992). A DNA clone that encodes a proline-rich protein (PRP) was isolated that is expressed at quantitatively different levels in juvenile and mature petioles *in vitro*. Five- to ten-fold greater steady state levels of this PRP mRNA accumulate in mature-phase petioles than juvenile-phase petioles cultured *in vitro*. There is no influence of auxin on the level of the PRP mRNA in the cultured juvenile- and mature-phase petioles. Chronologically young, mature-phase petioles not treated with auxin accumulate intermediate levels of the PRP mRNA, relative to the juvenile and mature petioles. However, treatment of chronologically young, mature-phase petioles with auxin reduces the accumulation of the PRP mRNA to a level that is comparable to that which accumulates in juvenile-phase petioles. The PRP mRNA is not detectable in intact shoot tissue of either phase unless it is wounded, but is detectable in petioles within one to three days of excision for *in vitro* treatment. These data indicate that expression of this PRP gene in petioles is induced by excision, and that the level of expression is inversely related to competence for adventitious root formation.

To determine whether the spatial expression pattern of the PRP gene in ivy petioles is related to the cell types involved in adventitious root formation, labeled probes of the PRP gene were hybridized to the RNA present in cross-sections of treated petioles. There is no PRP expression detected in petioles immediately after excision, and PRP expression is significantly greater in mature- than juvenile-phase petioles treated for five days, with no effect of auxin on the levels of expression. In both juvenile- and mature-phase petioles at day five of treatment, PRP expression is preferentially localized in the duct epithelial cells, cells surrounding the ducts, phloem parenchyma and inner cortical cells adjacent to the phloem. A high level of PRP expression is also observed in these cell types in chronologically young, mature-phase petioles treated for five days without auxin; however, PRP expression is reduced in chronologically young, mature-phase petioles five days after auxin treatment. Thus, in both juvenile- and mature-phase petioles the expression of the PRP gene is preferentially localized in the cell types that undergo early divisions leading to the formation of meristems and root initials in the competent juvenile-phase petioles. The greater level of expression of the PRP gene in mature- than juvenile-phase petioles is consistent with elevated levels of the PRP gene product having a preventive role in the morphogenesis of the root meristem. The reduced level of PRP expression in specific cells of chronologically young, mature-phase petioles treated with auxin and the competence of the cells to form root primordia are also consistent with a possible role of high levels of PRP preventing morphogenesis. DNA sequence analysis indicates that the ivy PRP gene has sequence similarity to a cell wall PRP gene of soybean. If the ivy PRP is a cell wall protein, it can be postulated that the higher level of wall PRP may disrupt the organized orientation of cell divisions required for the formation of a meristem. However, currently we cannot distinguish between the greater level of PRP expression having a causal role in preventing morphogenesis or being a consequence of altered developmental competence.

One experimental approach to assess the relationship between high PRP expression and a lack of competence for rooting is the use of antisense RNA technology. This approach would involve the introduction of a DNA construct of the PRP gene into the genome of mature-phase ivy. The construct would direct the transcription of the nonsense strand of the introduced PRP DNA, to yield antisense RNA that is complementary to PRP RNA transcribed from the native gene. The expression of antisense RNA in all cell types—or, more specifically, in the cell types of mature-phase petioles, where the native gene is expressed—would prevent or reduce the production of the PRP protein. If high levels of the PRP protein play a causal role in preventing morphogenesis of root meristems, a significant reduction in the level of PRP protein by antisense RNA should result in the formation of adventitious root in the mature-phase petioles expressing antisense RNA.

Alternatively, juvenile-phase ivy could be transformed with an appropriate construct of the PRP gene such that the there is an over-expression of the PRP protein in juvenile phase petioles. If the PRP protein is solely responsible for the lack of rooting response in mature-phase petioles, then juvenile petioles from the transformed, over-expressing plants should not root.

Both approaches of antisense RNA expression and over-expression may produce equivocal results if the PRP protein is only one component of several that are acting additively to inhibit rooting. This can only be determined by conducting the experiments. A major limitation to the use of this approach in ivy and many woody perennials is the lack of ability to transform the tissue with DNA constructs. In some woody perennial species, juvenile-, but not mature-phase tissue, is capable of being transformed, which would limit the utility of the approach. Additionally, the requirement for organogenesis to regenerate transformants in some transformation techniques can lead to the recovery of plantlets which are juvenile in phenotype. However, the "knockout" approach of antisense RNA has been very useful in a number of studies of the function of gene products and will continue to provide a means of further assessing the functional relationship between the expression of a gene and a developmental process. If a suitable experimental system can be identified, this approach could be useful for the analysis of competence for root initiation.

In summary, the histologically similar phloem and cortical parenchyma cells of petioles of juvenile- and mature-phase ivy are differentially competent of undergoing organized cell divisions. Both phases respond to auxin by undergoing cell divisions, but only juvenile-phase cells are competent for root meristem formation. Chronologically young, mature-phase petioles have an intermediate level of competence with respect to the percentage of petioles rooting and the number of roots per petiole. The greater expression of the PRP gene in phloem and cortical parenchyma of mature, rather than juvenile and chronologically young, mature-phase petioles, is consistent with it playing a role in preventing root formation. Further molecular genetic approaches would permit testing this relationship.

PIGMENTATION RESPONSE IN LAMINA AND STEM TISSUE

The competence of stem tissue to accumulate anthocyanin pigment has been shown to be correlated with ease of adventitious root formation in woody perennial species (Bachelard and Stowe, 1962; Mullins, 1985). It is unlikely that anthocyanin pigmentation is directly related to the easy-to-root character since the accumulation of anthocyanin in the vacuole of cells of the epidermis of apple (Mullins, 1985) and hypodermis of ivy (Murray and Hackett, 1991) is spatially separated from the cells involved in adventitious root formation. However, although anthocyanin expression and rooting expression may be merely coincidental in juvenile-phase tissue, it is possible that there is a common mechanism by which competence for anthocyanin accumulation and rooting are lost during maturation. Therefore, we have studied, in ivy, genes of anthocyanin biosynthesis in order to determine what prevents the expression of a specific anthocyanin-related gene in mature-phase tissue.

A hypodermal layer of collenchyma cells of stems and petioles of juvenile-phase ivy accumulates red anthocyanin pigment (one class of the flavonoids), whereas the histologically similar collenchyma cells of stems and petioles of mature-phase tissue do not. Leaf lamina tissues of both phases do not accumulate anthocyanin when grown at temperatures above 20 °C; however, at slightly lower temperatures juvenile- but not mature-phase laminae will accumulate anthocyanin. We have used lamina tissue treated *in vitro* to study the basis of this phase-dependent difference in pigmentation.

Juvenile- and mature-phase lamina tissues accumulate comparable levels of flavonols (another class of the flavonoids) in response to treatment with both light and sucrose (Murray and Hackett, 1991). Juvenile lamina tissue accumulates anthocyanin in response to this treatment in the mesophyll cells (palisade and spongy parenchyma), but not in the epidermal or vascular tissue. Mature-phase lamina tissue does not accumulate anthocyanin in response to light and sucrose, even in chronologically young tissue. The accumulation of flavonols but not anthocyanin in mature-phase tissue suggested that a limitation of dihydroflavonol reductase (DFR) activity prevented anthocyanin biosynthesis. DFR catalyzes a reaction late in the anthocyanin biosynthetic pathway, just beyond the production of flavonols. There is no detectable DFR activity in lamina tissue of either phase at the start of *in vitro* treatment, but there is an induction of DFR activity by light and sucrose treatment in juvenile- but not mature-phase lamina tissue. However, the accumulation of flavonols in mature-phase lamina tissue in response to light and sucrose demonstrates that mature-phase tissue is responsive to the inductive treatments. The accumulation of comparable levels of flavonols in both phases suggests that enzymes early in the flavonoid pathway, such as chalcone synthase (CHS), are not limiting in mature-phase tissue.

DNA clones of the DFR gene from *Antirrhinum* and the CHS gene from *Petroselinum*, were used to isolate clones of DFR and CHS from ivy. By use of nuclear "run-on" transcription assays and RNA blot analyses, we have demonstrated that the lack of induction of DFR activity in mature-phase lamina tissue is due to a lack of transcription of the DFR gene and the resultant lack of accumulation of detectable DFR mRNA. In contrast, both phases transcribe CHS and accumulate CHS mRNA in response to both light and sucrose. CHS mRNA does accumulate in juvenile- and mature-phase stem tissue, but DFR mRNA accumulates only in juvenile stem tissue.

These data indicate that the lack of competence to induce pigmentation in lamina tissue of mature-phase ivy is due to the lack of competence to express an enzyme late in the biosynthetic sequence, and is not due to an inactivation of many or all of the genes that encode enzymes of the pathway. The presence of histologically similar mesophyll cells, and the induction of expression of many of the genes of flavonoid synthesis in lamina of both phases to yield flavonols, suggest that the regulation of histogenesis and response to inductive stimuli may be very similar in both phases. The difference in competence for DFR expression is apparently superimposed on the fundamental pattern of expression to give a phase-dependent phenotype. This qualitative difference in expression of DFR between juvenile- and mature-phase ivy is regulated at the level of transcription. Further molecular genetic approaches will be used to determine whether the lack of transcription of the DFR locus in mature-phase tissue is due to an altered characteristic of the DFR locus, such as altered chromatin conformation, or due to the lack of expression of regulatory genes that control DFR expression.

CONCLUSION

The specialization of newly formed cells during primary differentiation to yield unique cell types and tissues is the result of the differential expression of genes in a cell-type or tissue-specific manner (Goldberg, 1987). In contrast to this, we have discussed the phase-dependent difference in expression of genes within similar cell types. The similarity of the cell types between phases suggests that the regulation of primary differentiation during histogenesis is fundamentally similar. The effect of phase on the competence of these cells to subsequently express a phenotype or developmental process, such as pigmentation or adventitious root formation, appears to be superimposed on the control of primary differentiation. However, the intermediate competence of histologically young, mature-phase petioles to form adventitious roots suggests that the effect of phase may vary or interact with the developmental stage of a tissue within a phase.

Competence to express a phenotype or process is operationally defined, that is, competence can be assessed by determining whether or not a tissue is capable to respond in a specific way to inductive treatments. In our histological studies we have not been able to identify anatomical or cytological markers for the phloem parenchyma or inner cortical parenchyma that would allow us to distinguish or predict which cells are competent or incompetent for root meristem formation. Additionally, although mature-phase ivy lacks competence for the specific responses of adventitious root formation or anthocyanin pigmentation, the tissue is responsive to the inductive stimuli. Mature-phase phloem and cortical parenchyma undergo divisions in response to excision and auxin, and lamina tissue accumulates flavonols in response to light and sucrose. Thus, the effect of phase on competence may be the result of subtle differences in gene expression. This is supported by the fact that to date only a few genes have been shown to be differentially expressed in juvenile- and mature-phase tissue (Hutchison et al., 1990; Murray et al., 1989; Woo, 1992), and for those that are differentially expressed, their differential expression is limited to specific cell types.

Although the competence to express a phenotype or process within a phase is relatively stable, there is evidence that competence is manipulable or even reversible (Brand and Lineberger, 1992; Huang et al., 1992a,b; Hackett and Murray, 1993;). For instance in ivy, chronologically young, mature-phase petioles do form a limited number of adventitious roots in response to excision and auxin. Occasionally, mature-phase stem tissue will have some anthocyanin pigmentation. However, all of the phase-related characteristics may not have the same ease of reversibility, and the ease of reversibility of any one characteristic may change over developmental time.

Etiolation induced change in rooting competence of mature-phase tissues is persistent and may be associated with changes in specific cell types. Analysis of expression of specific genes in these cells using *in situ* hybridization could give insight into this possibility.

REFERENCES

Bachelard, E.P., and Stowe, B.B., 1962, A possible link between root initiation and anthocyanin formation. *Nature* 194:209.

Brand, M.H., and Lineberger, R.D., 1992, *In vitro* rejuvenation of *Betula* (Betulaceae): morphological evaluation, *Amer. J. Bot.* 79:618.

Beakbane, A.B., 1961, Structure of the plant stem in relation to adventitious rooting. *Nature* 192:954.

Davies, F.T., Lazarte, J.F., and Joiner, J.N., 1982, Initiation and development of roots in juvenile and mature leaf bud cuttings of *Ficus pumila* L., *Amer. J. Bot.* 69:804.

Doss, R.P., Torre, L.C., and Barritt, B.H., 1980, An investigation of the influence of etiolation on the rooting of red raspberry (*Rubus idaeus* L., c.v. Meeker) root-shoots, *Acta Hortic.* 112:77.

Doud, S.L., and Carlson, R.F., 1977, Effects of etiolation, stem anatomy, and starch reserves on root initiation of layered *Malus* clones, *J. Amer. Soc. Hortic. Sci.* 102:487.

Frolich, E.F., 1961, Etiolation and the rooting of cuttings, *in:* "Proc. Int. Plant Prop. Soc.," 11:277.

Gardner, F.E., 1929, The relationship between tree age and rooting of cuttings, *Proc. Amer. Soc. Hortic. Sci.* 26:101.

Gardner, R.E., 1936, Etiolation as a method of rooting apple variety stem cuttings, *Proc. Amer. Soc Hortic. Sci.* 34:323.

Geneve, R.L., 1985, The role of ethylene in adventitious root initiation in de-bladed petioles of the juvenile and mature phase of *Hedera helix* L., Ph.D. Thesis, Univ. of Minn., St. Paul.

Geneve, R.L., Hackett, W.P., and Swanson, B.T., 1988, Adventitious root initiation in de-bladed petioles from juvenile and mature phases of English ivy, *J. Amer. Soc. Hortic. Sci.* 113:630.

Girouard, R.M., 1967a, Initiation and development of adventitious roots is stem cuttings of *Hedera helix*. Anatomical studies of the juvenile growth phase, *Can. J. Bot.* 45:1877.

Girouard, R.M., 1967b, Initiation and development of adventitious roots is stem cuttings of *Hedera helix*. Anatomical studies of the mature growth phase, *Can. J. Bot.* 45:1883.

Goldberg, R.B. 1987, Emerging patterns of plant development, *Cell* 49:298.

Hackett, W.P., 1985, Juvenility, maturation and rejuvenation in woody plants, *Hortic. Rev.* 7:109.

Hackett, W.P., and Murray J.R., 1993, Maturation and rejuvenation in woody species, *in*: "Micropropagation of Woody Plants," M.R. Ahuja, ed., Kluwer Academic Pubs., Dordrecht.

Harrison-Murray, R.S., 1982, Etiolation of stock plants for improved rooting of cuttings: I. Opportunities suggested by work with apple, *in*: "Proc. Int. Plant Prop. Soc.," 31:386.

Herman, D.E., and Hess, C.E., 1963, The effect of etiolation upon the rooting of cuttings, *in*: "Proc. Int. Plant Prop. Soc.," 13:42.

Hess, C.E., 1962, Characterization of rooting cofactors extracted from *Hedera helix* L. and *Hibiscus rosa-sinensis* L., *in*: "Proc. 16th Int. Hortic. Congr.," p. 328.

Hess, C.E., 1969, Internal and external factors regulating root initiation, *in*: "Root Growth," W.J. Whittington, ed., Butterworths, London, p. 42.

Howard, B.H., 1980, Plant propagation, *in*: "Rep. East Malling Res. Sta. for 1979," p. 67.

Howard, B.H., 1981, Plant propagation, *in*: "Rep. East Malling Res. Sta. for 1980," p. 59.

Howard, B.H., 1982, Plant propagation, *in*: "Rep. East Malling Res. Sta. for 1981," p. 57.

Howard, B.H., 1984, Plant propagation, *in*: "Rep. East Malling Res. Sta. for 1983," p. 77.

Howard, B.H., 1985, Plant propagation, *in*: "Rep. East Malling Res. Sta. for 1984," p. 131.

Howard, B.H., Harrison-Murray, R.S., Vasek, J., and Jones, O.P., 1988, Techniques to enhance rooting potential before cutting collection, *Acta Hortic.* 227:176.

Huang, L.-C., Hsiao, C.-K., Lee, S.-H., Huang, B.-L., and Murashige, T., 1992a, Restoration of vigor and rooting competence in stem tissues of mature citrus by repeated grafting of shoot apices onto freshly germinated seedlings *in vitro*, *In Vitro Cell Dev. Biol.*, 28P:30.

Huang, L.-C., Suwenza, L., Huang, B.-L., Murashige, T., Mahdi, E.F.M., and Gundy, R.V., 1992b, Rejuvenation of *Sequoia sempervirens* by repeated grafting of shoot tips onto juvenile rootstocks *in vitro*, *Plant Physiol.* 98:166.

Hutchison, K.W., Sherman, C.B., Weber, J., Smith, S.S., Singer, P.B., and Greenwood, M.S., 1990, Maturation in larch, II. Effects of age on photosynthesis and gene expression in developing foliage, *Plant Physiol.* 94:1308.

Kawase, M., 1965, Etiolation and rooting of cuttings, *Physiol. Plant.* 18:1066.

Kawase, M., and Matsui, H., 1980, Role of auxin in root primordium formation in etiolated 'Red Kidney' bean stems, *J. Amer. Soc. Hortic. Sci.* 105:898.

Maynard, B.K., and Bassuk, N.L., 1987, Stockplant etiolation and blanching of woody plants prior to cutting propagation, *J. Amer. Soc. Hortic. Sci.* 112:273.

Maynard, B.K., and Bassuk, N.L., 1988, Etiolation and banding effects on adventitious root formation, *in*: "Adventitious Root Formation by Cuttings," T.D. Davis, B.E. Haissig, and N. Sankhla, eds., Adv. in Plant Sci. Ser., vol 2, Dioscorides Press, Portland, p. 29.

McDaniel, C., 1982, Shoot meristem development, *in*: "Positional Controls in Plant Development," P. Barlow, and D.J. Carr, eds., Cambridge Univ. Press, Cambridge.

Mullins, M.G., 1985, Regulation of adventitious root formation in microcuttings, *Acta Hortic.* 166:53.

Murray, J.R., and Hackett, W.P., 1991, Dihydroflavonol reductase activity in relation to differential anthocyanin accumulation in juvenile and mature phase *Hedera helix* L., *Plant Physiol.* 97:343.

Murray, J.R., Smith, A.G., and Hackett, W.P., 1989, Phase-dependent competence for dihydroquercetin reductase and anthocyanin expression, *in*: "Abstracts of Hortic. Biotech. Symp.," Aug. 21-23, 1989, Univ. of Calif., Davis.

Poethig, R.S., 1990, Phase change and regulation of shoot morphogenesis in plants, *Science* 250:923.

Rogler, C.E., and Hackett, W.P., 1975, Phase change in *Hedera helix*: Induction of the mature to juvenile phase change by gibberellin A3, *Physiol. Plant.* 33:141.

Schreiber, L.R., and Kawase, M., 1975, Rooting of cuttings from tops and stumps of American Elm, *HortSci.* 10:615.

Stein, O.L., and Fosket, E.B., 1969, Comparative developmental anatomy of shoots of juvenile and adult *Hedera helix*, *Amer. J. Bot.* 56:546.

Stoutemyer, V.T., 1937, "Regeneration in Various Types of Apple Wood," Res. Bull. Iowa Agric. Exp. Sta., No. 220, p. 308.

Stoutemyer, V.T., 1961, Light and propagation, *in*: "Proc. Int. Plant Prop. Soc.," 11:252.

Woo, H.-H., 1992, Molecular analysis of phase variation in *Hedera helix* L., English ivy, Ph.D. Thesis, Univ. of Minn., St. Paul.

Zimmerman, R.H., 1973, Juvenility and flowering in fruit trees, *Acta Hortic.* 34:139.

Zimmerman, R.H., ed., 1976, "Symposium on Juvenility in Woody Perennials," *Acta Hortic.*, No. 56.

BIOCHEMICAL AND MOLECULAR MARKERS OF

CELLULAR COMPETENCE FOR ADVENTITIOUS ROOTING

Paul Hand

Molecular Biology Department
Horticulture Research International
Littlehampton
West Sussex, BN17 6LP, UK

INTRODUCTION

A key stage in adventitious rooting is the *de novo* formation of a root meristem. This involves the dedifferentiation of induced cells followed by cell division and enlargement to form a meristem. Subsequent primordium development leads to root emergence and growth. Cells are said to be competent for root formation when they are able to respond directly to an inducing stimulus (usually wounding and/or auxin) by the direct formation of root primordia [Geneve (1991); see also the chapter by Mohnen in this volume]. Cells which are not competent are unable to respond directly to the stimulus, but may attain the competent state indirectly via non-directed cell division. It is clear that such a major shift in the developmental fate of cells involves complex interacting changes at the biochemical level and at the level of gene expression. There is substantial evidence that auxins play a crucial role in the formation of adventitious roots, but they are not the sole determinant (see the chapter by Blakesley in this volume). Many other compounds are involved in the rooting process, as reviewed by Haissig and Davis in this volume.

There is, however, no direct evidence for the mode of action of auxins or cofactors in controlling or directing the rooting process (see the chapter by Palme and coworkers in this volume.) Most of the research to date has correlated qualitative or quantitative changes in a particular compound or group of compounds with the appearance or numbers of adventitious roots in the plant under study. Several biochemical markers of rooting have been proposed, but the supporting evidence is often contradictory [see reviews by Haissig (1986), Jarvis (1986), Davis et al. (1988), and Wilson and van Staden (1990)]. Problems arise from the difficulty of measuring events which are localized in a small number of cells, and of isolating key factors from an extremely complex set of biochemical interactions. Some researchers have called for more fundamental studies of adventitious rooting to identify the key markers and controls of the process (Davies and Hartmann, 1988; Haissig et al., 1992).

Biology of Adventitious Root Formation, Edited by
T.D. Davis and B.E. Haissig, Plenum Press, New York, 1994

In this chapter I will review the recent literature on potential biochemical and molecular markers and suggest some techniques of molecular biology which could be applied to provide a deeper understanding of adventitious root formation.

BIOCHEMICAL MARKERS

Polyamines

Polyamines are low molecular weight, aliphatic, nitrogenous polycations. They have been shown to play a role in a variety of biological processes in a wide range of organisms (Slocum et al., 1984; Tabor and Tabor, 1984; Galston and Kaur-Sawhney, 1987). In a review of the role of polyamines in adventitious root formation and root growth, Sankhla and Upadhyaya (1988) concluded that there was little evidence to link polyamines to the regulation of rooting. However, several studies up to that time had demonstrated that cell division and differentiation were correlated with early increases in endogenous levels of polyamines, especially putrescine and spermidine. Work on easy-to-root mung bean hypocotyls suggested that applications of exogenous polyamines could stimulate adventitious rooting (Jarvis et al., 1983).

Since then, studies in other plant systems have supported the correlation of elevated polyamine levels with cell division and root primordium development, suggesting a role for polyamines as markers of the rooting process (see the chapter by Tepfer and coworkers in this volume). Tiburcio et al. (1989), Biondi et al. (1990) and Geneve and Kester (1991), working with different species *in vitro*, all linked an increase in putrescine levels with the increase in mitotic activity during primordium development, after the induction phase. Geneve and Kester (1991) also found that a greater increase in putrescine occurred in juvenile- than mature-phase ivy petioles. Addition of polyamines to mature petioles did not, however, induce root formation. This suggested that the difference in putrescine levels was not directly controlling the process.

Attention has also focused on the specificity of the enzyme inhibitors commonly used in studies of polyamine metabolism. This has led in turn to an appraisal of the roles of free and bound forms of polyamines. Burtin et al. (1989) found that difluoromethyl ornithine, the inhibitor of ornithine decarboxylase, was more active in inhibiting the formation of putrescine conjugates, whereas difluoromethyl arginine, the arginine decarboxylase inhibitor, directly inhibited putrescine biosynthesis. The conjugated forms of polyamines were not found to be storage forms, but their function is unknown. This may vary according to the partner molecule. Studies of free and conjugated polyamines during rooting have produced contradictory results, and no clear pattern has yet emerged (Torrigiani et al., 1989; Altamura et al., 1991).

Peroxidases

Many studies have attempted to correlate peroxidase (PER) enzyme activity with rooting. Changes in PER isoenzyme levels associated with rooting have been observed, and differences in activity between easy- and difficult-to-root species (Molnar and LaCroix, 1972; Quoirin et al., 1974) suggested that they could have potential as biochemical markers. However, plant PERs are found in multiple forms (isoenzymes), and participate in auxin metabolism, respiration, cell wall synthesis and wound responses, among other processes. The specific role of PERs in rooting is, therefore, difficult to ascertain. PER activity has been linked to the oxidation of auxin. Many basic PERs have indole-3-acetic acid (IAA) oxidase activity and other PERs have been shown to be effective in IAA oxidation, at least *in vitro* (Hinman and Lang, 1965; Pressey, 1990). However, the reaction requires several cofactors including oxygen, Mn^+ and a monophenol. Hence, it has been suggested that PER-catalyzed IAA oxidations are efficient only in damaged cells, where all of the necessary reactants can come together. It has also been suggested that the consequent reduction in IAA levels could induce physiological changes in neighboring cells which elicit a wound response and possibly root induction (Pressey, 1990). The acidic PERs are believed to be involved in lignification; Gaspar et al. (1985) proposed a two-step system of PER response to wounding or stress, with a rapid activation of basic PERs (which affect auxin metabolism and lead to the induction of rooting) followed by a later activation of acidic PERs leading to lignification during root initiation and development. Competent (preinduced) tissues have initially high levels of PER activity. Auxin levels are said to mirror the fluctuations in basic PER activity. However, the physiological role of free auxin compared with oxidation products or conjugates is still far from understood, which complicates the interpretation of studies in this area.

The conflicting evidence from published studies was reviewed by Haissig (1986) and Bhattacharya (1988). More recent studies have also produced apparently contradictory results. Berthon et al. (1989) found that PER activity initially increased then decreased during rooting of *Sequoiadendron giganteum* cuttings *in vitro*. The changes in enzyme activity were largely due to changes in basic isoenzymes, which is in agreement with the model proposed by Gaspar et al. (1985). Moncousin et al. (1989) found a similar pattern of PER activity in *Vitis* shoots rooting *in vitro*. In addition, PER activity showed a transient decline in the first few hours of culture. This corresponded to a peak in IAA levels, and occurred before any visible cytological event. Using tissue-cultured *Prunus avium*, Dalet and Cornu (1989) also observed an initial delay in the increase in PER activity in shoots during rooting. The number of basic PER increased more rapidly in auxin-treated rooting shoots than in controls, but overall there was no correlation between rooting and PER activity. Whole shoots (stem and leaves) were sampled for this study, which could have masked localized changes in PERs at the site of root initiation. Pythoud and Buchala (1989) could find no correlation between total PER activity and rooting in *Populus tremula*. The relative contributions of basic and acidic PERs were not assessed, but a steady increase in total PER activity was observed in both rooting and nonrooting plants.

Polyphenol Oxidase

Despite years of research, the physiological role of this enzyme remains unknown [see reviews by Mayer (1987) and Vaughn et al. (1988)]. It is able to catalyze the hydroxylation of phenolic compounds *in vitro*, but *in vivo* studies using specific inhibitors have demonstrated that polyphenol oxidase (PPO) is not involved in synthesis of phenolic compounds in intact cells (Duke and Vaughn, 1982; Strack et al., 1986), being localized exclusively in plastids. However, cell damage by wounding allows enzyme and substrates to come into contact, thereby facilitating the oxidation of phenolics. Al Barazi and Schwabe (1984) suggested that PPO activity was positively correlated with rooting ability in *Pistacia vera*, and Bassuk et al. (1981) found that products of phenolic oxidation stimulated root induction in apple cuttings. PPO may, therefore, have an indirect role in adventitious root formation as a producer of rooting "cofactors" (see the chapter by Howard in this volume.) Little progress has been made in this area in recent years [the subject was reviewed recently by Wilson and Van Staden (1990)]. From the limited data available at this time, PPO cannot be considered a biochemical marker of rooting. More precise information is needed on the mode of action of PPO in wounded cells and the role of the products of enzyme action in stimulating rooting, especially in relation to auxin action or metabolism.

Flavonoids

The flavonoids are a class of phenolic compounds that are potential biochemical markers of rooting. Flavonoids include the anthocyanins, the red pigments commonly found in plants. They are produced via the shikimic acid, phenylpropanoid and flavonoid biosynthetic pathways. A branch of the phenylpropanoid pathway leads to lignin synthesis. These enzyme pathways are affected by a variety of environmental stimuli such as temperature, light and wounding [for a review see Hahlbrock and Scheel (1989)]. Several studies on woody plants have linked high endogenous flavonoid levels with ease of rooting (Table 1). Recently Curir et al. (1990) found that *in vitro* rooting of *Eucalyptus gunnii* could be stimulated by preconditioning of shoots on a medium containing the cytokinins kinetin or zeatin. The preconditioning was accompanied by an accumulation of two glycosides of quercetin in the shoots. The preconditioning could be replaced by addition of these compounds to the rooting medium. This effect was specific to the quercetin glycosides: addition of quercetin reduced the rooting response by 90%. Interestingly, Bachelard and Stowe (1962) identified the anthocyanins in easy-to-root *Acer rubrum* cuttings as glycosides of cyanidin. Curir and coworkers speculated that the flavonoids could act by inhibition of IAA oxidase activity or via a higher affinity as substrates for oxidase and PER enzymes, as suggested by Barz (1977). This might maintain high levels of endogenous IAA which in turn would stimulate rooting. However, endogenous IAA levels and IAA oxidase activity were not measured directly.

Flavonoid metabolism has also been studied with regard to another aspect of rooting. The potential of woody plant cuttings to root may vary between juvenile and adult phases of the same species. Juvenile-phase shoots generally root more readily than adult-phase shoots, and often contain significantly higher levels of anthocyanin and other flavonoids (Vazquez and Gesto, 1986).

Table 1. Examples of high flavonoid levels associated with ease of rooting.

Flavonoid	Reference
Cyanidin glycosides	Bachelard & Stowe (1962)
Anthocyanin	Vazquez & Gesto (1986)
Quercetin glycoside	Curir et al. (1990)
Cyanidin glycosides	Murray & Hackett (1991)

Haissig (1986) suggested that stock plant maturation may modify endogenous levels of anthocyanin precursors that are essential for the rooting of cuttings. *Hedera helix* shows phase-specific differences in rooting ability (Hackett, 1988). Murray and Hackett (1991) explored this theme in relation to flavonoid biosynthesis and enzyme activity in cultured leaf discs of juvenile- and mature-phase *Hedera helix*. They found several differences which could be markers of phase-state. When discs were cultured on sucrose medium in the light, mature-phase tissues developed a higher specific activity of phenylalanine ammonia lyase (PAL; the first enzyme in the phenylpropanoid pathway), and accumulated higher levels of phenylpropanoids, especially caffeic and ferulic acids. Juvenile-phase leaf discs accumulated anthocyanin and cyanidin glycoside, and activity of the enzyme dihydroflavonol reductase, whereas none of these were detectable in the mature-phase tissue. These specific differences are discussed by Murray and coworkers elsewhere in this volume. Caffeic and ferulic acids were found by James and Thurbon (1981) to inhibit root formation in apple, and the positive correlation between flavonoid glycosides and root formation has been noted by other workers, as described above.

NUCLEIC ACID MARKERS

Several studies have shown that nucleic acid (and protein) synthesis are necessary for adventitious root formation [for a review see Haissig (1986)]. This is perhaps not surprising, given that the process involves cell division and major changes in the metabolism of the cells which redifferentiate to form root primordia. Early studies were limited to an examination of total protein and nucleic acid levels. They could, therefore, only provide information on net synthesis and steady-state levels of the protein and messenger RNA (mRNA) populations in the tissues examined. Advances in the field of molecular biology mean that techniques are available for the isolation and quantitation of individual mRNA species. Hackett et al. (1990) found that the translation products of abundant mRNAs were very similar in auxin-treated juvenile and mature ivy petioles. They isolated two complementary DNA (cDNA) clones of differentially expressed mRNAs. The expression of one of these mRNAs was induced by wounding and displayed an inverse relationship to rooting potential of the tissues. Nucleotide sequencing showed this to have homology with cell wall proline-rich protein genes of soybean, pea and carrot (Woo, 1992). Genes with similar sequences have been shown to be associated with lateral root initiation in tobacco (Keller and Lamb, 1989) and the development of vascular tissue in maize (Stiefel et al., 1990).

The induction of roots in plants infected with the pathogenic bacterium *Agrobacterium rhizogenes* has attracted much interest at the molecular level. Studies of the expression of the *A. rhizogenes* genes responsible (the *rol* genes) and their interaction with plant genes may help to target genes involved in adventitious root formation. Aspects of this subject are reviewed in detail elsewhere in this volume by Hamill and Chandler, and by Tepfer and coworkers.

Work in other fields not directly related to adventitious root formation may provide useful information on potential molecular markers: examples are the genetics and molecular biology of apical meristems (Medford, 1992). Although most effort has been directed at the function of established meristems, some information has emerged on the *de novo* induction of meristematic activity. Carmen Martinez et al. (1992) have described the spatial pattern of expression of the *cdc2* gene in *Arabidopsis*. This has been found in several plant species and its protein kinase product is a key component of the cell cycle required for the G_1- to S-phase transition and entry into mitosis. mRNA transcripts of this gene accumulate in the root meristem and in the initial stages of the activation of a new meristem at sites of lateral root development. This suggests that *cdc2* expression is an early event in meristem activation. Expression of *cdc2* is also strong in root pericycle and stelar parenchymal cells, which are arrested in the G_2 stage of mitosis. The authors suggested that *cdc2* expression may, therefore, be involved in determining competence for proliferation, but additional (unknown) "signals" are required for cell division to proceed. Genes like *cdc2* could be good candidates for further study, both as markers of the competent state and for their function in the root induction process.

NEW APPROACHES AND TECHNOLOGIES

Some of the practical difficulties involved in defining biochemical and molecular markers of competence to root were mentioned earlier. Several problems need to be addressed in future studies: 1) the dilution effect from the tissues surrounding the relatively small number of cells which form the root initials and primordia, 2) heterogeneity of experimental plant material, 3) the direct measurement of compounds involved in root initiation and development, and 4) the fragmentary nature of much of the existing evidence.

Experimental Systems

Experimental systems are reviewed in depth elsewhere in this volume [see the chapter by Ernst, by Riemenschneider, and by Mohnen in this volume] but it is worth mentioning briefly the advantages of *in vitro* culture [see also the chapter by Howard in this volume]. This technique has been expanding rapidly, especially in its application to woody plants (Jones, 1979; Bonga and Durzan, 1987). Several elegant systems have been developed for the study of adventitious rooting, such as ivy petioles (Geneve et al., 1988), cotyledons of walnut (Jay-Allemand et al., 1991) and stem disks of apple (van der Krieken et al., 1991). *In vitro* cultures allow uniform clonal material to be used in rooting experiments, leading to more synchronized root formation. With difficult-to-root adult woody material, the phenomenon termed *in vitro* rejuvenation has been observed (see the chapter by Howard in this volume). Shoots from mature plants subcultured *in vitro* may regain the rooting ability (competence) associated with juvenile shoots (Webster and Jones, 1989; Hammatt and Grant, 1993). This allows comparisons between difficult-to-root and easy-to-root material of the same genotype under controlled experimental conditions. The ultimate test of any rooting promoter or marker is with whole cuttings *in vivo*, but *in vitro* systems such as those described above allow experimenters to probe further into the fundamental aspects of adventitious rooting.

Biochemical Approaches

The importance of auxin in adventitious root formation is well known, yet many of the studies on putative biochemical markers of the process have not involved direct measurements of auxin levels. Even studies of enzymes such as IAA oxidases have often simply assumed that levels of enzyme activity in extracts of tissue are negatively correlated with auxin levels. This may have been due to uncertainty about the physiological significance and role of free and conjugated forms of auxin or practical problems in measuring endogenous auxin levels. Auxins can now be measured directly by gas chromatography coupled to mass spectroscopy (GC-MS) (Schneider et al., 1985; Epstein et al., 1988; Sutter and Cohen, 1992), and the availability of radioactively labelled auxins allows their metabolism and distribution during rooting to be examined (van der Krieken et al., 1992). These techniques will undoubtedly become more widely used in the future, and should allow biochemical markers to be correlated more directly with auxin levels.

Antibody technology seems to have found little application in adventitious rooting research, yet is widely applied in other fields (Knox, 1982). Techniques are available for the quantitation of plant growth regulators and other low molecular weight metabolites by immunoassay [see review by Weiler (1984)] and results have been validated against GC-MS (Julliard et al., 1992). Antibodies can be used to study changes in steady-state levels of specific enzymes and proteins. Site-directed monoclonal antibodies can be produced, which as their name suggests are specific for a particular region on the antigen molecule. Such antibodies may have an application in the study of specific PER isoenzymes, for example. The use of immunocytolocalization is potentially of great interest. This technique would allow the distribution of enzymes to be determined within tissues and cells, as demonstrated by Stroobants et al. (1991). In this way, biochemical and cytological data can be brought together. It may also be possible to examine auxin-induced changes in cell wall characteristics, or sensitivity of cells to auxin using these methods; antibodies to an auxin binding protein have been produced (Napier et al., 1988).

Molecular Approaches

The control of adventitious rooting has received little attention from molecular biologists until recently, yet results from areas such as the control of flower development have shown what a powerful tool molecular techniques can be (Coen, 1991; Coen and Meyerowitz, 1991). Work has begun to isolate and identify genes whose expression is modified during the rooting process (van der Krieken et al., 1991; Woo, 1992). The strategy adopted by these workers is to compare gene expression in rooting and nonrooting tissues, by the construction and differential screening of cDNA libraries. Clones which show qualitative or quantitative differences between treatments can be isolated and their nucleic acid sequences determined. This information can provide clues to the identity and function of the genes and their putative protein products. This approach can be used to detect very rare transcripts by use of a subtractive technique. This involves subtracting out the majority of transcripts which are common to both tissues, utilizing the potential of complementary nucleic acid strands to combine or hybridize. The resulting "library" of transcripts is enriched for differential sequences. This method has been used to isolate sequences expressed in shoot apices during the transition to flowering (Kelly et al., 1990; Melzer et al., 1990). The sensitivity of the method is such that it is possible to detect transcripts which represent 0.01% of the total cellular mRNA. Even greater sensitivity can be obtained using the polymerase chain reaction (PCR). This technique was developed in the mid 1980s (Saiki et al., 1985) and has since become one of the most powerful tools in molecular biology. PCR is a rapid procedure for the *in vitro* amplification of a specific segment of DNA between two short sections of known nucleotide sequence. Sequences of interest can be amplified many million-fold in a few hours. In a modification of the original technique, short primer sequences can be attached to the ends of a heterologous population of DNA molecules for amplification. By this means, PCR based methods of differential screening have been developed (Duguid and Dinauer, 1990; Timblin et al., 1990). These can be used to isolate differentially expressed genes from very low starting populations of cells.

The PCR technique can be useful in an alternative molecular strategy. This uses information on the biochemical pathways and genes believed to be associated with rooting and cell division to focus on key enzymes and proteins. Promising candidates for this approach are the enzymes for flavonoid and polyamine biosynthesis and the PER enzymes. Genes or transcripts cloned from other species could be used to identify and clone target genes from the species under investigation, or sequence information could be used to design PCR primers for direct amplification of clones from DNA. Using RACE-PCR (Rapid Amplification of cDNA Ends), genes can be amplified when only a single section of nucleotide sequence is available (Frohman et al., 1988). The disadvantage of this strategy is that it depends upon the availability of at least some sequence information from the gene of interest. The first strategy is technically difficult, but does not require preexisting genetic information.

Once genes of interest have been isolated and sequenced, the next challenge is to examine their function and patterns of expression. Nucleic acid sequences can be labelled *in vitro* and used to probe extracts of tissue mRNA immobilized onto membranes. This allows the levels of a particular mRNA species to be determined in different tissues or with time in response to, for example, an inductive treatment. Hybridization of probes to tissue sections (*in situ* hybridization) enables the cells which are expressing a particular gene to be localized within a tissue.

The ability to introduce novel genetic material into many plant species through the use of techniques such as *Agrobacterium*-mediated transformation or microprojectile bombardment allows gene expression to be manipulated. Several woody plant species have already been stably transformed, and methodologies for others are currently under development. Transformation by microprojectile bombardment often leads to only transient expression of introduced genes. This approach could still be useful, for example, in studies on the induction of competence in difficult-to-root mature woody plants. Rapidly increasing knowledge of regulatory sequences which influence the expression of plant genes could allow quite elegant experiments to be performed to investigate the role of potential marker genes in the rooting process. A gene of interest could be fused to regulatory elements which control expression in an organ-specific manner or in response to a specific chemical or environmental stimulus. Alternatively, gene expression can be suppressed using "antisense" technology (van der Krol et al., 1988). This involves the introduction of a DNA sequence which codes for an mRNA complementary to the sequence of the target mRNA. Hybridization is believed to occur between the two RNA molecules, which has the effect of "down-regulating" the target gene.

CONCLUSIONS

To date there is no direct evidence linking a biochemical or molecular marker to competence to root. The accumulated evidence is largely correlative. Studies to date have concentrated on the search for markers, rather than on the rooting process itself. Nevertheless, they have provided important clues for the direction that future research should take.

The scale of the problem and the practical difficulties involved necessitate a multi-disciplinary approach. Molecular studies of rooting have only recently begun. The techniques available through molecular biology provide powerful additional tools for researchers. We need to bring together the developmental, cytological, physiological, biochemical and molecular evidence to elucidate the whole rooting process. The problem needs to be addressed from both ends: More precise information is needed about the links between specific markers, the competent state and root induction; and, also, the root induction process needs to be examined in detail to determine the key differences between competent and noncompetent cells. Only then will we be in a position to consider manipulating adventitious rooting in a controlled manner.

ACKNOWLEDGMENTS

The author is supported by the UK Agricultural and Food Research Council, and thanks Dr. B. Thomas and Dr. N. Hammatt for critical reading of the manuscript.

REFERENCES

Al Barazi, Z., and Schwabe, W.W., 1984, The possible involvement of polyphenol-oxidase and the auxin-oxidase system in root formation and development of cuttings of *Pistacia vera*, *J. Hortic. Sci.* 59:453.

Altamura, M.M., Torrigiani, P., Capitani, F., Scaramagli, S., and Bagni, N., 1991, *De novo* root formation in tobacco thin layers is affected by inhibition of polyamine biosynthesis, *J. Exp. Bot.* 42:1575.

Bachelard, E.P., and Stowe, B.B., 1962, A possible link between root initiation and anthocyanin formation, *Nature* 194:209.

Barz, W., 1977, Degradation of polyphenols in plants and plant cell suspension cultures, *Physiol. Vég.* 15:261.

Bassuk, N.L., Hunter, L.D., and Howard, B.H., 1981, The apparent involvement of polyphenol oxidase and phloridzin in the production of apple rooting co-factors, *J. Hortic. Sci.* 56:313.

Berthon, J.-Y., Maldiney, R., Sotta, B., Gaspar, T., and Boyer, N., 1989, Endogenous levels of plant hormones during the course of adventitious rooting in cuttings of *Sequoiadendron giganteum* (Lindl.) *in vitro*, *Biochem. Physiol. Pfl.* 184:405.

Bhattacharya, N.C., 1988, Enzyme activities during adventitious rooting, *in*: "Adventitious Root Formation in Cuttings," T.D. Davis, B.E. Haissig, and N. Sankhla, eds., Advances in Plant Sci. Series, vol. 2, Dioscorides Press, Portland.

Biondi, S., Diaz, T., Iglesias, I., Gamberini, G., and Bagni, N., 1990, Polyamines and ethylene in relation to adventitious root formation in *Prunus avium* shoot cultures, *Physiol. Plant.* 78:474.

Bonga, J.M., and Durzan, D.J., eds., 1987, "Cell and Tissue Culture in Forestry, General Principles and Biotechnology," Martinus Nijhoff, Boston.

Burtin, D., Martin-Tanguy, J., Paynot, M., and Rossin, N., 1989, Effects of the suicide inhibitors of arginine and ornithine decarboxylase activities on organogenesis, growth, free polyamine and hydroxycinnamoyl putrescine levels in leaf explants of *Nicotiana xanthi* M.C. cultivated *in vitro* in a medium producing callus formation, *Plant Physiol.* 89:104.

Carmen Martinez, M., Jorgensen, J.-E., Lawton, M.A., Lamb, C.J., and Doerner, P.W., 1992, Spatial patterns of *cdc2* expression in relation to meristem activity and cell proliferation during plant development, *Proc. Natl. Acad. Sci. USA* 89:7360.

Coen, E.S., 1991, The role of homeotic genes in flower development and evolution, *Annu. Rev. Plant Physiol. Plant Mol. Biol.* 42:241.

Coen, E.S., and Meyerowitz, E.M., 1991, The war of the whorls: genetic interactions controlling flower development, *Nature* 353:31.

Curir, P., van Sumere, C.F., Termini, A., Barthe, P., Marchesini, A., and Dolci, M., 1990, Flavonoid accumulation is correlated with adventitious root formation in *Eucalyptus gunnii* Hook micropropagated through axillary bud stimulation, *Plant Physiol.* 92:1148.

Dalet, F., and Cornu, D., 1989, Lignification level and peroxidase activity during *in vitro* rooting of *Prunus avium*, *Can. J. Bot.* 67:2182.

Davies, F.T. Jr., and Hartmann, H.T., 1988, The physiological basis of adventitious root formation, *Acta Hortic.* 227:113.

Davis, T.D., Haissig, B.E., and Sankhla, N., eds., 1988, "Adventitious Root Formation in Cuttings," Dioscorides Press, Portland.

Duguid, J.R., and Dinauer, M.C., 1990, Library subtraction of *in vitro* cDNA libraries to identify differentially expressed genes in scrapie infection, *Nucleic Acids Res.* 18:2789.

Duke, S.O., and Vaughn, K.C., 1982, Lack of involvement of polyphenol oxidase in ortho-hyroxylation of phenolic compounds in mung bean seedlings, *Physiol. Plant.* 54: 381.

Epstein, E., Muszkat, L., and Cohen, J.D., 1988, Identification of indole-3-butyric acid (IBA) in leaves of cypress and corn by gas chromatography - mass spectrometry, *Alon. Hanolea.* 42:917.

Frohman, M.A., Dush, M.K., and Martin, G.R., 1988, Rapid production of full-length cDNAs from rare transcripts: amplification using a single gene-specific oligonucleotide primer, *Proc. Natl. Acad. Sci. USA* 85:8998.

Galston, A.W., and Kaur-Sawhney, R., 1987, Polyamines as endogenous growth regulators, *in*: "Plant Hormones and their Role in Plant Growth and Development," P.J. Davis, ed., Martinus Nijhoff, Dordrecht.

Gaspar, T., Penel, C., Castillo, F.J., and Greppin, H., 1985, A two-step control of basic and acidic peroxidases and its significance for growth and development, *Physiol. Plant.* 64:418.

Geneve, R.L., 1991, Patterns of adventitious root formation in English Ivy, *J. Plant Growth Regul.* 10:215.

Geneve, R.L., Hackett, W.P., and Swanson, B.T., 1988, Adventitious root initiation in de-bladed petioles from the juvenile and mature phases of English Ivy, *J. Amer. Soc. Hortic. Sci.* 113:630.

Geneve, R.L., and Kester, S.T., 1991, Polyamines and adventitious root formation in the juvenile and mature phase of English Ivy, *J. Exp. Bot.* 42: 71.

Hackett, W.P., 1988, Donor plant maturation and adventitious root formation, *in*: "Adventitious Root Formation in Cuttings,", T.D. Davis, B.E. Haissig, and N. Sankhla, eds., Advances in Plant Sci. Series, vol. 2, Dioscorides Press, Portland.

Hackett, W.P., Murray, J.R., Woo, H.-H., Stapfer, R.E., and Geneve, R., 1990, Cellular, biochemical and molecular characteristics related to maturation and rejuvenation in woody species, *in*: "Plant Ageing: Basic and Applied Approaches", R. Rodriguez, R. Sanchez Tames, and D.J. Durzan, eds., NATO ASI Ser. A, vol. 186, Plenum Press, New York.

Hahlbrock, K., and Scheel, D., 1989, Physiology and molecular biology of phenylpropanoid metabolism, *Ann. Rev. Plant Physiol. Plant Mol. Biol.* 40:347.

Haissig, B.E., 1986, Metabolic processes in adventitious rooting of cuttings, *in*: New Root Formation in Plants and Cuttings", M.B. Jackson, ed., Martinus Nijhoff, Dordrecht.

Haissig, B.E., Davis, T.D., and Riemenschneider, D.E., 1992, Researching the controls of adventitious rooting, *Physiol. Plant.* 84:310.

Hammatt, N., and Grant, N., 1993, Apparent rejuvenation of wild cherry (*Prunus avium* L.) during micropropagation, *J. Plant Physiol.* (in press)

Hinman, R.L., and Lang, J., 1965, Peroxidase catalyzed oxidation of indole-3-acetic acid, *Biochem.* 4:144.

James, D.J., and Thurbon, I.J., 1981, Phenolic compounds and other factors controlling rhizogenesis *in vitro* in the apple rootstocks M.9 and M.26., *Z. Pflanzenphysiol.* 105:11.

Jarvis, B.C., 1986, Endogenous control of adventitious rooting in non-woody cuttings, *in*: "New Root Formation in Plants and Cuttings," M.B. Jackson, ed., Martinus Nijhoff, Dordrecht.

Jarvis, B.C., Shannon, P.R.M., and Yasmin, S., 1983, Involvement of polyamines with adventitious root development in stem cuttings of mung bean, *Plant Cell Physiol.* 24:677.

Jay-Allemand, C., De Pons, V., Doumas, P., Capelli, P., Sossountzov, L., and Cornu, D., 1991, *In vitro* root development from walnut cotyledons: a new model to study the rhizogenesis processes in woody plants, *C.R. Acad. Sci. Paris Ser. III* 312:369.

Jones, O.P., 1979, Propagation *in vitro* of apple and other woody fruit plants, *Sci. Hortic.* 30:44.

Julliard, J., Sotta, B., Pelletier, G., and Miginiac, E., 1992, Enhancement of naphthaleneacetic acid-induced rhizogenesis in T_L-DNA-transformed *Brassica napus* without significant modification of auxin levels and auxin sensitivity, *Plant Physiol.* 100:1277.

Keller, B., and Lamb, C.J., 1989, Specific expression of a novel cell wall hydroxyproline-rich glycoprotein gene in lateral root initiation, *Genes Dev.* 3:1639.

Kelly, A.J., Zagotta, M.T., White, R.A., Chang, C., and Meeks-Wagner, R., 1990, Identification of genes expressed in the tobacco shoot apex during the floral transition, *Plant Cell* 2:963.

Knox, R.B., 1982, Immunology and the study of plants, *in*: "Antibody as a Tool," J.J. Marchalonis, and G.W. Smith, eds., Wiley, Chichester.

Mayer, A.M., 1987, Polyphenol oxidases in plants - recent progress, *Phytochemistry* 26:11.

Medford, J.I., 1992, Vegetative apical meristems, *Plant Cell* 4:1029.

Melzer, S., Majewski, D.M., and Apel, K., 1990, Early changes in gene expression during the transition from vegetative to generative growth in the long-day plant *Sinapis alba*, *Plant Cell* 2:953.

Molnar, J.M., and LaCroix, L.J., 1972, Studies of the rooting of cuttings of *Hydrangea macrophylla*: enzyme changes, *Can. J. Bot.* 50:315.

Moncousin, C., Favre, J.M., and Gaspar, T., 1989, Changes in peroxidase activity and endogenous IAA levels during adventitious root formation in vine cuttings, *in*: "Physiology and Biochemistry of Auxins in Plants," M. Kutacek, R.S. Bandurski, and J. Krekule, eds., Academia, Praha.

Murray, J.R., and Hackett, W.P., 1991, Dihydroflavonol reductase activity in relation to differential anthocyanin accumulation in juvenile and mature phase *Hedera helix* L., *Plant Physiol.* 97:343.

Napier, R.M., Venis, M.A., Bolton, M.A., Richardson, L.I., and Butcher, G.W., 1988, Preparation and characterisation of monoclonal and polyclonal antibodies to maize membrane auxin-binding protein, *Planta* 176:519.

Pressey, R., 1990, Anions activate the oxidation of indoleacetic acid by peroxidases from tomato and other sources, *Plant Physiol.* 93:798.

Pythoud, F., and Buchala, A.J., 1989, Peroxidase activity and adventitious rooting in cuttings of *Populus tremula*, *Plant Physiol. Biochem.* 27:503.

Quoirin, M., Boxus, P., and Gaspar, T., 1974, Root initiation and isoperoxidases of stem tip cuttings from mature *Prunus* plants, *Physiol. Vég.* 12:165.

Saiki, R.K., Scharf, S., Faloona, F., Mullis, K., Horn, G., Erlich, H.A., and Arnheim, N., 1985, Enzymatic amplification of β-globin genomic sequences and restriction site analysis for diagnosis of sickle cell anemia, *Science* 230:1350.

Sankhla, N., and Upadhyaya, A., 1988, Polyamines and adventitious root formation, *in*: "Adventitious Root Formation in Cuttings," T.D Davis,, B.E. Haissig, and N. Sankhla, eds., Advances in Plant Sci. Series, vol. 2, Dioscorides Press, Portland.

Schneider, E., Kazakoff, C., and Wightman, F., 1985, Gas chromatography - mass spectrometry evidence for several endogenous auxins in pea seedling organs, *Planta* 165:232.

Slocum, R.D., Kaur-Sawhney, R., and Galston, A.W., 1984, The physiology and biochemistry of polyamines in plants, *Arch. Biochem. Biophys.* 235:283.

Stiefel, V., Ruiz-Avila, L., Raz, R., Valles, M.P., Gomez, J., Pages, M., Martinez-Izquierdo, J.A., Ludevid, M.D., Langdale, J.A., Nelson, T., and Puigdomenech, P., 1990, Expression of a maize cell wall hydroxyproline-rich glycoprotein gene in early leaf and root vascular differentiation, *Plant Cell* 2:785.

Strack, D., Ruhoff, R., and Grawe, W., 1986, Hydroxycinnamoylcoenzyme-A:tartronate hydroxycinnamoyltransferase in protein preparations from mung bean, *Phytochem.* 25:833.

Stroobants, C., Sossountzov, L., and Miginiac, E., 1991, Immunocytolocalization of zeatin riboside in rooting tobacco leaves during the early stages of rhizogenesis, *C.R. Acad. Sci. Paris Ser. III* 312:261.

Sutter, E.G., and Cohen, J.D., 1992, Measurement of indolebutyric acid in plant tissues by isotope dilution gas chromatography - mass spectrometry analysis, *Plant Physiol.* 99:1719.

Tabor, C.W., and Tabor, H., 1984, Polyamines, *Ann. Rev. Biochem.* 53:749.

Tiburcio, A.F., Gendy, C.A., and Tran Than van, K., 1989, Morphogenesis in tobacco subepidermal cells: putrescine as marker of root differentiation, *Plant Cell, Tissue and Organ Culture* 19:43.

Timblin, C., Battey, J., and Kuehl, W.M., 1990, Application of PCR technology to subtractive cDNA cloning: identification of genes expressed specifically in murine plasmacytoma cells, *Nucleic Acids Res.* 18:1587.

Torrigiani, P., Altamura, M.M., Capitani, F., Serafini-Fracassini, D., and Bagni, N., 1989, *De novo* root formation in thin cell layers of tobacco: changes in free and bound polyamines, *Physiol. Plant.* 77:294.

van der Krieken, W.M., Breteler, H., and Visser, M.H.M., 1991, Indolebutyric acid-induced root formation in apple tissue culture, *Acta Hortic.* 289:343.

van der Krieken, W.M., Breteler, H., and Visser, M.H.M., 1992, The effect of the conversion of indolebutyric acid into indoleacetic acid on root formation on microcuttings of *Malus*, *Plant Cell Physiol.* 33:709.

van der Krol, A.R., Mol, J.N.M., and Stuitje, A.R., 1988, Modulation of eukaryotic gene expression by complementary RNA or DNA sequences, *Biotechniques* 6:958.

Vaughn, K.C., Lax, A.R., and Duke, S.O., 1988, Polyphenol oxidase: the chloroplast oxidase with no established function, *Physiol. Plant.* 72:659.

Vazquez, A., and Gesto, D.V., 1986, Rooting, endogenous root-inducing cofactors and proanthocyanidins in Chestnut, *Biol. Plant.* 28:303.

Webster, C.A., and Jones, O.P., 1989, Micropropagation of the apple rootstock M.9: effect of sustained subculture on apparent rejuvenation *in vitro*, *J. Hortic. Sci.* 64:421.

Weiler, E.W., 1984, Immunoassay of plant growth regulators, *Ann. Rev. Plant Physiol.* 35:85.

Wilson, P.J., and van Staden, J., 1990, Rhizocaline, rooting co-factors and the concept of promoters and inhibitors of adventitious rooting - a review, *Ann. Bot.* 66:479.

Woo, H.-H., 1992, Molecular Analysis of Phase Variation in *Hedera helix* L., English ivy, Ph.D. thesis, Univ. of Minnesota, St. Paul.

MANIPULATING ROOTING POTENTIAL IN

STOCKPLANTS BEFORE COLLECTING CUTTINGS

Brian H. Howard

Propagation Science Section
Horticulture Research International
East Malling,
Kent ME19 6BJ, UK

INTRODUCTION

Vegetative propagation is the bridge between plant improvement and commercial exploitation of clonally produced plants. The ease and extent of commercialization is determined largely by readiness to root from cuttings, and ready-rooting has subsequent benefits for "weaning," "hardening," rapid growth and batch uniformity. These benefits do not apply when propagation is difficult and cuttings root in low numbers over a protracted period. Speed of adventitious rooting is important because rapid rooting minimizes cutting exposure to adverse environments and to the diseases to which unrooted cuttings are prone.

Plant breeders are not always able to place ease of propagation high on their list of selection criteria, and it is often subordinated to more market-related characteristics on the assumption that deficiencies will be made-up by improved propagation technology. Propagation ability cannot be predetermined in mutations and other naturally occurring selections whose commercialization, therefore, depends on finding effective methods for their multiplication.

The majority of vegetatively propagated crops are characterized by a wide varietal diversity, often with relatively few of any one type being required each year. The need for relatively small-scale production makes such varieties low priority candidates for improving rooting via genetic engineering, when considering limitations imposed by the present state of the science (see the chapter by Riemenschneider in this volume) and the accompanying legislation. A complementary approach is to raise the rooting potential of existing varieties by enhancing gene expression, rather than by changing their genetic make-up. The increasing use of stockplants to facilitate production, and to ensure correct identity and optimal health status, offers the opportunity of "manipulating" shoots to improve rooting potential in cuttings made from them. Improving the potential for rooting must be accompanied by treatments and environments that will ensure the full realization of that potential.

Biology of Adventitious Root Formation, Edited by
T.D. Davis and B.E. Haissig, Plenum Press, New York, 1994

In view of the importance of adventitious rooting to the introduction and production of improved plants, it will be the purpose of this review to reach a better understanding of those factors determining rooting potential in cuttings. This will be done especially in the context of improving practical propagation of relatively difficult species—challenging, where necessary, existing concepts in order to point the way forward. The literature describing techniques and general propagation background will be used illustratively, not treated exhaustively [see, instead, Macdonald (1986), Davis et al. (1988), Hartmann et al. (1990)].

STOCKPLANT TREATMENTS

The concept of treating stockplants, or individual shoots, to raise rooting potential before cutting collection is old [e.g., Gardner (1937)]. This approach recognizes that post-cutting-collection treatments, and propagation environments in which difficult-to-root cuttings are subsequently placed, are rarely optimal, and that cuttings with high rooting potential suffer least from any deficiencies. The many preconditioning techniques used to treat stockplants before cutting collection include:

• General light exclusion (often referred to as "etiolation," but often without the implied internode elongation) and localized light exclusion ["blanching," banding; reviewed by Maynard and Bassuk (1988)]. Darkness is usually confounded with high temperature when stockplants are etiolated under black polyethylene-clad structures (Harrison-Murray and Howard, 1983).

• Modifying light quality, photoperiod, CO_2, water and mineral nutrition—often with highly interactive and inconsistent effects between species [reviewed by Moe and Andersen (1988)].

• Severe annual winter pruning [hedging; Libby et al. (1972), Howard (1987)].

• Raising stockplants via micropropagation (Howard et al., 1989; Webster and Jones, 1992), which is often referred to as "rejuvenation," although enhanced rooting is not necessarily accompanied by marked or sustained increase in shoot vigor. Comparisons with cuttings from conventional sources are sometimes confounded with different age and type of stockplant, and different types of cuttings (Jones and Webster, 1989; Marks, 1991).

• Growing plants rapidly, often at higher-than-normal temperatures and humidity (Campen et al., 1990).

• Serial grafting of adult scions onto seedling rootstocks (Siniscalco and Pavolettoni, 1988).

• Girdling or ringing the mainstem, branch or individual shoot (Higdon and Westwood, 1963).

• Induction of adventitious shoots from roots (Garner and Hatcher, 1964) or sphaeroblasts (Hatcher and Garner, 1955).

• Treating stockplants with chemical growth regulators, notably gibberellins, auxins, cytokinins or growth retardants (Stoutemyer et al., 1961; Sadhu, 1979).

These pretreatments can be categorized in terms of their general effectiveness, especially with respect to wide species range and consistency of response, their practical usefulness and the opportunity for relatively large-scale uptake. Light exclusion (dark-preconditioning), severe pruning (hedging) and raising stockplants via micropropagation are reliable and practical, and will be considered here in greatest detail, with only passing reference to other preconditioning techniques where appropriate.

It is a marked feature of the most reliable preconditioning methods that their practical application has been demonstrated repeatedly with a wide range of plants, but that our understanding of the biological processes involved is lacking or is based on assumptions which have not been tested (see the chapter by Haissig and Davis in this volume). Therefore, our opportunities to increase the precision of the responses and to extend their practical applications are limited.

PRECONDITIONING TREATMENTS RELATED TO TYPE OF CUTTING

Before assessing progress in understanding preconditioning processes it is necessary to consider the significance of different types of cuttings in terms of their possible interplay with preconditioning treatments, and attempt to put into context views on "rejuvenation." The extent to which the mechanisms that control rooting potential have developed, and the effectiveness of different preconditioning treatments, may not be similar for leafy "softwood" cuttings and for leafless "hardwood" cuttings. Softwood cuttings are often propagated from the distal portion of the annual shoot as apical cuttings (Macdonald, 1986) containing the youngest, most recently developed tissue. Successful preconditioning treatments are likely to be those that influence dynamic processes of cell development in a relatively short time (days rather than months). Hardwood cuttings, on the other hand, often have to include the proximal part of the shoot (Howard et al., 1983) which contains the chronologically oldest tissue influenced by shoot development throughout the growing season, and which as a basal cutting (Howard, 1971) is often the only tissue capable of rooting. Furthermore, although hardwood cuttings are environmentally sensitive, leafy softwood cuttings are much more so, and the relative importance of the propagation environment in either contributing to or masking rooting potential is likely to be relatively greater for softwood than hardwood cuttings.

JUVENILITY

The free-rooting of cuttings from juvenile-phase plants [Hackett (1988); see the chapter by Murray and coworkers in this volume] is often used as a basis for attempting to develop scientific understanding of how to improve rooting potential in adult-phase cuttings. The implication is that when juvenile-phase characteristics such as absence of flowers, increased shoot vigor, increased spininess and delayed defoliation are produced by treatments to stockplants, the often observed improved rooting of cuttings is a consequence of temporary phase-reversal or "rejuvenation." The tendency to assume a causal link between certain of the juvenile shoot characters associated with phase-change, such as enhanced shoot vigor (Garner and Hatcher, 1964), and improved rooting of cuttings in adult plants, together with the concept of "rejuvenation," may be teleological and unhelpful in seeking an under-standing of rooting potential in cuttings that will lead to practical improvements. By the same token, it is useful to consider the relevance of the concept that in adult plants grown from seed the base of the trunk is permanently juvenile simply because the buds in that zone developed in the true juvenile phase and the shoots produced in response to pruning have characteristics similar to those of the juvenile phase. Chronologically, the base of the plant is older than the crown. Thus, it is perhaps less reasonable to suppose that the rooting potential of shoots from either zone is more affected by some inherent biological clock, determined by ontogenetic age, than by the contemporary interrelationship of shoots at different positions in the stockplant, and the physiological imbalance between these and the root system of the stockplant caused by pruning. The possibly greater importance of physiological age is underlined by the fact that the base of adult plants raised from cuttings through many vegetative cycles is also the zone of highest regenerative capacity, but it cannot be described as permanently juvenile. The situation is further confused by evidence from many workers (Hackett, 1985) that the length of the juvenile period is inversely related to tree size and, hence, that the early vegetative vigor associated with high rooting potential is also associated with early onset of flowering, sometimes facilitated by girdling, which can also enhance rooting.

In order to improve understanding of rooting potential in cuttings there may be benefit in developing a simple concept based on the observation that plants exhibit a range of rooting ability, from those that are impossible to root with current technology, to those that respond to treatment with auxin, to those that develop preformed roots in intact stems without any treatment (see the chapter by Barlow, and by Haissig and Davis in this volume). Preformed roots frequently emerge at proximal nodes (Howard, 1987), especially at the base of annual shoots (Fig. 1), which is the zone in "normal shoots" (i.e., without preformed roots) from which roots most often develop in hardwood cuttings. The opportunity for rooting potential to extend above the shoot base is enhanced by growing stockplants in humid, warm conditions (Campen et al., 1990) as seen by the emergence of preformed roots from non-basal nodes (Fig. 1) and by severe stockplant pruning (Howard et al., 1983). If root initiation processes occur or can be made to occur endogenously to varying degrees, it is reasonable to suppose that pretreatments to raise rooting potential operate, at least to some extent, by advancing these ongoing processes. This in turn raises questions as to which stages in the root initiation and development processes are affected by pretreatments. To what extent should treatments applied immediately after cutting collection be seen as part of the process of improving rooting potential? Does the application of exogenous auxin complete processes started by endogenous auxin? Does wounding in the course of collecting cuttings, or additional wounding in the form of incisions, slicing or splitting the cutting base, stimulate the production of a novel wound stimulus, or simply disrupt tissues leading to their regeneration under conditions conducive to root initiation, for example, in the presence of applied auxin? (see the chapter by Haissig and Davis in this volume).

Figure 1. Left: proximal development of preformed roots in field-grown shoots of *Prunus avium* x *P. pseudocerasus* 'Colt' cherry; Right: nodal development of preformed roots in MM.106 apple grown at high temperature and humidity.

Insights into the apparent *de novo* initiation and development of adventitious roots were obtained in experiments investigating wounding responses in non-basal internodal tissue of apple hardwood cuttings (Howard et al., 1984) and associated anatomical studies

(Mackenzie et al., 1986). The combination of splitting the stem and applying indole-3-butryic acid (IBA) induced callus to form in which a new cambium developed. When the separated halves of the cutting stem prevented the cambium from fusing, it turned back on itself to produce outward-pointing cambial projections or salients. These in turn appeared to determine the nearby location of new root initials in the surrounding callus, which subsequently developed vascular connections with the salients. Similarities were noted between the gaps produced in the phloem sclerenchyma sheath by wounding, and those caused naturally by the entry of leaf and bud traces (Pontikis et al., 1979), which are relatively numerous at the free-rooting bases of cuttings, and with which cambial salients are also associated. It is tempting to suggest that in both additionally wounded and normal cuttings the location and triggering of root initials are determined by factors focused by the salients in ways not yet understood. Such factors might include endogenous and possibly exogenous auxin.

DARK-PRECONDITIONING FOR LEAFY CUTTINGS

Methods

Growing stockplants in the dark or in heavy shade before collecting cuttings often results in large increases in root numbers, with associated improvement in the number of cuttings which root (Harrison-Murray, 1982). Deciduous stockplants are grown in the dark from bud-burst for about two weeks and then weaned into the light for about the same period before taking cuttings. The dark stimulus is weakened but not totally destroyed by prolonged light exposure before propagation. There is less need for light-weaning and less check to stockplants when heavy shade (one to two percent light) rather than complete darkness is used. An effective technique is to enclose stockplants in a structure covered with black polyethylene. Daytime temperatures inside the frame are higher than those outside, and both darkness and elevated temperature contribute to the enhanced rooting (Harrison-Murray and Howard, 1983). Precautions include the need to provide minimum ventilation to avoid heat stress from excessively high temperatures, and to limit periods of high relative humidity at night, which encourages botrytis disease. A prophylactic fungicide spray applied at the time of covering the plants is also helpful in this respect.

Local blanching enhances rooting less than total darkness (Harrison-Murray, 1982). To be effective, an opaque tape [or Velcro band, sometimes impregnated with IBA (Maynard and Bassuk, 1986)] needs to be applied as closely as possible behind the shoot apex without killing the meristem (Gardner, 1937; Harrison-Murray and Howard, 1982). The importance of excluding light from the shoot apex implies that the dark stimulus operates initially on undifferentiated tissue, a view supported by the resulting production of very large numbers of roots, often produced in vertical ranks. In practice, developing leaves are removed from a node within five cm of the apex and the black tape is applied with adhesive surfaces pressed together to allow for stem expansion. Local blanching in this way becomes less effective as the stem tissue ages (Harrison-Murray and Howard, 1982), as is the case when stoolbeds (mound-layers) of fruit rootstocks are earthed-up too late in summer—earthing-up being a practical form of localized blanching. Blanching can only achieve its effect in difficult-to-root clones such as the M.9 apple rootstock when accompanied by severe winter pruning; the combined stimulus manifests itself in improved rooting potential by mid summer (Howard et al., 1985). The gradual loss of rooting potential suggests that a secondary stimulus is negated in the light, or that in the light secondary stem development partly reverses the initial stimulus. The emergence of relatively few roots following taping suggests that blanching only completes the later stages of an ongoing root development process. Taping is particularly effective in maintaining the maximum dark-stimulus during the light-weaning process (Fig. 2), especially in terms of the numbers of roots produced (Harrison-Murray, 1982).

Figure 2. Left: heavy rooting of dark-preconditioned M.9 apple softwood cuttings, with the dark stimulus maintained by opaque tape during the light-weaning period; Right: dark-preconditioned cuttings of *Syringa vulgaris* 'Madame Lemoine' propagated in low-light conditions preventing the accumulation of photosynthates, leading to basal necrosis.

Table 1. Effects of dark pretreatment on variates of *Syringa vulgaris* 'Madame Lemoine' cuttings at collection. In the larger data set from which these examples are taken all treatment differences were significant ($P<0.001$) except for leaf area (n.s.).

	Control	Prior Darkness
Stem proximal diameter (mm)	3.7	3.1
Stem dry weight (g)	0.43	0.21
Stem dry weight content (%)	22.0	17.5
Leaf area per cutting (cm^2)	87.0	89.9
Ratio of leaf area to stem dry weight	202	428

Pointers to Rooting Enhancement Processes

A mechanism to explain at least part of the reason for enhanced rooting following dark-preconditioning has been proposed, based on recent experiments with the difficult-to-root *Syringa vulgaris* 'Madame Lemoine' (Howard and Ridout, 1992). Shoots which developed in the dark were thinner and had a lower dry weight:fresh weight ratio than normal light-grown shoots, but their leaf areas were not reduced (Table 1). Rooting was predicted accurately by a model based on stem proximal diameter, which was further improved by incorporating the area and number of leaves per cutting. This relationship, whereby cuttings with a high leaf area to stem diameter (or weight) ratio rooted best, held for both dark-preconditioned shoots compared to light-grown ones, and for normally thin-stemmed cuttings compared to thicker ones from any source. A similar inverse relationship between stem thickness and rooting was obtained in light-grown shoots that were defoliated in parallel with the dark-preconditioning period (Fig. 3). The rooting stimulus extended to stem tissue which developed during the light-weaning period and to that which developed after defoliation stopped. A clue to the possible mechanism involved comes from the fact that during the 10 days or so before roots emerged, the proximal portion of the dark-preconditioned stem increased in dry matter by a proportionately greater amount than that of control cuttings (Fig. 4), implying that, given equal opportunities for photosynthesis, the thinner-stemmed cuttings required less carbohydrates for maintaining stem tissue, thereby possibly making more carbohydrates available to "drive" the rooting process than in non-preconditioned cuttings, many of which showed a net dry weight loss in the rooting zone.

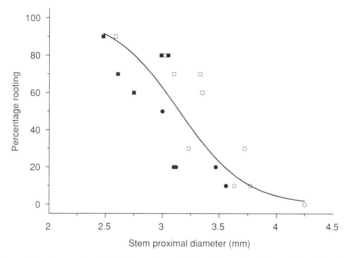

Figure 3. Rooting of *Syringa vulgaris* 'Madame Lemoine' softwood cuttings related to stem thickness as influenced by dark-preconditioning and predefoliation. (O = control; □ = dark; ● = defoliation; ■ = dark and defoliation combined).

If a key part of the process involves the stem being starved of carbohydrate during the dark period this is likely to be enhanced by high respiration rates caused by elevated ambient temperatures inside the polythene-covered structure, which was also shown to contribute to the rooting stimulus (Harrison-Murray and Howard, 1983). Furthermore, successful rooting of "semi-starved" cuttings would be totally dependent on adequate net photosynthesis during the rooting period. This view is supported by the fact that the benefits of dark-preconditioning are lost during dull weather, or when irradiation is experimentally reduced below the 20% available light that is considered optimal on a sunny day when propagating in highly supportive "wet fog," providing both leaf-wetting and high humidity. Under these conditions the bases of dark-preconditioned cuttings become necrotic and rot rather than root. In extreme cases the entire cutting collapses, and commonly the proximal few cm of stem rot, with roots sometimes developing belatedly in healthy tissue above (Fig. 2). Necrosis in these "starved" stems is no doubt exacerbated by enhanced respirational demands induced by IBA and by bottom heat applied to the cutting base. All types of responses in dark-preconditioned cuttings, from improved rooting to extensive stem necrosis, have been obtained in an environmentally controlled propagation facility by varying the intensity of sodium lighting.

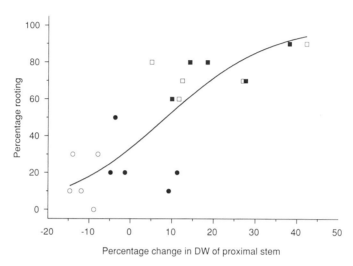

Figure 4. Rooting of *Syringa vulgaris* 'Madame Lemoine' softwood cuttings related to the percentage change in dry matter at the cutting base before root emergence (O = control;□ = dark; ● = defoliation;■ = dark and defoliation combined).

Put into the context of the review by Maynard and Bassuk (1988), the proposed carbohydrate redistribution mechanism is able to accommodate the conflict as to whether or not etiolation creates a mobile stimulus. Both shoot tissues which developed in the dark, and those which extended subsequently in the light, showed enhanced rooting associated with thinner stems and reduced dry matter content. Therefore, it is possible to obtain a rooting response in tissues distant from the darkened zone without a specific dark-generated stimulus being involved. Nevertheless, this mechanism is unlikely to account for the total dark-preconditioning effect. In common with many species, thin, dark-grown *Syringa* stems are also less lignified, including reduced sclerenchymatous fibre development in the cortex, which might facilitate the physical emergence of roots (Beakbane, 1970) or simply provide a larger number of parenchyma cells capable of dedifferentiation and formation of root initials. Although extending beyond the scope of this review, of possibly greater significance are the common features in the biosynthesis and destruction of both auxin and lignin, and the role of light in auxin breakdown, which suggest that auxin enrichment at the expense of lignification might be a major component of general dark-preconditioning, and to some extent of localized blanching processes. Working with pea stems, Baadsmand and Andersen (1984) proposed that under low irradiance IAA moved more slowly and was more uniformly distributed along the stem than at higher irradiance, which correlated with the production of more scattered and numerous roots. It is possibly relevant that a feature of many dark-preconditioned cuttings in experiments at East Malling is the large increase in root numbers and their emergence along the stem (Fig. 2), rather than mainly at the base, as in light-grown stems—a situation parallelled in dark-grown bean plants (Batten and Goodwin, 1981). This effect is greatly increased when dark-preconditioning and auxin application are combined (Harrison-Murray, 1982), although in this and other similar studies questions of synergism or substitution are difficult to answer.

The complexity of the biochemistry of events which determine whether stems develop in ways that enhance their adventitious rooting potential is underlined by the involvement of phenolic substances in both lignification and, apparently, in root initiation, especially in terms of seasonal changes in rooting ability (Bassuk and Howard, 1981; Curir et al., 1992). However, interpretation of studies which appeared to associate poor mid-winter rooting in the apple rootstock M.26 with low phenolic cofactor activity (Bassuk and Howard, 1981) was complicated by subsequently discovering that seasonal rooting trends could be reversed simply by controlling net water balance in the cutting after planting (Howard and Harrison-Murray, 1988).

SEVERE WINTER PRUNING AND HEDGING

General Considerations

Annual shoots are cut-back to relatively dormant proximal buds when pruned severely in winter. This reduces the number of shoots which grow from the crown, and encourages additional shoots to grow from older framework and from the trunk. Once established, a hedge produces a relatively large number of shoots which grow faster and for longer in the season than those on unpruned trees. The hardwood cuttings from a well-established hedge propagate more readily than those from newly planted stockplants (Sinha and Vyvyan, 1943; Howard and Harrison-Murray, 1982; see the chapter by Ritchie in this volume). Stoolbeds (which represent a very severely pruned hedge), and large hedges cut back into old branches, initially produce excessively vigorous shoots, often bearing many spines in appropriate species such as *Malus, Prunus* and *Pyrus*. These vigorous, dominant shoots do not root readily, but until now the possible significance of this in terms of understanding processes influencing rooting potential was not considered. Part of the reason for this was the lack of pressure to propagate these vigorous dominant shoots, which instead were used to produce framework for the following year's cuttings. For softwood cuttings, the readily observed effect of cutting-back to deeply dormant buds is to delay shoot development and reduce their number, while increasing their vigor and duration of growth, compared to shoots on less severely pruned plants.

Shoots not removed as softwood cuttings have the potential to become hardwood cuttings, and as the growing season progresses distal shoots assume dominance and subordinate shoots stop growing. The range of shoot sizes available for hardwood cuttings is often greater than for softwood cuttings, especially when the latter are prepared from only the distal part of the shoot to include the growing tip. Most research into the mechanism of hedging has been conducted with hardwood cuttings, especially for temperate fruit species. It is a generally long-held belief that improved rooting following severe stockplant pruning is in some way due to the enhanced shoot vigor, which is likened to the vigorous non-flowering shoots of seedlings, with associated high rooting potential during the juvenile phase (Garner and Hatcher, 1964). Hedging is sometimes referred to as "rejuvenation," and the lower part of the stockplant induced to sprout by severe pruning is sometimes described as being permanently juvenile in ontogenetic terms (Bonga, 1982), despite it being the chronologically oldest tissue. While it is undeniable that true juvenile-phase cuttings have high adventitious rooting potential and that many treatments to adult-phase plants enhance rooting ability along with shoot vigor, there are a variety of reasons why the concept of phase-change, and of tapping into a permanently juvenile zone, may be counterproductive in attempting to understand better the factors determining rooting potential in cuttings leading to improved practice. A direct link between shoot vigor and rooting potential in adult-phase plants runs counter to the poor rooting observed in the vigorous shoots of young stoolbeds and hedges—and as seen immediately after severely cutting back into old hedge framework. Furthermore, vigorous shoots from annually pruned M.9 apple stoolbeds will set flower buds (Garner and Hatcher, 1964) and flowers readily develop on severely pruned hedges of this rootstock if two-year-old shoots in the form of small spurs are not removed at pruning. The vigorous annual shoots on severely pruned hedges of *Syringa vulgaris* will set flower buds also. An artefact which prevents clear interpretation of many experiments is that most workers specify a particular cutting size and use only a proportion of the shoots available from each source, which may not represent the mode of the population, and certainly not the range. It follows that links between shoot morphology as a marker of rejuvenation and rooting will not be seen clearly, and they may be more assumed than real.

Alternative hypotheses need exploring which could explain improved rooting potential in specially treated adult-phase stockplants in terms of accepted shoot developmental physiology and which, in turn, may embrace also the mechanisms operating in truly juvenile shoots. Although hard-pruning invariably results in enhanced shoot vigor through a reduction in shoot numbers, this is confounded with their decreasing height from the ground and closer proximity to the stockplant root system. The annual removal of either entire shoots as cuttings, or all but the proximal few nodes which are left for renewed shoot growth the following year, creates an angular "multi-jointed" framework whose vascular system is clearly disorganized compared to normal extension growth from apical buds. It is not uncommon for very vigorous shoots in *Malus* and *Prunus* hedges and in young nursery trees to be blown-out in summer gales, giving the appearance of a ball and socket joint with weak vascular continuum. This has implications for hormone, photosynthate and nutrient translocation (see the chapter by Friend and coworkers in this volume).

Components of the Rooting Response to Hedging

Work with *Prunus insititia* 'Pixy' is taking a new look at hedging without preconceived ideas linked to rejuvenation. By following individually the fate of many thousands of shoots propagated as hardwood cuttings it has been shown that features associated with the juvenile condition such as very rapid and prolonged shoot growth, and frequent emergence of laterals, are not directly associated with enhanced rooting. Rooting was fastest and, hence, obtained the highest percentage at the end of the experiment in cuttings from the smallest and thinnest shoots within the bush.

It is likely that the inconsistencies in performance of thick and thin cuttings revealed in the literature are due to interacting factors and artefacts which cloud a critical assessment of rooting potential. Two prerequisites to studying the effects of within-plant shoot variation in *Prunus insititia* 'Pixy' have been identified. First, rooting must be disassociated from cutting survival by investigating the rooting process itself, and not by inferring rooting performance from the number of plants which eventually grow from cuttings planted directly into the field, as the literature shows is often the case. Second, the propagation conditions must give cuttings of all sizes an equal chance of success. Hardwood cuttings have a tendency to take up excessive amounts of water from the medium—which in the absence of leaves cannot be transpired rapidly—and basal tissues rot rather than root, particularly in thin cuttings. This was overcome by placing the bases of the cuttings on a 20-cm-deep bed of fine sand that rapidly removed excess water, and by using granulated pine bark as a mulch to cover the bases to a depth of eight cm, rather than planting cuttings into a peat-containing medium. Proximal (basal) cuttings were used and rooting was promoted by providing electrical bottom heat at 18° C, with the propagation beds contained in an insulated, cooled building with ambient air temperature at 10° C to retard bud development. Having taken these precautions, it was shown that rooting percentage increased with decreasing diameter of the cutting base (Fig. 5). Application of exogenous auxin (IBA at 2500 mg L^{-1} in 50% acetone for five seconds to a depth of eight mm) was a prerequisite for rooting.

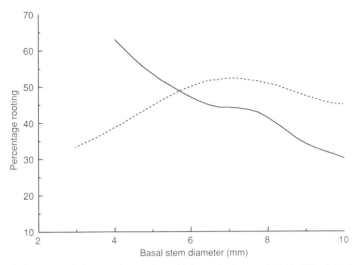

Figure 5. Smoothed curves relating rooting to shoot thickness in *Prunus insititia* 'Pixy' hardwood cuttings under rapidly draining conditions (———— , 252 cuttings), and under slowly draining conditions (- - - - -, 509 cuttings) leading to basal necrosis, especially in thin cuttings.

Although diameter of the shoot from which the cutting was made predicted rooting well, the length of shoots in the crown of the hedge could also do so, because shoot length and proximal diameter were positively and closely correlated (Howard and Ridout, 1991a). This was not so for lower shoots and those emerging from the trunk which grew up through the canopy. Their rooting potential was best described in terms of diameter:length ratio of the original shoot, with cuttings from relatively thin, long shoots rooting faster than those from relatively thick, short ones (Howard and Ridout, 1991b). These relationships between shoot morphology and rooting were obtained despite reducing shoots from all sources to 60 cm in length when preparing cuttings.

The significance of shoot morphology on the rooting of hardwood cuttings is not yet understood, but has been investigated further in work hitherto unpublished. Delaying hedge pruning from the usual dormant period in early March until mid-May, when the new season's growth was well advanced, increased the rooting of hardwood cuttings propagated the following autumn, especially those normally difficult-to-root from the distal crown zone. A major effect of delayed pruning was to reduce the thickness of the cuttings (Fig. 6).

In another experiment, established hedges were differentially pruned to produce crowns at different heights above ground, ranging from 1.1 m (normal) to 2.3 m (elevated). The height of the crown above ground had no effect on rooting, which was determined entirely by the relative position and, therefore, to a large extent size of shoots. In the following year a larger number of cuttings was produced on the elevated crowns, arising from the additional framework developed the previous year, and their rooting percentage was increased by prolonging the propagation period. A 35 mm-wide ring of cortex removed in August from the main vertical branch of the elevated hedges below the cluster of crown shoots depressed rooting in November of the shoots above the ring, in association with a marked increase in shoot thickness, compared to shoots from non-ringed controls (Fig. 7). There was no effect of ringing on either size of shoots or rooting below the ring.

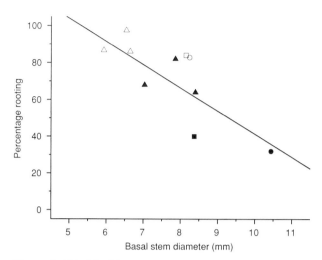

Figure 6. Rooting of Prunus insititia 'Pixy' hardwood cuttings propagated from hedges pruned in late winter (solid symbols) or early summer (open symbols) in relation to shoot thickness (○ = distal crown cuttings; □ = proximal crown cuttings; ▲ = trunk-derived cuttings).

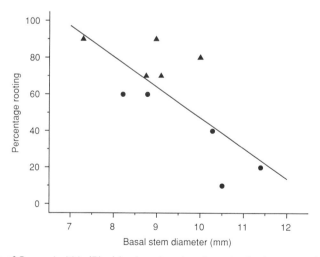

Figure 7. Rooting of *Prunus insititia* 'Pixy' hardwood cuttings from the distal crown position of ringed (●) or non-ringed (▲) branches.

These results suggest that the developmental physiology of individual shoots, which manifests itself in terms of shoot thickness, is an important component of their rooting potential when propagated as basal hardwood cuttings. When propagated earlier as softwood cuttings, the thin, subordinate crown shoots, and all shoots from the trunk zone of 'Pixy' hedges, also rooted better than the thick, dominant crown shoots. However, a direct comparison with hardwood cuttings is confounded by the fact that, in summer, failure of thick cuttings to root was accompanied by rotting of the stem base (Table 2), as was the case for dark-preconditioned *Syringa* cuttings when propagated with inadequate light intensity. Ability to survive and, hence, the opportunity to root, will be determined by the interaction of shoot morphology and cutting environment. Thick hardwood cuttings survive better than thin ones in adverse environments, but fail to root, whereas thick softwood cuttings survive less well than thin ones and so have less opportunity to root.

During the course of these experiments it was observed that the ratio of "bark" (mainly epicambial tissues, but including the most recently formed secondary xylem) to "wood" (mainly xylem) increased with decreasing diameter of shoot. It could be significant, therefore, that fast-rooting, thinner cuttings have a relatively high proportion of tissues in which roots initiate (Table 3).

ENHANCED ROOTING VIA MICROPROPAGATION

Shoot vigor and rooting potential often increase *in vitro* with successive subcultures, with the rate of improvement being variety-specific (Sriskandarajah et al., 1982; Economou and Read, 1986; Webster and Jones, 1989). Cultures of juvenile origin with high rooting potential can tolerate higher levels of auxin in the medium than those of adult origin (Pliego-Alfaro and Murashige, 1987). When micropropagated plants are containerized or planted in the field as stockplants it is often found that the improved rooting potential is carried through to the benefit of conventional cuttings (Howard et al., 1989; Marks, 1991; Webster and Jones, 1992). This may last for a few months or many years depending on the species, but when the improvement appears to be long-term it is often associated with severe winter pruning or hedging of the *ex vitro* source. Generally, it is considered that these are temporary epigenetic effects, but in seeking an explanation it must also be acknowledged that clones with improved growth and rooting potential may have been selected unknowingly via somaclonal variation, whose visual characteristics still conform to the type. These *ex vitro* benefits are expressed in both softwood and hardwood cuttings and are often referred to as "rejuvenation," although the implied vegetative invigoration is often short-lived and flowering occurs within two to three years of propagation.

Progress in understanding the reasons for the *ex vitro* improvements is hampered by not understanding the processes which lead to the preceding improvements during progressive subculturing *in vitro*. Also, as with stockplant hedging, progress has probably been hampered by the use of subsamples of cuttings which meet predetermined criteria, rather than by experimenting with entire shoot-populations or representative samples, and following the fate of individual cuttings.

Using an improved rooting *ex vitro* source of the same *Prunus insititia* 'Pixy' rootstock that was used for studies of stockplant pruning and hedging, it was found that not only did the *ex vitro* source produce faster growing, more vigorous shoots with earlier and more profuse lateral production than the conventional source, but the *ex vitro* source also had a higher proportion of the relatively ready-rooting, thin cuttings which contributed to its easy-propagating characteristic (Fig. 8). Additionally, however, shoots of any particular diameter from the *ex vitro* source rooted better than did those of similar thickness from the conventional source, an effect not associated directly with their enhanced lateral production. The combination of *ex vitro* source, and of making cuttings from the shoots arising from the favored trunk zone, raised rooting of all cuttings to virtually 100%, irrespective of thickness (Howard and Ridout,1991b) (Fig. 9). The cumulative effects of *ex vitro* source material, hedging and trunk-derived cuttings possibly indicate a common mechanism for improving rooting potential.

It is not clear whether the carry-over of the *in vitro* improvement in vigor and rooting is limited to the *in vitro*-generated material (V_o generation) or whether the improvement passes into the V_1 generation and possibly beyond. If limited to the V_o generation, enhanced rooting might be a direct response to the carry-over of the *in vitro* influence such that the first *ex vitro* generation is behaving as the final subculture *in vitro*. *Ex vitro* fruit plants which are not severely pruned in the V_o generation lose vigor and flower relatively quickly. A key to understanding the *ex vitro* effect may lie in the fact that it is maintained by regular severe winter pruning, which also improves conventional source material.

Table 2. Rooting and basal necrosis of *Prunus insititia* 'Pixy' softwood cuttings propagated during summer.

Type and Position of Shoot	Thick Dominant, Crown	Thin Subordinate, Crown	Thick, Trunk Zone	Thin, Trunk Zone	P	s.e.d.	d.f.
Rooting (%)	15	85	75	95	<0.01	11.6	9
Roots per rooted cutting	3.8	4.9	6.9	5.2	<0.01	0.61	7
Proximal necrosis ≥ 5 mm (%)	82	13	43	18	<0.01	12.3	9

Table 3. "Bark:wood" fresh weight ratio of *Prunus insititia* 'Pixy' crown shoots related to shoot proximal diameter (mm).

Shoot Position	Diameter	Ratio
Distal-dominant	7.05	0.46
Median-subordinate	5.55	0.58
Proximal-subordinate	4.97	0.62

CONCLUSION AND FORWARD LOOK

New initiatives are needed to understand the physiological and developmental processes in shoots that influence rooting, so as to identify key mechanisms in gene expression and obtain reliable experimental systems free from the inconsistencies that often plague propagation research, as a basis for studying genetic control (see the chapter by Ernst, and by Riemenschneider in this volume).

The aim of much research is to improve the propagation of hitherto difficult adult-phase subjects, and it takes the form of empirical reassessments of key factors such as propagation season, type of cutting, auxin treatment, type of compost and ways of managing the aerial environment. The resulting lack of understanding of the processes involved is exacerbated by assumptions that need challenging. These include questioning the extent to which "juvenility" and "rejuvenation" are helpful or distracting concepts in attempting to increase practical success, and to speculate on possible alternative processes. Potentially important results from current studies include the following.

In both softwood and hardwood cuttings of difficult-to-propagate subjects, rooting potential is correlated inversely with shoot thickness—so the link between enhanced rooting and increased stockplant vigor in hedged, *ex vitro* and "etiolated" source plants is, at most, indirect. The stem diameter:length ratio of a shoot sometimes describes rooting potential in hardwood cuttings better than thickness alone, even though cuttings are reduced to a standard length for propagation. This implies that rooting potential is determined by physiological and developmental events during the preceding growing season, rather than by only some aspect of shoot thickness at the base of the cutting during propagation. This view is supported by the fact that rooting differences in softwood cuttings of *Prunus insititia* 'Pixy' parallel those of hardwood cuttings propagated later in terms of the effect of shoot size and location in the stockplant, although confounded with necrosis. On the other hand, shoot thickness is of little importance when rooting potential is naturally high, as in subjects producing preformed roots and in trunk-derived cuttings from *in vitro* sources.

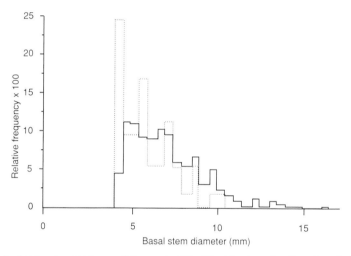

Figure 8. Distribution by stem thickness of *Prunus insititia* 'Pixy' hardwood cuttings from hedges produced by micropropagation (••••••) or conventionally (——).

In softwood cuttings the effect on rooting potential of differences in shoot morphology operate, at least in part, via the different accumulation of dry matter in the rooting zone before roots emerge. A stress-free photosynthetically conducive environment is essential to exploit dark-preconditioning. Ultra-severe pruning creates conditions in shoots that enhance the response to localized light exclusion (blanching), so interactive processes appear likely. The effect of shoot morphology on the rooting of hardwood cuttings is not yet understood. There is no absolute relationship between shoot thickness and rooting potential, but these are often closely and inversely linked within sources of cuttings, and frequently confounded with the position of the shoot in the stockplant. Relative shoot position is more important than the height of the crown, and physiological ageing appears more relevant than distance of shoots from the putative ontogenetic juvenile zone. Although it is often difficult to disentangle effects on rooting due to cutting thickness from those associated with shoot position in the bush, thickness can affect rooting independently of shoot position as shown by responses to different severity of pruning, to different times of pruning and to branch ringing. Where treatments did not affect shoot thickness, rooting was not affected either.

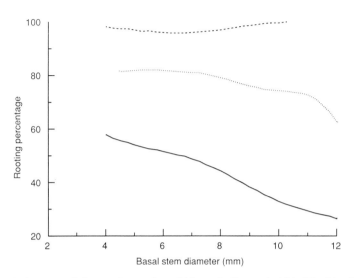

Figure 9. Smoothed curves relating rooting to shoot thickness in *Prunus insititia* 'Pixy' hardwood cuttings from different sources (- - - - - - = 34 trunk-derived cuttings from a micropropagated hedge; ••••• = 146 crown-derived cuttings from a micropropagated hedge; ——— = 1643 crown-derived cuttings from a conventional hedge).

Exogenous auxin enhances the effectiveness of dark-preconditioning in softwood cuttings and it is a prerequisite for expressing rooting potential in difficult-to-propagate hardwood cuttings. *Endogenous* auxin may have a major role in determining rooting potential (sometimes leading to preformed roots), and phenolic metabolism appears to also be implicated (see the chapter by Blakesley in this volume). New data make it clear that assumptions and conventional wisdom linking shoot vigor with rooting potential, and the implication that this has for our understanding of stockplant management, juvenility and "rejuvenation," must be challenged. Taking together new information, accepted concepts of the role of auxin in root initiation and, for difficult-to-propagate subjects, the readily observed advantages of the shoot base and of severe annual pruning, the following points appear relevant: Rooting potential is a direct response of processes within the developing shoot; chronological and physiological age, but not ontogenetic age, are important. In distal softwood cuttings rooting potential is minimally developed, but can be enhanced markedly immediately prior to cutting collection. In hardwood cuttings rooting potential is often highly developed and is mostly restricted to the proximal part of the shoot. There is little opportunity to modify potential immediately before cutting collection, but apparent stimulation of *de novo* rooting in hardwood cuttings can be induced by additional wounding after cutting collection in the presence of exogenous auxin. Rooting potential might be determined by exposure of proximal tissues to endogenous auxin, and the disjointed anatomy of an annually pruned hedge could delay polar transport to the benefit of that potential. Shoot-thickness often describes relative rooting potential among a population of shoots, but the relevance of shoot thickness diminishes as overall rooting potential increases, so it seems sensible for the time being to investigate processes which determine rooting potential, and which themselves are regulated by a shoot thickness factor.

The following are among the specific questions that need to be addressed around the central need to understand the relationship between shoot morphology and rooting potential in softwood and hardwood cuttings in the context of stockplant management and manipulation: Does "hormonal imprinting" (Csaba, 1986) occur in plants? (Guern, 1987)— whereby exposure of tissue to endogenous auxin might induce hormone receptors which sensitize that tissue to subsequent auxin application to an extent determined by the intensity of the initial hormone imprinting (see the chapter by Palme and coworkers in this volume). To what extent is hormone imprinting a key factor in developing rooting potential? and is it enhanced during dark-preconditioning of softwood cuttings and in hardwood cuttings by the "multi-jointed" nature of the hedge framework? What is the relationship between the putative hormone imprinting and shoot thickness that results in weak shoots having the highest rooting potential?—when minimal auxin imprinting might be expected in subordinate shoots. Can other data be interpreted in terms of hormone imprinting? For example, oak cuttings maintained a higher capacity to respond to exogenous auxin treatment when grown vegetatively under glass than when taken from the same genotype in a reproductive condition in the field (Morgan et al., 1980); and difficult-to-root pear shoots rooted readily when planted around established pear trees and grafted by inarching into their trunks, suggesting that their rooting potential was improved by factors from the growing tree (Higdon and Westwood, 1963).

When ringing or girdling enhances rooting potential does the process parallel that proposed for severely pruned hedges?—where the rate of polar transport of auxin might be delayed to the benefit of shoots destined for cuttings. What is the process which causes the rooting enhancement which follows girdling in many plants to be reversed in *Prunus insititia* 'Pixy' in association with a marked increase in shoot thickness? Are other negative rooting responses to girdling [e.g., Evert and Smittle (1990)] associated with increased thickness of cuttings? Given that these workers showed that carbohydrate increase above a girdle at the base of the annual shoot need not be associated with improved rooting, is the role for carbohydrate availability in hardwood cuttings (Cheffins and Howard, 1982a,b) and in softwood cuttings (Howard and Ridout, 1992) only indirectly linked to the development of rooting potential?—by enabling or enhancing the realization of that potential created by other processes. What is the effect of dark-preconditioning on stem structure with respect to possibly altering the requirement for maintenance carbohydrate?

Given the many examples [e.g., Doorenbos (1954), Garner and Hatcher (1957), Stoutemyer et al. (1961), Howard and Ridout (1992)] of improved rooting in cuttings taken from a scion grafted onto vigorous and/or seedling rootstocks, would this not be a useful system with which to investigate possible hormone imprinting?—given the opportunity to alter shoot vigor and possibly impede polar translocation at the graft union. When serial grafting to seedling rootstocks enhances rooting [e.g., Muzik and Cruzado (1958), Siniscalco and Pavolettoni (1988)] might this be because the graft union delays polar movement of auxin? and this enhances hormonal imprinting progressively with time?—similar to that proposed in space within a hedge plant. Is it necessary to infer the accumulation of a juvenile factor? Does *ex vitro* improvement in rooting extend beyond the V_0 generation? and, if not, what are the processes that enhance rooting of V_0 conventional cuttings that interact with source plant pruning level?

Answers to these and similar questions will bring together onto a firmer footing attempts to understand rooting potential from both a physiological and a genetical standpoint, with important positive consequences for practical propagation.

ACKNOWLEDGMENTS

Work reported from HRI East Malling is part of a program funded by the Ministry of Agriculture, Fisheries and Food. Figures 3, 4 and 8 were reproduced with the kind permission of the *Journal of Horticultural Science*.

REFERENCES

Baadsmand, S. and Andersen, A.S., 1984, Transport and accumulation of indole-3-acetic acid in pea cuttings under two levels of irradiance, *Physiol. Plant.* 61:107.

Bassuk, N.L., and Howard, B.H., 1981, A positive correlation between endogenous root-inducing cofactor activity in vacuum-extracted sap, and seasonal changes in rooting of M.26 winter apple cuttings, *J. Hortic. Sci.* 56:301.

Batten, D.J., and Goodwin, P.B., 1981, Auxin transport inhibitors and the rooting of hypocotyl cuttings from etiolated mung-bean *Vigna radiata* L. Wilczek seedlings, *Ann. Bot.* 47:497.

Beakbane, A.B., 1970, Relationships between structure and adventitious rooting, *Comb. Proc. Int. Plant Prop. Soc. for 1969* 19:192.

Bonga, J.M., 1982, Vegetative propagation in relation to juvenility, maturity and rejuvenation, *in*: "Tissue Culture in Forestry," J.M. Bonga and D.J. Durzan, eds., Martinus Nijhoff/Dr. W. Junk, The Hague.

Campen, R., Weston, G.D., Howard, B.H., and Harrison-Murray, R.S., 1990, Enhanced rooting potential in MM.106 apple rootstock shoots grown in a polythene tunnel, *J. Hortic. Sci.* 65:367.

Cheffins, N.J., and Howard, B.H., 1982a, Carbohydrate changes in leafless winter cuttings, I. The influence of level and duration of bottom heat, *J. Hortic. Sci.*, 57:1.

Cheffins, N.J., and Howard, B.H., 1982b, Carbohydrate changes in leafless winter cuttings, II. Effects of ambient air temperature during rooting, *J. Hortic. Sci.*, 57:9.

Csaba, G., 1986, Receptor ontogeny and hormonal imprinting, *Experientia* 42:750.

Curir, P., Sulis, S., Bianchini, P., Marchesini, A., Guglieri, L., and Dolci, M., 1992, Rooting herbaceous cuttings of *Genista monosperma* Lam., Seasonal fluctuations in phenols affecting rooting ability, *J. Hortic. Sci.*, 67:301.

Davis, T.D., Haissig, B.E., and Sankhla, N., 1988, "Adventitious Root Formation in Cuttings," Dioscorides Press, Portland.

Doorenbos, J., 1954, "Rejuvenation" of *Hedera helix* in graft combinations, *Proc. Kon. Ned. Akad.Wetensch. Amst. Ser. C* 57:99.

Economou, A.S., and Read, P.E., 1986, Microcutting production from sequential reculturing of hardy deciduous azalea shoot tips, *HortSci.* 21:137.

Evert, D.R., and Smittle, D.A., 1990, Limb girdling influences rooting, survival, total sugar, and starch of dormant hardwood peach cuttings, *HortSci.* 25:1224.

Gardner, F.E., 1937, Etiolation as a method of rooting apple variety stem cuttings, *Proc. Amer. Soc. Hortic. Sci.* 34:323.

Garner, R.J., and Hatcher, E.S.J., 1957, Rootstock effect on the regeneration of apple cuttings, *Rep. E. Malling Res. Sta. for 1956*:60.

Garner, R.J., and Hatcher, E.S.J., 1964, Regeneration in relation to vegetative vigour and flowering, *Proc. 16th Int. Hortic. Cong. 1962, Belgium*, 3:105.

Guern, J., 1987, Regulation from within: The hormone dilemma, *Ann. Bot.*, 60 (suppl.) 4:75.

Hackett, W.P., 1985, Juvenility, maturation and rejuvenation in woody plants, *Hortic.Revs.*, 7:109.

Hackett, W.P., 1988, Donor plant maturation and adventitious root formation, *in*: "Adventitious Root Formation in Cuttings," T.D. Davis, B.E. Haissig and N. Sankhla, eds., Dioscorides Press, Portland.

Harrison-Murray, R.S., 1982, Etiolation of stock plants for improved rooting of cuttings: 1. Opportunities suggested by work with apple, *Comb. Proc. Int. Plant Prop. Soc. for 1981* 31:386.

Harrison-Murray, R.S., and Howard, B.H., 1982. Preconditioning shoots for rooting-Blanching, *Rep. E. Malling Res. Sta. for 1981*:57.

Harrison-Murray, R.S., and Howard, B.H., 1983, Pre-etiolation of M.9 - Environmental factors, *Rep. E. Malling Res. Sta. for 1982*:60.

Hartmann, H.T., Kester, D.E., and Davies, F.T., 1990, "Plant Propagation - Principles and Practices," Prentice-Hall, Englewood Cliffs.

Hatcher, E.S.J., and Garner, R.J., 1955, The production of sphaeroblast shoots of apple for cuttings, *Rep. E. Malling Res. Sta. for 1954*:73.

Higdon, R.J., and Westwood, M.N., 1963, Some factors affecting the rooting of hardwood pear cuttings, *Proc. Amer. Soc. Hortic. Sci.*, 83:193.

Howard, B.H., 1971, Propagation techniques, *Sci. Hortic.*, 23:116.

Howard, B.H., 1987, Propagation, *in*: "Rootstocks for Fruit Crops," R.C. Rom and R.F. Carlson, eds., John Wiley and Sons, New York.

Howard, B.H., and Harrison-Murray, R.S., 1982, Cutting techniques - Interaction with cutting source and season of collection, *Rep. E. Malling Res. Sta. for 1981*:61.

Howard, B.H., and Harrison-Murray, R.S., 1988, Effects of water status on rooting and establishment of leafless winter (hardwood) cuttings, *Acta Hortic.*, 227:134.

Howard, B.H., and Ridout, M.S., 1991a, Rooting potential in plum hardwood cuttings: 1. Relationship with shoot diameter, *J. Hortic. Sci.*, 66:673.

Howard, B.H., and Ridout, M.S., 1991b, Rooting potential in plum hardwood cuttings: II. Relationships between shoot variables and rooting in cuttings from different sources, *J. Hortic. Sci.*, 66:681.

Howard, B.H., and Ridout, M.S., 1992, A mechanism to explain increased rooting in leafy cuttings of *Syringa vulgaris* 'Madame Lemoine' following dark-treatment of the stockplant, *J. Hortic. Sci.*, 67:103.

Howard, B.H., Harrison-Murray, R.S., and Arjyal, S.B., 1985, Responses of apple summer cuttings to severity of stockplant pruning and to stem blanching, *J. Hortic. Sci.*, 60:145.

Howard, B.H., Harrison-Murray, R.S., and Fenlon, C.A., 1983, Effective auxin treatment of leafless winter cuttings, *in*: "Growth Regulators in Root Development," M.B. Jackson, and A.D. Stead, eds., Monograph 10, British Plant Regulator Group, Wantage, Oxford.

Howard, B.H., Harrison-Murray, R.S., and Mackenzie, K.A.D., 1984, Rooting responses to wounding winter cuttings of M.26 apple rootstock, *J. Hortic. Sci.*, 59:131.

Howard, B.H., Jones, O.P., and Vasek, J., 1989, Long-term improvement in the rooting of plum cuttings following apparent rejuvenation, *J. Hortic. Sci.*, 64:147.

Jones, O.P., and Webster, C.A., 1989, Improved rooting from conventional cuttings taken from micropropagated plants of *Pyrus communis* rootstocks, *J. Hortic. Sci.*, 64:429.

Libby, W.J., Brown, A.G., and Fielding, J.M., 1972, Effects of hedging radiata pine on production, rooting, and early growth of cuttings, *N.Z. J. For. Sci.*, 2:263.

Macdonald, B., 1986, "Practical Woody Plant Propagation for Nursery Growers," Timber Press, Portland.

Mackenzie, K.A.D., Howard, B.H., and Harrison-Murray, R.H., 1986, The anatomical relationship between cambial regeneration and root initiation in wounded winter cuttings of the apple rootstock M.26, *Ann. Bot.*, 58:649.

Marks, T.R., 1991, Rhododendron cuttings. I. Improved rooting following 'rejuvenation' *in vitro*, *J. Hortic. Sci.*, 66:103.

Maynard, B., and Bassuk, N., 1986, Etiolation as a tool for rooting cuttings of difficult-to-root woody plants, *Comb. Proc. Int. Plant Prop. Soc. for 1985*, 35:488.

Maynard, B.K., and Bassuk, N.L., 1988, Etiolation and banding effects on adventitious root formation, *in*: "Adventitious Root Formation in Cuttings," T.D. Davis, B.E. Haissig and N. Sankhla, eds., Dioscorides Press, Portland.

Moe, R., and Andersen, A.S., 1988, Stock plant environment and subsequent adventitious rooting, *in*: "Adventitious Root Formation in Cuttings," T.D. Davis, B.E. Haissig and N. Sankhla, eds., Dioscorides Press, Portland.

Morgan, D.L., McWilliams, E.L., and Parr, W.C., 1980, Maintaining juvenility in live oak, *HortSci.*, 15:493.

Muzik, T.J., and Cruzado, H.J., 1958, Transmission of juvenile rooting ability from seedlings to adults of *Hevea brasiliensis*, *Nature*, 181:1288.

Pliego-Alfaro, F., and Murashige, T., 1987, Possible rejuvenation of adult avocado by graftage onto juvenile rootstocks *in vitro*, *HortSci.*, 22:1321.

Pontikis, C.A., Mackenzie, K.A.D., and Howard, B.H., 1979, Establishment of initially unrooted stool shoots of M.27 apple rootstock, *J. Hortic. Sci.*, 54:79.

Sadhu, M.K., 1979, Effect of pretreatment of stock plants of mango with cycocel, ethrel and morphactin on the rooting of cuttings and air layers, *Sci. Hortic.*, 10:363.

Sinha, A.C., and Vyvyan, M.C., 1943, Studies on the vegetative propagation of fruit tree rootstocks, II. By hardwood cuttings, *J. Pomol.*, 20:127.

Siniscalco, C., and Pavolettoni, L., 1988, Rejuvenation of *Eucalyptus* x *trabutii* by successive grafting, *Acta Hortic.*, 227:98.

Sriskandarajah, S., Mullins, M.G., and Nair, Y., 1982, Induction of adventitious rooting *in vitro* in difficult-to-propagate cultivars of apple, *Plant Sci. Lett.*, 24:1.

Stoutemyer, V.T., Britt, O.K., and Goodwin, J.R., 1961, The influence of chemical treatments, understocks, and environment on growth phase changes and propagation of *Hedera canariensis*, *Proc. Amer. Soc. Hortic. Sci.*, 77:552.

Webster, C.A., and Jones, O.P., 1989, Micropropagation of the apple rootstock M.9: effect of sustained subculture on apparent rejuvenation *in vitro*, *J. Hortic. Sci.*, 64:421.

Webster, C.A., and Jones, O.P., 1992, Performance of field hedge and stoolbed plants of micropropagated dwarfing apple rootstock clones with different degrees of apparent rejuvenation, *J. Hortic. Sci.*, 67:521.

AUXIN METABOLISM

AND ADVENTITIOUS ROOT INITIATION

David Blakesley

School of Biological Sciences
University of Bath
Claverton Down
Bath BA2 AY, UK

INTRODUCTION

In this chapter I will report on and discuss the role of auxins in adventitious root initiation, particularly the relations between endogenous indole-3-acetic acid (IAA) and the early events of adventitious rooting. The fact that IAA is involved is well established, although much of the data to support this is circumstantial. It will be not possible in this review paper to describe all the work on auxin application, transport and metabolism that might be relevant to studies of adventitious root initiation. Evidence derived from studies on auxin application has been reviewed many times [e.g., Audus (1959) and Blakesley et al. (1991b)] and will not be considered in detail in this paper. Evidence from more recent work on the analysis of endogenous auxin, and from studies on transgenic plant tissue, will be germane to the present paper. From these studies it will be apparent that we still do not have a clear understanding of the exact role of auxin in the process of adventitious root initiation. The aim of the latter part of this paper will be to describe briefly the newer technologies which are available to plant developmental physiologists, and to indicate new directions for researchers to approach the problem of auxin involvement in adventitious root initiation.

It is pertinent to consider the terminology used in the present paper to describe the process of adventitious root initiation, as this varies considerably in the literature and may be interpreted differently by scientists working in different fields [see the chapter by Barlow, and by Haissig and Davis in this volume]. Aside from preformed root primordia, Lovell and White (1986) have identified three categories of adventitious root primordia: those that develop *in situ*, those that develop towards the base of a cutting but not at the site of the initial cell divisions and those that develop in basal callus. Therefore, in the latter two cases, initial cell divisions are required before the first anatomical event of root initiation occurs. The lag phase may be hours or days, depending on the nature of the potential root initiation site. Anatomically and biochemically a number of different stages in the formation of an adventitious root initial have been identified. However, once these events are underway it is still difficult to be sure at what point the cellular events reach a state of determination. Clearly this can be established with certainty once organized root primordium initials are identifiable.

In those cases where primordia develop *in situ*, several workers have attempted to define distinct stages in root initiation (Doré, 1965; Girouard, 1967; Smith and Thorpe, 1975), all of which essentially include:

- Induction of a new meristematic locus.
- Early cell divisions.
- Later cell divisions to form an organized, determined root meristem
- Development of the root by extension growth of cells produced by the meristem, including the vascular union between primordium and stem.

The first visible sign of a new meristematic locus in *Pinus radiata* seedling cuttings is the expansion of a single cell, with simultaneous nuclear swelling and formation of dense cytoplasm (Smith and Thorpe, 1975). These early events seem characteristic of a meristematic locus and have been reported in other species (Coleman and Greyson, 1977; Oppenoorth, 1979). It is important to stress that if auxin is involved in the induction of root initiation, then one should consider examining auxin biosynthesis, metabolism and localization during the time prior to these first cellular events. This should be in addition to determining the role of auxin in each phase of adventitious root initiation.

ANALYSIS OF ENDOGENOUS AUXIN

Earlier work on the measurement of endogenous auxin largely utilized bioassays and often considered the auxin status of the cutting at the time of excision. Results were contradictory, for example, no differences were found in the auxin levels in an easy- and a hard-to-root cultivar of *Dahlia* (Biran and Halevy, 1973), sugar maple (Greenwood et al., 1976), *Dendranthema* (Stolz, 1968) and, more recently, *Rhododendron* (Wu and Barnes, 1981). In contrast a positive correlation between the ease of rooting and auxin levels has been reported with two cultivars of grapevine rootstock (Kracke et al., 1981). Much higher levels of auxin were found in the easy-to-root cultivar, although the bioassay used for this estimation was not reported. A better experimental system may be to compare rooting differences within the same species. Hengst (1959) showed that the variation in rooting of *Streptocarpus* leaves of different ages was closely correlated with the variation in auxin content of the whole plant. Smith and Wareing (1972) reported a decline in auxin levels in the stem of *Populus* x *robusta* cuttings taken over the growing season, and that this decline correlated well with a gradual decrease in the rooting ability of the cuttings. Blakesley et al. (1991a) reported a correlation between the rooting percentage of *Cotinus coggygria* cuttings and their endogenous auxin levels. *Cotinus* shoots which subsequently rooted well contained significantly higher levels of free IAA in the stem tissue than shoots which failed to root. Further, the level of conjugated IAA was significantly higher in the cuttings which failed to root.

The problems of determining the association between endogenous plant growth regulators and root initiation in woody plant material are considerable. More recently workers have concentrated more on nonwoody and hypocotyl explants to investigate these problems, examining the changes in the levels of endogenous auxin during the process of adventitious root initiation. This approach should give the opportunity to measure auxin levels in the rooting zone at specific stages such as induction, early cell divisions, primordium organization and emergence. Root initiation and emergence in 6-day-old *Phaseolus aureus* hypocotyl cuttings is very rapid. The first cell divisions take place within 30 hours, well developed root primordia are visible within 60 hours and roots emerge after 100 hours. If the cotyledons are removed from such cuttings, three to four root primordia develop at the base of the hypocotyl. Elevated levels of IAA occur in this zone within hours of cutting excision, but decline before the first cell division takes place (Blakesley et al., 1985). No such increase in IAA level was found in adjacent parts of the hypocotyl where adventitious root initiation does not occur. These results show that a transient rise in the levels of IAA is associated with the induction period of adventitious root initiation in this system, and that a decline in auxin is associated with the first cytological events. They do not prove that this rise and fall in IAA levels is causal. More recently a similar transient rise in the level of IAA has been reported in *in vitro* cuttings of grape (Gaspar et al., 1990). This early peak of IAA was not found in the apical parts of the shoot. Gaspar et al. (1990) also predicted from earlier work that early cytological events such as nuclear swelling would be visible at the time of the auxin decline, but not before. This led them to speculate that the enhanced IAA level was responsible for cell reactivation in the interfasicular cambium (Moncousin et al., 1989). Further, they also demonstrated that the decline in auxin was associated with a rise in the levels of peroxidases (PER) (Gaspar et al., 1990). These studies indicate that reduced levels of IAA are associated with early cell divisions and the organization of root primordia. Further support for the requirement of an early peak of IAA comes from the work of Maldiney et al. (1986) with hypocotyl cuttings of Craigella and Craigella lateral suppressor tomatoes. They also detected a rise in IAA, although the peak occurred after 72 hours. Organized root primordia were detectable in cleared hypocotyl sections after 96 hours. The authors suggested that the induction of adventitious root primordia occurred concomitant with the rise in IAA level. A similar finding was reported by Label et al. (1989) during *in vitro* rooting of *Prunus avium* explants. Growth of the primordia occurred as the level of IAA declined. Berthon et al. (1989) reported that adventitious root initiation in *Sequoiadendron giganteum* cuttings *in vitro* was accompanied by a decrease in IAA level. However the timing of the sampling was such that a transient rise in IAA could have been missed. Nordström and Eliasson (1991) found no rise in endogenous IAA levels in the base of leafy pea cuttings, but their first sample time was 24 hours. Although this harvest time was prior to any cell division, it could be after the actual induction of adventitious root primordia.

Further circumstantial evidence comes from an exciting study carried out by Prinsen et al. (1992), who attempted to correlate morphological changes in *Agrobacterium rhizogenes*-transformed pea epicotyls with altered endogenous levels of IAA and indole-3-acetamide (IAM). They suggested that callus and subsequent root formation were dependent upon the accumulation of endogenous IAA that resulted from expression of the T_R *aux* genes, concomitant with IAM accumulation. This is the first assay of endogenous IAA in tissue inoculated with *Agrobacterium*-derived strains containing these T_R DNA *aux* genes. Although root primordia were not present at the times when IAA and IAM were measured, it seems likely that the elevated IAA levels were indeed involved in the adventitious root initiation process.

AUXIN METABOLISM

Auxin Conjugation

The preceding discussion has concentrated on free IAA in plant tissue. IAA is also known to be present in plant tissue covalently bonded (conjugated) through the carboxyl group by an amide or an ester linkage to an amino acid or sugar/inositol, respectively. Bandurski and coworkers have carried out an extensive research program to elucidate the identity and specific roles of these conjugates in *Zea mays*. This research led Bandurski (1980) to propose that the conjugation and hydrolysis of IAA are important mechanisms in the regulation of free IAA levels in plant tissue. Certainly plant tissues show an increased capacity to accumulate bound auxins following auxin application (Andreae and van Ysselstein, 1956; Venis, 1972; Nordström et al., 1991). The conjugation of auxin has been reviewed previously [e.g., Bandurski (1980)], but it is important to summarize the nature of these compounds.

Indole-3-acetylaspartate (IAAsp) has been identified as the predominate amide conjugate formed after IAA application (Andreae and Good, 1955; Venis, 1972) and is also known to occur naturally (Law and Hamilton, 1982; Nordström and Eliasson, 1991). Zenk (1964) and Andreae (1967) regarded IAAsp as a detoxification product of applied auxin. Feung et al. (1977) suggested that IAAsp may also be involved in the regulation of endogenous auxin because of a reversible conjugation. Although amide conjugation has been regarded as a reversible process (Bialek and Cohen, 1989), other workers have reported very little hydrolysis of IAAsp to free IAA (Plüss et al., 1989; Nordström and Eliasson, 1991). The second type of IAA conjugate involves an ester link with a sugar or *myo*-inositol such as indole-3-acetylglucose. Although esters with a number of sugars have been reported, glucose is the most commonly reported component (Cohen and Bandurski, 1982). The potential roles of the various IAA conjugates has been discussed previously (Cohen and Bandurski, 1982).

A number of workers have implicated auxin conjugates in the control of adventitious root initiation. Blakesley et al. (1991a) examined the levels of IAA and IAA conjugates in cuttings of the woody shrub *Cotinus coggygria* taken at different times of the year. Cuttings taken at the time of bud break rooted well, but cuttings taken much later in the growing season rooted very poorly. Endogenous IAA conjugates were cleaved by a mild hydrolysis treatment, which would be expected to release IAA esters. At the time of harvest, free IAA levels in young shoots were significantly higher than those of IAA conjugates. Later in the season, when rooting was poor, the reverse was found. Although the "pool" of IAA was similar on both occasions, the ratio of free IAA to total IAA in the rooting zone was 0.94 in young shoots which rooted, and 0.02 in older shoots which failed to root. Nordström and coworkers published data on the levels of bound IAA in young pea cuttings during the rooting period. They found a build-up of IAAsp in the rooting zone (Nordström and Eliasson, 1991), which they postulated prevented a longer term build-up of free IAA. Following the application of indole-3-butyric acid (IBA) and IAA to these pea cuttings, Nordström et al. (1991) again found no long term accumulation of IAA, but IBA was much more persistent. The number of roots on cuttings treated with IBA was an order of magnitude greater than on cuttings treated with IAA. Conversion of the exogenous IAA to IAAsp was the predominant route of metabolism for the applied IAA. A similar metabolic pathway was reported by Wiesman et al. (1988, 1989), although in *Phaseolus aureus* cuttings they found much more rapid metabolism of applied IBA. In contrast to the results of Nordström et al. (1991), they were not able to attribute the significantly higher root number in *P. aureus* cuttings treated with IBA to persistence of free IBA in the rooting zone. IAAsp applied to the rooting solution did not stimulate rooting (Nordström et al., 1991), which supports the assumption that once endogenous IAA (or IBA) is conjugated to IAAsp it does not stimulate adventitious root initiation.

Auxin Catabolism

In addition to the biosynthesis and conjugation of IAA, plant tissues are also capable of non-decarboxylative IAA degradation and decarboxylative IAA oxidation. Oxidative IAA catabolism is catalyzed by various PERs (E.C. 1.11.1.7) and has been recently reviewed [Gaspar et al. (1985); see also the chapter by Hand in this volume]. Several pathways of IAA degradation have been identified (Grambow and Langenbeck-Schwich, 1983) which lead to the major products of indole-3-aldehyde and 3-methyleneoxindole. More recently attention has been drawn to an intermediate in these pathways, indole-3-methanol (Sabater et al., 1983) which was identified as the main IAA oxidative product catalyzed by extracellular PERs from lupin (Ros Barcelo et al., 1990). Basic isoPERs localized in the cytoplasm of lupin phloem parenchyma cells are apparently involved in the oxidation of polarly transported IAA (Ros Barcelo et al., 1990). Based on *in vitro* studies they suggested that the basic PER system, functioning with monophenols and Mn^{2+} might exist *in vivo*. They further suggested that although *in vitro* IAA oxidation can also be carried out by acidic isoPERs, *in vivo* oxidation is mainly by basic isoPERs, as proposed by Gaspar et al. (1985). More recently, Ferrer et al. (1990, 1992) have reported that in cell walls of lupin hypocotyls, the acidic isoenzyme group of isoPERs was responsible for oxidizing IAA through oxidative and peroxidative cycles. Two research groups established that IAA catabolism in *Z. mays* seedlings can also take place without the loss of the carboxyl carbon of IAA (Reinecke and Bandurski, 1981; Nonhebel et al., 1983). In comparison to the work on oxidative decarboxylation, however, the enzyme systems involved in the oxidation of IAA without decarboxylation have received less attention, although Reinecke and Bandurski (1988) have recently oxidized IAA to oxindole-3-acetic acid *in vitro* with a partially characterized enzyme from *Z. mays*. The actual function and diversity of this IAA catabolism system are not understood.

Gaspar et al. (1990) reported a transient decrease in the levels of PERs in the first few hours following the excision of *in vitro* cuttings of grapevine. This decline coincided with a transient rise in the levels of IAA in the rooting zone. Levels of PERs subsequently increased during the development of the root initials, similar to observations in other systems [e.g., Berthon et al. (1987) and Mato et al. (1988)]. This supports the notion raised earlier in this chapter that elevated levels of IAA are important during root induction, but subsequently lower levels of auxin enable the development of the root primordium.

HYPOTHESIS BASED ON PAST WORK

I have recently restated the hypothesis (Blakesley et al., 1985) that, based on all the evidence to date, absolute levels of auxin are important for the stimulation of the primary events of root initiation (Blakesley and Chaldecott, 1993). For primordia that develop *in situ*, an elevated level of auxin may be required to induce the anatomical and biochemical events of root initiation which precede the first cell divisions. This notion has also been expressed by Jarvis (1986) and Gaspar et al. (1990). Several studies have demonstrated such a rise in IAA levels in the rooting zone during adventitious root induction (Blakesley et al., 1985; Gaspar et al., 1990). However, the necessity for such a rise may be partially dependent on the levels of cytokinins in the tissue. In an intact plant cytokinins are transported from the root to the shoot. Excision of the root system removes this source of cytokinins and might, therefore, be expected to lead to a reduced cytokinin concentration at the base of the shoot. I have shown that cytokinin levels in cuttings of *Cotinus coggygria* and *Phaseolus aureus* were easily detectable at the time of cutting excision, but fell dramatically to very low levels on subsequent harvest times (unpublished date). The decline in cytokinins in the rooting zone of *P. aureus* hypocotyl cuttings was concomitant with a rise in IAA (Blakesley et al., 1985). Maldiney et al. (1986) also found that cytokinin levels decreased dramatically in the rooting zone of young tomato cuttings within 24 hours of excision. It is possible that this reduction in the level of endogenous cytokinins, a result of the excision of the root system, may be necessary before adventitious root initiation, stimulated by free IAA, can take place.

FUTURE RESEARCH: DIRECTION AND TECHNIQUES

It is important that future work on endogenous IAA metabolism in the rooting zone is correlated with the developmental stages of root initiation. Analysis of IAA metabolism should be accompanied by detailed anatomical and molecular studies to establish the precise stage of root initiation at which the analysis is carried out. Rarely in past work has this been achieved. I believe that such work would greatly enhance our understanding of the involvement of IAA in root initiation. However, this approach still utilizes the rooting zone as a whole, as analysis of IAA and its metabolism in the actual cells involved in the early stages of root initiation would be very difficult, if not impossible, using conventional dissection and analytical techniques. Advances have been made in the actual analysis of IAA in recent years, particularly the accuracy and sensitivity of physicochemical assays. We are employing highly sensitive methods of analysis for IAA in small pieces of tissue. Using combined gas chromatography-mass spectrometry (GC-MS), for example, the mass of IAA in the rooting zone of a single *P. aureus* cutting can be accurately measured in the selected ion-monitoring (SIM) mode, as the limit of sensitivity lies around 100 fmol (Blakesley, unpublished). It is further possible using methane-chemical-ionization mass spectrometry to increase this level of sensitivity by several orders of magnitude (Netting and Milborrow, 1988). In theory this would allow the assay of IAA in individual root initial cells, if suitable microdissection techniques were available. Harris and Outlaw (1990) have recently published a low volume enzyme-amplified immunoassay with sub-fmol sensitivity. This has been applied to the measurement of abscisic acid in individual stomatal guard cells. Such techniques are being developed for IAA and could further enhance the sensitivity of GC-MS techniques.

Immunocytolocalization

Immunocytolocalization techniques should also be considered to provide evidence of the role of IAA in root initiation. In brief, the technique enables the localization of plant growth regulators in tissues and cells using antiplant growth regulator antibodies, which are subsequently detected using anti-serum to the primary antibody linked to the antigen. These can be visualized by light or electron microscopy depending on the detail required. A good example of the potential of these techniques is provided by the work of Stroobants et al. (1991) who used immunocytolocalization to locate zeatin riboside within the cells of tobacco leaves transformed with *Agrobacterium rhizogenes*. It should be noted that these techniques are difficult and require careful control treatments to avoid artifacts. Nevertheless their potential is such that the limitations are likely to be overcome as scientists appreciate the opportunity which they afford to further our understanding of the role of plant growth regulators in plant growth and development.

Photoaffinity Labelling

A further exciting microtechnique that could enhance our understanding of the role of IAA in root initiation is photoaffinity labelling. In a hypocotyl, for example, it is possible to locate the cell type(s) which carry polar auxin transport in the intact plant using microautoradiography. It should also be possible to identify lateral migration of IAA away from the cells involved in polar IAA transport in an intact stem/hypocotyl. Following cutting excision, any changes in the lateral diffusion and redistribution of IAA between different cell types could also be monitored, and the actual location within the cells determined. An elegant study of the location of transported auxin in etiolated shoots of *Z. mays* illustrates the potential for this technique very well (Jones, 1990). Microautoradiography has been utilized to examine the location of exogenously applied [^{3}H]IAA in buds, coleoptiles and epicotyls. Jones (1990) has examined the problem of polar auxin transport in 3.5-day-old *Z. mays* seedling shoots using a newly developed technique involving photoaffinity labelling. In this technique, [^{3}H],5-N$_3$IAA, an analog of IAA, is photolytically fixed *in situ*. Covalent fixation facilitates the identification of cells containing the labelling reagent at the time of UV radiation by autoradiography. Jones found that [^{3}H],5-N$_3$IAA was present in every cell, although at different levels. The highest concentration was in the epidermis, whereas cells subtending the epidermis and parenchyma cells in the stele contained the next highest level. These data are relevant to the present discussion on auxin and root initiation, because in *Z. mays* it shows that auxin is available to all cells, and that internal pathways for auxin movement are also possible and significant. This work indicates that polar auxin transport can occur in the epidermal cells and the vascular tissue. However, these studies were carried out in excised mesocotyl segments over four hours, and it is not clear whether the same distribution of $[^3H],5\text{-}N_3IAA$ would have been found in an intact seedling.

Sensitivity and Evidence from Transgenic Plant Tissues

In much of this section I will describe and discuss techniques and approaches concerned with elucidating the localization of auxin within tissues and individual cells, and the measurement of absolute levels of IAA concerned with adventitious root initiation. It will not address the manipulation of endogenous auxin levels, or the issue of differential cellular sensitivity to IAA. Trewavas (1981) fostered the theory that in many cases sensitivity (responsiveness) to plant growth substances is more important than absolute concentrations of the plant growth substances. This was echoed recently by Gaspar and Hofinger (1988) who stressed that data on auxin biosynthesis and metabolism should be interpreted in the context of differential sensitivity to auxin. More recently Trewavas (1991) has discussed its actual measurement and theoretical basis. His criteria for an unambiguous measurement of growth substance sensitivity are:
- The avoidance of artifact by manipulation at the endogenous level.
- Preferable use of whole plants.
- Methods which assess the plant growth substance contribution to control when a number of other factors also contribute.

This would allow the measurement of sensitivity at the plant growth substance level, and is defined by Trewavas (1991) as "control strength." Control strength requires knowledge of the fractional change in response and the fractional change in endogenous concentration. The change in response (R) is divided by the change in endogenous concentration of the plant growth substance (C) to give a value of the control strength. Very few existing plant growth substance studies allow the calculation of control strength. Studies of auxin receptors and auxin binding may provide useful information towards understanding this issue, as discussed elsewhere in the present volume.

One of the most exciting developments for plant physiologists is the exploitation of various gene *loci* in *Agrobacterium* species; hairy root disease and crown gall disease are growths on plant tissue that are incited by *Agrobacterium rhizogenes* and *A. tumefaciens*, respectively. Hairy root disease is characterized by prolific adventitious root development. Crown gall tumors are characterized by amorphous, disorganized callus or by shoot teratomas. In the case of virulent strains of these bacteria, a small segment of DNA (T-DNA) contained on a large extrachromosomal plasmid is transferred, integrated and expressed in the plant genome. Virulent *A. rhizogenes* bacteria harbor an Ri (root-inducing) plasmid [recently reviewed by Blakesley and Chaldecott (1993)] and virulent strains of *A. tumefaciens* harbor a Ti (tumor-inducing) plasmid. The expression of Ri T-DNA results in an abundant proliferation of roots at the site of inoculation. Ri plasmids have conveniently been characterized by the type of opine which is synthesized in transformed tissue. Agropine-type Ri plasmids, for example, have two T-DNA regions, T_L and T_R. DNA hybridization studies on the T_R region show considerable homology to the *tms1* and *tms2* genes of *A. tumefaciens* Ti plasmids (Huffman et al., 1984; Jouanin, 1984; White et al., 1985; Offringa et al., 1986). Hairy root disease is also a characteristic of plant tissue transformed by the mannopine-type *A. rhizogenes* which does not contain the T_R region.

The T_L region of the Ri plasmid stimulates root formation independently of the transfer and expression of the T_R T-DNA genes. White et al. (1985) were the first group to characterize the *A. rhizogenes* T-DNA agropine-type Ri plasmid pRiA4. They identified four genetic loci (*rolA, rolB, rolC* and *rolD*) which were defined according to the tumor root morphology observed when insertions were made into each gene locus. *Rol* gene loci have since been investigated by many research groups [e.g., Shen et al. (1988), Estruch et al. (1991), Maurel et al., 1991)].

These plasmids could be useful because they contain genes involved in the biosynthesis of plant growth substances (see the chapter by Hamill and Chandler, and by Tepfer and coworkers in this volume). Further, transfer of certain regions of the Ri plasmid has been reported to confer increased sensitivity to auxin in plant tissue (Shen et al., 1988; Maurel et al., 1991). Enhanced auxin levels may be important for auxin sensitivity, according to the recent work of Estruch et al. (1991) which reports the *rolB* gene coding an enzyme capable of hydrolysing indole-ß-glucosides. It should be noted, however, that the substrate tested for the *rolB* gene enzyme by Estruch et al. (1991) was indoxyl-ß-glucoside, which is not an auxin. The study of the structure and function of these genes involved in the biosynthesis and metabolism of plant growth substances, and in apparent sensitivity to plant growth substances, could, therefore, offer considerable insight into the role of auxin in root initiation. They may allow the manipulation of endogenous IAA levels in the rooting zone through the incorporation of auxin biosynthetic genes from *A. tumefaciens* or *A. rhizogenes*. The work of Prinsen et al. (1992), described earlier, is a good example of the potential of this technique. These workers were able to transform decapitated *Pisum sativum* epicotyls with *pA-303,d16Agrobacterium*-derived strains containing T_R DNA and *aux* genes, among others, and to compare morphological and histological changes with altered auxin biosynthesis. I believe that controlled incorporation and expression of these genes offers a powerful tool to plant developmental physiologists in their quest to understand the role of plant growth substances in adventitious root initiation. If the ß-glucosidase enzyme encoded by *rolB* (Estruch et al., 1991) is shown to hydrolyse auxin conjugates *in vivo*, then this too could be a very useful technique for manipulating endogenous auxin metabolism.

In a carrot disc transformation system described by Ryder et al. (1985) and Cardarelli et al. (1987a,b), the Ri T-DNA apparently sensitizes the plant cells to auxin. Shen et al. (1988) investigated various physiological properties of protoplasts derived from root tips of *Lotus corniculatus* that were transformed by *A. rhizogenes*. These properties were compared with their non-transformed counterparts. Hairy root transformed cells were 10^2 to 10^3 times more sensitive to the effects of auxin than were untransformed cells. The authors postulated that this increased sensitivity to auxin by the transformed cells was an early cellular event, possibly involving the reception or transduction of the hormone signal. A similar enhanced sensitivity to auxin was found in subsequent studies by these authors (Shen et al., 1990) and also by Barbier-Brygoo et al. (1990) who later reported that each of the single genes *rolA, rolB* and *rolC* was able to confer increased sensitivity to auxin to transformed protoplasts, *rolB* being the most effective (Maurel et al., 1991). Maurel et al. (1991) assumed that a high sensitivity to auxin could be a major determinant for root differentiation. They further speculated that the increased sensitivity to auxin conferred by the *rolB* gene might direct the transformed cell into root organogenesis.

CONCLUSION

The literature reviewed in this paper has widened the discussion of the involvement of auxin in root initiation to consider the future direction of work in this area. I have presented the data of a variety of workers which support the notion that in young seedling cuttings elevated levels of IAA are associated with the initiation of adventitious roots. In the rooting zone of *Phaseolus aureus*, for example, an early rise in free IAA during the induction of adventitious root initiation was accompanied by a decline in the level of cytokinins (Blakesley et al., 1985). A hypothesis raised previously by these and other authors stated that these plant growth substance events are necessary to "trigger" the start of the formation of a root initial at a potential root initiation site. Further work is being undertaken in my laboratory to test this hypothesis using *P. aureus* and *Eucalyptus* species. The issue of sensitivity is a complex one which has been discussed widely in the literature [see references in Trewavas (1991)] and has been equated with plant growth substance reception and transduction. There has been some discussion of high sensitivity to auxin which is reported to be a common feature of hairy root cultures. Indeed, incorporation of the *rol*B gene alone can induce root initiation on tobacco stems (Cardarelli et al., 1987a). It seems possible that hydrolysis of IAA conjugates may be involved in this phenomenon.

REFERENCES

Andreae, W.A., and Van Ysselstein, M.W.H., 1956, Studies on 3-indoleacetic acid metabolism, III. The uptake of 3-indoleacetic acid by pea epicotyls and its conversion to 3-indoleacetylaspartic acid, *Plant Physiol.* 31:235.

Andreae, W.A., 1967, Uptake and metabolism of indoleacetic acid, napthaleneacetic acid and 2,4-dichlorophenoxyacetic acid by pea root segments in relation to growth inhibition during and after auxin application, *Can. J. Bot.* 45:737.

Andreae, W.A., and Good,N.E., 1955, The formation of indoleacetylaspartic acid in pea seedlings, Plant Physiol. 30:380.

Audus, L.J., 1959, "Plant Growth Substances," Leonard Hill, London.

Bandurski, R.S., 1980, Homeostatic control of concentrations of indole-3-acetic acid, *in*: "Plant Growth Substances 1979," F. Skoog, ed., Springer-Verlag, Berlin.

Barbier-Brygoo, H., Guern, J., Ephritikhine, G., Shen, W.H., Maurel, C., and Klämbt, D., 1990, The sensitivity of plant protoplasts to auxin: modulation of receptors at the plasmalemma, *in*: "Plant Gene Transfer," UCLA Symp. on Mol. and Cell. Biol., new ser., C.Lamb, and R.Beachy, eds., Liss, New York.

Berthon, J.Y., Boyer, N., and Gaspar, T., 1987, Sequential rooting media and rooting of *Sequoiadendron giganteum in vitro*, Peroxidase activity as a marker, *Plant Cell Rep.* 6:341.

Berthon, J.Y., Maldiney, R., Sotta, B., Gaspar, T., and Boyer, B., 1989, Endogenous levels of plant hormones during the course of adventitious rooting in cuttings of *Sequoiadendron giganteum* (Lindl.) in vitro, *Biochem. Physiol. Pfl.*184:405.

Bialek, K., and Cohen, J.D., 1989, Free and conjugated indole-3-acetic acid and its derivatives in plants, *Plant Physiol.* 91:775.

Biran, I., and Halevy, A.H., 1973, Stock plant shading and rooting of dahlia cuttings, *Sci. Hortic.* 1:125.

Blakesley, D., Hall, J.F., Weston, G.D., and Elliott, M.C., 1985, Endogenous plant growth substances and the rooting of *Phaseolus aureus* cuttings, *in*: "Abst. 12th Int. Conf. Plant Growth Subs.," Heidelberg.

Blakesley, D., Weston, G.D., and Elliott, M.C., 1991a, Endogenous levels of indole-3-acetic acid and abscisic acid during the rooting of *Cotinus coggygria* cuttings taken at different times of the year, *Plant Growth Reg.* 10:1.

Blakesley, D., Weston, G.D., and Hall, J.F., 1991b, The role of endogenous auxin in root initiation. Part I: Evidence from studies on auxin application, and analysis of endogenous levels, *Plant Growth Reg.* 10:341.

Blakesley, D., and Chaldecott, M.C., 1993, The role of endogenous auxin in root initiation, Part II: Sensitivity, and evidence from studies on transgenic plant tissue, *Plant Growth Reg.* (in press)

Cardarelli, M., Mariotti, D., Pomponi, M., Spanò, L., Capone, I., and Costantino, P., 1987a, *Agrobacterium rhizogenies* T-DNA genes capable of inducing hairy root phenotype, *Mol. Gen. Genet.* 209:475.

Cardarelli, M., Spanò, L., Marriotti, D., Mauro, M.L., Van Sluys, M.A., and Costantino, P., 1987b, The role of auxin in hairy root induction, *Mol. Gen. Genet.* 208:457.

Cohen, J.D., and Bandurski, R.S., 1982, Chemistry and physiology of bound auxins, *Annu. Rev. Plant Physiol.* 33:403.

Coleman, W.K., and Greyson, R.I., 1977, Analysis of root formation in leaf discs of *Lycopersicon esculentum* Mill. cultured *in vitro*, *Ann. Bot.* 41:307.

Doré, J., 1965, Physiology and regeneration of cormophytes, *in*: "Encyc. of Plant Physiol.," W. Rhuland, ed., Springer-Verlag, Berlin.

Estruch, J.J., Schell, J., and Spena, A., 1991, The protein encoded by the *rolB* plant oncogene hydrolyses indole glucosides, *EMBO J.* 10:3125.

Ferrer, M.A., Pedreño, M.A., Muñoz, R., and Ros Barcelo, A., 1990, Oxidation of coniferyl alcohol by cell wall peroxidases at the expense of indole-3-acetic acid and O_2. A model for the lignification of plant cell walls in the absence of H_2O_2, *FEBS* 276:127.

Ferrer, M.A., Pedreño, M.A., Ros Barcelo, A., and Muñoz, R., 1992, The cell wall localization of two strongly basic isoperoxidases in etiolated *Lupinus albus* hypocotyls and its significance in coniferyl alcohol oxidation and indole-3-acetic acid catabolism, *J. Plant Physiol.* 139:611.

Feung, C.S., Hamilton, R.H., and Mumma, R.O., 1977, Metabolism of indole-3-acetic acid. IV. Biological properties of amino acid conjugates, *Plant Physiol.* 59:91.

Gaspar, T., Penel, C., Castilllo, F.J., and Greppin, H., 1985, A two step control of basic and acidic peroxidases and its significance for growth and development, *Physiol. Plant.* 64:418.

Gaspar, T., and Hofinger, H., 1988, Auxin metabolism during adventitious rooting, *in*: "Adventitious Root Formation in Cuttings," T.D.Davis, B.E.Haissig, and N.Sankhla, eds., Dioscorides Press, Portland.

Gaspar, T., Moncousin, C., and Greppin, H., 1990, The place and role of exogenous and endogenous auxin in adventitious root formation, *in*: "Intracellular Communications in Plants," B.Millet and H.Greppin, eds, INRA, Paris.

Girouard, R.M, 1967, Initiation and development of adventitious roots in stem cuttings of *Hedera helix*, *Can. J. Bot.* 45:1883.

Grambow, H.J., and Langenbeck-Schwich, B., 1983, The relationship between oxidase activity, peroxidase activity, hydrogen peroxide and phenolic compounds in the degradation of indole-3-acetic acid *in vitro*, *Planta*. 157:131.

Greenwood, M.S., Atkinson, O.R., and Yawney, H.W., 1976, Studies of hard- and easy-to-root ortets of sugar maple: Differences not due to endogenous auxin content, *Plant Prop.* 22:3.

Harris, M.J., and Outlaw, W.H., 1990, Histochemical techniques: a low volume, enzyme-amplified immunoassay with sub-fmol sensitivity, Application to measurement of abscisic acid in stomatal guard cells, *Physiol Plant.* 78:495.

Hengst, K.H., 1959, Untersuchungen zur physiologie der regeneration in der guttang Streptocarpus, II. Korrelationsersheinungen and polarität, *Z. Bot.* 47:383.

Huffman, G.A., White, F.F., Gordon, M.P., and Nester, E.W., 1984, Hairy-root-inducing plasmid: physical map and homology to tumor-inducing plasmids, *J. Bacteriol.* 157:269.

Jarvis, B.C., 1986, Endogenous control of adventitious rooting in non-woody cuttings, *in*: "New Root Formation in Plants and Cuttings," M.B. Jackson, ed, Martinus Nijhoff, Dordrecht.

Jones, A.M., 1990, Location of transported auxin in etiolated maize shoots using 5-azidoindole-3-acetic acid, *Plant Physiol.* 93:1154.

Jouanin, L., 1984, Restriction map of an agropine-type Ri plasmid and its homologies with Ti plasmids, *Plasmid.* 12:91.

Kracke, H., Cristoferi, G., and Marangoni, B., 1981, Hormonal changes during the rooting of hardwood cuttings of grapevine rootstocks, *Amer. J. Enol. Vitic.* 32:135.

Label, P.H., Sotta, B., and Miginiac, E., 1989, Endogenous levels of abscisic acid and indole-3-acetic acid during in vitro rooting of wild cherry explants produced by micropropagation, *Plant Growth Reg.* 8:325.

Law, D.M., and Hamilton, R.A., 1982, A rapid isotope dilution method for analysis of indole-3-acetic acid and indoleacetyl aspartic acid from small amounts of plant tissue, *Biophys. Res. Commun.* 106:1035.

Lovell, P.H., and White, J., 1986, Anatomical changes during adventitious root formation, *in*: "New Root Formation in Plants and Cuttings," M.B. Jackson, ed., Martinus Nijhoff, Dordrecht.

Maldiney, R., Pelèse, F., Pilate, G., Sotta, B., Sossountzov, L., and Miginiac, E., 1986, Endogenous levels of abscisic acid, indole-3-acetic acid, zeatin and zeatin riboside during the course of adventitious root formation on cuttings of Craigella and Craigella lateral suppressor tomatoes, *Physiol. Plant.* 68:426.

Mato, M.C., Rua, M.L., and Ferro, E., 1988, Changes in levels of peroxidases and phenolics during root formation in *Vitis* cultured *in vitro*, *Physiol. Plant.* 72:84.

Maurel, C., Barbier-Brygoo, H., Brevet, J., Spena, A., Tempé, J., and Guern, J., 1991, *Agrobacterium rhizogenes* T-DNA genes and sensitivity of plant protoplasts to auxins, *in*: "Advances in Mol. Genet. of Plant-Microbe Interactions," vol. 1, H.Hennecke, and D.P.S.Verma, eds., Kluwer Academic Pubs., Dordrecht.

Moncousin, C., Favre, J-M., and Gaspar, T., 1989, Early changes in auxin and ethylene production in vine cuttings before adventitious rooting, *Plant Cell Tissue Organ Cult.* 19:235.

152

Netting, A.G., and Milborrow, B.V., 1988, Methane chemical ionization mass spectrometry of the pentafluorobenzyl derivatives of abscisic acid, its metabolites and other plant growth regulators, *Biomed. Env. Mass Spectrom.* 17:281.

Nonhebel, H.M., Crozier, A., and Hillman, J.R., 1983, Analysis of [^{14}C] indole-3-acetic acid metabolites from the roots of *Zea mays* seedlings using reverse-phase high-performance liquid chromatography, *Physiol. Plant.* 57:129.

Nordström, A.-C., Alvarado Jacobs, F., and Eliasson, L., 1991, Effect of exogenous indole-3-acetic acid and indole-3-butyric acid on internal levels of the respective auxins and their conjugation with aspartic acid during adventitious root formation in pea cuttings, *Plant. Physiol.* 96:856.

Nordström, A.C., and Eliasson, L., 1991, Levels of endogenous indole-3-acetic acid and indole-3-acetylaspartic acid during adventitious root formation in pea cuttings, *Physiol Plant.* 82:599.

Offringa, I.A., Melchers, L.S., Regensburg-Tuink, A.J.G., Costantino, P., Schilperoort, R.A., and Hooykaas, P.J.J., 1986, Complimentation of *Agrobacterium tumefaciens* tumor-inducing aux mutants by genes from the TR region of the Ri plasmid of *Agrobacterium rhizogenes*, *Proc. Natl. Acad. Sci. USA.* 83:6935.

Oppenoorth, J.M., 1979, Influence of cylohexamide and actinomycin D on initiation and early development of adventitious roots. *Physiol. Plant.* 47:134.

Plüss, R., Titus, J., and Meier, H., 1989, IAA-induced adventitious root formation in greenwood cuttings of *Populus tremula* and formation of 2-indolone-3-acetylaspartic acid, a new metabolite of exogenously applied indole-3-acetic acid, *Physiol. Plant.* 75:89.

Prinsen, E., Bercetche, J., Chriqui, D., and van Onckelen, H., 1992, *Pisum sativum* epicotyls inoculated with *Agrobacterium rhizogenes* agropine strains harbouring various T-DNA fragments: Morphology, histology and endogenous indole-3-acetic acid and indole-3-acetamide content, *J. Plant Physiol.* 140:75.

Reinecke, D.M., and Bandurski, R.S., 1981, Metabolic conversion of ^{14}C-indole-3-acetic acid to ^{14}C-oxindole-3-acetic acid, *Biochem. Biophys. Res Commun.* 103:429.

Reinecke, D.M., and Bandurski, R.S., 1988, Oxidation of indole-3-acetic acid to oxindole-3-acetic acid by an enzyme preparation from *Zea mays*, *Plant Physiol.* 86:868.

Ros Barcelo, A., Pedreño, M.A., Ferrer, M.A., Sabater, F., and Muñoz, R., 1990, Indole-3-methanol is the main product of the oxidation of indole-3-acetic acid catalyzed by two cytosolic basic isoperoxidases from *Lupinus*, *Planta.* 181:448.

Ryder, M.H., Tate, M.E, and Kerr, A., 1985, Virulence properties of strains of *Agrobacterium* on the apical and basal surfaces of carrot root discs, *Plant Physiol.* 77:215.

Sabater, F., Acosta, M., Sanchez-Bravo, J., Cuello, J., and del Rio, J.A., 1983, Indole-3-methanol as an intermediate of the oxidation of indole-3-acetic acid by peroxidase, *Physiol. Plant.* 57:75.

Shen, W.H., Davioud, E., David, C., Barbier-Brygoo, H., Tempé, J., and Guern, J., 1990, High sensitivity to auxin is a common feature of hairy root, *Plant Physiol.* 94:554.

Shen, W.H., Petit, A., Guern, J., and Tempé, J., 1988, Hairy roots are more sensitive to auxin than normal roots, *Proc. Natl. Acad. Sci. USA.* 85:3417.

Smith, D.R., and Thorpe, T.A., 1975, Root initiation in cuttings of *Pinus radiata* seedlings, I. Developmental sequence, *J. Exp. Bot.* 26:184.

Smith, N.G., and Wareing, P.F., 1972, The rooting of actively growing and dormant leafy cuttings in relation to the endogenous hormone levels and photoperiod, *New Phytol.* 71:483.

Stolz, L.P., 1968, Factors influencing root initiation in an easy and a difficult-to-root Chrysanthemum, *Proc. Amer. Soc. Hortic. Sci.* 92:622.

Stroobants, C., Sossountzov, L., and Miginiac, E., 1991, Immunocytolocalisation du riboside de la zéatine dans des feuilles de tabac isolées et bouturées, cours des phases initiales de la rhizogénèse, *C.R. Acad. Sci. Paris.* 312:261.

Trewavas, A.J., 1981, How do plant growth substances work? I, *Plant, Cell and Env.* 4:203.

Trewavas, A.J., 1991, How do plant growth substances work? II, *Plant, Cell and Env.* 14:1.

Venis, M.A., 1972, Auxin-induced conjugation system in peas, *Plant Physiol.* 49:24.

Wiesman, Z., Riov, J., and Epstein, E., 1988, Comparison of movement and metabolism of indole-3-acetic acid and indole-3-butyric acid in mung bean cuttings, *Physiol. Plant.* 74:556.

Wiesman, Z., Riov, J., and Epstein, E., 1989, Characterization and rooting ability of indole-3-butyric acid conjugates formed during rooting of mung bean cuttings, *Plant. Physiol.* 91:1080.

White, F.F., Taylor, B.H., Huffman, G.A., Gordon, M.P., and Nester, E.W., 1985, Molecular and genetic analysis of the transferred DNA regions of the root-inducing plasmid of *Agrobacterium rhizogenes*, *J Bacteriol.* 164:33.

Wu, F.T., and Barnes, M.F., 1981, The hormone levels in the stems of difficult-to-root and easy-to-root rhododendrons, *Biochem. Biophys. Pfl.* 176:13.

Zenk, M.H., 1964, Isolation, biosynthesis and function of indoleacetic acid conjugates, *in*: "Rég. Nat. Croiss. Vég.", J.P. Nitsch, ed., C.N.R.S., Paris.

THE *ERabp* GENE FAMILY:

STRUCTURAL AND PHYSIOLOGICAL ANALYSES

Klaus Palme[1], Thomas Hesse[1], Christine Garbers[1,2],
Carl Simmons[2], and Dieter Söll[2]

[1]Max-Planck-Institut für Züchtungsforschung
D-5000 Köln 30, FRG

[2]Department of Molecular Biophysics and Biochemistry
Yale University,
New Haven, USA

INTRODUCTION

Auxins are a group of phytohormones that influence a wide range of growth and developmental responses in plants. Effects induced by auxins include a stimulation of cell enlargement and stem growth, cell division, vascular tissue differentiation, initiation of roots on stem cuttings, the development of branch roots and the differentiation of roots in tissue culture (Davies, 1987). Although auxin can inhibit the growth of a primary root at rather low concentrations, probably due to the induction of ethylene production, lateral branch roots and adventitious roots are stimulated by high auxin levels, an effect that has been very useful in horticultural practice for plant propagation by cuttings (see chapter by Blakesley, by Haissig and Davis, and by Howard in this volume).

Lateral roots or root hairs originate from small groups of cells in the pericycle which are activated by auxin to proliferate. More diverse cell types from which adventitious roots may form, are also activated by auxin. How can auxin stimulate such cells to divide? and what are the intracellular target sites for auxins? In the numerous physiological responses of plants to auxins, the specific structure/function requirements of natural and synthetic auxins suggest that the first step of the response must be an interaction of auxin with a protein. Such proteins may be operationally termed receptors, but should better be termed auxin binding proteins as long as definitive proof for their receptor function is missing. Many auxin binding proteins have been identified over the past decades, but biochemical progress in the characterization of these proteins has been slow due to the many difficulties associated with classical ligand binding techniques. This has been overcome by the introduction of photoaffinity labelling techniques as well as by improvements in other biochemical and immunological tools available for structural and functional studies. Hence the list of auxin binding proteins has been growing. Initially, based on binding of radiolabelled auxins to plant membrane fractions, the search for auxin binding proteins led to the identification of three binding sites that were located on the endoplasmic reticulum (ER; site I), the tonoplast (site II) and the plasma membrane (site III) (for recent reviews see Palme et al., 1991; Jones and Prasad, 1992). Due to its relative abundance, the site I-specific auxin binding site was for many years most actively studied but, with the help of photoaffinity labeling techniques using photolabile auxins, more proteins were identified. These photolabile and auxin-specific ligands include 5'-azido[7-^{3}H]indole-3-acetic acid [azido-IAA] and azido[5,6-^{3}H$_2$]naphthylphthalamic acid [azido-NPA] (Melhado et al., 1981; Jones et al., 1984; Hicks et al., 1989a; Campos et al., 1991; Zettl et al., 1991, 1992). Using these ligands a plasma membrane-associated 23 kDa protein was identified which is thought to be part of the auxin efflux carrier system (Feldwisch et al., 1992; Zettl et al., 1992). With the same ligand (i.e., azido-IAA), a 60 kDa ß-glucosidase was identified and purified to homogeneity; this enzyme is thought to play a role in the metabolism of phytohormone conjugates (Campos et al., 1992; Brzobohaty and Palme, unpublished). There is also a 31 kDa endo-1,3-ß-glucanase, however, its role in auxin physiology is at present largely unknown (MacDonald et al., 1991). In addition, a 65 kDa protein was identified using antibodies "raised" against auxin (Prasad and Jones, 1991). Other proteins which were identified by Hicks et al. (1989a,b) await further characterization.

In this review we will focus on "site I-specific" auxin binding proteins and recent progress on ER-located auxin binding proteins (ERabps). In particular, we will examine their structural, molecular and physiological features.

THE "SITE I" CLASS OF AUXIN BINDING PROTEINS

The "site I" class of auxin binding proteins was first characterized in detail in maize (*Zea mays*) coleoptiles (Hertel et al. 1972; Dohrmann et al., 1978). Corresponding auxin binding sites were later identified in many other mono- and dicotyledonous plant species [for a review see Venis, (1985)]. This site I-specific auxin binding moiety was found to be of low abundance in most plants except barnyard grass and etiolated maize coleoptiles. The relative abundance of this auxin binding moiety in maize coleoptiles allowed solubilization and subsequent purification of the corresponding protein (Venis, 1977; Shimomura et al., 1986; Venis, 1987; Palme et al., 1990). Equilibrium auxin binding assays as well as auxin-specific photoaffinity labeling techniques demonstrated that this protein binds various auxins including naphthaleneacetic acid (NAA) and IAA (Palme et al., 1990; Campos et al., 1991). The protein that was isolated by chromatography on NAA bound to aminohexyl-Sepharose has an apparent molecular mass of 20 kDa. Equilibrium dialysis showed that the protein binds NAA with a K_D of 2.4 x 10^{-7} mol. Scatchard analysis revealed one auxin binding site per protein for this phytohormone (Shimomura et al., 1986; Palme et al., 1990). Biochemical experiments established that this protein is a luminal component of the ER (Hesse et al., 1989).

STRUCTURE OF ERabps

Several isoforms of this protein were identified by Western-blotting analysis. Three of these proteins were purified and partial amino acid sequences determined (Hesse et al., 1989). The primary structures of several of these auxin binding proteins, termed ERabp, for ER-localized auxin binding protein, were deduced from cDNAs isolated from maize and *Arabidopsis* (Hesse et al., 1989; Inohara et al., 1989; Tillmann et al., 1989; Palme et al., 1992). It was found that all predicted amino acid sequences of the auxin binding proteins share common structural features: 1) a N-terminal signal peptide responsible for uptake into the lumen of the ER (Campos et al., unpublished), 2) a C-terminal-located ER retention signal composed of the amino acids Lys-Asp-Glu-Leu [KDEL] (Pelham, 1990),

(3) an Asn-Xxx-Thr glycosylation signal, 4) three cysteine residues at conserved positions, and 5) a sequence composed of residues His-Arg-His-Ser-Cys-Glu which is thought to be involved in auxin binding or transduction (Venis et al., 1992). The proteins are processed when entering the lumen of the ER and the mature proteins have a size of about 163 to 164 amino acid residues. A comparison of the primary amino acid sequences of several ERabps is shown in Fig. 1. The sequences illustrate that despite the high degree of variability found in the DNA sequences encoding these proteins, evolutionary pressure obviously resulted in the conservation of various blocks of amino acid residues within the mature proteins.

```
                                                        |
Zm-ERabp1    MAPDLSELAAAAAARGAYLAGVGVA---VLL-AASFLPVAES/-SCV-RDNSLVRDI    50
Zm-ERabp4    .VRRRPATG..PRPHL.-AV.R.LLLAS..AA...S....../-..P-.........    53
Zm-ERabp5    .VRRRPATG..QRPQL.-AV.R.LLLAS..AA...S....../-..P-.........    53
At-ERabp1    ----MIV.SVGS.SSSPIVVVFS..--L-..FYF.-ETSLG/-AP.PINGLPI..N.    47

                              |                          AUXIN
Zm-ERabp1    SQMPQSSYGIEGLSHITVAGALNHGMKEVEVWLQTISPGQRTPIHRHSCEEVFTVL    106
Zm-ERabp4    .R.Q.RN..R..F.....T...A..T.........FG............I..      109
Zm-ERabp5    .R.Q.RN..R..F.....T...A..T.........FG............I..      109
At-ERabp1    .D...DN..RP....M....SVL.......I....FA..SE...........T..    103

                            *               |
Zm-ERabp1    KGKGTLLMGSSSLKYPGQPQEIPFFONTTFSIPVNDPHQVWNSDEHEDLQVLVIIS    162
Zm-ERabp4    .......L...............V.V.....................N.........    165
Zm-ERabp5    .......L...............V.V.....................N.........    165
At-ERabp1    ..S...YLAETHGNF..K.I.F.I.A.S.IH..I..A...K.TG-...........    158

             |
Zm-ERabp1    RPPAKIFLYDDWSMPHTAAVLKFPFVWDEDCFEAA-KDEL       [163]    201
Zm-ERabp4    ...V...I..........K....Y......LP.P-....       [163]    204
Zm-ERabp5    ...V...I..........K....Y......LP.P-....       [163]    204
At-ERabp1    ...I...I.E..F......R....YY...Q.IQESQ....       [165]    198
```

Figure 1. Amino acid alignments of *Zea mays* and *Arabidopsis* ERabp protein sequences.

THE *ERabp* GENE FAMILY

ERabp genes have been isolated from maize and *Arabidopsis*. DNA gel blot analysis suggests the presence of a small gene family in maize, whereas in *Arabidopsis* probably only one gene exists (Hesse et al., 1989; Palme et al., 1992). The *Arabidopsis* gene AtERabp1 was RFLP-mapped to chromosome 4 on a RFLP containing the *hy4* and the *ga1* locus flanking the *AtERabp1* gene within a distance of approximately 100 kb (Palme et al., 1992). Recent studies indicate that, in *Arabidopsis*, at least one more pseudo gene related to *AtERabp1* exists which contains the C-terminal part of the coding sequence but does not contain the 5' part of the gene. From analysis of the regions flanking this pseudo gene it seems likely that this genomic region arose by duplication. However, from the large number of stop codons found in this sequence it seems unlikely that the ORF found in this region encodes a functional protein (Marques and Palme, unpublished).

In maize, five ERabp genes have been identified, three of which (i.e., the *ZmERabp1*, *ZmERabp4*, and *ZmERabp5* gene) have been characterized in detail (Schwob et al., 1993). The restriction map of these genes is shown in Fig. 2. *ZmERabp3* which is very similar to *ZmERabp1* has also been reported by Lazarus et al. (1991) and by Yu and Lazarus (1991), whereas the *ZmERabp2* gene has not yet been cloned although a partial amino acid sequence for the protein encoded by this gene has been reported (Hesse et al., 1989). All maize *ERabp* genes sequenced up to now encode proteins related in size and primary sequence, particularly in those domains thought to be important for its function. These regions are the presumed auxin binding domain, the glycosylation site, three cysteines that probably play a role in oligomerization of the ERabps and the C-terminal KDEL motif. Differences in the 5'- and 3'-flanking regions as well as in the regions encoding the signal sequences suggest that these regions may play a role in control of tissue-specific expression of their genes, whereas protein sequence diversity in the N-terminal signal sequences could be responsible for differential cellular targeting.

Intron size is often used to analyze relationships between members of a gene family. It has been found that intron 1 is conserved between *ZmERabp4* and *ZmERabp5*, but is much larger for *ZmERabp1* or *ZmERabp3*. Intron 2 is conserved between *ZmERabp4* and *ZmERabp5*, and at different size between *ZmERabp1* and *ZmERabp3*, whereas the size of intron 3 is conserved for all three genes. Only the *ZmERabp3* gene reported by Lazarus et al. (1991) seems to have an intron of much larger size (i.e., intron I, 5.2 kb in size).

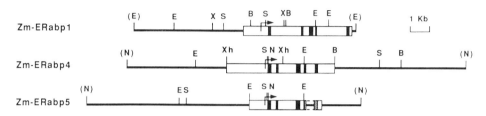

Figure 2. Restriction map of three maize ERabp genes. The exon and intron positions are indicated by black and white boxes, respectively. The arrow points to the transcription start site. Abbreviations: B, *Bam* HI; E, *Eco* RI; N, *Not* I; S, *Sal* I; X, *Xba* I; Xh, *Xho* I.

ERabp GENE EXPRESSION

Maize *ERabp* genes have been found to be expressed in a wide variety of tissues including callus tissue cultures, coleoptiles, leaves, ears and tassels. Expression levels of *ERabp* genes are under developmental control. It has been shown that although the *ZmERabp1* and *ZmERabp4* genes are expressed to apparently similar levels in maize coleoptiles, their expression levels differ widely in ears and tassels (Hesse et al., 1989). *ZmERabp5* seems to be expressed at a much lower level. Similar relative expression levels were reported for the *ZmERabp4* and *ZmERabp5* genes in protoplast transient expression experiments using the full length promoters fused to the *chloramphenicol acetyl transferase* gene as a promoter, whereas the *ZmERabp1* promoter consistently showed higher expression (Schwob et al., 1993).

To localize the expression more precisely, *in situ* hybridization experiments have been performed. In three-day-old etiolated maize coleoptiles, expression of *ZmERabp1* and *ZmERabp4* genes were found in both the coleoptile and the primary leaf. Further detailed analysis of *in situ* hybridization patterns by reflection confocal laser scanning microscopy revealed 1.7-fold higher expression for the *ZmERabp1* gene in the coleoptile as compared to the primary leaf (Hesse et al., unpublished).

ERabp PHYSIOLOGY

Since 1977, it has been argued that ERabp1 functions as an auxin receptor (Ray et al., 1977). Early evidence in favor of this protein being an auxin receptor was at best circumstantial and, hence, its role as a receptor has been questioned. However, recent electrophysiological results suggest a more direct involvement of this protein in control of plasma membrane polarization. It was shown that IAA causes two distinct electrical responses. An immediate depolarization was observed after external application of auxins to maize coleoptiles, probably caused by an H^+/IAA-symport (Felle et al., 1991). After depolarization of the plasma membrane, a hyperpolarization occurred resulting from proton extrusion, probably due to activation of the plasma membrane H$^+$ATPase. In many experiments it was observed that this latter response was delayed when auxins were added (Felle et al., 1986; Peters and Felle, 1991). Using classical electrophysiology, it was demonstrated that antibodies raised against *ZmERabp1* prevented an auxin-induced variation of transmembrane potential difference of tobacco mesophyll protoplasts (Barbier-Brygoo et al., 1989, 1991). These observations were questioned because mesophyll protoplasts can be electrically "leaky" and thus make the interpretation of voltage recordings difficult. Hence it was necessary to apply the patch-clamp technique (Hedrich and Schroeder, 1989) to determine the physical basis for the observed potential differences. Using this technique it could indeed be demonstrated that natural auxin and synthetic analogues enhance a H$^+$ATPase driven membrane current. Furthermore, antibodies directed against the *ZmERabp1* protein prevented this auxin-induced current. Most interestingly, antibodies raised against a peptide part of the presumed auxin binding site (Venis et al., 1992), acting like an auxin antagonist, induced this current in the absence of auxin (Rück et al., 1993).

Because antibodies raised against *ZmERabp1* almost completely prevent the auxin-induced H$^+$ATPase current, the auxin receptor mediating this activation of the plasma membrane proton pump is apparently localized at the plasma membrane. Evidence is now accumulating suggesting a relocalization of *ZmERabp1* from the ER to the plasma membrane (Vente and Palme, unpublished). Hence this protein could be directly involved in the primary control of the membrane responses controlling the current. At present it is not clear whether *ZmERabp1* directly interacts with the H$^+$ATPase or whether, and similar to other cellular signal perception systems described in mammals or yeast, it could interact with a plasma membrane protein that mediates these effects. Recent results using transgenic tobacco plants suggest that *ZmERabp1*, together with another gene product identified by gene tagging, can influence root elongation (our unpublished data).

While most of the *ZmERabp1* is located within the ER, it is unlikely that this fraction of *ZmERabp1* plays a significant role in the electrophysiological responses measured by various electrophysiological techniques. Instead, a trafficking route may exist that allows translocation of ERabps through the Golgi apparatus to the plasma membrane. It is well known that some ER proteins containing C-terminal KDEL sequences can be secreted when cellular calcium levels have been perturbed and so calcium or other cellular components could perturb *ZmERabp1* retention by the cellular KDEL retention system (Booth and Koch, 1989). Moreover, the observation of Napier and Venis (1990) that binding of a monoclonal antibody to the C-terminus of the ERabp1 was blocked by the addition of auxin raise the possibility that an auxin-induced conformational shift could affect the C-terminal interaction of *ZmERabp1* with the KDEL retention system. Loss of ER retention and thus passage of ERabp to the plasma membrane through the secretory pathway would be the result.

PERSPECTIVE

For a long time the site I-specific auxin binding site was the only one amenable to biochemical and physiological analysis. This site has been analyzed by molecular and electrophysiological techniques and these studies provided surprising evidence that this protein must be present at the cell surface where it apparently plays a role in controlling ion fluxes. We expect that analysis of the genes encoding these proteins will contribute to the understanding of some of the aspects of auxin perception, in particular how this hormone is linked to the cellular signal transduction machinery. This, in turn may provide new insight into the control of auxin-induced developmental phenomena such as adventitious root formation.

ACKNOWLEDGEMENT

This work was supported by the Human Science Frontier Programme Organization and the BRIDGE programme of the European Economic Community.

REFERENCES

Barbier-Brygoo, H., Ephritikhine, G., Klämbt, D., Ghislain, M., and Guern, J., 1989, Functional evidence for an auxin receptor at the plasmalemma of tobacco mesophyll protoplasts, *Proc. Natl. Acad. Sci. USA* 86:891.

Barbier-Brygoo, H., Ephritikhine, G., Klämbt, D., Maurel, C., Palme, K., Schell, J., and Guern, J., 1991, Perception of the auxin signal at the plasma membrane of tobacco mesophyll protoplasts. *Plant J.* 1:83.

Booth, C., and Koch, G.L.E., 1989, Perturbation of cellular calcium induces secretion of luminal ER proteins, *Cell* 59:729.

Campos, N., Feldwisch, J., Zettl, R., Boland, W., Schell, J., and Palme, K., 1991, Identification of auxin binding proteins using an improved assay for photoaffinity labeling with 5-N$_3$-[7-^{3}H]-indole-3-acetic acid, *Technique* 3:69.

Campos, N., Bako, L., Feldwisch, J., Schell, J., and Palme, K., 1992, A protein from maize labeled with azido-IAA has novel ß-glucosidase activity, *Plant J.* 2:675.

Davies, P.J., 1987, "Plant Hormones and their Role in Plant Growth and Development," Martinus Nijhoff Pubs., Dordrecht.

Dohrmann, U., Hertel, R., and Kowalik, W., 1978, Properties of auxin binding sites in different subcellular fractions from maize coleoptiles, *Planta* 140:97.

Feldwisch, J., Zettl, R., Hesse, F., Schell, J., and Palme, K., 1992, An auxin binding protein is localised to the plasma membrane of maize coleoptile cells: Identification by photoaffinity labeling and purification of a 23 kDa polypeptide, *Proc. Natl. Acad. Sci. USA* 89:475.

Felle, H., Brummer, B., Bertl, A., and Parish, R.W., 1986, Indole-3-acetic acid and fusicoccin cause cytosolic acidification of corn coleoptile cells, *Proc. Natl. Acad. Sci. USA.* 83:8992.

Felle, H., Peters, W., and Palme, K., 1991, The electrical response of maize to auxins, *Biochim. Biophys. Acta* 1064:199.

Hedrich, R., and Schroeder J.I., 1989, The physiology of ion channels and electrogenic pumps in higher plants, *Annu. Rev. Plant Physiol.* 40:539.

Hertel, R., Thomson, K.S., and Russo, V.E.A., 1972, In vitro auxin binding to particulate cell fractions from corn coleoptiles, *Planta* 107:325.

Hesse, T., Feldwisch, J., Balshüsemann, D., Bauw, G., Puype, M., Vandekerckhove, J., Löbler, M., Klämbt, D.,Schell, J., and Palme, K., 1989, Molecular cloning and structural analysis of a gene from *Zea mays* (L.) coding for a putative receptor for the plant hormone auxin, *EMBO J.* 8:2453.

Hicks, G.R., Rayle, D.L., Jones, A.M., and Lomax, T.L., 1989a, Specific photoaffinity labeling of two plasma membrane polypeptides with an azido auxin, *Proc. Natl. Acad. Sci. USA* 86:4948.

Hicks, G.R., Rayle, D.L., and Lomax, T.L., 1989b, The *diageotropica* mutant of tomato lacks high specific activity auxin binding sites, *Science* 245:52.

Inohara, N., Shimomura, S., Fukui, T., and Futai, M., 1989, Auxin-binding protein located in the endoplasmic reticulum of maize shoots: Molecular cloning and complete structure, *Proc. Natl. Acad. Sci. USA* 86:3564-3568.

Jones, A.M., Melhado, L.L., Ho, T.-H., and Leonhard, N.J., 1984, Azido auxins, Quantitative binding data in maize, *Plant Physiol.* 74:295.

Jones, A. M., and Prasad, P. V., 1992, Auxin binding proteins and their possible roles in auxin-mediated plant cell growth, *BioEssays* 14:43.

Lazarus, C.M., Napier, R.M., Yu, L.-X., Lynas, C., and Venis, M.A., 1991, Auxin binding protein antibodies and genes, *in* "Molecular Biology of Plant Development," G.I. Jenkins, and W. Schuch., eds., Company of Biologists Ltd., Cambridge.

Macdonald, H., Jones, A.M., and King, P., 1991, Photoaffinity labeling of soluble auxin-binding proteins, *J. Biol. Chem.* 266:7393.

Melhado, L.L., Jones, A.M., Leonard, N.J., and Vanderhoef, L., 1981, Azido auxins: synthesis and biological activity of fluorescent photoaffinity labeling agents, *Plant Physiol.* 68:469.

Napier, R., and Venis, M. (1990) Monoclonal antibodies detect an auxin-induced conformational change in the maize auxin-binding protein, *Planta* 182:313-318.

Palme, K., Feldwisch, J., Hesse, T., Bauw, G., Puype, M., Vandekerckhove, J., and Schell, J., 1990, Auxin binding proteins from maize coleoptiles: Purification and molecular properties, *in* "Hormone

Perception and Signal Transduction in Animals and Plants," vol. XLIV, J.A. Roberts, C. Kirk, and M. Venis, eds., The Company of Biologists Ltd., Cambridge.

Palme, K., Hesse, T., Moore, I., Campos, N., Feldwisch, J., Garbers, C., Hesse, F., and Schell, J., 1991, Hormonal modulation of plant growth: The role of auxin perception, *Mechanisms of Development* 33:97.

Palme, K., Hesse, T., Campos, N., Garbers, C., Yanofsky, M.F., and Schell, J., 1992, Molecular analysis of an auxin binding protein gene located on chromosome 4 of arabidopsis, *Plant Cell* 4:193.

Pelham, H.R.B., 1990, The retention signal for the soluble proteins of the endoplasmic reticulum, *Trends Biochem. Sci.* 15:483.

Peters, W.S., and Felle, H., 1991, Control of apoplast pH in corn coleoptile segments, II. The effect of various auxins and auxin analogues, *J. Plant Physiol.* 137:691.

Prasad, P.V., and Jones, A.M., 1992, Putative receptor for the plant growth hormone auxin identified and characterized by anti-idiotypic antibodies, *Proc. Natl. Acad. Sci. USA* 88:5479.

Ray, P.M., Dohrmann, U., and Hertel, R., 1977, Specificity of auxin-binding sites on maize coleoptile membranes as possible receptor sites of auxin action, *Plant Physiol.* 60:585.

Rück, A., Palme, K., Venis, M.A., Napier, R., and Felle, H.H., 1993, Patch-clamp analysis establishes a role for an auxin binding protein in the auxin stimulation of plasma current in *Zea mays* protoplasts, Plant J. (in press)

Shimomura, S., Sotobayashi, T., Futai, M., and Fukui, T., 1986, Purification and properties of an auxin-binding protein from maize shoot membranes, *J. Biochem.* 99:1513.

Schwob, E., Choi, S.-Y., Simmons, C., Migliaccio, F., Ilag, L., Hesse, T., Palme, K., and Söll, D., 1993, Molecular analysis of three maize 22 kDa auxin binding protein genes - transient promoter expression and regulatory regions, *Plant J.* (in press)

Tillmann, U., Viola, G., Kayser, B., Siemeister, G., Hesse, H., Palme, K., Löbler, M., and Klämbt, D., 1989, cDNA clones of the auxin-binding protein from corn coleoptiles (*Zea mays* L.): Isolation and characterization by immunological methods, *EMBO J.* 8:2463.

Venis, M.A., 1977, Solubilisation and partial purification of auxin-binding sites of corn membranes, *Nature* 266:268.

Venis, M., 1985, "Hormone Binding Sites in Plants," Longman, New York.

Venis, M. , 1987, Can auxin receptors be purified by affinity chromatography? *in* "Plant Hormone Receptors," D. Klämbt, ed.,. NATO ASI Series H, vol. 10., Springer-Verlag, Berlin.

Venis, M.A., Napier, R.M., Barbier-Brygoo, H., Maurel, C., Perrot-Rechenmann, C., and Guern, J., 1992, Antibodies to a peptide from the maize auxin-binding protein have auxin agonist activity, *Proc. Natl. Acad. Sci. USA* 89:7208.

Yu, L.-X., and Lazarus, C.M., 1991, Structure and sequence of an auxin-binding protein gene from maize (*Zea mays* L.), *Plant. Mol. Biol.* 16:925.

Zettl, R., Campos, N., Boland, W., Schell, J., and Palme, K., 1991, 5'-azido-[3,6-^{3}H$_2$]-naphthylphtalamic acid, a photoactivatable probe for auxin efflux carrier proteins, *Technique* 3:151.

Zettl, R., Feldwisch, J., Boland, W., Schell, J., and Palme, K., 1992, Azido-[3,6-3H2]-N-1-naphthylphtalamic acid, a novel photo-activatable probe for auxin efflux carrier proteins from higher plants: Identification of a 23 kDa protein from maize coleoptile plasma membranes, *Proc. Natl. Acad. Sci.USA* 89:480.

USE OF TRANSFORMED ROOTS FOR

ROOT DEVELOPMENT AND METABOLISM STUDIES AND

PROGRESS IN CHARACTERIZING ROOT-SPECIFIC GENE EXPRESSION

John D. Hamill[1] and Stephen F. Chandler[2]

[1]Department of Genetics and Developmental Biology
Monash University
Clayton, Melbourne, Victoria 3168, Australia

[2]Calgene Pacific Pty Ltd
16 Gipps Street
Collingwood, Melbourne, Victoria 3066, Australia

INTRODUCTION

Following reports in the early 1980s that adventitious roots that were transformed with *Agrobacterium rhizogenes* often grow rapidly in hormone-free medium *in vitro*, there has been an upsurge in the use of transformed root cultures as a system to study root metabolism and associated root biology. As DNA transfer is involved in the interaction between *A. rhizogenes* and higher plants, there has also been a great deal of interest in using this bacterium as a vector to genetically manipulate plants. More than 460 plant species from over 100 families have been transformed by *A. rhizogenes* (Porter, 1991) and in many cases axenic transformed roots have been reported to grow rapidly *in vitro* (Hamill and Rhodes, 1992). In general, such transformed root cultures are quite stable at the gross chromosomal level, so long as the integrity of the meristem is not disturbed, e.g., by the addition of phytohormones to the medium (Aird et al., 1988a,b). Transformed root cultures have been used for secondary metabolite studies and also, to a lesser extent, to examine the interactions between soil organisms and roots of higher plants (Hamill and Rhodes, 1992). The recovery and growth of plants in soil demonstrates that roots containing Ri T-DNA can provide nutrient and water uptake requirements, though few studies have critically evaluated the differences between normal and transformed roots with respect to these functions.

In this chapter we will consider some of the opportunities offered by transformation of higher plants with *A. rhizogenes* with respect to root development studies and also biotechnology. We also will review recent literature concerned with root-specific gene expression in plants. Such information may provide opportunities to utilize genes such as those contained in the T-DNA of *A. rhizogenes* to specifically alter the growth patterns of

Biology of Adventitious Root Formation, Edited by
T.D. Davis and B.E. Haissig, Plenum Press, New York, 1994

roots without expression of the genes in the upper parts of the plant which may perturb normal shoot development. Details on the effects of incorporation of specific genes from *A. rhizogenes* into the genomes of higher plants, together with consideration of the biochemical mechanisms involved in gene action *in vivo*, are the subject of another chapter in this volume (see the chapter by Tepfer and coworkers in this volume).

GROWTH PATTERNS OF TRANSFORMED ROOTS

Numerous reports in the literature allude to the observation that transformed roots grow very rapidly *in vitro* or that roots of transformed plants containing Ri T-DNA are very prolific compared to nontransformed controls (e.g., Hamill and Rhodes, 1992; Hauth and Beiderbeck, 1992). However, few reports substantiate these observations with experimental details regarding rates and patterns of growth. Using *Arabidopsis thaliana* as an experimental model we have undertaken a more detailed comparison between roots of regenerants containing or lacking one copy of the Ri T-DNA from an agropine strain of *A. rhizogenes*.

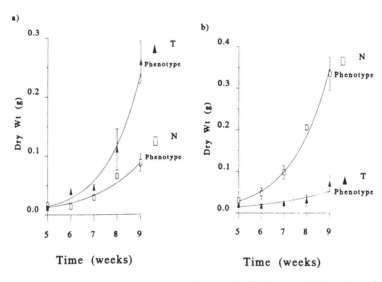

Figure 1. Dry weight analysis of root (a) and shoot (b) growth of N (□) and T (▲) plants of *Arabidopsis thaliana* over a nine week period following germination. Plants were grown *in vitro* in 50 mL of MS medium containing 30 g L^{-1} sucrose without hormones. Medium was solidified with 1.8 g L^{-1} phytagel (Sigma). Points represent means of duplicates with ± S.E. bars.

Before we consider these results it would be beneficial to consider briefly the role of auxins and cytokinins in controlling root growth in normal, nontransformed plants. Growth of roots occurs by lateral branch differentiation and growth of the main root and lateral root systems. In dichotomous root systems, initial laterals differentiate from the main root in acropetal sequence and often themselves produce laterals (Fitter, 1991). Lateral root primordia initiation is stimulated by auxin (mostly derived from the shoot *in vivo*) resulting in cell division which originates in the pericycle. Lateral root emergence consists largely of cell elongation and is inhibited by the high levels of auxin that stimulate lateral root primordia initiation (Wightman et al., 1980; MacIsaac et al., 1989). Root apices are a major source of cytokinin which also plays an important role in lateral root formation and growth.

A cytokinin gradient is established along the root and a high cytokinin concentration inhibits both the initiation of lateral root primordia and the emergence and growth of lateral roots. However, at a lower cytokinin concentration lateral root primordia initiation and growth are promoted, providing the appropriate concentration of auxin is present (Wightman et al., 1980).

We have considered the question of whether the more prolific nature of transformed roots compared to nontransformed controls is due to faster rates of root-tip elongation or increased rates of lateral root initiation and growth. To begin to answer this question, we compared the growth pattern of seedlings of *A. thaliana* Ri T-DNA segregants which contained one copy of Ri T-DNA (T plants) compared to those which did not contain T-DNA (N plants). Transformation of *A. thaliana* ecotype Landsberg *erecta* was carried out using the protocol of Valveekens et al. (1988) using the agropine strain *A. rhizogenes* LBA 9402. Transformed roots were selected on the basis of their ability to grow rapidly in hormone-free MS medium (Murashige and Skoog, 1962) followed by plant regeneration and seed set *in vitro* using the Valveekens et al. (1988) protocol (Hamill, unpublished).

Dry weight analysis of roots of T plants compared to N plants revealed a three-fold increase in dry weight of the T root system after nine weeks growth (Fig. 1a). Interestingly, the shoot dry weight of T plants was significantly lower than that of N plants (Fig. 1b) suggesting a significant alteration in the source:sink relationship between the shoot and root due to incorporation of Ri T-DNA. As has been reported for other plants containing Ri T-DNA (Tepfer, 1984; Spena et al., 1987; Handa, 1992), T plants of *A. thaliana* had an altered shoot morphology compared to N plants.

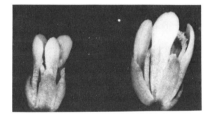

Figure 2. Phenotype of N (left) and T (right) plants of *Arabidopsis thaliana* after five weeks growth *in vitro* in MS medium without phytohormones. Note the more prolific root system and altered shoot phenotype in the T plant. Diameter of petri dish = 50 mm. **Inset:** Phenotype of a normal-sized flower from an N plant (left) and the smaller flower from a T plant (right) of *A. thaliana*.

Figure 2 shows the phenotype of N and T plants five weeks after germination and growth *in vitro*. The stem of T plants was dwarfed, with reduced internodal distances and slightly wrinkled, epinastic leaves. The inflorescence was compact and flowers were slightly smaller than those from N plants. Flowers of T plants had some self-fertility. The root system was much more prolific than that of N plants.

Isolated roots of both N and T plants grew in hormone-free MS liquid medium, with the T cultures characteristically being more prolific. However, whereas detached T roots grew on the surface of hormone-free medium, detached roots of N seedlings showed very poor growth on the surface of solidified MS medium (1.8 g L^{-1} phytagel, Sigma). To compare the root branching pattern of both N and T roots in hormone-free medium it was decided therefore to use intact seedlings placed singly after germination into the center of a 140 mm-wide petri dish containing 50 mL of MS medium solidified with 1.8 g L^{-1} phytagel. As the medium was only four mm deep, it was possible to measure root growth and branching patterns accurately by marking the bottom of the plates at regular time intervals.

Under the conditions noted above, the main root of both N and T seedlings produced primary (1°) roots, which produced secondary roots (2°), which in turn produced tertiary roots (3°) and so on. The rate of growth of the main root was similar for N and T plants, suggesting that the increased overall dry weight exhibited by the T root systems when cultured *in vitro* was not due to an inherent difference in growth rate *per se* (data not shown). However, a detailed study of the number of laterals formed showed that there was a difference between roots of N and T plants when seedlings were monitored over a 27 day period (after this time roots became so interwoven it was impossible to distinguish their origin with accuracy). During the first two weeks of growth, N and T plants produced lateral roots at the same rate; but, from day 15, T plants began producing more 2° laterals, compared to N plants. This phenomenon continued up to day 27, by which time T plants had twice as many 2° laterals, five times as many 3° laterals and nine times as many 4° laterals, compared to root systems of N plants. The total number of lateral root tips per plant averaged 262 for N plants and 678 for T plants at day 27 (Fig. 3). Thus, in intact transformed plants at least, incorporation of Ri T-DNA genes results in a stimulation of lateral root differentiation and growth. In axenic root culture this difference is even more pronounced as auxin cannot be obtained from the shoot system and must be entirely synthesized in the roots. In comparison to axenic roots of N plants, the auxin supply in roots containing Ri T-DNA is increased due to the action of T-DNA genes (see below).

In the light of the auxin-cytokinin balance model for root growth (noted above) it appears that the effect of Ri T-DNA is to maximize the number of branched roots by finely adjusting the auxin:cytokinin hormone balance within the root system to stimulate lateral root differentiation and growth. The *rolB* gene causes an increase in auxin sensitivity and a release of auxin from conjugates in the plant cells in which it is expressed (Estruch et al., 1991a; Maurel et al., 1991). The promoter of the *rolB* gene is regulated by auxin (Maurel et al., 1990) and the gene is expressed early in the meristem initial cells when a high auxin concentration will be most effective in stimulating lateral root primordia differentiation (Altamura et al., 1991). After a meristem initial cell is formed, the expression of *rolB* diminishes and thus the level of auxin probably is also reduced. At the same time, the expression of the *rolC* gene from the Ri T-DNA may cause an increase in the localized cytokinin concentration to a level where the auxin-cytokinin balance is optimal for the primordia to develop into an actively growing lateral root. [The *rolC* gene is induced weakly by auxin (Maurel et al., 1990) and encodes an enzyme which causes release of active cytokinin from inactive conjugates (Estruch et al., 1991b)]. Cytokinin produced in the newly growing root tips may inhibit further lateral formation until the gradient within the growing root has adjusted to a level which allows a new lateral root to differentiate again. The biochemical role of other T-DNA genes is not entirely clear but the involvement of genes such as *rolA* (which promotes rooting when over-expressed in tobacco plants) and ORF13 and 14 [which together with the *rolABC* genes are needed for rapid growth of carrot roots *in vitro* (Capone et al., 1989)] is likely to be important in ensuring that the hormone balance of the cells of the plant species that the bacterium has infected are optimal for lateral root emergence and root growth. In nature, transformation with Ri T-DNA results in a large surface area of opine-secreting roots to provide nutrition for agrobacteria in the rhizosphere.

In summary, the introduction of genes into plants which alter the auxin-cytokinin balance to change the topology of root systems is thus a real possibility and may provide benefits to agriculture by improving the nutrient and water uptake capacity due to a more extensive root system. The infection of the basal regions of stem cuttings by *A. rhizogenes* or insertion of isolated genes from *A. rhizogenes* are attractive propositions for the manipulation of the root system of certain vegetatively propagated crops [e.g., *Corylus avellana* (hazel nut) (Bassil et al., 1991), *Malus pumila* (apple) (Lambert and Tepfer, 1991)]; and we are considering this approach to stimulate root initiation and improve the rooting potential of micropropagated eucalypts of commercial importance (see below). In addition it may be beneficial to incorporate genes from *A. rhizogenes* into the genome of annual crop plants and in such experiments it would be beneficial to ensure that these genes were predominantly or exclusively expressed in the root system. Consideration of some of the promoters available for this purpose, and strategies to obtain additional promoters, will be presented below.

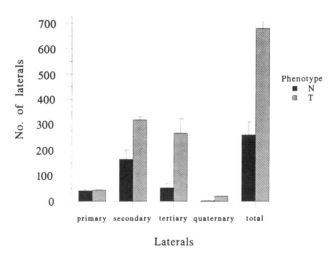

Figure 3. Number and position (1°, 2°, 3°, 4°) of root laterals in N and T plants of *Arabidopsis thaliana* after 27 days growth *in vitro* in MS medium without hormones. Data represent means with ± S.E. bars

MANIPULATION OF THE ROOTING POTENTIAL OF MICROPROPAGATED EUCALYPTS USING *AGROBACTERIUM RHIZOGENES*

The control of root initiation and development *in vitro* has particular importance for hard-to-root woody plants. For example, the ability to initiate root formation in vegetative, micropropagated material of commercially valuable trees would be of enormous value to the forestry industry, where breeding programs are in their infancy and wide genetic variation exists (see the chapter by Ritchie in this volume). We have recently started a program to use the *rol* genes from *A. rhizogenes* to initiate roots on stems of shoots from *in vitro* cultures of three temperate eucalypt species, *Eucalyptus globulus* (Tasmanian Blue Gum), *E. nitens* (Shining Gum) and *E. regnans* (Mountain Ash). These species dominate hardwood plantations in southern Australia, Chile, Spain, Portugal and other countries, yet over 80% of seedling clones cannot be induced to root *in vitro* or only root very poorly (Willyams et al., 1992). The consequent need to impose selection for rooting ability within a vegetative propagation program severely restricts possible genetic gain.

Infection of tissue culture-derived shoots with wild-type strains of *A. rhizogenes* has been shown to result in adventitious root initiation in the woody plants *Olea europeae* (olive) (Rugini, 1992) and *M. pumila* (apple) (Patena et al., 1988; Lambert and Tepfer, 1991). As is illustrated in Figure 4, wild-type *A. rhizogenes* strains can also be used to induce adventitious roots on shoots of micropropagated *E. globulus*. Application of auxin is not necessary for root formation by infected shoots, though it is a prerequisite for noninfected shoots. Shoots isolated from cultures of *E. nitens*, *E. grandis* and *E. dunnii* have also been shown to form adventitious roots after inoculation with wild-type *A. rhizogenes* (MacRae, 1991).

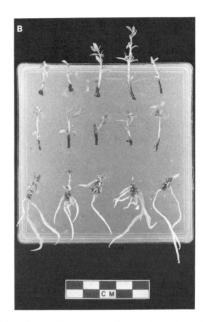

Figure 4. (A) Micropropagated plantlets of a clone of *Eucalyptus globulus* growing *in vitro* prior to infection with *Agrobacterium rhizogenes*. (B) Effect of inoculating micropropagated plantlets of a clone of *E. globulus* with: (i) Upper row—Nutrient broth not containing agrobacteria; (ii) Middle row—A two-day-old suspension of disarmed *A. tumefaciens* diluted to 10^8 bacteria mL^{-1} in nutrient broth; and (iii) Bottom row—A-two-day old suspension of *Agrobacterium rhizogenes* A4 diluted to 10^8 bacteria mL^{-1} in nutrient broth. Treatments (i) and (ii) resulted in blackening of the stem but no roots formed. Treatment (iii) resulted in prolific rooting from the stem of inoculated plantlets. Cultures were maintained in modified MS medium without hormones for five weeks prior to photographing.

Our preliminary experiments have shown (Fig. 4) that A4 is a particularly virulent *A. rhizogenes* strain for initiation of transformed roots in eucalypts. The virulence of A4 has been shown for a number of plants (Porter, 1991) and may relate to the fact that A4 is an agropine-type strain, which contains auxin biosynthesis genes as well as *rol* genes in the T-DNA. Additional virulence factors, encoded by the bacterial chromosome or Ri-plasmid, are also likely to be very important for infection and cultivar-specific virulence as has been reported for some plants infected with *A. rhizogenes* (Porter, 1991). It is critical that the genetic modification of rooting ability using *A. rhizogenes* genes is applicable to any eucalypt clone as hundreds of clones would typically be employed in a large-scale tree improvement program and any restriction imposed by strain infectivity would compromise potential genetic gain. Using the GUS reporter gene on a binary vector to measure infection rate, we have screened a number of "disarmed" *Agrobacterium* strains and have identified one *A. tumefaciens* strain which gives reproducibly high infection rates in a large number of micropropagated eucalypt clones (data not shown). Using binary vectors in this strain of *A. tumefaciens*, various combinations of *rol* and/or *aux* genes can, therefore, be transferred to eucalypt stem tissue and our hope is that it will be possible for root formation to occur after inoculation at least as effectively as with wild type *A. rhizogenes*. If necessary, this may be achieved in conjunction with exogenous application of auxin.

It is unlikely that the dramatic effects on shoot phenotype observed in transformed plants containing Ri T-DNA (e.g., Fig. 2) will be observed in chimeric eucalypt trees having a normal stem and a transformed root system. Nevertheless it will be important to assess these chimeric eucalypt clones with respect to the topology of their root systems, the root:shoot dry weight ratio of the chimeric plants and the ability of the root system to provide the shoot with the appropriate balance of water and nutrients, to enable wood of high quality to be produced when trees are grown in a plantation.

ROOT METABOLISM AND GENE EXPRESSION STUDIES

Transformed Roots and Secondary Metabolite Biosynthesis

As transformed roots possess the combined attributes of rapid growth *in vitro* and long-term genetic stability, they have been increasingly utilized for root metabolism studies. The number of reports involving secondary metabolism in transformed roots continues to grow as illustrated in Table 1.

Transformed root cultures have also been used for precursor feeding and biochemical inhibitor experiments to elucidate biosynthetic pathways (e.g., Walton et al., 1990; Robins et al., 1991), genetic manipulation experiments aimed at altering flux through biochemical pathways (e.g., Hamill et al., 1990; Berlin et al., 1991a), elicitation experiments aimed at increasing secondary metabolite production (Mukundan and Hjortso, 1990; Furze et al., 1991) and experiments investigating the growth of transformed roots in bioreactors with the long-term aim of producing expensive phytochemicals in an industrial setting (Rhodes et al., 1986; Wilson et al., 1990; Shimomura et al., 1991).

Many secondary metabolites are synthesized mainly or exclusively in the roots of the intact plant, so it is likely that at least some of the key genes encoding biosynthetic enzymes are root-specific in their expression profile. Relatively few secondary metabolite genes have been isolated to date but there is evidence that this assumption is at least partly true. For example, the enzyme 6β-hydroxylase (6β-H) converts hyoscyamine to 6-hydroxy hyoscyamine, the precursor of the important muscle relaxant scopolamine which is produced commercially from *Datura* and *Duboisia* species. The 6β-H enzyme has been purified and antibody detection followed by gene expression studies demonstrated that it is expressed specifically in roots (see also below). Similarly, the gene for strictosidine synthase, an important enzyme in the synthesis of indole alkaloids has been cloned and its expression profile suggested it is active mainly, though not exclusively, in the root of *Rauvolfia serpentina* (Bracher and Kutchan, 1992) (see also below). Transformed root cultures, therefore, represent an excellent source of enzymes and mRNA for the isolation of genes involved in plant secondary metabolism which may lead to the recovery of promoters capable of conferring root-specific or root-predominant expression profiles upon heterologous genes.

Table 1. Recent reports of secondary metabolites produced by transformed roots.

Secondary Metabolite	Genus	Reference
Alkamide alkaloids	*Echinacea*	Trypsteen et al., 1991
Anthraquinones	*Cassia*	Asamizu et al., 1988; Soo et al., 1988
	Rubia	Sato et al., 1991
β-carboline alkaloids	*Peganum*	Berlin et al., 1992
Betalain pigments	*Beta*	Hamill et al., 1986
Cardioactive glycosides	*Digitalis*	Saito et al., 1990
Diterpenoids	*Salvia*	Hu & Alferman, 1993
Flavonoids	*Raphanus*	Ahn et al., 1992.
	Lotus	Morris & Robbins, 1992
	Lupinus	Berlin et. al., 1991b
Indole alkaloids	*Catharanthus*	Parr et al., 1988; Davioud et al., 1989; Toivonen et al., 1989
	Cinchona	Hamill et al., 1989
	Amsonia	Sauerwein et al., 1991a
Lobetyolin	*Lobelia*	Ishimaru et al., 1991
Napthoquinones	*Lithospermum*	Shimomura et al., 1991
Piperdine alkaloids	*Lobelia*	Yonemitsu et al., 1990
	Hyoscyamus	Sauerwein et al., 1991b
Polyacetylenes	*Coreopsis, Bidens,*	Marchant, 1988
	Lobelia	Ishimaru et al., 1992
Polyines	*Chaenactis*	Constabel & Towers, 1988
Pyridine alkaloids	*Nicotiana*	Hamill et al., 1986; Parr & Hamill, 1987
Saponines	*Panax*	Yoshikawa & Furuya, 1987; Hwang et al., 1992
Sesquiterpenes	*Lippia*	Sauerwein et al., 1991c
	Datura	Furze et al., 1991
Steroids	*Ajuga*	Matsumoto & Tanaka, 1991
Steviol glucosides	*Stevia*	Yamazaki et al., 1991
Terpenoids	*Daucus*	Chamberlain et al., 1991
Thiophenes	*Tagetes*	Westcott, 1988; Croes et al., 1989
	Ambrosia, Carthamus, Rudbeckia	Flores et al., 1988
Tropane alkaloids	*Duboisia*	Deno et al., 1987; Mano et al., 1989
	Atropa	Kamada et al., 1986; Jung & Tepfer, 1987; Sharp & Doran, 1990
	Datura	Payne et al., 1987; Christen et al., 1989; Robins et al., 1990; Parr et al., 1991
	Hyoscyamus	Flores & Filner, 1985; Parr et al., 1991; Sauerwein & Shimomura, 1991
	Scopolia	Mano et al., 1986; Nabeshima et al., 1986; Parr et al., 1990
	Nicandra	Parr, 1992

Root-Specific or Root-Predominant Gene Expression

As has been noted, the isolation of genes and regulatory regions which lead to root-specific or root-predominant gene expression is highly desirable. A number of genes have been reported to be either root-specific or, more commonly, root-predominant in their expression profile and in several cases the promoter of these genes has been shown to confer a root-specific or -predominant expression profile on reporter genes such as β-glucuronidase (GUS) in transgenic plant tissues. With the increased understanding of promoter modularity (Dynan, 1989; Benfey et al., 1990), deletion analysis of promoters which do not lead to root-specific expression when fused to reporter genes has identified domains or modules in these promoters which are responsible for expression specifically in the root system. We can, for the sake of convenience, divide these sequences into two types depending upon their origin, as we will discuss below.

Promoters of Genes from Plant Pathogens. The so-called "constitutive" promoter of the CaMV35S gene has been dissected and shown to be composed of domains A and B which act in synergy to promote expression in a wide range of tissues of many plant species (Benfey et al., 1989, 1990). Domain A, extending to -90, led to expression of the GUS reporter gene predominantly in the root, though some plants with high expression in roots (due presumably to position effects) did exhibit some expression in the shoot tip (Benfey et al., 1989). Expression from domain A was mainly observed in the meristematic region and cortex of the root tip and also in the pericycle. Less frequently, expression was seen in the root cap and root hairs (Benfey et al., 1989). A *cis* element, termed activation sequence 1 [(as)-1], located between -85 and -64, was found to be primarily responsible for root-specific or -predominant expression. The element contains a tandem repeat TGACG motif and binds a nuclear "activation sequence" factor, (ASF)-1. Mutation of the TGACG motif produced a decrease in expression of the reporter gene in the root (Lam et al., 1989).

A cDNA clone for a factor that binds to (as)-1 has been isolated and shown to be expressed in roots at five to 10 times the level present in shoot tissue (Katagiri et al., 1989). A tetrameric (as)-1 sequence, fused to a minimal promoter (-72 deletion of CaMV35S with TATA box intact), was sufficient to confer expression of GUS in shoot tissue due to the increased target for binding of the (ASF)-1 protein (Lam et al., 1989). An element with a strong similarity to (as)-1 has been found in the promoter of several other genes from plant pathogens including the octopine synthase (OCS) and the mannopine synthase genes (mas 1' 2') from *A. tumefaciens* (Bouchez et al., 1989). The mas 1' 2' promoter showed predominant expression in the root system of young plants, particularly the root tip (Langridge et al., 1989; Saito et al., 1991). Interestingly, the mas 1' 2' promoter also showed significant expression in germinated pollen (Langridge et al., 1989). Fromm et al. (1989) have shown that the promoter from the OCS gene contains a palindromic sequence that binds the (ASF)-1 protein. When inserted into the promoter of the ribulose bisphosphate carboxylase gene, which normally does not express in roots, this sequence caused that promoter to be active in root tips. When fused to a minimal 35S TATA CaMV35S promoter (-46 to +8), ligated to the coding sequence of the GUS gene, the OCS element directed expression to the root tips and, as in the -90 deletion of CaMV35S promoter, those transgenics with strong expression in the root also showed expression in the shoot tip (Fromm et al., 1989).

The *rolD* (ORF15) gene of *A. rhizogenes* was reported to be expressed specifically in the root system of transgenic plants containing Ri T-DNA (Durand-Tardif et al., 1985) though the *rolD* promoter did enable the GUS gene to be expressed in shoots as well as roots (Leach and Aoyagi, 1991). The latter authors carried out preliminary deletion analysis of the *rolD* promoter and found one construct (pD-02), containing 373 bp of flanking DNA 5' to the *rolD* gene, in which GUS activity was greatly reduced in the leaves but remained high in the roots (ca. 20-fold greater, compared to leaves). By contrast, trans-genics containing 1222 bp of 5' flanking region fused to GUS showed a three-fold greater level of GUS in the leaf than in the root. Further deletion of the promoter to 132 bp dramatically reduced expression in the roots while increasing expression in leaf tissue slightly. It thus appears a root-specific or -predominant expression module lies between -132 and -373 of the 5' region of the *rolD* gene (Leach and Aoyagi, 1991).

Similar analysis has identified a root-specific or -predominant expression module within the 5' region of the promoter of the *rolB* (ORF11) gene of *A. rhizogenes*. Deletion of a 35 bp sequence lying between -306 and -341 of this promoter abolished expression of the GUS reporter gene in the meristematic zone of root tips of transgenic tobacco and carrot tissues (Capone et al., 1991). Interestingly, deletion of this sequence also prevented the promoter from being induced by auxin suggesting that at least one of the auxin responsive domains of the *rolB* promoter is responsible for expression of the *rolB* gene in the initial cells of the root meristem (Altamura et al., 1991; Capone et al., 1991).

Promoters from Endogenous Plant Genes. As previously stated, a number of plant genes have been identified as being root-specific or root-predominant in their expression profiles. For example, a haemoglobin cDNA isolated from *Trema tomentosa* (a non-N_2 fixing, non-nodulating tree species) was found to have a root-specific expression profile (Bogusz et al., 1988). A homolog of this gene was also isolated from the non-leguminous N_2 fixing, nodulating tree *Paranasponia andersonii* (Bogusz et al., 1988). Fusion of the promoter regions of these genes to the coding sequence of GUS resulted in expression in the meristematic region of roots of transgenic tobacco and the *Paranasponia* promoter also caused the GUS gene to be expressed in the differentiating vascular system. Though both promoters gave rise to GUS activity predominantly in the root system of transgenic tobacco, the root:leaf expression ratio was significantly higher, on average, for the promoter from the *Paranasponia* species compared to that of the *Trema* species. This was due largely to increased expression of the latter in leaf tissue (Bogusz et al., 1990).

A hydroxyproline-rich glycoprotein (HRGP) gene from tobacco was found to be expressed specifically in roots, though at low levels. Fusion of its promoter to the GUS coding sequence indicated that the promoter was active transiently in the pericycle and endodermis during the induction of lateral root differentiation (Keller and Lamb, 1989). The role of the protein encoded by this gene was suggested to be in strengthening the cell wall of the newly forming lateral root, enabling it to withstand the pressure involved in breaking through the cortical and epidermal cells of the main root. This assumption was strengthened by the observation that the GUS activity in emerging laterals was located almost exclusively in the tip of the emerging root (Keller and Lamb, 1989). The promoter of a gene coding for another cell wall protein, the glycine-rich (GRP) cell wall protein GRP 1.8, was found to be vascular tissue-specific in its expression profile and to contain a root-specific element located between -94 bp and -76 bp (Keller and Baumgartner, 1991). This element was found to be responsible for expression mainly in a band of tissue approximately corresponding to the cell expansion-vascular differentiation zone of root tips (Keller and Baumgartner, 1991). As in the (as)-1 sequence of the CaMV35S promoter (Benfey et al., 1989), this root module interacts synergistically with other "upstream" modules to ensure optimal expression of the gene in shoot vascular tissue.

As was noted previously, the enzyme 6β-H has been localized to the pericycle of young roots of *H. niger* (Hashimoto et al., 1991). Using a cDNA for 6β-H as a probe, analysis of mRNA from different tissues demonstrated that the gene was active only in roots of intact plants (Matsuda et al., 1991). A higher level of transcript was observed in cultured roots than in roots from intact plants, presumably due to the higher percentage of young, branching roots without secondary growth, which is characteristic of these cultures (Matsuda et al., 1991). A cDNA clone for strictosidine synthase was isolated and found to be most highly expressed in roots of *Rauvolfia serpentina*, though some expression was also observed in leaves and flowers (Bracher and Kutchan, 1992). A series of strictosidine synthase promoter deletions, fused to the GUS gene coding sequence, was constructed by those authors. Initially these constructs have been tested for expression in leaf protoplasts but it would not be too surprising if root-specific module(s) were found in this promoter.

An extensin cDNA from *Brassica napus* was found to be homologous to two transcripts from *B. napus* root tissue (1.45 and 1.26 kb) but the transcripts were not present in RNA from *B. napus* leaf or stem tissue (Shirsat et al., 1991). Fusion of 1 Kb of promoter sequence of the extensin gene to the GUS coding sequence, with subsequent introduction into *B. napus*, showed the promoter was active in the phloem of the vascular tissues of roots but not in root meristem, epidermal, cortex or pericycle cells (Shirsat et al. 1991).

A structural gene encoding nitrate reductase has been isolated from *Phaseolus vulgaris* and found to be nitrate-inducible and only expressed in root tissues (Hoff et al., 1991). As nitrate reductase enzyme is predominantly found in leaves in *P. vulgaris*, Hoff et al. (1991) suggested that another nitrate reductase gene which is expressed in leaves remains to be isolated from *P. vulgaris*.

There have been a number of studies which have identified genes and/or enhancer sequences of plant origin responsible for root-specific expression but where the biochemical function of the gene product is not known. Ligation of fragments of genomic *Arabidopsis thaliana* (200-600 bp) DNA to a minimal CaMV35S promoter (-72 to +8), which was fused to the GUS coding sequence, allowed detection of transgenic tobacco plants containing an enhancer module in the genomic fragments (Ott and Chua, 1990). Over 500 transgenic plants were examined for expression of GUS in roots; 11% exhibited GUS activity in some cell types. About 50% of plants exhibiting expression in the root also showed expression in the leaf or stem. Several enhancer sequences were identified which conferred expression specifically to root tissues—including two localized in the root meristem; two in the meristem and the mature epidermis; one in the meristem and root cap; and one in the root meristem, mature root epidermis and the root cap (Ott and Chua, 1990). Tagging of the genomic sequences of *Arabidopsis*, potato and tobacco, which enable expression of the GUS gene coding sequence by activation of a promoter-trap following transformation, has been reported recently (Lindsey et al., 1993). A number of regulatory sequences that confer root-specific or root-predominant expression have been identified using this technique (Lindsey et al., 1993). Clearly this approach of promoter-enhancer tagging represents a powerful strategy for identification and isolation of DNA sequences which control expression in a specific manner in a range of plant organs, including different cell types within the root. An alternative strategy is to use a differential screening approach to identify cDNAs which are expressed specifically or predominantly in a particular tissue or cell type. Conkling et al. (1990) used this strategy to identify and isolate four root-specific cDNA clones, and subsequently their corresponding genomic clones, from *N. tabacum*. Characterization of one of these genes (TobRB7) by *in situ* hybridization, followed by analysis of transgenics containing promoter-GUS fusions, revealed that expression was localized to root meristem and immature central cylinder regions. Promoter deletion analysis showed that sequences between -299 and -636 were essential for root-specific expression. A negative regulatory element was found between -636 and -813 (Yamamoto et al., 1991). Further analysis will reveal the precise sequences responsible for root-specific expression.

Several mRNAs accumulate rapidly after addition of 2,4-dichlorophenoxyacetic acid (2,4-D) to auxin-starved tobacco cells grown *in vitro*. The promoter of one cDNA clone of this class of genes, GNT1, resulted in expression of the GUS gene only in the meristematic and vascular differentiation zone of root tips of transgenic plants (van der Zaal et al., 1991). GUS activity was not detected in root cap cells. Treatment of these transgenic plant tissues with 2.2 μmol 2,4-D increased the intensity of the GUS activity as detected by histochemical staining, and activity was detected in the vascular system. No significant homology was apparent between the promoter of this gene and the auxin-inducible promoter of the *rolB* gene (van der Zaal et al., 1991).

Differential screening of a rice cDNA library enabled the recovery of two cDNA clones (COS 6 and COS 9) which showed predominant expression in the roots of young rice plants (de Pater and Schilperoort, 1992). Preliminary analysis of the promoter of COS 9 has commenced with the aim of recovering a strong and specific expression sequence to enable expression of foreign genes only in the roots of transgenic rice plants (de Pater and Schilperoort, 1992).

Recently, Dietrich et al. (1992) have analyzed the relative importance of 5' and 3' regions from a gene (AX92) which shows predominant expression in the root cortex of *B. napus* plants. They found that DNA sequences in the 3' region of the gene were sufficient to direct expression in the seedling roots and the embryonic axis in developing seed. As yet it is unclear whether the 3' regions affect gene activity at the transcriptional or post-transcriptional level (Dietrich et al., 1992).

In conclusion, therefore, the number of reports of root-specific cDNAs and also promoters and modules within 5' and 3' regulatory regions, which allow expression specifically or predominantly in the root system of higher plants, has increased significantly in the past two to three years. Sequences have been identified (and no doubt many more will be) which are responsible for expression in a wide range of cell types within the root (root cap, meristem, pericycle, vascular system, cortex and epidermis) and at least two classes of promoters have been identified which are inducible by externally applied agents such as auxin or nitrate. The ability to fuse modules from different promoters to form new promoters with hybrid expression patterns (Fromm et al., 1989; Comai et al., 1990) suggests that there is a real prospect of being able to direct expression of "foreign" genes at high levels to the appropriate cell type within the root system, at the appropriate time in development. Genes of interest include the *rol* genes noted previously, but also pathogen, herbicide and heavy metal resistance genes and genes facilitating beneficial interactions between plants and organisms in the rhizosphere.

ACKNOWLEDGMENTS

We thank Marcus Lee (*Arabidopsis* transformations) and Assunta Pelosi, Richard Harrison, Paula Moolhuijzen and Cecilia O'Dwyer (*Eucalyptus* transformations) for discussions and also providing experimental data prior to publication. We acknowledge the financial support of the Australian Research Council and the Industrial Research and Development Board of Australia to enable research in these areas. We also thank Mrs. J. Elliston for preparing the manuscript.

REFERENCES

Ahn, J.C., Paek, Y.W., Kang, Y.H., and Hwang, B., 1992, Production of anthocyanin by culture of hairy roots of *Raphanus sativus*, *Korean J. Bot.* 35:37.

Aird, E.L.H., Hamill, J.D., and Rhodes, M.J.C., 1988a, Cytogenetic analysis of hairy root cultures from a number of plant species transformed by *Agrobacterium rhizogenes*, *Plant Cell Tissue Organ Cult.* 15:47.

Aird, E.L.H., Hamill, J.D., Robins, R.J., and Rhodes, M.J.C., 1988b, Chromosome stability in transformed hairy root cultures and the properties of variant lines of *Nicotiana rustica* hairy roots, *in*: "Manipulating Secondary Metabolism in Culture," R.J. Robins, and M.J.C. Rhodes, eds., Cambridge Univ. Press, Cambridge.

Altamura, M.M., Archiletti, T., Capone, I., and Costantino, P., 1991, Histological analysis of the expression of *Agrobacterium rhizogenes rol* B-GUS gene fusions in transgenic tobacco, *New Phytol.* 118:69.

Asamizu, T., Akiyama, K., and Yasuda, I., 1988, Anthoquinones production by hairy root culture in *Cassia obtusifolia*, *Yakagaku zasshi* 108:1215.

Bassil, N.V., Proebsting, W.M., Moore, L.W., and Lightfoot, D.A., 1991, Propagation of Hazelnut stem cuttings using *Agrobacterium rhizogenes*, *Hortic. Sci.* 26:1058.

Benfey, P.N., Ling, R., and Chua, N.H., 1990, Combinatorial and synergistic properties of CaMV35S enhancer subdomains, *EMBO J.* 9:1685.

Benfey, P.N., Ren, L., and Chua, N.H., 1989, The CaMV35S enhancer contains at least two domains which can confer different developmental and tissue specific expression patterns, *EMBO J.* 8:2195.

Berlin, J., Dietze, P., Fecker, L., Goddijn, O., and Hoge, J.H.C., 1991a, Production of high levels of serotonin in *Peganum* cultures by expression of a foreign plant tryptophan decarboxylase, *in*: "Abstracts 3rd Int. Cong. Plant Mol. Biol.," R.B. Hallick, ed., Tucson.

Berlin, J., Fecker, L., Ruegenhagen, C., Sator, C., Strack, D., Witte, L., and Wray, V., 1991b, Isoflavone glycoside formation in transformed and non-transformed suspension and hairy root cultures of *Lupinus polyphyllus* and *Lupinus hartwegii*, *Z. Naturforsch. (C)* 46:725.

Berlin, J., Kuzokina, I.N., Ruegenhagen, C., Fecker, L., Commandeur, U., and Wray, V.L., 1992, Hairy root cultures of *Peganum harmala*: characterization of cell lines and effects of culture conditions on the accumulation of beta-carboline alkaloids and seratonin, *Z. Naturforsch. (C)* 47:222.

Bogusz, D., Appleby, C.A., Landsmann, J., Dennis, E.S., Trinick, M.J., and Peacock, W.J., 1988, Functioning hemoglobin genes in a non-nodulating plant, *Nature* 331:178.

Bogusz, D., Llewellyn, D.J., Craig, S., Dennis, E.S., Appleby, C.A., and Peacock, W.J., 1990, Non-legume hemoglobin genes retain organ specific expression in heterologous transgenic plants, *Plant Cell* 2:633.

Bouchez, D., Tokuhisa, J.G., Llewellyn, D.J., Dennis, E.S., and Ellis, J.G., 1989, The *ocs*-element is a component of the promoters of several T-DNA and plant viral genes, *EMBO J.* 8:4197.

Bracher, D., and Kutchan, T.M., 1992, Strictosidine synthase from *Rauvolfia serpentina*: analysis of a gene involved in indole alkaloid biosynthesis, *Arch. Biochem. Biophys.* 294:717.

Capone, I., Spano, L., Cardarelli, M., Bellincampi, D., Petit, A., and Costantino, P., 1989, Induction and growth properties of carrot roots with different complements of *Agrobacterium rhizogenes* T-DNA, *Plant Mol. Biol.* 13:43.

Capone, I., Cardarelli, M., Mariotti, D., Pomponi, M., De Paolis, A., and Costantino, P., 1991, Different promoter regions control level and tissue specificity of expression of *Agrobacterium rhizogenes rol* B gene in plants, *Plant Mol. Biol.* 16:427.

Chamberlain, D.A., Wilson, G., and Ryan, M.F., 1991, Trans-2-nonenal insect repellent, insecticide and flavour compound in carrot roots, cell suspension and hairy root cultures, *J. Chem. Ecol.* 17:615.

Christen, P., Roberts, M.F., Phillipson, J.D., and Evans, W.C., 1989, High yield production of tropane alkaloids by hairy-root cultures of a *Datura candida* hybrid, *Plant Cell Rep.* 8:75.

Comai, L., Moran, P., and Maslyar, D., 1990, Novel and useful properties of a chimeric plant promoter combining CaMV35S and MAS elements, *Plant Mol. Biol.* 15:373.

Conkling, M.A., Cheng, C.L., Yamamoto, Y.T., and Goodman, H.M., 1990, Isolation of transcriptionally regulated root specific genes from tobacco, *Plant Physiol.* 93:1203.

Constabel, C.P., and Towers, G.H.N., 1988, Thiarubrine accumulation in hairy root cultures of *Chaenactis douglasii*, *J. Plant Physiol.* 133:67.

Croes, A.F., van den Berg, A.J.R., Bosveld, M., Breteler, H., and Wullems, G.J., 1989, Thiophene accumulation in relation to morphology in roots of *Tagetes patula*: Effects of auxin and transformation by *Agrobacterium*, *Planta* 179:43.

Davioud, E., Kan, C., Hamon, J., Tempé, J., and Husson, H.P., 1989, Production of indole alkaloids by *in vitro* root cultures from *Catharanthus trichophyllus*, *Phytochem.* 28:2675.

Deno, H., Yamagata, T., Emoto, T., Yoshioka, T., Yamada, Y., and Fijita, Y., 1987, Scopalamine production by root cultures of *Duboisa myoporoides*, II. Establishment of a hairy root culture by infection with *Agrobacterium rhizogenes*, *J. Plant Physiol.* 131:315.

de Pater, B.S., and Schilperoort, R.A., 1992, Structure and expression of a root-specific rice gene, *Plant Mol. Biol.* 18:161.

Dietrich, R.A., Radke, S.E., and Harada, J.T., 1992, Downstream DNA sequences are required to activate a gene expressed in the root cortex of embryos and seedlings, *Plant Cell* 4:1371.

Dynan, W.S., 1989, Modularity in promoters and enhancers, *Cell* 58:1.

Durand-Tardif, M., Broglie, R., Slighton, J., and Tepfer, D., 1985, Structure and expression of Ri T-DNA from *Agrobacterium rhizogenes* in *Nicotiana tabaccum*: organ and phenotypic specificity, *J. Mol. Biol.* 186:557.

Estruch, J.J., Schell, J., and Spena, A., 1991a, The protein encoded by the *rol* B plant oncogene hydrolyses indole glucosides, *EMBO J.*, 10:3125.

Estruch, J.J., Chriqui, D., Grossman, K., Schell, J., and Spena, A., 1991b, The plant oncogene *rol* C is responsible for the release of cytokinins from glucoside conjugates, *EMBO J.* 10:2889.

Fitter, A.H., 1991, Characteristics and functions of root systems, *in*: "Plant Roots: The Hidden Half," Y. Waisel, A. Eshel, and U. Kafkafi, eds., Marcel Dekker, New York.

Flores, H.E., Pickard, J.J., and Hoy, M.W., 1988, Production of polyacetylenes and thiophenes in heterotrophic and photosynthetic root cultures of Asteraceae, *in*: "Chemistry and Biology of

Naturally Occurring Acetylenes and Related Compounds (NOARC). Bioactive Molecules," J. Lam, H. Breheler, T. Arnason, and L. Hansen, eds., 7:233.

Flores, H.E., and Filner, P., 1985, Metabolic relationships of putrescine, GABA and alkaloids in cell and root cultures of Solanaceae, in: "Primary and Secondary Metabolism of Plant Cell Cultures, K.H. Nuemann, W., Barz, and E. Reinhard, eds., Springer-Verlag, Berlin.

Fromm, H., Katagiri, F., and Chua, N.H., 1989, An octopine synthase enhancer element directs tissue specific expression and binds ASF-1, a factor from tobacco nuclear extracts, *Plant Cell* 1:977.

Furze, J.M., Rhodes, M.J.C., Parr, A.J., Robins, R.J., Whitehead, J.M., and Threlfall, D.R., 1991, Abiotic factors elicit sesquiterpenoid phytoalexin production but not alkaloid production in transformed roots of *Datura stramonium, Plant Cell Rep.* 10:111.

Hamill, J.D., and Rhodes, M.J.C., 1992, Manipulating secondary metabolism in culture, in: "Biosynthesis and Manipulation of Plant Products, Plant Biotechnology," vol. 3, D. Grierson, ed., Chapman and Hall, London.

Hamill, J.D., Robins, R.J., and Rhodes, M.J.C., 1989, Alkaloid production by transformed root cultures of *Cinchona ledgeriana, Planta Med.* 55:354.

Hamill, J.D., Parr, A.J., Robins, R.J., and Rhodes, M.J.C., 1986, Secondary product formation by cultures of *Beta vulgaris* and *Nicotiana rustica* transformed with *Agrobacterium rhizogenes, Plant Cell Rep.* 5:111.

Hamill, J.D., Robins, R.J., Parr, A.J., Evans, D.M., Furze, J.M., and Rhodes, M.J.C., 1990, Overexpressing a yeast ornithine decarboxylase gene in transgenic roots of *Nicotiana rustica* can lead to enhanced nicotine accumulation, *Plant Mol. Biol.* 15:27.

Handa, T., 1992, Genetic transformation of *Antirrhinum majus* L. and inheritance of altered phenotype induced by Ri T-DNA, *Plant Sci.* 81:199.

Hashimoto, T., Hayashi, A., Amano, Y., Kohno, J., Iwanari, H., Usuda, S., and Yamada, Y., 1991, Hyosycamine 6β-hydroxylase, an enzyme involved in tropane alkaloid biosynthesis, is localised at the pericycle of the root, *J. Biol. Chem.* 266:4648.

Hauth, S., and Beiderbeck, R., 1992, *In vitro* culture of *Agrobacterium rhizogenes* induced hairy roots by *Salix alba, Silvae Genetica* 41:46.

Hoff, T., Stummann, B.M., and Henningsen, K.W., 1991, Cloning and expression of a gene encoding a root specific nitrate reductase in bean (*Phaseolus vulgaris*), *Physiol. Plant.* 82:197.

Hu, Z.B., and Alfermann, A.W., 1993, Diterpenoid production in hairy root cultures of *Salvia miltiorrhiza, Phytochem.* 32:699.

Hwang, B., Ko, K.M., Hwang, K.H., Hwang, S.J., and Kang, Y.H., 1992, Production of saponin by hairy root cultures of ginseng (*Panax ginseng*), *Korean J. Bot.* 34:289.

Ishimaru, K., Yonemitsu, H., and Shimomura, K., 1991, Lobetyolin and lobetyol from hairy root culture of *Lobelia inflata, Phytochem.* 30:2255.

Ishimaru, K., Sadoshima, S., Neer, S., Koyama, K., Takahashi, K., and Shimomura, K., 1992, A polyacetylene gentiobioside from hairy roots of *Lobelia inflata, Phytochem.* 31:1577.

Jung, G., and Tepfer, D., 1987, Use of genetic transformation by the Ri T-DNA of *Agrobacterium rhizogenes* to stimulate biomass and tropane alkaloid production in *Atropa belladonna* and *Calystegia sepium* roots grown *in vitro, Plant Sci.* 50:145.

Kamada, H., Okamura, N., Satake, M., Harada, M., and Shimomura, K., 1986, Alkaloid production by hairy root cultures in *Atropa belladonna, Plant Cell Rep.* 5:239.

Katagiri, F., Lam, E., and Chua, N.H., 1989, Two tobacco DNA-binding proteins with homology to the nuclear factor CREB, *Nature* 340:727.

Keller, B., and Baumgartner, C., 1991, Vascular specific expression of the bean GRP 1.8 gene is negatively regulated, *The Plant Cell* 3:1051.

Keller, B., and Lamb, C.J., 1989, Specific expression of a novel cell wall hydroxyproline rich glycoprotein gene in lateral root induction, *Genes Devel.* 3:1639.

Lam, E., Benfey, P., Gilmartin, P., Fang, R., and Chua, N.H., 1989, Site specific mutations alter *in vitro* factor binding and change promoter expression pattern in transgenic plants, *Proc. Natl. Acad. Sci. (USA)* 86:7890.

Lambert, C., and Tepfer, D., 1991, Use of *Agrobacterium rhizogenes* to create chimeric apple trees through genetic grafting, *Bio/Technol.* 9:80.

Langridge, W.H.R., Fitzgerald, K.J., Koncz, C., Schell, J., and Szalay, A.A., 1989, Dual promoter of *Agrobacterium tumefaciens* mannopine synthase genes is regulated by plant growth hormones, *Proc. Natl. Acad. Sci. (USA)* 86:3219.

Leach, F., and Aoyagi, K., 1991, Promoter analysis of the highly expressed *rol* C and *rol* D root inducing genes of *Agrobacterium rhizogenes*: enhancer and tissue specific DNA determinants are dissociated, *Plant Sci.* 79:69.

Linsey, K., Wei, W., Clarke, M.C., McArdle, H.F., Rooke, L.M., and Topping, J.F., 1993, Tagging genomic sequences that direct transgene expression by activation of a promoter trap in plants, *Trangenic Res.* 2:33.

MacIsaac, S.A., Sawhney, V.K., and Pohorecky, Y., 1989, Regulation of lateral root formation in lettuce (*Lactuca sativa*) seedling roots: Interacting effects of -naphthalene acetic acid and kinetin, *Physiol. Plant.* 77:287.

MacRae, S., 1991, *Agrobacterium*-mediated transformation of eucalypts to improve rooting ability, *in*: "IUFRO Symp. on Intensive Forestry: The Role of Eucalypts," September, 1991, Durban.

Mano, Y., Ohkawa, H., and Yamada, Y., 1989, Production of tropane alkaloids by hairy root cultures of *Duboisia leichhardtii* transformed by *Agrobacterium rhizogenes, Plant Sci.* 59:191.

Mano, Y., Nabeshima, S., Matsui, C., and Ohkawa, H., 1986, Production of tropane alkaloids by hairy root cultures of *Scopolia japonica, Agric. Biol. Chem.* 50:2715.

Marchant, Y.Y., 1988, *Agrobacterium rhizogenes* - transformed root cultures for the study of polyacetylene metabolism and biosynthesis, *in*: "Chemistry and Biology of Naturally-occurring Acetylenes and Related Compounds (NOARC): Bioactive Molecules," J. Lam, H. Breheler, T. Arnason, and L. Hansen, eds., 7:217.

Matsuda, J., Okabe, S., Hashimoto, J., and Yamada, Y., 1991, Molecular cloning of hyoscyamine 6 β-hydroxylase, a 2-oxoglutarate-dependent dioxygenase from cultured roots of *Hyoscyamus niger, J. Biol. Chem.* 266:9460.

Matsumoto, T., and Tanaka, N., 1991, Production of phytoecdysteroids by hairy root cultures of *Ajuga reptans, Agric. Biol. Chem.* 55:1019.

Maurel, C., Brevet, J., Barbier-Brygoo, H., Guern, J., and Tempé, J., 1990, Auxin regulates the promoter of the root inducing rol B gene of *Agrobacterium rhizogenes* in transgenic tobacco, *Mol. Gen. Genet.* 223:58.

Maurel, C., Barbier-Brygoo, H., Brevet, J., Spena, A., Tempé, J., and Guern, J., 1991, Single *rol* genes from the *Agrobacterium rhizogenes* T_L T-DNA alter some of the cellular responses to auxin in *Nicotiana tabacum, Plant Physiol.* 97:212.

Morris, P., and Robbins, M.P., 1992, Condensed tannin formation by *Agrobacterium rhizogenes* transformed root and shoot organ cultures of *Lotus corniculatus, J. Exp. Bot.* 43:221.

Mukundan, U., and Hjortso, M., 1990, Thiophene accumulation in hairy roots of *Tagetes patula* in response to fungal elicitors, *Biotech. Lett.* 12:609.

Murashige, T., and Skoog, F., 1962, A revised medium for rapid growth and bioassays with tobacco tissue cultures, *Physiol. Plant.* 15:473.

Nabeshima, S., Maro, Y., and Ohkawa, H., 1986, Production of tropane alkaloids of hairy root cultures of *Scopolia japonica, Symbiosis* 2:11.

Ott, R.W., and Chua, N.H., 1990, Enhancer sequences from *Arabidopsis thaliana* obtained by library transformation of *Nicotiana tabacum, Mol. Gen. Genet.* 223:169.

Parr, A.J., 1992, Alternative metabolic fates of hygrine in transformed root cultures of *Nicandra physaloides, Plant Cell Rep.* 11:270.

Parr, A.J., and Hamill, J.D., 1987, Relationship between *Agrobacterium rhizogenes* transformed hairy roots and intact uninfected *Nicotiana* plants, *Phytochem.* 26:3241.

Parr, A.J., Peerless, A.C.J., Hamill, J.D., Walton, N.J., Robins, R.J., and Rhodes, M.J.C., 1988, Alkaloid production by transformed root cultures of *Catharanthus roseus, Plant Cell Rep.* 7:309.

Parr, A.J., Payne, J., Eagles, J., Chapman, B.T., Robins, R.J., and Rhodes, M.J.C., 1990, Variation in tropane alkaloid accumulation within the Solanaceae and strategies for its exploitation, *Phytochem.* 29:2545.

Patena, I., Sutter, E., and Dandekar, A.M., 1988, Root induction by *Agrobacterium rhizogenes* in a difficult-to-root woody species, *Acta Hortic.* 227:324.

Payne, J., Hamill, J.D., Robins, R.J., and Rhodes, M.J.C., 1987, Production of hyoscyamine by hairy root cultures of *Datura stramonium, Planta Med.* 53:474.

Porter, J., 1991, Host range and implications of plant infection by *Agrobacterium rhizogenes, Crit. Rev. Plant Sci.* 10:387.

Rhodes, M.J.C., Hilton, M., Parr, A.J., Hamill, J.D., and Robins, R.J., 1986, Nicotine production by hairy root cultures of *Nicotiana rustica*: fermentation and product recovery, *Biotech. Lett.* 8:415.

Robins, R.J., Parr, A.J., Payne, J., Walton, N.J., and Rhodes, M.J.C., 1990, Factors affecting tropane-alkaloid production in a transformed root culture of a *Datura candida* X *D. aurea* hybrid, *Planta* 18:414.

Robins, R.J., Parr, A.J., Bent, E.G., and Rhodes, M.J.C., 1991, Studies on the biosynthesis of tropane-alkaloids in *Datura stramonium* L. transformed root cultures. I. The kinetics of alkaloid production and the influence of feeding intermediate metabolites, *Planta* 183:185.

Rugini, E., 1992, Involvement of polyamines in auxin and *Agrobacterium rhizogenes* induced rooting of fruit trees *in vitro, J. Amer. Soc. Hortic. Sci.* 117:532.

Saito, K., Yamazaki, M., Shimomura, K., Yoshimatsu, K., and Murakoshi, T., 1990, Genetic transformation of foxglove (*Digitalis purpurea*) by chimeric foreign genes and production of cardioactive glycosides, *Plant Cell Rep.* 9:121.

Saito, K., Yamazaki, M., Kaneko, H., Murakoshi, I., Fukuda, Y., and Van Montagu, M., 1991, Tissue-specific and stress-enhancing expression of the TR promoter for mannopine synthase in transgenic medicinal plants, *Planta* 184:40.

Sato, K., Yamazaki, T., Okugawa, E., Yoshihira, K., and Shimomura, K., 1991, Anthraquinone production by transformed root cultures of *Rubia tinctorum*, *Phytochem.* 30:1507.

Sauerwein, M., and Shimomura, K., 1991, Alkaloid production in hairy roots of *Hyoscyamus albus* transformed by *Agrobacterium rhizogenes*, *Phytochem.* 30:3277.

Sauerwein, M., Ishimaru, K., and Shimomura, K., 1991a, Indole alkaloids in hairy roots of *Amsonia elliptica*, *Phytochem.* 30:1153.

Sauerwein, M., Ishimaru, K., and Shimomura, K., 1991b, A piperidone alkaloid from *Hyoscyamus albus* roots transformed with *Agrobacterium rhizogenes*, *Phytochem.* 30:2977.

Sauerwein, M., Yamazaki, T., and Shimomura, K., 1991c Hernandulcin in hairy root cultures of *Lippia dulcis*, *Plant Cell Rep.* 9:579.

Sharp, J.M., and Doran, P.M., 1990, Characteristics of growth and tropane alkaloid synthesis in *Atropa belladonna* roots transformed by *Agrobacterium rhizogenes*, *J. Biotechnol.* 16:171.

Shimomura, K., Sudo, H., Saga, H., and Kamada, H., 1991, Shikonin production and secretion by hairy root cultures of *Lithospermum erythrorhizon*, *Plant Cell Rep.* 10:282.

Shirsat, A.H., Wilford, N., Evans, I.M., Gatehouse, L.N., and Croy, R.D.N., 1991, Expression of a *Brassica napus* extensin gene in the vascular system of transgenic tobacco and rape plants, *Plant Mol. Biol.* 17:701.

Soo, K.K., Ebizuka, Y., Noguchi, H., and Sankawa, U., 1988, Production of secondary metabolites by hairy roots and regenerated plants transformed with Ri plasmids, *Chem. Pharm. Bull.* 36:4217.

Spena, A., Schmlling, T., Koncz, C., and Schell, J.S., 1987, Independent and synergistic activity of *rol* A, B and C loci in stimulating abnormal growth in plants, *EMBO J.* 6:3891.

Tepfer, D., 1984, Transformation of several species of higher plants by *Agrobacterium rhizogenes*: sexual transmission of the transformed genotype and phenotype, *Cell* 37:959.

Toivonen, L., Balsevich, J., and Kurz, G.W., 1989, Indole alkaloid production by hairy root cultures of *Catharanthus roseus*, *Plant Cell, Tissue Organ Cult.* 18:79.

Trypsteen, M., van Lijsebettens, M., van Severen, R., and van Montagu, M., 1991, *Agrobacterium rhizogenes*-mediated transformation of *Echinacea purpurea*, Plant Cell Rept. 10:85.

Valveekens, D., Van Montagu, M., and Van Lijsebettens, M., 1988, *Agrobacterium tumefaciens* mediated transformation of *Arabidopsis thaliana* root explants by using kanamycin selection, *Proc. Natl. Acad. Sci. (USA)* 85:5536.

van der Zaal, E.J., Droog, F.N.J., Boot, C.J.M., Hensgens, L.A.M., Hoge, J.H.C., Schilperoort, R.A., and Libbenga, K.R., 1991, Promoters of auxin-induced genes from tobacco can lead to auxin-inducible and root tip-specific expression, *Plant Mol. Biol.* 16:983.

Walton, N.J., Robins, R.J., and Peerless, A.C.J., 1990, Enzymes of *N*-methylputrescine biosynthesis in relation to hyoscyamine formation in transformed root cultures of *Datura stramonium* and *Atropa belladonna*, *Planta* 182:136.

Westcott, R., 1988, Thiophene production from *Tagetes* hairy roots, *in*: "Manipulating Secondary Metabolism in Culture," R.J. Robins, and M.J.C Rhodes, eds., Cambridge Univ. Press, Cambridge.

Wightman, F., Schneider, E.A., and Thimann, V.K., 1980, Hormonal factors controlling the initiation and development of lateral roots. II. Effects of exogenous growth factors on lateral root formation in pea roots, *Physiol. Plant* 49:304.

Willyams, D., Whiteman, P., Cameron, J., and Chandler, S., 1992, Inter- and intra-family variability for rooting capacity in micropropagated *Eucalyptus globulus* and *Eucalyptus nitens*, *in*: "AFOCEL/IUFRO Symp. on Mass Production Technology for Genetically Improved Fast Growing Forest Trees," 1992, Bordeaux.

Wilson, P.D.G., Hilton, M.G., Meehan, P.T.H., Waspe, C.R., and Rhodes, M.J.C., 1990, The cultivation of transformed roots from laboratory to pilot plant, *in*: "Progress in Plant Cellular and Molecular Biology," H.J.J. Nijkamp, L.H.W. van der Plas, and J. van Aartrijk, eds., Kluwer Academic Pubs., Dordrecht.

Yamamoto, Y.T., Taylor, C.G., Acedo, G.N., Cheng, C.L., and Conkling, M.A., 1991, Characterization of *cis*-acting sequences regulating root-specific gene expression in tobacco, *Plant Cell* 3:371.

Yamazaki, T., Flores, H.E., Shimomura, K., and Yoshihira, K., 1991, Examination of steviol glucosides production by hairy root and shoot cultures of *Stevia rebaudiana*, *J. Nat. Prod. (Lloydia)* 54:986.

Yonemitsu, H., Shimomura, K., Satake, M., Mochida, S., Tanaka, M., Endo, T., and Kaji, A., 1990, Lobeline production by hairy root culture of *Lobelia inflata L., Plant Cell Rep.* 9:307.

Yoshikawa, T., and Furuya, T., 1987, Saponin production by cultures of *Panax ginseng* transformed with *Agrobacterium rhizogenes, Plant Cell Rep.* 6:449.

CONTROL OF ROOT SYSTEM ARCHITECTURE THROUGH CHEMICAL

AND GENETIC ALTERATIONS OF POLYAMINE METABOLISM

David Tepfer[1], Jean-Pierre Damon[1], Gozal Ben-Hayyim[2],
Alessandro Pellegrineschi[1], Daniel Burtin[3], and Josette Martin-Tanguy[3]

[1]INRA, F-78026 Cedex Versailles, France

[2]The Volcani Center Institute of Horticulture
Bet-Dagan, Israel

[3]INRA, Dijon, France

INTRODUCTION

Plant growth is plastic, responding to a variety of environmental cues and conditions; and plant growth has been altered using genetics, biochemistry, horticulture and agronomy. However, efforts to modify growth have been mostly been aimed at the aerial parts of the plant. In the present paper we will retrace and discuss our efforts to alter root system architecture, first through genetic transformation, then through chemical and physiological means.

CHANGING ROOT SYSTEMS THROUGH GENETIC TRANSFORMATION

Since soil microorganisms compete for nutrients supplied through plant photosynthesis, they have acquired, in many cases, the ability to inhibit, stimulate and re-direct plant growth. Such microorganisms have thus provided models for human attempts to understand and manipulate plant growth. We have chosen *Agrobacterium rhizogenes* as a model and a source of genes for studying root system development. The ability of this bacterium to induce the formation of adventitious roots in dicotyledonous plants has long been recognized (Riker et al., 1930). The roots produced by *A. rhizogenes* were shown to be genetically transformed, first by the finding that they contain opines (Tepfer and Tempé, 1981), which are metabolites whose synthesis is a marker for genetic transformation by *Agrobacterium tumefaciens*, then by the direct demonstration of DNA from *A. rhizogenes* in axenic cultures of roots formed as a result of bacterial inoculation (Chilton et al., 1982; Tepfer, 1982; White et al., 1982; Willmitzer et al., 1982). The transferred DNA (T-DNA) is termed *Ri* (root-inducing), and often exists in two parts, one of which, the left-hand or TL-DNA, carries the primary responsibility for root induction.

Biology of Adventitious Root Formation, Edited by
T.D. Davis and B.E. Haissig, Plenum Press, New York, 1994

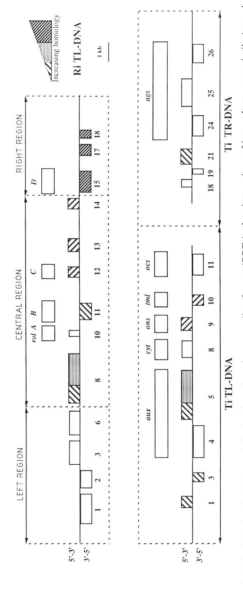

Figure 1. The Ri (root-inducing) TL-DNA with genetic loci and numbered open reading frames (ORFs), showing regions of internal sequence similarity and similarity to *Ti* (tumor-inducing) T-DNA from *Agrobacterium tumefaciens*. The *Ri* TL-DNA is divided into the left, central and right regions, according to the level of internal sequence similarity (Levesque *et al.*, 1988); study based on the sequence of Slightom *et al.*, 1986. The central and right regions carry the *rol* loci, which are defined as regions in which transposon insertion alters root production by *A. rhizogenes* (White *et al.*, 1985). ORFs 9 and 16 are not shown. In the *Ti* T-DNA *aux* refers to auxin synthesis and *cyt* to cytokinin synthesis.

Agrobacterium rhizogenes engages in genetic transformation by using the same basic DNA transfer mechanisms as *Agrobacterium tumefaciens*. The principal difference is in the information transferred. The *Ti* (tumor-inducing) T-DNA of *A. tumefaciens* contains genes that encode the synthesis of auxins and cytokinins; hence, tumor formation. The mechanism of root induction by the *Ri* T-DNA is not fully understood. The roots induced by *A. rhizogenes* have an unusual ability to grow in culture (Tepfer and Tempé, 1981; Tepfer, 1984), where they can be propagated indefinitely as organ clones. Transformed root cultures provide an *in vitro* model for the rhizosphere, permitting the production of secondary metabolites and the study of plant-microorganism interactions. The advantage of this model is that cultured roots are not subject to influence from the plant's aerial parts (Tepfer, 1989). These roots are also a source of metabolites important in plant/microorganism interactions (Tepfer et al., 1988a,b), and these interactions can sometimes be modelled *in vitro* using transformed roots (Mugnier and Mosse, 1987; Bécard and Fortin, 1988; Tepfer, 1989). Not only are transformed roots able to grow in culture, but their pattern of growth is altered. They exhibit reduced apical dominance, i.e., they are highly branched, which explains (at least in part) their altered growth properties.

When transformed roots are allowed (or induced) to regenerate shoots, one obtains whole plants containing *Ri* T-DNA in all of their cells, and the alterations observed in the pattern of growth of transformed roots in culture are found in the root systems of these transformed plants (Tepfer, 1982, 1983a,b, 1984). Roots become more branched, and tend to be plagiotropic. (These changes may differ according to species and variety.) In tobacco, *Nicotiana tabacum*, which we have used as our primary species model, the root/shoot ratio is not altered by this genetic transformation. Other features appear, such as the regeneration of shoots on roots; and numerous changes occur in the aerial parts, such as the formation of roots along the stems (Ackermann, 1977; Tepfer, 1984). These changes in the aerial parts can be used as markers for the activity of genes carried by the *Ri* TL-DNA, and they can be correlated with biochemical changes. They are collectively referred to as "the transformed phenotype" (Tepfer, 1984).

The *Ri* TL-DNA contains 18 stretches of DNA sequence devoid of stop codons and encoding putative proteins of 100 amino acids or more (Slightom et al., 1986). Such regions are termed ORFs (open reading frames). Some of these ORFs have been recognized, through insertional mutagenesis, as being important in root induction (White et al., 1985). These are the "*rol*" genes (*rol* designates "root locus"), and they are designated *A-D* (Fig. 1). ORFs 8, 13 and 14 are of interest because they are similar in sequence to *rol* genes or to genes encoded in the *Ti* T-DNA (Fig. 1) (Levesque et al., 1988). Messenger RNAs have been attributed to several ORFs (Durand-Tardif et al., 1985; Taylor et al., 1985; Ooms et al., 1986).

The phenotypic consequences of *Ri* TL-DNA genes have been examined by expressing them either under the control of their endogenous promoters or the more generally expressed 35S promoter from the Cauliflower Mosaic Virus (CaMV), or by establishing correlations between phenotypic changes and changes in the expression of single genes (Durand-Tardif et al., 1985; Taylor et al., 1985; Cardarelli et al., 1987; Jouanin et al., 1987; Oono et al., 1987; Spena et al., 1987; Vilaine et al., 1987; Schmülling et al., 1988; Sinkar et al., 1988a,b). Although the picture that emerges from this work is sometimes unclear. But, at least in tobacco, dwarfing of the aerial parts can generally be assigned to both *rolA* and *rolC*; wrinkled leaves are primarily due to *rolA*; and reduced apical dominance is due to *rolC*. Other ORFs can also cause changes in phenotype, e.g., ORF 13 (unpublished results) produces stunting and leaf deformation. Curiously, in all of this work little attention has been paid to effects on root system architecture.

In the present research we have used aeroponic culture to examine the effects of *rolA* and *rolC* on root system development in tobacco (unpublished results). Both genes were expressed using the 35S CaMV promoter. We found that *rolA* had no detectable effect on root system architecture, nor did it change the root/shoot ratio. One the other hand *rolC* was associated with the formation of a thick, short tap root, from which a large number of laterals emerged (Fig. 2). Furthermore, the root/shoot ratio was increased by about 60% in plants carrying 35S-*rolC*.

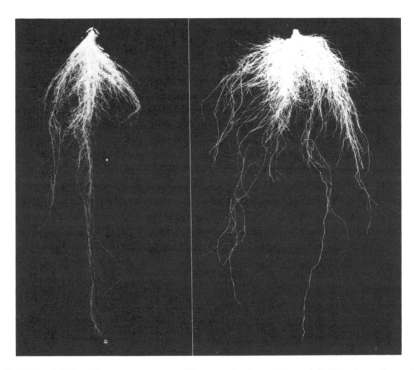

Figure 2. Effect of 35S-rolC on root system architecture of tobacco. Control (left) and transformed (right) plants were grown in aeroponic culture, and their root systems were excised for photography.

FUNCTIONS OF GENES ENCODED IN THE *RI* TL-DNA: THE CASE FOR POLYAMINE INVOLVEMENT

The functions of the genes that alter the phenotype of both the shoot and root systems have come under three types of scrutiny, involving hormone sensitivity, hormone production and polyamine accumulation. We shall discuss the first two only briefly. An increase in sensitivity to auxin was attributed primarily to *rolB* (ORF 11) (Shen et al., 1988; Maurel et al., 1991a,b). The activity of the protein encoded by *rolB* was defined as hydrolyzing indole glucosides (Estruch et al., 1991a), and that encoded by *rolC* as releasing cytokinins from glucose conjugates (Estruch et al., 1991b). These activities are not necessarily reflected in increases in the amounts of plant hormones that can be extracted from transformed tissues (Julliard et al., 1992; Spena et al., 1992; Schaerer and Pilet, 1993).

In studying the chain of events that begins with gene expression and results in the transformed phenotype, we have concentrated on the polyamines as possible key agents in effecting the transformed phenotype (Maliga et al., 1973; Martin-Tanguy et al., 1990; Burtin et al., 1991; Sun et al., 1991). The impetus for this research was accumulating evidence that polyamines are growth regulators in a variety of biological systems. Unlike hormones, they are metabolites present in high enough concentrations to be important in intermediary metabolism. We first looked for correlations between changes in polyamine profiles over time and the degree of phenotypic alteration due to the *Ri* TL-DNA. The transformed phenotype exists in states designated T or T', with the latter being an exaggerated form of the former. The T and T' states are correlated, respectively, with the homozygous and heterozygous states of the foreign DNA, and the T' is unstable, reverting to the T state in lateral branches or sometimes in the apical meristem (Tepfer, 1984; Durand-Tardif et al., 1985). This reversion is correlated with repression of the expression of *rolA* (Sinkar et al., 1988b), which is due to methylation of sequences adjacent to the gene (L.-Y. Sun and D. Tepfer, unpublished results). Polyamines and their conjugates were assayed in sibling tobacco plants, having the control (N), T or T' phenotypes, over a time-course starting 40 days after sowing, and ending at day 115 (Martin-Tanguy et al., 1990). Profiles for free amines and conjugates were established for both roots and shoot tops. In all cases transformation was associated with a reduction in amine titers, and, for most of the substances measured, this reduction was greater in the T' than in the T phenotype. These experiments were repeated using plants carrying either *rolA* or *rolC*, and both genes were shown to be capable of inhibiting polyamine accumulation [(Sun et al. (1991); Martin-Tanguy et al., unpublished results].

Since the transformed genotype and phenotype appeared to be associated with changes in polyamine metabolism, we sought to determine whether reducing polyamine titers through other than genetic means would produce a similar phenotype. We treated tobacco plants with a chemical inhibitor of putrescine synthesis, α-DL-difluoromethylornithine (DFMO). DFMO specifically binds to and inactivates ornithine decarboxylase (ODC), one of the enzymes involved in putrescine synthesis. This inhibitor is widely used in yeast and animal systems. In tobacco, it produces a phenotype that closely resembles the transformed phenotype attributed to the *Ri* TL-DNA (Burtin et al., 1991), supporting the hypothesis that the transformed phenotype is produced through interference with the ornithine pathway of putrescine synthesis.

Table 1. Effects of α-DL-difluoromethylornithine (DFMO) and 35S-*rolA* on growth (increase in fresh weight) of tobacco roots excised from young seedlings.

Treatment	Fresh Weight (% of Control)
Control	100
DFMO, 1.0 mmol	205
Putrescine, 1.0 mmol	106
DFMO+Putrescine	111
35S-*rolA*	191
35S-*rolA*+DFMO	349

CHANGING ROOT SYSTEMS BY INHIBITING POLYAMINE ACCUMULATION

The ability to simulate the transformed phenotype in the aerial parts of the plant by specifically inhibiting an enzyme involved in putrescine synthesis suggested that the effects of *Ri* T-DNA on root architecture might also be mimicked by inhibiting putrescine synthesis. This conjecture was tested *in vitro* using roots excised from normal (non-transformed) tobacco seedlings (G. Ben-Hayyim et al., unpublished results). The advantage of such a model system derives, as with the transformed roots described above, from its independence from the aerial parts of the plant. Excised, non-transformed roots exposed to DFMO grew faster and their morphology changed. Instead of producing several roots of similar length, the initial excised roots exposed to DFMO continued to grow as a primary root, from which numerous laterals emerged (Fig. 3). A similar phenotype was produced when roots were excised from seedlings transformed by 35S-*rolA*, and the effects of DFMO and *rolA* on overall growth were cumulative (Table 1). Furthermore, treatment of whole, untransformed plants with DFMO caused an increase, by a factor of two, in the root/shoot ratio. Thus, both the extent and design of root system development varies with polyamine titers, which can be altered both genetically and chemically.

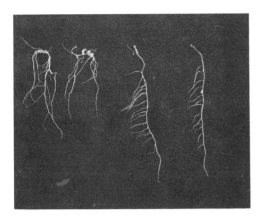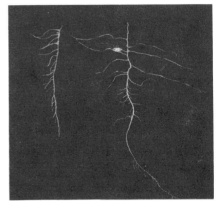

Figure 3. Effects of α-DL-difluoromethylornithine (DFMO) and *rolA* on the growth of excised tobacco roots *in vitro*. Left photograph: from left, control (two roots) and DFMO-treated (two roots). Right photograph: from left, *rolA* and *rolA*+DFMO. Roots excised from control, non-transformed tobacco plants were grown in the absence or presence of DFMO, as were roots excised from seedlings transformed by 35S-*rolA* .

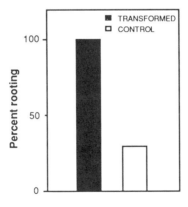

Fig. 4. Effects of transformation by *A. rhizogenes* on the rooting of cuttings of scented geranium.

STIMULATING ADVENTITIOUS ROOT FORMATION ON CUTTINGS

Insertion of wild type *Ri* T-DNA into a plant genome can stimulate the rooting of cuttings. *Pelargonium graveolens* (scented geranium) is propagated, through cutting, for the lemon-like odor it produces. The ability of cuttings to root is stimulated by incorporating T-DNA from *A. rhizogenes* into the whole plant, via root induction and regeneration from transformed root cultures (Fig. 4). In addition, the general shape of the plant's aerial parts is improved, due to a shortening of internodes and an increase in leaf production (Fig. 5).

Fig. 5. Effects of transformation by *A. rhizogenes* on habit in scented geranium. Control (left), transformed plant (right).

DISCUSSION

The *Ri* TL-DNA carries genes that induce the formation of altered roots, and these roots regenerate into plants having a changed phenotype. The biochemistry of the action of genes carried by this T-DNA is not clear, but it would seem that among the events that link the transformed genotype to the transformed phenotype, changes in titers of polyamines must be important. Genes from the *Ri* TL-DNA are thus useful tools for probing the mechanisms that control plant development, in particular those that govern root growth. We have shown that the transformed phenotype is correlated with a reduction in the titers of polyamines and their derivatives, and a similar phenotype can be obtained using a chemical inhibitor of putrescine synthesis. We thus believe that the *Ri* T-DNA causes the transformed phenotype (at least in part) by interfering with polyamine metabolism. This contention is further supported by the ability of *rolA*, *rolC* and DFMO to change the extent and pattern of root system growth. These results lead us to the conclusion that polyamines are important in regulating root system development.

The future of this research lies in causing localized changes in polyamine metabolism by regulating the expression of *Ri* TL-DNA genes so that they are active only in certain organs (e.g., just in roots) or just in parts of organs (e.g., just in lateral roots) or just under certain conditions (e.g., drought). We hope to tailor root system architecture to specific needs. The use of organ- and condition-specific promoters to control foreign gene activity is just beginning, but we are confident that redesigning root systems will lead to improvements in plant productivity. Just increasing the ability to root cuttings is useful, but producing drought resistance in crops like rice could be more beneficial to general human welfare. The *Ri* TL-DNA genes are clearly candidates for such applications, but other genes known to be involved in polyamine metabolism are equally promising. Such genes are well characterized in other systems, and are becoming available in plants (Malmberg et al., 1992).

REFERENCES

Ackermann, C., 1977, Pflanzen aus *Agrobacterium rhizogenes* tumoren aus *Nicotiana tabacum, Plant Sci. Lett.* 8:23.

Bécard, G., and Fortin, J., 1988, Early events of vesicular-arbuscular mycorrhiza formation on Ri T-DNA transformed roots, *New Phytol.* 108:211.

Burtin, D., Martin-Tanguy, J., and Tepfer, D., 1991, α-DL-difluoromethylornithine, a specific, irreversible inhibitor of putrescine biosynthesis, induces a phenotype in tobacco similar to that ascribed to the root-inducing, left-hand transferred DNA of *Agrobacterium rhizogenes*, *Plant Physiol.* 95:461.

Cardarelli, M., Mariotti, D., Pomponi, M., Spano, L., Capone, I., and Costantino, P., 1987, *Agrobacterium rhizogenes* T-DNA genes capable of inducing hairy root phenotype, *Mol. Gen. Genet.* 209:475.

Chilton, M.-D., Tepfer, D., Petit, A., David, C., Casse-Delbart, F., and Tempé, J., 1982, *Agrobacterium rhizogenes* inserts T-DNA into the genomes of host plant root cells, *Nature* 295:432.

Durand-Tardif, M., Broglie, R., Slightom, J., and Tepfer, D., 1985, Structure and expression of Ri T-DNA from *Agrobacterium rhizogenes* in *Nicotiana tabacum*: organ and phenotypic specificity, *J. Mol. Biol.* 186:557.

Estruch, J., Schell, J., and Spena, A., 1991a, The protein encoded by the *rolB* plant oncogene hydrolyses indole glucosides, *EMBO J.* 10:3125.

Estruch, J., Chriqui, D., Grossmann, K., Schell, J., and Spena, A., 1991b, The plant oncogene *rolC* is responsible for the release of cytokinins from glucoside conjugates, *EMBO J.* 10:2889.

Jouanin, L., Vilaine, F., Tourneur, J., Pautot, V., Muller, J.-F., and Caboche, M., 1987, Transfer of a 4.3 kb fragment of the TL-DNA of *Agrobacterium rhizogenes* strain A4 confers the pRi transformed phenotype to regenerated plants, *Plant Sci.* 53:53.

Julliard, J., Sotta, B., Pelletier, G., and Miginiac, E., 1992, Enhancement of naphthaleneacetic acid-induced rhizogenesis in TL-DNA-transformed *Brassica napus* without significant modification of auxin levels and auxin sensitivity, *Plant Physiol.* 100:1277.

Levesque, H., Delepelaire, P., Rouzé, P., Slightom, J., and Tepfer, D., 1988, Common evolutionary origin of the central portions of the Ri TL-DNA of *Agrobacterium rhizogenes* and the Ti T-DNAs of *Agrobacterium tumefaciens*, *Plant Mol. Biol.* 11:731.

Maliga, P., Sz.-Breznovits, A., and Márton, L., 1973, Streptomycin-resistant plants from callus culture of haploid tobacco, *Nature New Biol.* 244:29.

Malmberg, R.L., Smith, K.E., Bell, E., and Cellino, M.L., 1992, Arginine decarboxylase of oats is clipped from a precursor into 2 polypeptides found in the soluble enzyme, *Plant Physiol.* 100:146.

Martin-Tanguy, J., Tepfer, D., Paynot, M., Burtin, D., Heisler, L., and Martin, C., 1990, Inverse relationship between polyamine levels and the degree of phenotypic alteration induced by the Ri TL-DNA from *Agrobacterium rhizogenes*, *Plant Physiol.* 92:912.

Maurel, C., Barbierbrygoo, H., Brevet, J., Spena, A., Tempé, J., and Guern, J., 1991a, *Agrobacterium rhizogenes* T-DNA genes and sensitivity of plant protoplasts to auxins, *in:* "Advances in Molecular Genetics of Plant-Microbe Interactions," vol. 1, H. Hennecke, and D.P.S. Verma, eds., Kluwer Academic Pubs., Dordrecht.

Maurel, C., Barbierbrygoo, H., Spena, A., Tempé, J., and Guern, J., 1991b, Single *rol* genes from the *Agrobacterium rhizogenes* TL-DNA alter some of the cellular responses to auxin in *Nicotiana tabacum*, *Plant Physiol.* 97:212.

Mugnier, J., and Mosse, B., 1987, Vesicular-arbuscular mycorrhizal infection in transformed root-inducing T- DNA roots grown axenically, *Phytopath.* 77:1045.

Ooms, G., Twell, D., Bossen, M., Hoge, J., and Murrel, M., 1986, Developmental regulation of Ri TL-DNA gene expression in roots, shoots and tubers of transformed potato (*Solanum tuberosum* cv. Desiree), *Plant Mol. Biol.* 6:321.

Oono, Y., Handa, T., Kanaya, K., and Uchimiya, H., 1987, The TL-DNA gene of Ri plasmids responsible for dwarfness of tobacco plants, *Jpn. J. Genet.* 62:501.

Riker, A., Banfield, W., Wright, W., Keitt, G., and Sagen, H., 1930, Studies on infectious hairy root of nursery apple trees, *J. Agric. Res.* 41:887.

Schaerer, S., and Pilet, P.-E., 1993, Quantification of indole-3-acetic acid in untransformed and *Agrobacterium rhizogenes*-transformed pea roots using gas chromatography mass spectrometry, *Planta* 189:55.

Schmülling, T., Schell, J., and Spena, A., 1988, Single genes from *Agrobacterium rhizogenes* influence plant development, *EMBO J.* 7:2621.

Shen, W., Petit, A., Guern, J., and Tempé, J., 1988, Hairy roots are more sensitive to auxin than normal roots, *Proc. Natl. Acad. Sci. USA* 85:3417.

Sinkar, V., Pythoud, F., White, F., Nester, E., and Gordon, M., 1988a, *rol A* locus of the Ri plasmid directs developmental abnormalities in transgenic plants, *Genes and Dev.* 2:688.

Sinkar, V., White, F., Furner, I., Abrahamsen, M., Pythoud, F., and Gordon, M., 1988b, Reversion of aberrant plants transformed with *Agrobacterium rhizogenes* is associated with the transcriptional inactivation of the TL-DNA genes, *Plant Physiol.* 86:47.

Slightom, J., Durand-Tardif, M., Jouanin, L., and Tepfer, D., 1986, Nucleotide sequence analysis of TL-DNA of *Agrobacterium rhizogenes* agropine type plasmid, *J. Biol. Chem.* 261:108.

Spena, A., Schmülling, T., Koncz, C., and Schell, J., 1987, Independent and synergistic activity of *rol A*, *B* and *C* loci in stimulating abnormal growth in plants, *EMBO J.* 6:3891.

Spena, A., Estruch, J.J., Prinsen, E., Nacken, W., Vanonckelen, H., and Sommer, H., 1992, Anther-specific expression of the *rolB* gene of *Agrobacterium rhizogenes* increases IAA content in anthers and alters anther development and whole flower growth, *Theor. Appl. Genet.* 84:520.

Sun, L.-Y., Monneuse, M.-O., Martin-Tanguy, J., and Tepfer, D., 1991, Changes in flowering and accumulation of polyamines and hydroxycinnamic acid-polyamine conjugates in tobacco plants transformed by the *rolA* locus from the Ri TL-DNA of *Agrobacterium rhizogenes*, *Plant Sci.* 80:145.

Taylor, B., Amasino, R., White, F., Nester, E., and Gordon, M., 1985, T-DNA analysis of plants regenerated from hairy root tumors, *Mol. Gen. Genet.* 201:554.

Tepfer, D., 1982, La transformation génétique de plantes supérieures par *Agrobacterium rhizogenes*, In: "2e Colloq. Recher. Fruitières," Cent. Tech. Interprof. des Fruits et Légumes, Bordeaux.

Tepfer, D., 1983a, The biology of genetic transformation of higher plants by *Agrobacterium rhizogenes*, *in*: "Molecular Genetics of the Bacteria Plant Interaction," A. Puhler, ed., Springer-Verlag, Berlin.

Tepfer, D., 1983b, The potential uses of *Agrobacterium rhizogenes* in the genetic engineering of higher plants: nature got there first, *in*: "Genet. Eng. in Eukaryotes," P. Lurquin, and A. Kleinhofs, eds., Plenum Press, New York.

Tepfer, D., 1984, Transformation of several species of higher plants by *Agrobacterium rhizogenes*: sexual transmission of the transformed genotype and phenotype, *Cell* 47:959.

Tepfer, D., 1989, Ri T-DNA from *Agrobacterium rhizogenes*: a source of genes having applications in rhizosphere biology and plant development, ecology and evolution, *in*: "Plant-Microbe Interactions," T. Kosuge, and E. Nester, eds., McGraw Hill, New York.

Tepfer, D., and Tempé, J., 1981, Production d'agropine par des racines formées sous l'action d'*Agrobacterium rhizogenes*, souche A4, *C. R. Acad. Sci. Paris* 292:153.

Tepfer, D., Goldmann, A., Fleury, V., Maille, M., Message, B., Pamboukdjian, N., Boivin, C., Dénarié, J., Rosenberg, C., Lallemand, J.Y., Descoins, C., Charpin, I., and Amarger, N., 1988a, Calystegins, nutritional mediators in plant-microbe interactions, *in*: "Molecular Genetics of Plant-Microbe Interactions," R. Palacios, and D. Verma, eds., APS Press, St. Paul.

Tepfer, D., Goldmann, A., Pamboukdjian, N., Maille, M., Lépingle, A., Chevalier, D., Dénarié, J., and Rosenberg, C., 1988b, A plasmid of *Rhizobium meliloti* 41 encodes catabolism of two compounds from root exudate of *Calystegia sepium.*, *J. Bacteriol.* 170:1153.

Vilaine, F., Charbonnier, C., and Casse-Delbart, F., 1987, Further insight concerning the TL-region of the Ri plasmid of *Agrobacterium rhizogenes* strain A4: transfer of a 1.9 kb fragment is sufficient to induce transformed roots on tobacco leaf fragments, *Mol. Gen. Genet.* 210:111.

White, F., Ghidossi, G., Gordon, M., and Nester, E., 1982, Tumor induction by *Agrobacterium rhizogenes* involves the transfer of plasmid DNA to the plant genome., *Proc. Natl. Acad. Sci. U.S.A.* 79:3193.

White, F., Taylor, B., Huffman, G., Gordon, M., and Nester, E., 1985, Molecular and genetic analysis of the transferred DNA regions of the root inducing plasmid of *Agrobacterium rhizogenes*, *J. Bacteriol.* 164:33.

Willmitzer, L., Sanchez-Serrano, J., Bushfeld, E., and Schell, J., 1982, DNA from *Agrobacterium rhizogenes* is transferred to and expressed in axenic hairy root plant tissues, *Mol. Gen. Genet.* 186:16.

MODELING ROOT SYSTEM MORPHOLOGY IN RICE

Shigenori Morita and Jun Abe

Faculty of Agriculture
The University of Tokyo
Tokyo 113, Japan

INTRODUCTION

Although rice is one of the most important crops in the world, especially in Asia, and many agronomic studies have been done, information on development of the root system is limited compared with the shoot. However, most agricultural practices for growing rice (e.g., fertilization and water management) affect root growth. Several researches, in fact, suggest that the dimension and distribution of the root system have possible relations to yield in rice (Kawata et al., 1978c; Yamazaki et al., 1980; Morita et al., 1988b; Mawaki et al., 1990). Therefore, knowledge of structure and development of the root system is essential for selecting effective practices in rice cultivation, though it is quite tedious and time-consuming to study the root system of rice plants grown under flooded conditions. In this chapter, I will briefly review the structure and development of the rice root system; then I will discuss the mathematical modeling of root system morphology and the applications of such modeling (see chapter by Dickmann and Hendrick in this volume).

THE RICE ROOT SYSTEM

Structure of the Root System

The fundamental structure of the rice root system is almost the same as that of other cereal crops (Fig. 1). The root system of a rice plant consists of two kinds of main axes: one seminal root originating from the seed and hundreds of adventitious roots formed from the stem. The seminal root emerges first, followed by the adventitious roots. Formation, emergence, elongation and branching of the adventitious roots proceed acropetally along both the main stem and tillers, to keep pace with successive leaf emergence (Fujii, 1961; Kawata et al., 1963). Both seminal and adventitious roots form first-order lateral roots, second-order lateral roots, and so on, in order. Sometimes up to fifth-order lateral roots are found (Kawata and Soejima, 1974). Moreover, the main axes of seminal and adventitious roots, as well as most of their laterals, bear root hairs (Kawata and Chung, 1976; Kawata and Ishihara, 1959).

Biology of Adventitious Root Formation, Edited by
T.D. Davis and B.E. Haissig. Plenum Press, New York. 1994

Development of the Root System

Using the monolith method (Bohm, 1979), Kawata et al. (1963) observed root system development of rice plants grown in a farmer's paddy field. They excavated the soil monolith at several growth stages and carefully washed-out the root system by spraying water. It was shown that the root system of a rice plant enlarged concurrently with shoot development, whereas the relative shape of the root system was almost constant throughout the developmental stages (Fig. 1).

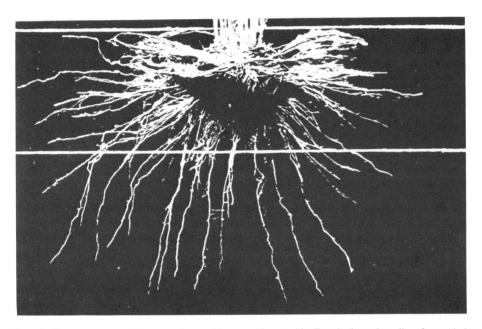

Figure 1. Rice root system at the harvest stage. Upper and lower white lines indicate the soil surface and plow pan, respectively; the distance between them is about 15 cm.

Because root system development proceeds in harmony with shoot development, root-shoot relationships were examined. Periodical analyses on the relationships between shoot weight and root weight showed the following allometry between them, until the heading stage (Mori, 1960):

$$\log_{10}(root\ weight) = a[\log_{10}(shoot\ weight)] + b \tag{1}$$

where "a" is the "relative growth coefficient," which is a significant index to characterize the developmental pattern, and "b" is a constant. Recently a similar relationship was found between root length and leaf area. At that time it was also shown that the relative growth coefficient changed with the speed of leaf emergence (Suga and Yamazaki, 1988).

Although root weight is a useful parameter to describe the root system dimension, it is necessary to study the root system in terms of structure—because the adventitious roots, which emerge and elongate successively, are different from one another in their number, diameter, length and branching, depending on the exact position of the root formation on the stem (Fujii, 1961; Kawata et al. 1963; Kawata et al., 1980; Yamazaki and Nemoto, 1986).

Dimensions of the Root System

The total numbers and total lengths of adventitious roots of rice increase with shoot growth (Mori, 1957; Inada, 1967). The total number of adventitious roots reaches a maximum value at about the heading stage, and does not increase thereafter. The root system at the heading stage consists of hundreds of elongated adventitious roots with thousands of lateral roots and hundreds of stunted, short roots which are less than five cm in length (Kawata et al, 1978a). The number of the stunted roots increases in the later growth stages. Moreover, some primordia of the adventitious roots formed in the stem are dormant and they do not emerge from stem (Kawata et al., 1978b).

It is known that the main axes of most adventitious roots are < 40 cm in length (Kawata et al., 1963). The total length of adventitious root axes also reaches a maximum value at about the heading stage, but the developmental stage at which the maximum length is reached differs slightly in each case (Mori, 1957; Inada, 1967). It is reported that the time when emergence and elongation of the main axes ends depends on the total leaf number of a main stem (Kawashima, 1983). Kawata and Soejima (1974), on the other hand, reported that some adventitious roots and many lateral roots continue to elongate after heading and through the ripening stage. Therefore, it is probable that the total length of roots, including lateral roots, increases after heading. There is a need for further research on the growth and development of rice roots after the heading stage, including decreases in total root length due to root death, because root turnover in rice has seldom been studied.

Distribution of the Root System

Root length density is often used to describe the root system morphology of upland crops [e.g., Gregory et al. (1978)]. This is defined as the total root length (including lateral roots of any order) per unit soil volume and is usually expressed as cm/cm^3 or cm^{-2}. Studies that address the rice root system with reference to root length density, however, are quite limited (Anon., 1978, 1979). Recently several researches on the rice root system have been done in Japan (Morita et al., 1988a; Mawaki et al., 1990) and in USA (Beyrouty et al., 1988; Slaton et al., 1990) by using root length density. However, it is not always easy to compare such studies because the environmental conditions and the cultivation methods are often quite different.

In Japan, rice seedlings were transplanted experimentally to make hills of three plants. Then soil cores were taken with a core sampler of 15 cm diameter below and between hills at the ripening stage (Fig. 2). The total root length in each soil layer was determined by a root length scanner (Morita et al., 1988c) and then were divided by the corresponding soil volumes to calculate root length density. The analyses indicated that root length density decreased exponentially with soil depth, though there was some variation in the surface soil layers. In the surface soil layers, the root length density below hills was higher than that between hills. In the deeper soil layers at both geographic locations, the root length density was comparable. Most parts of a root system were distributed in the upper 30 cm of soil (Morita et al., 1988c; Mawaki et al., 1990).

In USA, the root system of rice plants grown in rows was examined using a minirhizotron (Sanders and Brown, 1978) with reference to developmental stages including the ratoon stage (Beyrouty et al., 1988; Slaton et al., 1990). Total root length, in general, increases to reach a maximum at panicle initiation to the booting stage. Decrease in root length, however, occurred at several growth stages, which might be due to the change in the partitioning. Distribution of root length density showed that most parts of the root system were distributed in the upper 40 cm of soil throughout the growing period.

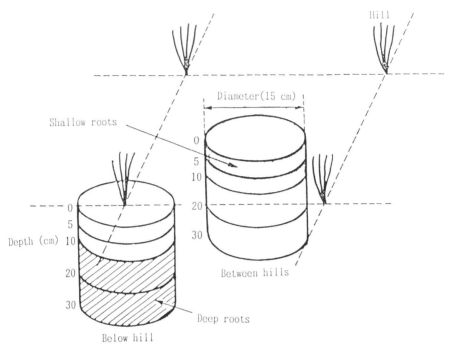

Figure 2. Schematic showing positions of sample collection for root length density measurements.

MODELLING THE ROOT SYSTEM

The Growth Direction of Adventitious Roots

The distribution of the rice root system is mainly regulated by the growth direction of adventitious roots, because adventitious roots comprise the bulk of the root system, as in other upland cereal crops (Oyanagi et al., 1993). It is often written in textbooks that plant roots show positive gravitropism and elongate vertically. Because the primary seminal root of maize usually shows orthogravitropism and elongates in the vertical direction, it is often used as material to study the gravitropic response of roots (Jackson and Barlow, 1981). It has been shown, however, that adventitious roots composing the rice root system show plagiogravitropism and elongate at various angles (Fig. 1) (Kawata et al., 1963; Oyanagi et al., 1993). Therefore, there is a need to study the growth direction of each adventitious root to characterize the distribution of the whole root system.

Several researchers have attempted to quantitatively describe the growth direction of adventitious roots. Methods to estimate the growth direction of individual adventitious roots have been developed and improved (Fig. 3) (Kawata and Katano, 1976; Yamazaki et al., 1981; Morita et al., 1986, 1987). The principle of the method is as follows: Root samples are collected with a core sampler of 15 cm diameter below the hill. After washing out the root system, the radius of the hill (r_0) and the length of each adventitious root (l) are measured. Because the adventitious roots of rice plants elongate almost straight in various directions, the length of an adventitious root plus the radius of the hill ($L = 1 + r_0$) should be linear to the cosine of its growth direction (θ). Therefore, the growth direction of each adventitious can be calculated with the equation:

$$\theta = \arccos(R/L) \tag{2}$$

where R is the radius of the core sampler.

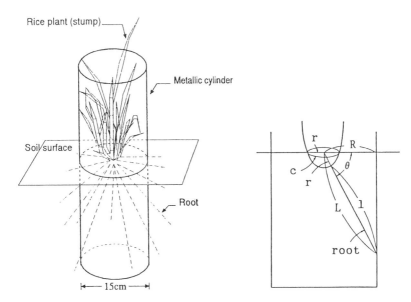

Figure 3. Schematic of the method for estimating the growth direction of rice adventitious roots. The method includes the parameters θ, growth direction of the adventitious root; 1, length of adventitious root; c, circle of the crown (= $2\pi r_0$); r, radius of the hill; L, corrected root length (= $1 + r_0$); R, radius of the sampler.

Basic Model of the Root System

After the method to estimate the growth direction of adventitious roots has been established, there is a need to define a standard root system with which comparisons can be made. A basic model of the root system was proposed with the assumption that adventitious roots elongate uniformly in the soil (Morita et al., 1984). From the assumption, P_θ, the probability of adventitious roots with the growth direction of θ is proportional to $2\pi\cos\theta$. That means that the relative frequency of adventitious roots with θ is $\cos\theta$ (Fig. 4). Then,

$$\int_0^{\pi/2} P_\theta d\theta = [\theta\sin\theta + \cos\theta]_0^{\pi/2} \tag{3}$$

$$= (\pi/2 - 1) \tag{4}$$

195

Therefore, the average growth angle of all the adventitious root is about 32.7° in the basic model. In addition,

$$\int_0^\theta P_\theta dx = \sin\theta \quad (because \quad \int_0^{\pi/2} P_\theta dx = 1)$$ (5)

Since $\sin\theta$ equals 0.5 when q is 30°, the number of adventitious roots with growth angle of 0° through 30° and those of 30° through 90° should be the same in the basic model.

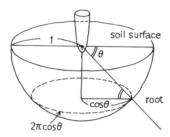

Figure 4. Schematic of the basic model for a root system.

This model is a kind of scale to analyze root system morphology with reference to growth direction of adventitious roots. The spatial distribution of adventitious roots in a root system could be analyzed quantitatively, if compared with this model. A case study showed that the spatial distribution of adventitious roots in a root system was similar to the basic model (Harada et al., 1986).

Root Length Density Model

Assumptions for the Model. Recently a root length density model was constructed based on three simple assumptions (Suga et al., 1988). Because the root length density model evolved from the basic model mentioned above, the first assumption is the same as the basic model of root system: all the adventitious roots of any hill, H_i, elongate uniformly in various directions from the central point of the hill, O_i, which is located at the soil surface level (Fig. 5). The second assumption is that root branching degree is constant along the main axis of any adventitious root. That means that the total length of lateral roots formed on a unit length of adventitious roots is constant. The third assumption is that any root system has a hemispheric rooting zone. The rooting zone of H_i is a double-hemispheric space from the hemisphere with radius of r_0, to the hemisphere with radius of r_{max} (Fig. 5).

196

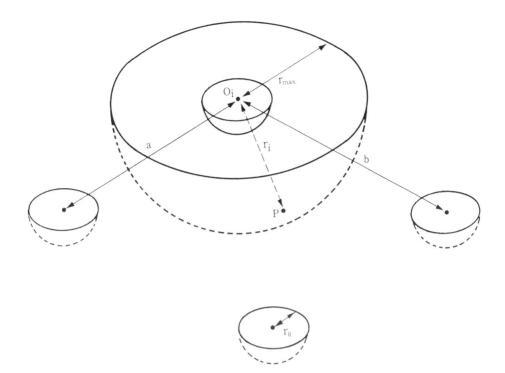

Figure 5. Schematic of the root length density model. Parameters include O_i, central point of the hill H_i; r_0, radius of hill; r_{max}, radius of rooting zone; P, point in soil; r_i, distance between O_i and P; a, distance between rows; b, distance between hills.

Fundamental Formulae. Consider a small soil volume, DV, around any point, P, in soil. Let DRL be the total root length derived from all the hills in the population in DV, and let DRL_i be its part from the single hill H_i, so that:

$$\Delta RL = \sum_i \Delta RL_i \tag{6}$$

Root length density at P, ρ, and its part derived from a single hill H_i, ρ_i, then are defined as follows:

$$\rho_i = \lim_{\Delta V \to 0} (\Delta RL_i / \Delta V) \tag{7}$$

$$\rho = \lim_{\Delta V \to 0} (\Delta RL / \Delta V) \tag{8}$$

$$= \lim_{\Delta V \to 0} \{(\sum_i \Delta RL_i) / \Delta V\} \tag{9}$$

$$= \sum_i \rho_i \tag{10}$$

From the first assumption, the number of adventitious roots elongating from O_i to ΔV is inversely proportional to the square of the distance between O_i and P. From the second assumption, ΔRL_i which is the total length of roots from H_i in ΔV, is also inversely proportional to the square of the distance between O_i and P. From the third assumption, when P is not in the rooting zone of H_i, ΔRL_i is zero. Thus, let r_i be the distance between O_i and P, and ρ_i can be expressed as:

$$\rho_i = \lim(\Delta RL_i / \Delta V) \tag{11}$$

$$= k/r_i^2 \ (r_0 \leq r_i \leq r_{max}) \tag{12}$$

$$= 0 \quad (r_i < r_0 , \ or \ r_i > r_{max}) \tag{13}$$

where k is a constant of proportionality, termed a root length density constant, which is related to the branching degree of adventitious roots. Equations (10), (11) and (12) indicate that the variable ρ is a function of the position of P. If one could know the distance between rows and hills, and if the values of r_0, r_{max} and k were known, it would be possible to calculate values of ρ with equation (10), (12) and (13). In the simulation program, values of r_{max} and k were evaluated so that simulated values of ρ_{model} fit the actual field data of ρ_{actual}.

Table 1. Parameters included in the simulation program. See text for details.

Category	Symbol	Name
Input	a	Distance between rows (cm)
	b	Distance between hills (cm)
	r_0	Radius of hill (cm)
	ρ_{actual}	Measured root length densities (cm/cm^3)
Output	k	Root length density constant
	r_{max}	Radius of rooting zone (cm)
	ρ_{model}	Simulated root length densities (cm/cm^3)

Simulation Program and Parameters. The parameters of our simulation program are listed in Table 1. The value of k is evaluated so that the average of ρ_{model} is equal to the average of ρ_{actual}. The best value of r_{max} is selected with the method of least squares, i.e., the sum of squared differences between ρ_{actual} and ρ_{model} is minimized to give the best value of r_{max}. The flow chart of the simulation program is shown in Figure 6 (Suga et al., 1988). After the values of k and r_{max} are determined, several indices of root morphology can be calculated from simulation parameters and the number of adventitious roots (Table 2) (Suga et al., 1988; Morita et al., 1988b).

Table 2. Parameters and indices in control and shaded plots.

Parameters and Indices	Symbol	Unit	Control Plot	Shaded Plot
Distance between rows	a	cm	26.4	26.4
Distance between hills	b	cm	18.1	18.1
Radius of hill	r	cm	1.80	1.67
Total number of elongated roots per hill	N_r		539	498
Actual root length density	ρ_{actual}	cm^{-2}	16.4	12.7
Root length density constant	k		1211	985
Radius of rooting zone	r_{max}	cm	34	30
Model root length density	ρ_{model}	cm^{-2}	16.4	12.7
Model root length	$r_{max}.r_0$	cm	32	28
Coefficient of branching	$2\pi k/N_r$		14	12
Adventitious root length per hill	$Nr(r_{max}.r_0)$	km hill^{-1}	0.17	0.14
Total root length per hill	$2\pi k(r_{max}.r_0)$	km hill^{-1}	2.4	1.8
Adventitious root length per unit area	$Nr(r_{max}.r_0)/ab$	km ha^{-1}	360	300
Total root length per unit area	$2\pi k(r_{max}.r_0)/ab$	km ha^{-1}	5100	3700

An Example Analysis. Hokuriku is one of the most famous growing areas of rice in Japan, but recently the yield of rice has varied and not increased. It has been reported that meteorological conditions after the panicle initiation state may have been detrimental to yields. Thus, in order to examine the effect of solar radiation on the root system development of rice plants, cv. Koshihikari was grown under control and shaded conditions after the heading stage. At the grain-filling stage, root samples in two plots were taken with a core sampler. After measuring root length with a Root Length Scanner, root length densities were calculate for an analysis by means of the root length density model. The root length densities were simulated under various combinations of the root length density constant (k) and the radius of rooting zone (r_{max}) in the model. The most suitable values of both parameters were selected, so that the simulated root length densities would fit the actual densities. The values of k and r_{max} thus obtained, as well as other measured data, made it possible to calculate indices relating to the root system morphology; e.g., root length, coefficient of branching and total root length per hill or per unit area—which are hardly accessible in field studies. The comparison of these parameters and indices between two plots suggested that the root system development was clearly poor in the shaded plot as had been expected (Table 2). The simulated root length densities were substantially close to the actual densities (Fig. 7) (Morita et al., 1988b).

Modelling as a Method of Studying the Root System. In summary, the goal of root system modeling is not always to increase the fit of the model to the dimension or distribution of the actual root system. In the case of the root length density model described above, the model is a tool to examine, compare and evaluate root system morphology by determining the root length density constant and radius of rooting zone. In addition, several other morphological indices of the root system (e.g., total root length per hill and per unit area) can be obtained through the simulation. For that purpose, a model based on fewer and simpler assumptions is better. Using the root length density model, root system morphology is presently being examined in different rice cultivars (genotypes). Ultimately, this effort should aid in formulating more effective cultivation practices for rice. In addition, more information on root system morphology could aid genetic improvement programs in the development of root systems which better utilize resources such as water and nutrients.

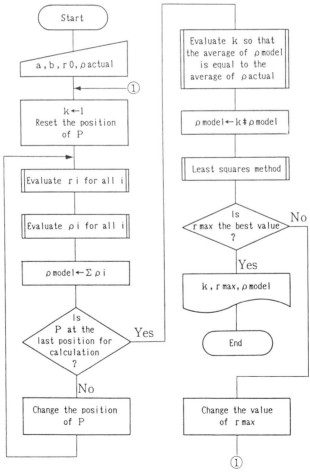

Figure 6. Flow chart of simulation by the root length density model.

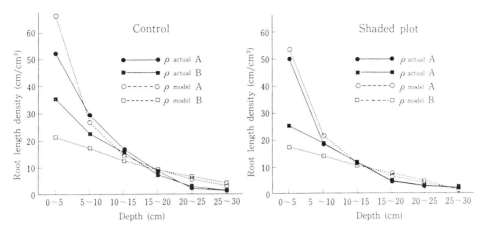

Figure 7. Comparisons between ρ_{actual} and ρ_{model} in the control and shaded plots. $\rho_{actual\ A}$, measured root length densities below hills; $\rho_{actual\ B}$, measured root length densities between hills; $\rho_{model\ A}$, simulated root length densities below hills; $\rho_{model\ B}$, simulated root length densities between hills.

REFERENCES

Anon., 1978, "IRRI Annual Report" for 1977," IRRI, Los Banos.

Anon., 1979, "IRRI Annual Report for 1978," IRRI, Los Banos.

Beyrouty, C.A., Wells, B.R.,Norman, R.J., Marvel, J.N., and Pillow, J.S., 1988, Root growth dynamics of a rice cultivar grown at two locations, Agron. J. 80:1001.

Bohm, W., 1979, Methods of studying root systems, Springer-Verlag, Berlin.

Fujii, Y., 1961, Studies on the regular growth of the roots in rice plants and wheats, *Agric. Bull. Saga Univ.* 12:1.*

Gregory, P.J., McGowan, P.J., Biscoe, P.V., and Hunter, R., 1978, Water relations of winter wheat, 1. Growth of the root system, J. Agric. Sci. Camb. 91:91.

Harada, J., Maeda, T., and Yamazaki, K., 1986, Spatial distribution of primary roots of rice hills with different plant numbers, *Jpn. J. Crop Sci. 55* (Extra Issue 1):62.**

Inada, K., 1967, Physiological characteristics of rice roots, especially with the viewpoint of plant growth stage and root age, *Bull. Nat. Inst. Agric. Sci.* D16:19.*

Jackson, M.B., and Barlow, P.W., 1981, Root geotropism and the role of growth regulators from the cap: a re-examination, *Plant Cell Environ.* 4:107.

Kawashima, C., 1983, Difference of the finishing time of crown roots elongation among rice cultivars with different number of leaves on the main stem, *Jpn. J. Crop Sci.*52:475.*

Kawata, S., and Chung, W., 1976, Formation of root hair in lateral roots of rice plants, *Proc. Crop Sci. Soc. Jpn.* 45:436.*

Kawata, S., and Ishihara, K., 1959, Studies on the root hairs in rice plants, *Proc. Crop Sci. Soc. Jap.* 27:341.*

Kawata, S., and Katano, M., 1976, On the direction of the crown root growth of rice plants, *Proc. Crop Sci. Soc. Jpn.* 45:471.*

Kawata, S., and Soejima, M., 1974, On superficial root formation in rice plants, *Proc.Crop Sci. Soc. Jpn.* 43:354.*

Kawata, S., El-Aishy, S.M., and Yamazaki, K., 1978a, The position of the formation of "stunted roots" in rice plants taken from the actual paddy fields, *Jpn. J. Crop Sci.* 47:609.*

Kawata, S., Harada, J., and Yamazaki, K., 1978b, On the number and the diameter of crown root primordia in rice plant, *Jpn. J. Crop Sci.* 47:644.*

Kawata, S., Sasaki, O., and Yamazaki, K., 1980, On the lateral root formation in relation to the diameter of crown roots in rice plants, Jpn. J. Crop Sci. 49:10.

Kawata, S., Soejima, M., and Yamazaki, K., 1978c, The superficial root formation and yield of hulled rice, *Jpn. J. Crop Sci.*47:617.*

Kawata S., Yamazaki, K., Ishihara, K., Shibayama, H., and Lai, K.-L., 1963, Studies on root system formation in rice plants in a paddy, *Proc. Crop Sci. Soc. Jpn.*32:163.*

Mawaki, M., Morita, S., Suga, T., Iwata, T., and Yamazaki, K., 1990, Effects of shading on root system morphology and grain yield of rice plants (*Oryza sativa* L.), I. An analysis on root length density, Jpn. J.Crop Sci. 59:89.*

Mori, T., 1957, Studies on ecological characters of rice root, Part 3. Development of root system with reference to the top growth, *Bull. Inst. Agric. Res. Tohoku Univ.*8:265.*

Mori, T., 1960, The relative growth of the root and shoot in rice plants, *Proc. Crop Sci. Soc. Jpn.* 29:69.*

Morita, S., Iwabuchi, A., and Yamazaki, K., 1986, Relationships between the growth direction of primary roots and yield in rice plants, *Jpn. J. Crop Sci.* 55:520.*

Morita, S., Iwabuchi, A., and Yamazaki, K., 1987, Relationships between the growth direction of primary roots and shoot growth in rice plants, *Jpn. J. Crop Sci.* 56:530*

Morita, S., Suga, T., and Yamazaki, K., 1988a, The relationship between root length density and yield in rice plants, *Jpn. J. Crop Sci.* 57:438.*

Morita, S., Suga, T., and Nemoto, K., 1988b, Analysis on root system morphology using a root length density model, U. Examples of analysis on rice root systems, *Jpn. J. Crop Sci.* 57:755.

Morita, S., Nemoto, K., Nakamoto, T., and Yamazaki, K., 1984, A method to estimate and to evaluate the spatial distribution of primary roots in rice, *Jpn. J. Crop Sci.*53 (extra issue 2):226.**

Morita, S., Suga, T., Haruki, Y., and Yamazaki, K., 1988c, Morphological characters of rice roots estimated with a root length scanner, *Jpn. J. Crop Sci.*57:371.

Oyanagi, A., Nakamoto, T., and Morita, S., 1993, The gravitropic response of roots and the shaping of the root system in cereal plants, *Envrn. Exp. Bot.* 33:141.

Sanders, J.L., and Brown, D.A., 1978, A new fiber optic technique for measuring root growth of soybeans under field conditions, *Agron. J.*70:1073.

Slaton , N.A., Beyrouty, C.A., Wells, B.R., Norman, R.J., and Gbur, E.E., 1990, Root growth and distribution of two short-season rice genotypes, *Plant and Soil* 121:269.

Suga, T., and Yamazaki, K., 1988, Developmental changes in root characters and growth correlation between root mass and leaf mass in rice, *Jpn. J. Crop Sci.*57:671.*

Suga, T., Nemoto, K., Abe, J., and Morita, S., 1988, Analysis on root system morphology using a root length density model, I. The model, *Jpn. J. Crop Sci.* 57:749.

Yamazaki, K., and Nemoto, K., 1986, Morphological trends and interrelationships of leaves, stem parts and roots along the main axes of rice plants, *Jpn. J.Crop Sci.* 55:236.*

Yamazaki, K., Katano, M., and Kawata, S., 1980, The relationship between the number of ears and the number of crown roots on a hill of rice plants, *Jpn. J. Crop Sci.* 49:317.*

Yamazaki, K., Morita, S., and Kawata, S., 1981, Correlations between the growth angles of crown roots and their diameters in rice plants, *Jpn. J.Crop Sci.* 50:452.*

---

*In Japanese with English summary, **in Japanese.

MODELING ADVENTITIOUS ROOT SYSTEM

DEVELOPMENT IN TREES: CLONAL POPLARS

Donald I. Dickmann and Ronald L. Hendrick

Department of Forestry
Michigan State University
East Lansing, MI 48824-1222 USA

INTRODUCTION

Mechanistic growth modelers have recently become interested in exploring the mysteries of tree root systems. At the outset of these modeling efforts, some understanding of the morphology, ecology and physiology of roots, and how they correlate with the aerial parts of the tree is essential. The growing tree is an integrated system, with water, minerals, nitrogenous compounds, carbohydrates, growth regulators and other organic substances moving freely, though often phasically, between the roots and the shoots. A perturbation or stress in one part of the tree is sensed and reacted to in all others. In addition, roots grow in a complex, heterogeneous soil environment. If we are to improve upon existing mechanistic and predictive models of tree growth, or to build more responsive new models, the physiology and ecology of this integrated shoot-root-soil system must be better understood.

In this paper we do not attempt to outline the procedures involved in building a root model. No algorithms are presented. Nor do we describe a successful mechanistic or morphological root model for poplars (genus *Populus*) because, to our knowledge, there are none. Rather, we will present a summary of our current knowledge of poplar root morphology, ecology and physiology. We also will present some inferences and conceptual models which should help modelers develop dynamic root models. Our premise is that a root model is no better than the biological foundation upon which it rests, both in terms of process flow within the model and its predictive output.

Most of the research to date on forest tree root systems has been done with conifers, so the literature on roots of hardwood forest trees is fairly depauperate. However, among hardwood forest tree taxa, poplars are fairly well represented (Dickmann and Pregitzer, 1992). There also is an extensive literature on the roots of horticultural woody crops, which we will draw upon when necessary to amplify a point. The work on root ecophysiology of short-rotation poplar hybrids and other hardwood taxa that has been done at Michigan State University also will be heavily drawn upon in our discussion.

Throughout our discussion we will use certain terms related to tree root systems in

Biology of Adventitious Root Formation, Edited by
T.D. Davis and B.E. Haissig, Plenum Press, New York, 1994

a consistent manner. **Fine roots** are those < 2 mm in diameter. These fine roots may be white, indicating they are primary growing points, or they may be brown, indicating suberization and the inception of secondary thickening. **Coarse roots** are > 2 mm in diameter and usually have undergone some secondary thickening. The literature about roots usually contains the aforementioned somewhat arbitrary terms. **Root architecture** refers to the morphology and physical structure of the root system of a tree. With reference to the physiology of root and shoot growth, **allocation** refers to a process of distribution of carbon (C), nitrogen (N) and other substances within the plant; the end product of this process is **biomass accumulation**.

EXISTING MODELS OF WOODY PLANT ROOT SYSTEMS

There are two types of mechanistic growth models (Landsberg, 1986) which can be applied to roots. Detailed "bottom-up" models synthesize growth by calculating the actions and interactions of the physiological processes contributing to it. Although some attempt has been made to construct bottom-up models for forest trees, we feel that more efforts in this area are warranted, especially if a root system component is included. The power of current computer technology certainly opens the way for additional bottom-up modeling, which can be quite complex because of the multitude of process flows and interactions among variables. Empirical "top-down" models use some simplified formulation(s) of major physiological processes contributing to growth and their response to major "driving" variables. Yield data, as the expression of physiological processes, are analyzed in terms of the driving variable (e.g. solar energy) and modifying factors. Top-down models, though useful in certain contexts, are less satisfying to the physiological ecologist because they are not very explanatory of the action and interaction of the multifarious physiological processes that determine growth in a particular environmental context.

There are several models that simulate tree root growth, ranging in scale from single-tree to stand and to ecosystem-level models. Top-down, species-specific, individual-tree computer models include the TREGRO model of red spruce sapling growth (Weinstein and Beloin, 1990; Weinstein et al., 1991) and the ECOPHYS model of first-year poplar growth (Host et al., 1990; Isebrands et al., 1990). TREGRO simulates red spruce growth under acid rain, nutrient or ozone stress. Biomass growth and non-structural C are allocated to various tree components (including roots) based on the proximity of the component sink to the source and on the time of the growing season. The effects of changes in solar radiation and temperature on total hybrid poplar growth can be simulated by ECOPHYS. The allocation of growth to roots is determined by a fixed set of transfer coefficients that can be altered initially by the user, but that do not change seasonally.

Of a more generic nature are the CARBON model (Bassow et al., 1990) and the Plant-Growth-Stress model of tree response to drought, ozone and stem density (Chen and Gomez, 1990). Root mass is maintained as a fixed proportion of foliage mass in CARBON, and a fixed proportion of fine roots are also assumed to turn over every year. Total tree, including root, growth in the Plant-Growth-Stress model is sensitive to changes in soil and atmospheric properties, and to the vertical distribution of tree roots within the soil changes according to fluctuations in soil moisture status.

A number of stand-level models of forest growth contain a root component. An allometric stand development model of *Pinus sylvestris* growth has been constructed in which root growth is driven by total needle area and by the ratio between nutrient demand and root uptake efficiency (Makela and Hari, 1986; Nikinmaa and Hari, 1990). Ewel and Gholz (1991) have developed a rather comprehensive model of below-ground dynamics for a *Pinus elliottii* plantation which incorporates root growth, mortality, decomposition and respiration. The model simulates C, P and water fluxes to, within and from the forest, and it is sensitive to changes in precipitation, soil and air temperature, solar radiation and relative humidity. King's (1993) model relates growth of even-aged conifer stands to N uptake and the accumulation of biomass among leaves, fine roots and woody tissues.

The computer model of Rastetter et al. (1991) is designed to predict the response of ecosystem C and N cycles to changes in climate, atmospheric CO_2 and pollutant N inputs. The model's fine root component includes the processes of growth, respiration, turnover and nutrient uptake. FOREST-BGC is an ecosystem model that uses stand water and N limitations to alter leaf/root/stem C allocation dynamically at each annual iteration (Running and Gower, 1991). Root maintenance respiration rates change with temperature, but the root turnover rate (proportion per year) is fixed.

The models discussed above are representative of the relatively few that consider below-ground growth. Many of these models are primarily allocation models, designed to study the effects of the environment on biomass accumulation, or on C and nutrient allocation and litter production. However, most lack a root-based physiological component and are less useful for studying short-term changes in root production and mortality, or the transfer of stored resources between the root system and above-ground plant parts. The absence of models describing these processes largely reflects the difficulty of studying and characterizing the root systems of trees. However, advances in both technology and methods for studying roots—e.g., minirhizotrons and video imaging (Hendrick and Pregitzer, 1992) and dual isotope labelling (Horwath, 1993)—have facilitated our understanding of root form and function, and will eventually permit the construction of more detailed, mechanistic models of below-ground processes.

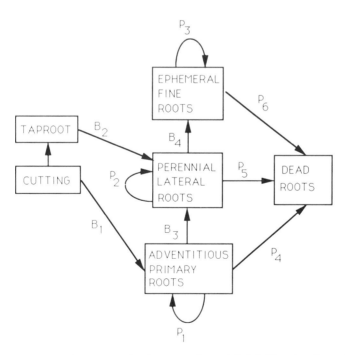

Figure 1. Conceptual model of root system development and growth in hybrid poplars. Boxes (state variables) represent pools of biomass or length in a given class of roots. Branching probabilities or rates are denoted by B_is, while P_is are the probability of a root (or proportion of a total pool) undergoing transition from one stage of development to another during some time interval.

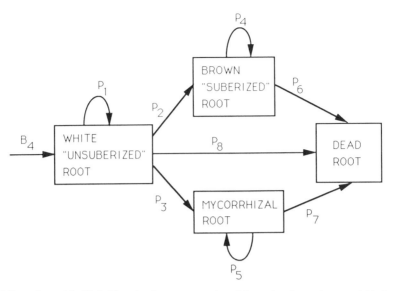

Figure 2. Life cycle model of hybrid poplar fine root growth and dynamics. Boxes (state variables) represent pools of root length or mass in a given stage of development. Probabilities of a root (or proportion of all roots) undergoing transition from one stage to another are denoted by P_is. The rate of new fine root production is represented by B_4. [cf. Fig. 1]

While not mathematical modelers, we (along with our colleague, Dr. Kurt Pregitzer), have developed a series of conceptual life-cycle models of root growth, mortality and internal resource transfers in poplars that lend themselves to formal modeling efforts (Figs. 1 and 2). These models are an extension of the one presented by Hendrick and Pregitzer (1992) for fine roots in a sugar maple forest. Boxes (state variables) in the diagram represent pools (biomass or length) of roots in different stages of development. Arrows represent transition rates of biomass (or length) during root system development, or fluxes of carbohydrates, nutrients, growth regulators, etc. from one pool to another.

There are distinct changes in root chemistry, form and function as new roots age, senesce and die. Mycorrhizal status also can affect fine root C costs and nutrient uptake capabilities (Harley and Smith, 1983; Allen, 1991), and possibly root longevity. The rate at which roots pass from one stage to another, and the proportion of total root system length or biomass within a given stage, can have significant effects on the balance of C and nutrients within the tree, as well as on whole-tree physiology. If chemical and functional changes associated with the state transitions depicted in Figs. 1 and 2 can be established, quantifying the relative or absolute amount of fine roots within a particular stage of development should lead to better understanding of both resource allocation and uptake capabilities within the root system.

ARCHITECTURE OF THE ADVENTITIOUS ROOT SYSTEM

Species and hybrids in the Aigeiros and Tacamahaca sections of the genus *Populus* are among the few tree taxa that readily produce adventitious roots on hardwood cuttings (i.e., segments of dormant stem wood, usually one-year-old in practice). This remarkable predisposition to adventitious rooting (see chapters by Barlow and by Haissig and Davis in this volume) is one of the primary reasons why poplars are used extensively throughout the world in clonal plantation forestry (see chapter by Ritchie in this volume). That clonal plantations of poplars derived from cuttings are supported on root systems that are, in origin, completely adventitious is a point worth emphasizing. The ease with which poplars can be clonally propagated has also lead to the development and commercial production of numerous genetically improved cultivars (Dickmann and Stuart, 1983).

Anatomically, most of the adventitious roots formed by clonally propagated poplars are derived from preformed root primordia (also called latent root primordia) in the periderm of the stem (Luxova and Lux, 1981a,b), with the number of primordia declining from the base to the tip of a one-year-old shoot (Bloomberg, 1959, 1963; Smith and Wareing, 1974). Induced adventitious roots also will form in the callus that proliferates at the base of a planted cutting. Adventitious rooting in poplars, as in all clonally propagated plants, is under strong genetic control, although it can be modified by environmental conditions (Haissig et al., 1992). Stem cuttings of species and hybrids from section Leuce, subsection Trepidae (aspens), generally do not adventitiously root because they do not produce preformed root primordia. Even in sections Tacamahaca and Aigieros, where stem cuttings usually root well, considerable genotypic variation in the extent and vigor of adventitious rooting occurs (Bloomberg, 1959, Ying and Bagley, 1977).

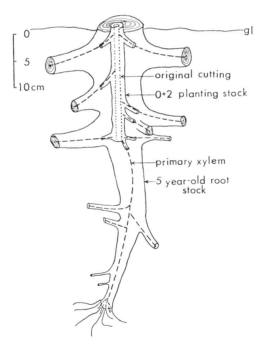

Figure 3. Five-year development of the tap root of a *Populus* x *euramericana* tree established from a hardwood cutting in the nursery and then transplanted as a two-year-old rooted plant to the field. Ground level is indicated by "gl." [from Faulkner, 1976]

The adventitious roots produced by a poplar cutting form the initial absorbing surface for the establishing plant. These roots vigorously expand into the soil during the period when buds at the apical end of the cutting are flushing to form the initial leaf surface. Although some of these initial roots may die, most continue to elongate, then become suberized and undergo secondary thickening to form the structural architecture of the root network. The cambium of the cutting continues to lay down secondary thickening after planting, while vertical sinker roots from basal callus elongate and thicken, forming a pronounced taproot (Fig. 3) (Faulkner, 1976). The development of this taproot is highly dependent upon the characteristics of the soil pedon, especially the presence or absence of impermeable layers or a water table (Sprackling and Read, 1979).

Strongly developed horizontal coarse roots grow radially away from the taproot. Most horizontal roots are generally found within the top 50 cm of the soil surface (Fig. 4) (Faulkner, 1976; Sprackling and Read, 1979). As trees grow in height there is a proportional extension of the horizontal roots. Both Faulkner (1976) and Hansen (1981) found significant linear relationships between tree height and length of horizontal roots. Horizontal roots can be found several tree-lengths away from the base of the stem. Vertical "sinker" roots branch from the horizontal roots and explore the soil to depths of one to three meters or more (Sprackling and Read, 1979; Graham et al., 1963). Poplar trees have at least four or more orders of roots, although the maximum number of developmental orders reported for trees is seven (Sutton and Tinus, 1983). We found that most of the roots of young hybrid poplar trees >1 mm in diameter were woody (Pregitzer et al., 1990).

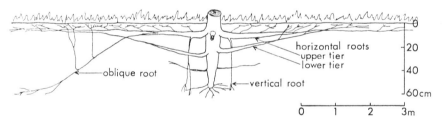

Figure 4. General appearance of the root system of a *Populus* x *euramericana* tree established from a two-year-old rooted cutting after about five years growth in the field. [from Faulkner, 1976]

Although species and cultivars within the genus *Populus* have similar root morphology, genotypic differences occur (Faulkner, 1976). These differences may be related to both the "C cost" of the root system as well as to the longevity of individual roots (Eissenstat, 1991). We found that two poplar genotypes from different sections of the genus differed radically in their allocation of C to the root system (Pregitzer et al., 1990). Fitter et al. (1991) pointed out that root architecture is of functional significance, but there is much yet to be learned about how genotypic differences in root morphology influence the acquisition of water and nutrients from the soil.

It is well known that most roots, especially fine roots, are concentrated near the soil surface or in areas of high availability of water and nutrients, and poplars are no exception. Both Elowson and Rytter (1984) and Ericsson (1984) showed that most roots in young *Salix*, *Betula* and *Alnus* plantations were in the top 10 cm of the soil. Soil drainage, texture, profile characteristics and genesis, as well as management practices such as fertilization or irrigation, interact to influence the distribution of tree roots. Because heavy equipment is often used in the establishment and tending of poplar plantations, there is a real possibility that cultural practices can negatively impact both soil porosity and root distribution.

Plantation density may also influence coarse root architecture just as it does crown architecture. Atkinson et al. (1976) demonstrated that the architecture of root systems of apple (*Malus* spp.) trees was dependent upon orchard spacing; roots of trees grown at high density were less wide-spreading than those grown at lower density. Since poplar plantations are usually composed of one or a few genotypes, root grafting is probably common (DeByle, 1964; Tew et al., 1969), so in older plantations the entire root system may be considered to be a single functional unit. It is also likely that mycorrhizal hyphae represent a continuum between the roots of individual trees, but the physiological importance of root-hyphal grafting remains largely unexplored. Accounting for these symbiotic interactions certainly would introduce another dimension of complexity into root models.

A major problem with modeling the architecture of poplar root systems is the complex, heterogeneous and competitive nature of the rhizosphere. Distinct horizons, some impermeable or inhospitable to root growth, exist in many forest soils. Rocks, gravel, dead and live roots, animal burrows etc. are distributed throughout the soil profile. Substantial spatial variation can occur within a small area. The extent and rate of root growth and development is decidedly impacted by these barriers, discontinuities and pedon changes. One approach is to assume a homogeneous soil matrix with a constant resistance to root penetration; yet it is better to set the distribution of discontinuities and barriers in a systematic or random way to mimic a particular pedon. From the standpoint of realistically modeling the dynamics of tree root systems, it is as important to realistically represent the milieu in which roots grow as it is to represent the functions and architecture of the roots themselves.

MORPHOLOGY AND DYNAMICS OF FINE ROOTS

Although coarse and fine roots are both physically and physiologically interconnected, the comparatively short functional lives of fine roots facilitates separate consideration of their growth, activity and mortality. Fine roots are small diameter (ca. one to two mm or less), relatively short-lived roots that are considered to function primarily in the uptake of water and nutrients. Defined as such, fine roots constitute only a small percentage of total tree or stand biomass (Pregitzer et al., 1990), yet can account for more than 50% of annual net biomass production (Grier et al., 1981, Keyes and Grier, 1981, Vogt et al., 1986). Substantial amounts of C and nutrients are also used for maintenance respiration, maintaining mycorrhizal fungi, uptake and transport of nutrients and production of growth regulators. Although the inclusion of a fine root component in models of tree growth and physiology should be regarded as essential, we know relatively little about the factors regulating fine root dynamics in trees. This makes the construction of mechanistic models of fine root growth and activity rather difficult.

Our experiences and the observations of other researchers support the hypothesis that there is a series of relatively discrete stages that fine roots pass through between root initiation and death (Fig. 2). Therefore, a life history, or demographic, framework categorizing different developmental stages in a fine root's "life" provides a useful conceptual model of root system development and dynamics which can be integrated into formal, mathematical models as well. Elucidating the effects of internal and external controlling factors on fine root demography should ultimately help in the construction of better models of root system behavior.

Fine Root Biomass and Nutrient Dynamics

To model biomass and nutrient allocation in the developing fine root systems of poplars and other trees, it is necessary to quantify both the rate at which fine roots are produced (B_4, Fig. 2) and the rate at which they die (P_{6-8}, Fig. 2). Measurements of fine root production and mortality rates are probably best generated from direct observations of root production and subsequent death using minirhizotrons (Hendrick and Pregitzer, 1992) or other *in situ* rhizotrons. The length of new roots produced during a given time interval can be expressed as a proportion of initial length, and a time-series of production rates can be generated for an entire season or more of growth. By monitoring individual roots, mortality can also be measured directly, and mortality rates calculated. These rates are relative measures, but they can be converted to biomass production and mortality estimates by multiplying the fine root length production and mortality rates for a given time interval by initial biomass of standing crops. Changes in fine root biomass and nutrient contents (and presumably the intervening C and nutrient fluxes) in mature forests can be accurately predicted (within ±10%) by using length dynamics from rhizotrons as indices of biomass dynamics (Hendrick and Pregitzer, unpublished). The same should be true on an individual tree basis as well.

Changes in root chemistry associated with root development must be quantified to accurately model changes in root nutrient content and nutrient fluxes to and from fine roots. For example, N concentration of white roots can be as much as 40% greater than that of brown roots (Goldfarb et al., 1990); this fact should be considered when attempting to model N allocation among fine roots during plant or stand development. If N losses from brown roots are due to retranslocation into the primary root system via a mechanism similar to that operating in the canopy, it would have profound implications for models of plant nutrition and nutrient cycling. There also can be large differences in the nutrient concentrations of fine roots varying only slightly in diameter (Pregitzer et al., 1990). Although commonly done, care should be taken when assigning "fine roots" to an arbitrary diameter class.

The soil factors affecting fine root dynamics are not well understood, which makes assigning conditional probabilities to the pathways in Fig. 2 problematic, from a modeling perspective. For example, there is general agreement that increased microsite N levels in the soil can cause a local proliferation of fine root biomass, although on a whole-tree level high soil N levels decrease relative accumulation of biomass in roots (see below). However, the relationship between soil fertility and fine root longevity (i.e., mortality or turnover) is the subject of much debate (Chapin, 1980; Grier et al., 1981; Keyes and Grier, 1981; Alexander and Fairley, 1983; Aber et al., 1985; Nadelhoffer et al., 1985; Sibley and Grime, 1986). Decreases in soil moisture and increases in soil temperature can lead to reductions in fine root growth, increases in rates of root browning (P_2) and mortality (P_{6-8}) (Rogers, 1939; Atkinson, 1980; Kuhns et al., 1985; Marshall and Waring, 1985; Bartsch, 1987; Hendrick and Pregitzer, 1993). There has been little research establishing either empirical or mechanistic relationships between fine root dynamics and soil factors. However, once these cause-effect relationships have been quantified for a particular species or forest type, it should be possible to modify the transition (or flux) pathways in Fig. 2 to account for changing soil conditions. This could be done using a multiple regression approach, with transition probabilities as dependent variables which vary according to changes in environmental factors (independent variables).

There are inherent temporal and ontogenetic changes in root system development and dynamics that can be moderated but not obviated by changes in the environment; these should be considered when attempting to model the development of the fine root system of poplars. For example, fine root production rates (B_4) would likely be quite high as root absorbing area was rapidly established during the juvenile growth phase, but rates would decline and equilibrate to some extent as the tree or stand matured. Seasonal variation in production rates also are evident. About 50% of annual root production occurs by the time the canopy is fully flushed in sugar maple forests, while < 10% occurs during the period between leaf fall and budbreak (Hendrick and Pregitzer, unpublished). A substantial amount of fine root production also occurs before and soon after budbreak in hybrid poplars (Dickmann and Pregitzer, unpublished). Other tree species vary from poplars in the timing of fine root growth (Lyr and Hoffmann, 1967), but, nonetheless, they have rather distinct seasonal peaks and troughs in production. Considerable work remains to be done elucidating the endogenous factors controlling phenological patterns of fine root production before mechanistic models of this process can be constructed. But intuition and the limited available evidence suggest a close correlation between canopy and fine root phenology, although there appears to be wide latitude in phenological complementarity among species (Lyr and Hoffmann, 1967).

Fine roots of most trees growing in natural conditions are infected to some extent by mycorrhizal fungi. This condition results in additional C costs to the tree above and beyond the energy expended in host tissue construction and maintenance, and it needs to be considered in any model of resource allocation to fine roots. Although the additional cost of maintaining mycorrhizal versus non-mycorrhizal roots has yet to be fully established for most species, respiration of heavily infected herbaceous root systems may be as much as 75 to 100% higher than non-infected roots (Pang and Paul, 1980; Harley and Smith, 1983). Three to four times as much fixed C can be allocated to the root systems of mycorrhizal tree seedlings in comparison to non-mycorrhizal seedlings (Harley and Smith, 1983). Quantifying and modeling C allocation to mycorrhizae is complicated by the difficulty of recovering all of the plant-derived C in fungal tissue (e.g., hyphae, spores) and the separation of fungal from host tissue in the mycorrhizal root.

Functional Changes in Fine Roots

From a physiological perspective, modeling changes in the quantity of resources allocated to, and residing in, fine roots in various stages of development is of limited interest unless the function of roots in various stages of development is known. Young, white roots are thought to be the most important class of roots for absorbing water and actively taking up nutrients. Older, brown (suberized) roots are generally considered to be less important for water and nutrient absorption. Theoretically, then, the balance of total root length (or surface area) in these two stages of development could have significant effects on the ability of the tree to acquire water and nutrients. The validity of these assumptions has been challenged by a number of researchers who claim that the older parts of the root system also may be important in water and nutrient uptake (e.g., Chung and Kramer, 1975; Atkinson, 1980, 1983, 1985; MacFall et al., 1991). However, many studies support the hypothesis that, for at least some species, young fine roots are most the important organs for both water and nutrient uptake (Passioura, 1980; Robinson et al., 1991). Furthermore, total root length (old plus young roots) and water absorption are poorly correlated (Hamblin and Tennant, 1987). Therefore, it seems likely that separating roots by development stage should prove useful in modeling shifts in root function during development, although better data on uptake efficiencies are certainly needed. The rate at which roots pass from one stage of development into another, and the proportion of the root system in any given stage, can be studied by direct observation. If development stage-related differences in uptake rates are found, then the transition rates of white to brown roots (and total relative pool sizes), which are relatively easy to determine and measure by direct observation (Hendrick and Pregitzer, 1992), can be the basis for assigning uptake coefficients.

The formation of a mycorrhizae by a plant root and an associated fungus will inevitably lead to an increase in the ability of the root to acquire nutrients and water, due largely to the exploitation of a greater soil volume by the mycorrhizal hyphae. Mycorrhizal hyphae transport water and a variety of minerals to the host plant, including P, N, Ca, Zn, S and K (Harley and Smith, 1983; Allen, 1991). It is difficult to quantify the absolute effect of mycorrhizal formation on uptake, but available data show that uptake rates of mycorrhizal roots can be up to six times those of uninfected roots (Harley and Smith, 1983).

THE DYNAMICS OF CARBON AND NITROGEN

The perennial growth habit of woody plants is dependent on a closely regulated system of resource acquisition, allocation, storage and reallocation, all of which involve root systems (Fig. 5). Fluxes of carbohydrates and N are very seasonally dependent and are regulated by endogenous factors established by the genome. The complexity of this system is especially revealed preceding and during the spring growth flush, and in the late summer and autumn during and after bud set. A mechanistic model of poplar growth and development must account for both the mobility and transitory nature of carbohydrate and N compounds in tree tissues. Whereas the general yearly pattern of these compounds in perennial woody plants has been well established, the exact timing and pathway of travel from place to place, and the regulatory mechanisms involved, are not.

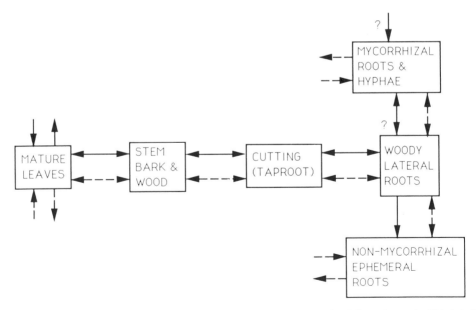

Figure 5. Conceptual whole-tree model of carbon and nutrient pools (boxes) and fluxes (arrows) within hybrid poplars. Carbon flows are designated by solid lines and nutrient flows by dashed lines. Question marks by arrows refer to possible uptake of carbon skeletons by mycorrhizal fungi. Movement between pools is regulated by seasonal environmental and endogenous factors.

Biomass Accumulation

Mechanistic models will have to accurately distribute biomass growth among various plant parts temporally, as well as account for environmental changes and genotype. The following observations indicate the complexity of these reponses.

The proportion of total-tree biomass accumulated in tree roots systems declines during the first year of growth, and this decline continues for a number of years thereafter. In poplars, root/shoot ratios during the first year vary from 0.2 to 0.6; leafless root/shoot ratios in late fall of the first year range from 0.5 to 1.7 (Pregitzer et al., 1990; Dickmann and Pregitzer, 1992). Root/shoot ratios of two- to three-year-old poplars generally fall below 0.25, although there is large variability in published data (Dickmann and Pregitzer, 1992). In older natural stands the proportion of standing biomass in coarse roots relative to the tops reaches a constant level (Zavitkovski and Stevens, 1972).

There is a strong genetic component to biomass accumulation in roots of poplars, especially among unrelated genotypes. For example, the two hybrid poplar clones that we have studied intensively at Michigan State University varied substantially in accumulation patterns after one and two years' growth in large pots in the field. *Populus* x *euramericana* cv. Eugenei had lower root/shoot ratios and accumulated less biomass into fine roots than did *P. tristis* x *P. balsamifera* cv. Tristis (Pregitzer et al., 1990). In every environment in which these two clones have been grown, the relative accumulation of biomass in roots of Tristis has exceeded that of Eugenei. Likewise, Ceulemans (1990) reported that root/shoot ratios in the *P. trichocarpa* x *P. deltoides* clones 44-136 and 11-11 were 0.59 and 0.27, respectively, after the first year of growth. At the end of the second year, ratios had declined to 0.26 for 44-136 and 0.20 for 11-11 (Friend et al., 1991). Clones of *P. trichocarpa* from four different provenances, ranging from 46° to 59° North latitude, showed leafless root/shoot ratios that varied from 0.5 for the southernmost sources to 1.7 for the northernmost after one year of growth in Scotland (Cannell and Willett, 1976). Gordon and Promnitz (1976), however, found little difference in root/shoot ratios among four related *P.* x *euramericana* clones after three years growth. Thus, it appears that C allocation to roots in any model will have to be driven by an independent genotypic variable.

High levels of soil resources, particularly N and other nutrients, promote higher relative growth rates in the shoot than the roots (Agren and Ingestad, 1987), a factor that must be built into environmentally driven models. In addition, our own data on hybrid poplars showed that accumulation responses to variation in soil N differed between genotypes (Pregitzer et al., 1990). At the end of the first year of growth in pots in the field, low-N Tristis trees had a leafless root/shoot ratio of 1.42, compared to 1.07 for high-N plants. In Eugenei low-N and high-N root/shoot ratios were 1.03 and 0.90, respectively. The response of biomass accumulation to water can be similar to the response to N (Ledig, 1983; Cannell, 1985; Pereira and Pallardy, 1989); water deficits tend to promote root growth relative to shoot growth, although this effect is not nearly so universal as the effect of high soil N.

Storage and Mobilization of Root Reserves

Prominent among the functions of roots is their role in the storage of carbohydrate and N reserves for subsequent utilization in maintenance respiration and growth. The specific roles of root reserves in tree survival, growth and development, however, become difficult to define (Loescher et al., 1990). There is abundant data available on the chemistry of root reserves and where they are stored, but much less on the physiology of their remobilization and use.

Starch is the carbohydrate most often associated with reserve pools in tree roots, but roots also contain considerable quantities of soluble sugars and, in certain genera such as *Fraxinus* or *Malus*, sugar alcohols. The relative proportion of starch and sugars in roots is dependent on taxon and season. In young *Populus* cv. Eugenei (Nguyen et al., 1990), starch predominated during the winter season, with concentrations exceeding 30% of root dry weight, but soluble sugar levels also comprised 10 to 15% of dry weight. In contrast, sugar levels in *P. trichocarpa* x *P. deltoides* cv. Raspalje (Bonicel et al., 1987) and *P.* cv. Tristis (Nguyen et al., 1990), which peaked at 80 mg g^{-1} dry weight during the dormant season, exceeded starch levels by 50%. In the trees studied by Bonicel et al., (1987), sucrose was the major storage sugar.

The total nonstructural carbohydrate (TNC) content of poplar roots can constitute a considerable part of their total dry weight, usually in the range of 10 to 20% (Schier and Zasada, 1973; Bonicel et al., 1987), but occasionally over 40% (Nguyen et al., 1990). Clonal differences in TNC content of roots also occurs; for example, 18 *P. tremuloides* clones varied from 10 to 26% (Schier and Johnston, 1971). The root system of hardwood trees typically contains higher concentrations of reserve carbohydrates than the stem system (Bonicel et al., 1987, Nguyen et al., 1990). However, on an absolute basis, the total weight of reserves in the roots is sometimes less than in the aerial portions, because the root system weighs less. Concentration of carbohydrates in the bark of roots is generally quite high, although not as high as in the wood (Keller and Loescher, 1989).

There appears to be an inverse relationship between root diameter and TNC or starch concentration, largely because small roots contain a larger proportion of living ray tissue, where most carbohydrate is stored (Wargo, 1979). However, since the total weight of large-diameter roots is usually so much greater than small roots, large roots constitute the principal underground reserve pool. In smaller trees this relationship may not hold; we found a positive relationship between carbohydrate concentration and root diameter in two poplar clones (Nguyen et al., 1990).

The general seasonal pattern is for root carbohydrate reserves in poplar trees to decline, often quite rapidly, just before or with the onset of the growing season, when shoots and roots are rapidly expanding. Then, when trees are fully refoliated, reserves begin to build back up to preflush levels, reaching a maximum early in the dormant season (Schier and Zasada, 1973; Bonicel et al., 1987; Nguyen et al., 1990). The implication of this pattern is that carbohydrate reserves are used for the early-season growth flush, but the extent to which carbohydrates are mobilized for root growth versus shoot growth is not known. In interpreting these trends it is important to consider that the concentration (percent of dry weight) of reserve carbohydrates in any rapidly growing tissue could decline simply due to dilution.

The strongest evidence for the seasonal cycling of root carbohydrate reserves comes from ^{14}C-studies. If leaves are exposed to $^{14}CO_2$ late in the growing season, the tracer is transported to the root system, as well as to sites of branch and stem storage (Priestly et al., 1976). The key regulatory event that shifts translocation of carbohydrate on a particular shoot axis basipetally to the root system appears to be bud set (Priestly et al., 1976; Isebrands and Nelson, 1983). During rapid shoot elongation little carbohydrate buildup occurs in the roots. As a particular shoot axis ceases growth and sets buds, particularly if it is removed from still active vegetative sinks, basipetal translocation to the lower stem and roots predominates. Genotypic variation in this pattern does occur, however. For example, *Populus trichocarpa* x *P. deltoides* hybrids 11-11 and 44-136 both showed increased translocation of ^{14}C assimilated by mature upper stem leaves to the roots following budset. But flow of the tracer to the roots of hybrid 44-136 was much stronger, both before and after budset, than in hybrid 11-11 (Ceulemans, 1990; Friend et al., 1991).

In several studies, though not with poplars, ^{14}C assimilated in the autumn and then transported to roots was followed through the subsequent growing season. These studies provide direct evidence of the mobilization of carbohydrates stored below ground for subsequent growth. For example, Lockwood and Sparks (1978a,b) studied remobilization of ^{14}C-labeled reserves in one- and three-year-old *Carya illinoensis* trees. In the spring following labeling, ^{14}C was detected in new shoots, inflorescences and roots. They concluded that only a small percentage of reserve materials are stored in the aerial portion of young pecan seedlings, and that they are rapidly supplemented by root reserves once rapid regrowth begins.

Our studies with poplars have shown that considerable loading of starch and sugars in fine roots (< 0.5 mm diameter) also can occur during autumn (Nguyen et al., 1990; Horwath, 1993). Starch concentrations in fine roots increased by over three times in Tristis, but an astonishing 75 times in Eugenei. During the same period, sugar concentrations doubled. Since no increase in the biomass of this root fraction occurred from August to November, and since late-season fine root turnover is minimal in these clones, fine roots apparently are sinks for nonstructural carbohydrate reserves during this period.

Roots also are important reservoirs of N, although much less is known about N reserves. Total N concentrations in coarse roots are generally less than one percent, but in fine roots (less than one 1 mm diameter) concentrations are higher, falling in the range of 1.0-1.8% (Goldfarb et al., 1990; Pregitzer et al., 1990). In young trees the amount of N stored in the roots during the dormant season may be substantial. In our two hybrid poplar clones 50 to 60% of total-tree N was found in the root system in November of the first year of growth (Pregitzer et al., 1990). The bark and newest layers of sapwood in the roots may be the most important reservoirs of N (Titus and Kang, 1982). Proteins are the primary storage forms of N (Titus and Kang, 1982), although certain amino acids like arginine and asparagine may also be important. Thus, the partitioning of N into various chemical storage forms in roots is similar to that of shoots (Sauter et al., 1989; Titus and Kang, 1982).

The annual cycle of N transformations is as follows: 1) N is mobilized in the autumn from senescing leaves and stored in stems and roots, largely as protein; 2) N is remobilized in the spring through hydrolysis of storage proteins and translocated as amino acids and amides via the xylem sap stream to developing tissues; and, 3) at the end of the growing season the cycle renews (Titus and Kang, 1982). Dickson (1979) demonstrated that *Populus* plants possess a system for recycling N from roots to shoots. In temperate climates, especially where soil N supplies are low, this recycling system is prominent, although the quantitative uptake of soil N, especially that mineralized from leaf litter and other sources, should not be underestimated (Titus and Kang, 1982; Horwath 1993).

CONCLUSIONS

We have presented both conceptual models of the function and dynamics of poplar adventitious root systems, as well as a review of what is known of certain aspects of root architecture, physiology and ecology. It is upon this foundation that modelers will have to build; in fact, some of them have already achieved a modicum of success. The poplar tree, like all woody plants, is an integrated above- and below-ground system (Fig. 5). We hope that in the future the co-dependency of the shoots and roots will be realized by modelers seeking to simulate the functioning and architecture of trees and stands as they really exist.

Such modeling efforts can have practical significance. Foresters managing poplar stands, who typically deal with the part of the tree protruding above the soil surface, must consider the effects that their silvicultural practices have on root system form and function (Dickmann and Pregitzer, 1992). Responsive models that include a root system component can help predict these effects. This knowledge may then evolve into revised silvicultural practices which modify root structure and function in such a way that product yield is maximized and environmental damage is minimized. In such a scenario, everyone would benefit.

REFERENCES

Aber, J.D., Melillo, J.M., Nadelhoffer, K.J., McClaugherty, C.A., and Pastor, J., 1985, Fine root turnover in forest ecosystems in relation to quantity and form of nitrogen availability: a comparison of two methods, *Oecologia* 66:317.

Agren, G.I., and Ingestad, T., 1987, Root:shoot ratio as a balance between nitrogen productivity and photosynthesis, *Plant Envir.* 10:579.

Alexander, I.J., and Fairley, R.I., 1983, Effects of N fertilization on populations of fine roots and mycorrhizae in spruce humus, *Plant Soil* 71:49.

Allen, M.F. 1991, "The Ecology of Mycorrhizae," Cambridge Univ. Press, Cambridge.

Atkinson, D., 1980, The distribution and effectiveness of roots of tree crops, *Hortic. Rev.* 2:425.

Atkinson, D., 1983, The growth, activity and distribution of the fruit tree root system, *Plant Soil* 71:23.

Atkinson, D., 1985, Spatial and temporal aspects of root distribution as indicated by the use of a root observation laboratory, *in:* "Ecological Interactions in Soil," A.H. Fitter, D. Atkinson, D.J. Read, and M. Usher, ed., Spec. Pub. No. 4, Blackwell Sci. Pubs., Oxford.

Atkinson, D., Naylor, D., and Coldrick, G.A., 1976, The effect of tree spacing on the apple root system, *Hortic. Res.* 16:89.

Bartsch, N, 1987, Response of root systems of young *Pinus sylvestris* and *Picea abies* plants to water deficits and soil acidity, *Can. J. For. Res.* 17:805.

Bassow, S.L., Ford, E.D., and Kiester, A.R., 1990, A critique of carbon-based tree growth models, *in:* "Process Modeling of Forest Growth Responses to Environmental Stress," R.K. Dixon, R.S. Meldahl, G.A. Ruark, and W.G. Warren, eds., Timber Press, Portland.

Bloomberg, W.J., 1959, Root formation of black cottonwood cuttings in relation to region of the parent shoot, *For. Chron.* 35:13.

Bloomberg, W.J., 1963, The significance of initial adventitious roots in poplar cuttings and the effect of certain factors on their development, *For. Chron.* 39:279.

Bonicel, A., Haddad, G., and Gagnaire, J., 1987, Seasonal variations of starch and major soluble sugars in the different organs of young poplars, *Plant Physiol. Biochem.* 25:451.

Cannell, M.G.R., 1985, Dry matter partitioning in tree crops, *in:* "Attributes of Trees as Crop Plants," M.G.R. Cannell, and J.E. Jackson, eds., Instit. Terrestrial Ecol., Huntingdon.

Cannell, M.G.R., and Willett, S.C., 1976, Shoot growth phenology, dry matter distribution and root:shoot ratios of provenances of *Populus trichocarpa*, *Picea sitchensis* and *Pinus contorta* growing in Scotland, *Silvae Genet.* 25:49.

Ceulemans, R., 1990, "Genetic Variation in Functional and Structural Productivity Determinants in Poplar, University of Antwerp," Dissertation, Thesis Pubs., Amsterdam.

Chapin, F.S., 1980, The mineral nutrition of higher plants, *Annu. Rev. Ecol. Syst.* 11:233.

Chen, C.W., and Gomez, L.E., 1990, Modeling tree responses to interacting stresses, *in:* "Process Modeling of Forest Growth Responses to Environmental Stress," R.K. Dixon, R.S. Meldahl, G.A. Ruark, and W.G. Warren, eds., Timber Press, Portland.

Chung, H.H., and Kramer, P.J., 1975, Absorption of water and ^{32}P through suberized and unsuberized roots of loblolly pine, *Can. J. For. Res.* 5:229.

DeByle, N.V., 1964, Detection of functional intraclonal aspen root connections by tracers and excavation, *For. Sci.* 10:386.

Dickmann, D.I., and Pregitzer, K.S., 1992, The structure and dynamics of woody plant root systems, *in* "Ecophysiology of Short Rotation Forest Crops," C.P. Mitchell, J.B. Ford-Robertson, T. Hinckley, and L. Sennerby-Forsse, ed., Elsevier Applied Sci., New York.

Dickmann, D.I., and Stuart, K.W., 1983, "The Culture of Poplars," Michigan State Univ., East Lansing.

Dickson, R.E., 1979, Xylem translocation of amino acids from roots to shoots in cottonwood plants, *Can.J. For. Res.* 9:374.

Eissenstat, D.M., 1991, On the relationship between specific root length and the rate of root proliferation: A field study using citrus rootstocks, *New Phytol.* 118:63.

Elowson, S., and Rytter, L., 1984, Biomass distribution within willow plants growing on a peat bog, *Swed. Univ. Agric. Sci. Rep.* 15:325.

Ericsson, T., 1984, Root biomass distribution in willow stands grown on a bog, *Swed. Univ. Agric. Sci. Rep.* 15:335-348.

Ewel, K.C., and Gholz, H.K., 1991, A simulation model of the role of belowground dynamics in a Florida pine plantation, *For. Sci.* 37:397.

Faulkner, H.G., 1976, Root Distribution, Amount, and Development from 5-Year-Old *Populus* x *euramericana* (Dode) Guinier., M.S.F. Thesis, Univ. Toronto, Canada.

Fitter, A.H., Stickland, T.R., Harvey, M.L., and Wilson, G.W., 1991, Architectural analysis of plant root systems, 1. Architectural correlates of exploitation efficiency, *New Phytol.* 118:375.

Friend, A.L., Scarascia-Mugnozza, G., Isebrands, J.G., and Heilman, P.E., 1991, Quantification of two-year-old hybrid poplar root systems: morphology, biomass, and ^{14}C distribution, *Tree Physiol.* 8:109.

Goldfarb, D., Hendrick, R., and Pregitzer, K.S., 1990, Seasonal nitrogen and carbon concentrations in white, brown and woody fine roots of sugar maple (*Acer saccharum* Marsh), *Plant Soil* 126:144.

Gordon, J.C., and Promnitz, L.C., 1976, A physiological approach to cottonwood yield improvement, in: "Symp. on East. Cottonwood and Related Species Proc., "B.A. Thielges, and S.B. Land, Jr., eds., Louisiana State Univ., Baton Rouge.

Graham, S.A., Harrison, R.P., and Westell, C.E., 1963, "Aspens: Phoenix trees of the Great Lakes Region," Univ. Michigan Press, Ann Arbor.

Grier, C.C., Vogt, K.A., Keyes, M.R., and Edmonds, R.L., 1981, Biomass distribution and above- and belowground production in young and mature *Abies amabilis* zone ecosystems of the western Cascades, *Can. J. For. Res. 11:155.*

Haissig, B.E., Davis, T.D., and Riemenschneider, D.E., 1992, Researching controls of adventitious rooting, *Physiol. Plant.* 84:310.

Hamblin, A., and Tennant, D., 1987, Root length density and water uptake in cereals and legumes: how well are they correlated? *Aust. J. Agric. Res.* 38:513-527.

Hansen, E.A., 1981, Root length in young hybrid *Populus* plantations: its implication for border width of research plots, *For. Sci.* 27:808.

Harley, J. L., and Smith, S.E., 1983, "Mycorrhizal Symbiosis," Academic Press, London.

Hendrick, R.L., and Pregitzer, K.S., 1992, The demography of fine roots in northern hardwood forests, *Ecology* 73:1094.

Hendrick, R.L., and Pregitzer, K.S., 1993, Patterns of fine root mortality in two sugar maple forests, *Nature* 361:59-61.

Host, G.E., Rauscher, H.M., Isebrands, J.G., Dickmann, D.I., Dickson, R.E., Crow, T.R., and Michael, D.A., 1990, "The ECOPHYS User's Manual," USDA For. Ser. Gen. Tech. Rep. NC-141, North Central Forest Experiment Station, St. Paul.

Horwath, W., 1993, "The Dynamics of Carbon, Nitrogen and Soil Organic Matter on *Populus* Plantations," Ph.D. Dissertation, Michigan State Univ., East Lansing.

Isebrands, J.G., and Nelson, N.D., 1983, Distribution of ^{14}C-labeled photosynthates within intensively cultured *Populus* clones during the establishment year, *Physiol. Plant.* 59:9.

Isebrands, J.G., Rauscher, H.M., Crow, T.R., and Dickmann, D.I., 1990, Whole-tree growth process models based on structural-functional relationships, in: "Process Modeling of Forest Growth Responses to Environmental Stress," R.K. Dixon, R.S. Meldahl, G.A. Ruark, and W.G. Warren, eds., Timber Press, Portland.

Keller, J.D., and Loescher, W.H., 1989, Nonstructural carbohydrate partitioning in perennial parts of sweet cherry, *J. Amer. Soc. Hortic. Sci.* 114: 969.

Keyes, M.R., and Grier, C.C., 1981, Above- and below-ground net production in 40-year-old Douglas-fir stands on low and high productivity sites, *Can. J. For. Res.* 11:599.

King, D.A., 1993, A model analysis of the influence of root and foliage allocation on forest production and competition between trees, *Tree Physiol.* 12:119.

Kuhns, M.R., Garrett, H.E., Teskey, R.O., and Hinckley, T.M., 1985, Root growth of black walnut trees related to soil temperature, soil water potential and leaf water potential, *For. Sci.* 31:617.

Landsberg, J.J., 1986, "Physiological Ecology of Forest Production," Academic Press, New York.

Ledig, F.T., 1983, The influence of genotype and environment on dry matter distribution in plants, in: "Plant Research and Agroforestry," P.A. Huxley, ed., Int. Coun. for Res. in Agroforestry, Nairobi.

Lockwood, D.W., and Sparks, D., 1978a, Translocation of ^{14}C in 'Stuart' pecan in the spring following assimilation of $^{14}CO_2$ during the previous growing season, *J. Amer. Soc. Hortic. Sci.* 103:38.

Lockwood, D.E., and Sparks, D., 1978b, Translocation of ^{14}C from tops and roots of pecans in the spring following assimilation of $^{14}CO_2$ during the previous growing season, *J. Amer. Soc. Hortic. Sci.* 103:45.

Loescher, W.H., McCamant, T., and Keller, J.D., 1990, Carbohydrate reserves, translocation, and storage of woody plant roots, *Hortic. Sci.* 25:274.

Lyr, H., and Hoffman, G., 1967, Growth rates and growth periodicity of tree roots, *Int. Rev. For. Res.* 2: 181.

Luxova, M., and Lux, A., 1981a, Latent root primordia in poplar stems, *Biol. Plant.* (Praha) 23:285.

Luxova, M., and Lux, A., 1981b, The course of root differentiation from root primordia in poplar stems, *Biol. Plant.* (Praha) 23:401.

MacFall, J.S., Johnson, G.A., and Kramer, P.J., 1991, Comparative water uptake by roots of different ages in seedlings of loblolly pine, *New Phytol.* 119:551.

Makela, A., and Hari, P., 1986, Stand growth model based on carbon uptake and allocation in individual trees, *Ecol. Model.* 33:204.

Marshall, J.D., and Waring, R.H., 1985, Predicting fine root production and turnover by monitoring root starch and soil temperature, *Can. J. For. Res.* 15:791.

Nadelhoffer, K.J., Aber, J.D., and Melillo, J.M., 1985, Fine roots, net primary production and nitrogen availability: a new hypothesis, *Ecology* 66:1377.

Nguyen, P.V., Dickmann, D.I., Pregitzer, K.S., and Hendrick, R., 1990, Late-season changes in allocation of starch and sugar to shoots, coarse roots, and fine roots in two hybrid poplar clones, *Tree Physiol.* 7:95.

Nikinmaa, E., and Hari, P., 1990, A simplified carbon partitioning model for Scots pine to address the effects of altered needle longevity and nutrient uptake on stand development, *in*: "Process Modeling of Forest Growth Responses to Environmental Stress," R.K. Dixon, R.S. Meldahl, G.A. Ruark, and W.G. Warren, eds., Timber Press, Portland.

Pang, P.C., and Paul, E.A., 1980, Effects of vesicular-arbuscular mycorrhizae on ^{14}C and ^{15}N distribution within nodulated fababeans, *Can. J. Soil Sci.* 60:241.

Passioura, J.B., 1980, The transport of water from root to shoot in wheat seedlings, *J. Exp. Bot.* 31:333.

Pereira, J.S., and Pallardy, S., 1989, Water stress limitations to tree productivity, *in*: "Biomass Production by Fast-Growing Trees," J.S. Pereira, and J.J. Landsberg, eds., Kluwer Acad. Pubs., Boston.

Pregitzer, K.S., Dickmann, D.I., Hendrick, R., and Nguyen, P.V., 1990, Whole-tree carbon and nitrogen partitioning in young hybrid poplars, *Tree Physiol.* 7:79.

Priestly, C.A., Catlin, P.B., and Olsson, E.A., 1976, The distribution of ^{14}C-labelled assimilates in young apple trees as influenced by doses of supplementary nitrogen, I. Total ^{14}C radioactivity in extracts, *Ann. Bot.* 40:1163.

Rastetter, E.B., Ryan, M.G., Shaver, G.R., Mellilo, J.M., Nadelhoffer, K.J., Hobbie, J.E.. and Aber, J.D., 1991, A general biogeochemical model describing the responses of the C and N cycles in terrestrial ecosystems to changes in CO_2, climate and N deposition, *Tree Physiol.* 9:101.

Robinson, D., Linehan, D.J., and Caul, S., 1991, What limits nitrate uptake from soil? *Plant Cell Envir.* 14:77.

Rogers, W.S., 1939, Root studies, VII. Apple root growth in relation to rootstock, soil, seasonal and climatic factors, *J. Pomol.* 17:99.

Running, S.W., and Gower, S.T., 1991, FOREST-BGC, A general model of forest ecosystem processes for regional applications, II. Dynamic carbon allocation and nitrogen budgets, *Tree Physiol.* 9:147.

Sauter, J.J., VanCleve, B., and Wellenkamp, S., 1989, Ultrastructural and biochemical results on the localization and distribution of storage proteins in a poplar tree and in twigs of other tree species, *Holzforsch.* 43:1.

Schier, G.A., and Johnston, R.S., 1971, Clonal variation in total nonstructural carbohydrates of trembling aspen roots in three Utah areas, *Can. J. For. Res.* 1:252.

Schier, G.A., and Zasada, J.C., 1973, Role of carbohydrate reserves in the development of root suckers in *Populus tremuloides, Can. J. For. Res.* 3:243.

Sibley, R.M., and Grime, J.P., 1986, Strategies of resource capture by plants - evidence for adversity selection, *J. Theor. Biol.* 118:247.

Sprackling, J.A., and Read, R.A., 1979, "Tree Root Systems in Eastern Nebraska," Nebraska Conserv. Bull. No. 37, Lincoln.

Sutton, R.F., and Tinus, R.W., 1983, "Root and Root System Terminology," For. Sci. Monograph 24.

Tew, R.K., Debyle, N.V., and Schultz, J.D., 1969, Intraclonal root connections among quaking aspen trees, *Ecology* 50:920.

Titus, J.S., and Kang, S.M., 1982, Nitrogen metabolism, translocation, and recycling in apple trees, *Hortic. Rev.* 4: 204.

Vogt, K.A, Grier, C.C., and Vogt, D.J., 1986, Production, turnover and nutritional dynamics of above-and below-ground detritus of world forests, *Adv. Ecol. Res.* 15:303.

Wargo, P.M., 1979, Starch storage and radical growth in woody roots of sugar maple, *Can. J. For. Res.* 9:45.

Weinstein, D.A., and Beloin, R., 1990, Evaluating effects of pollutants on integrated tree processes: a model of carbon, water and nutrient balances, *in*: "Process Modeling of Forest Growth Responses to Environmental Stress," R.K. Dixon, R.S. Meldahl, G.A. Ruark, and W.G. Warren, eds., Timber Press, Portland.

Weinstein, D.A, Beloin, R., and Yanai, R.D., 1991, Modeling changes in red spruce carbon balance and allocation in response to interacting ozone and nutrient stresses, *Tree Physiol.* 9:127.

Ying, C.C., and Bagley, W.T., 1977, Variation in rooting capability of *Populus deltoides, Silvae Genet.* 26:204.

Zavitkovski, J., and Stevens, R.D., 1972, Primary productivity of red alder ecosystems, *Ecology* 53:235.

THE ROLE OF EXPERT AND HYPERTEXT SYSTEMS IN MODELING

ROOT-SHOOT INTERACTIONS AND CARBON ALLOCATION

H. Michael Rauscher

USDA Forest Service
North Central Forest Experiment Station
Forestry Sciences Laboratory
1831 Highway 169 East
Grand Rapids, MN 55744 USA

INTRODUCTION

Access to knowledge and the ability to use it wisely has always been the hallmark of successful individuals, companies and nations. Scientific progress also depends upon accessible and organized knowledge (Bauer, 1992). Over the last 50 years, an increasing number of scientists have called attention to the deteriorating condition of our scientific knowledge infrastructure, i.e., the technical literature that supports scientific progress. Vannevar Bush (1945) was among the first influential scientists to point out that management of scientific knowledge has not essentially changed for more than 200 years. He summarized the situation as follows:

...There is a growing mountain of research. But there is increased evidence that we are being bogged down today as specialization extends. The investigator is staggered by the findings and conclusions of thousands of other workers --conclusions which he cannot find time to grasp, much less to remember, as they appear. Yet specialization becomes increasingly necessary for progress, and the effort to bridge between disciplines is correspondingly superficial....

As President Roosevelt's chief science advisor, Bush saw that the progress of scientific research was being impeded by the inability of researchers to efficiently access and manipulate the growing body of scientific knowledge. Forscher (1963) reinforced this view of the critical state of scientific knowledge management by writing a wonderful parable entitled, *Chaos in the Brickyard*. In this parable, Forscher pointed to an unending flood of peer-reviewed journal articles, which he called **bricks** of information, being produced and simply dumped onto a growing heap, the scientific literature. The hard work of providing context and meaning for these bricks—building compact edifices of knowledge by organizing and synthesizing theories—is not being done.

The increasing volume of scientific research results in all fields is staggering.

Scientists are buried under mountains of technical reports, and decision/policy makers are bewildered by the chaotic state of scientific knowledge. Under such conditions it is no wonder that knowledge rarely gets to many of the places it needs to be in a form that is of any use. The consequences of these knowledge shortfalls can range from minor inconveniences to personal, national or global tragedy (Herring, 1968; James, 1985; McRoberts et al., 1991; Price, 1963; Ziman, 1976).

Recent developments in hypertext and expert systems technology have provided powerful tools that can be used to improve the management of scientific knowledge. The purpose of this chapter is to illustrate how hypertext and expert systems can be used to organize knowledge about root-shoot interactions and carbon (C) allocation and, furthermore, how this organized knowledge can be used to improve a whole plant ecophysiological model of hybrid poplar growth and development, named ECOPHYS (Rauscher et al., 1990). Poplars are deployed worldwide in intensively cultured plantations, which are started from hardwood cuttings that are usually rooted directly in the field (see the chapter by Ritchie in this volume). Hence, root-shoot interactions and C allocation are especially important subjects for those physiologists and geneticists who are improving the efficiency of poplars for, e.g., production of biomass for conversion into energy (see the chapter by Dickmann and Hendrick in this volume).

I will begin this chapter by defining relevant terminology, followed by a brief overview of expert and hypertext systems. The methods of symbolically representing knowledge in electronic format will then be outlined. After establishing this technical background, a theory of C allocation and root-shoot interaction, taken from the literature, will be presented and used as an illustration. The theory will then be rendered into a small example hypertext document. Expert system rules will be extracted from the hypertext document and formulated so that they can be implemented in an expert system shell. Finally, the expert system will be conceptually placed within ECOPHYS.

TERMINOLOGY

This multidisciplinary chapter necessarily introduces numerous concepts from research in knowledge management, and in hypertext and expert systems that are likely to be new for many natural resource scientists. Definitions of these new concepts are, therefore, needed to ensure the best possible transfer of knowledge.

Knowledge Management Terminology

In this section I will described the differences between data, information and knowledge as used in this chapter. **Data** is defined as unevaluated observations of facts and relationships. Data are clearly unorganized and unevaluated, for example, numerical data tabulated for two quantities: the heights of hybrid poplars in some population, H; and the tree's dry weights, W. These two columns of numbers represent corresponding measurements from members of a population, but otherwise convey no more information or knowledge about that population.

In contrast, **information** is data that have been fitted into categories, classification schemes or numerical relationships. To continue with the hybrid poplar example, we could relate the two columns of data by fitting a mathematical function, e.g., $H = aW^{1.5}$. Information remains unevaluated, however, in the sense that we do not know whether to believe in its validity (truth value). Information turns into knowledge when we develop a justified belief in it. In terms of this height-dry weight relationship, justification would consist of some causal or correlational validation of our belief for this particular relation.

Traditional cognitive science view defines **knowledge** as justified true belief (Giere, 1984). Knowledge exists when: 1) a person believes that a fact or relationship is true, 2) the person is justified in so believing, and 3) the fact or relationship is actually true. A fact may be false in the sense that our observational abilities are not always faithful to us; occasionally we perceive something as true fact when in actuality it is not. False facts are one possible breakdown in our knowledge. If a person believes a true fact without proper justification, knowledge also does not exist because it violates part (2) of the definition of knowledge. We may have **belief** without independently verifiable justification, but we do not have knowledge. Finally, because we rarely know whether a fact or relationship is actually true, (3) above, with absolute certainty, we can only claim conditional knowledge. Absolute faith is possible; absolute knowledge is not. **Knowledge management**, then, is the art and science of manipulating human knowledge so that it can be easily accessed, compactly organized, objectively evaluated, and widely distributed.

Hypertext Terminology

A **Hypertext** document, or **hyperdocument**, consists of a network of chunks connected by electronic links. A **chunk** is an organized collection of information on a single topic roughly comparable to a paragraph of text (Seyer, 1991). Chunks are internally self-contained and independently understandable. A **link** is an electronic cross reference used to connect logically related chunks (Berk, 1991). By moving a "mouse" pointer to a link on a computer screen and clicking the mouse button, the user jumps from one chunk to another. The act of linking chunks creates the hyperdocument and at the same time creates one set of navigational jumps for readers to follow (Slatin, 1991). Links simulate the mental associations between chunks in the mind of the author.

Hyper in hypertext means extending into another dimension as in hyperspace, hypersphere or hyperdimensional. **Text** is a generic term denoting any written material, in particular, in printed form. The **structure** of a hyperdocument refers to those elements that deal only with the organization of the chunks. Text structure commonly takes the form of a table of content, an outline of the chunks, graphical diagrams of chunk relationships, indices, etc. The **content** of a hyperdocument refers to the domain-specific material that is the subject matter. **Hypermedia** is often used to refer to hypertext documents where a chunk can take many forms, such as music, maps, photographs, video, voice or animation, in addition to text. Hypertext, however, is commonly used to refer to both types of systems as the more generic term.

Expert Systems Terminology

Expert systems are defined as computer programs that use knowledge and symbolic inference (reasoning) procedures to solve problems that are difficult enough to otherwise require significant human expertise for their solution (Harmon and King, 1985). When most of the available knowledge is of a non-mathematical nature and when it exists in the form of heuristics, often encoded in "IF ... THEN ..." constructs, we can use expert systems methods to encode and manipulate it. A **heuristic** is: "...A rule of thumb or a technique based on experience and for which our knowledge is incomplete. A heuristic rule works with useful regularity but not necessarily all the time..." (Kurzweil, 1990).

HYPERTEXT SYSTEMS

Efforts have been underway since at least 1945 to devise a way to manage large masses of information (Parsaye et al., 1989). The basic idea has been to depart from the sequential, linear storage and retrieval of text to a random access, nonlinear method. The dominant analogue has been the familiar 3- x 5-inch card. The card represents a chunk or node of text, and the "trick" has been to devise systems that can easily and comfortably structure knowledge by linking these chunks together. The links, which represent knowledge structure, become as real and as important to the overall system as the chunks, which represent knowledge content. A hypertext document can be thought of as a 3-dimensional network of chunks and links (Fig. 1). The goal of hypertext is to organize information into systems of knowledge that are intuitive for all users.

Hypertext software systems are emerging as powerful knowledge management tools. Much work has been done to use the power of computers to organize and manage the facts and theories that constitute our scientific knowledge. An excellent survey of knowledge management systems, including database systems, bibliographic reference systems and

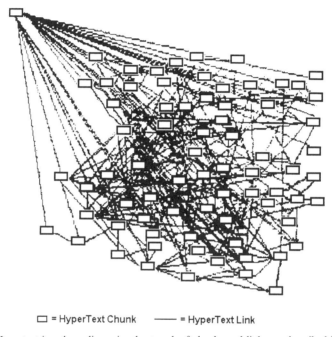

☐ = HyperText Chunk　　——— = HyperText Link

Figure 1. Hypertext is a three-dimensional network of chunks and links, as described in the text.

hypertext systems, can be found in Parsaye et al. (1989). The development of hypertext systems, first set into motion by Vannevar Bush's vision for a "Memex" system, may be edging us closer to the kind of dramatic breakthrough in knowledge management technology that he envisioned almost 50 years ago (Bush, 1945). The first functional hypertext systems appeared in the 1960s. The 1970s saw the development of a dozen research hypertext systems for mainframe computers and minicomputers (Conklin, 1987). During the 1980s, hypertext systems began to appear on microcomputers in commercial quantities and at microcomputer prices (Fersko-Weiss ,1991). Rauscher and Host (1990) discuss the intimate connection between hypertext and expert systems to improve knowledge management. A good general introduction to hypertext methods and techniques can be found in Shneiderman and Kearsley (1989) and Horn (1989). Rauscher et al. (in press) present a brief overview of hypertext system authoring. Two hypertext systems have been published in the field of natural resource management (Rauscher et al,. 1991; Thomson et al., 1993).

Benefits of Hypertext Systems

Hypertext technology is the most powerful tool currently available for the organization, synthesis and communication of scientific knowledge. Hypertext technology offers a greater range of media to capture and provide information than print technology. For example, both printed text and hypertext can use text and graphics; but, only hyperdocuments can include video, animation and sound as well as execute computer simulation models (Nielsen, 1989; Slatin, 1991). Some information is more naturally represented as sound, animation or video rather than text or still pictures. Hypertext fosters easy and quick access to immensely large and complex knowledge-bases that in printed form would be daunting (Schlumlienzer, 1989). Hyperdocuments are likely to contain more information than books and yet maintain ease of access. Hypertext takes up much less physical space than books and journals, and offers significantly lower publishing costs.

Hypertext systems allow us to view and evaluate the content and structure of our scientific knowledge explicitly. We can send copies to our peers and debate the additions, deletions and modifications to both content and structure of the knowledge base. We can identify agreements, as well as disagreements, with an ease never before possible. Scientific progress depends upon recognizing anomalies (Allen and Hoekstra, 1992) and the recognition of anomalies depends upon collecting what we know in a single place in an organized structure. Hypertext methods give us the powers to pull together what we know and to effectively manage our large and ever-expanding body of scientific knowledge.

Limitations of Hypertext Systems

Hypertext technology is still young and the implementations of the concepts are more crude than elegant. Recall how crude, from today's vantage point, wordprocessors were in 1980. Hypertext development systems are roughly at a similar state of development today. This reality translates into several limitations in comparison with print technology.

Printed text has some advantages that current hypertext systems are going to have to overcome before a technological shift is likely to occur. First, printed text provides an excellent high-resolution output, presenting text and color graphics in large and easy to read characters. Books are also easy to access, easy to annotate by writing on the margin and, usually, extremely portable (Shneiderman et al., 1991). Navigating through the knowledge in a book is something we learn early in life, and we are rarely conscious of this acquired skill. Because predictability of information flow is important to readers, authors use great skill to lead readers through every section of the book in a predetermined, fixed manner (Slatin, 1991). Another advantage of printed text is that, in the end, a printed book is an unchanging product. Books can and do remain unchanged for centuries (Slatin, 1991). This static, eternal quality confers on books great authority regardless of their content. Hypertext may have difficulty attaining such an authoritative status because of its propensity to be constantly changing.

Beyond these technological limitations lie some conceptual limitations of hypertext. Hypertext produces documents that are electronic but not executable in the sense that computer models are executable. We cannot "write" a hyperdocument, run it on a computer and have it automatically do any work for us. Whatever we want from the hyperdocument, as in producing a wordprocessing document or a printed book, we must actively do ourselves. This is in contrast to electronic models, which can be written and executed, to perform work without our direct control. Finally, computerized searches of realistically large topics such as root-shoot interactions or C allocation will yield vast numbers of citations. Assimilating, organizing and synthesizing this material is a task at least as daunting as writing a scientific textbook. Such an effort more realistically resembles the multi-author, interdisciplinary teamwork needed to write an update to a major encyclopedia.

Developing the first hyperdocument in each domain will be a large undertaking. Updating an existing hyperdocument to keep it current should be significantly easier.

EXPERT SYSTEMS

In structure, an expert system, also known as a knowledge-based system, may consist of many components (Fig. 2). Not all of these elements are necessarily included in any one system. Rather, I have illustrated a relatively comprehensive collection of components. At the heart of any expert system is its knowledge-base. The **knowledge-base** contains the domain-pertinent knowledge. The expert knowledge encoded in the knowledge-base of expert systems comes from human experts, either directly through interviews or indirectly from the scientific literature. Experts perform better than non-experts because they have a large amount of compiled, domain-specific knowledge stored in long-term memory and they have become skilled at using that knowledge to solve specific problems. A world-class expert is said to have between 50,000 and 100,000 chunks of knowledge stored in long-term memory. When young, we begin to organize our experiences into chunks centered around objects, either concrete or abstract. As we learn more, we cluster more and more knowledge around successively more abstract chunks. But no matter how complex our chunks become, our ability to manipulate any one chunk seems to stay about the same. Experts ordinarily emerge after about 10 years or more of study and work in their fields because it takes that long to acquire, chunk and organize the necessary knowledge and experience.

The process of running an expert system is commonly referred to as a **consultation**, alluding to the analogous scenario of consulting a human expert. The **reasoning engine** is a computer program that navigates through the knowledge-base using inference methods and control strategies to solve problems. To make the best decisions possible in an imperfect world, we must often use the best available resources at our disposal. Sometimes certain desirable resources are unavailable, sometimes the ones we do have are unreliable,

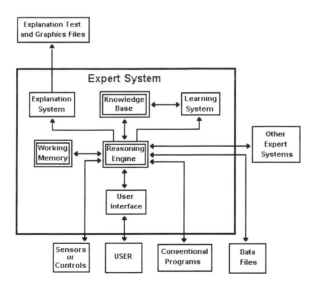

Figure 2. A complete expert system architecture showing the major structural components.

and sometimes our understanding is incomplete and can only provide us with vague answers. For expert systems to be useful, they, like us, also must possess an ability to reason in this type of "dirty" environment. Therefore, most expert systems allow for inexact reasoning or approximate reasoning in the course of the problem solving consultation (McNeill and Freiberger, 1993). A **working memory** serves as the work area ("scratch pad") of the **inference engine**, where all currently known, problem-specific information is stored. **User interface** software enables the system to elicit answers to questions and to display screen reports. Consultation with the expert system usually occurs through some user-interface program, which may be menu-driven or which may incorporate some sophisticated natural language capability or, possibly, visual-aid graphics. Facts about a current consultation problem are provided by the user and entered into the working memory. A reasoning engine then attempts to validate other "facts" (hypotheses) from working-memory facts and from the knowledge-base "understanding" of the application area.

The **explanation system** along with the user interface constitutes the human-machine interface, and is often in the form of a hypertext system. Explanations may answer "why" a question is asked of the user, "how" an answer has been derived, the meaning of terms, and the intent and source of reference for the knowledge in the knowledge-base. An expert system may input data from simulation models (in external programs) or database/spreadsheet files—as well as output data to simulation models and change database/spreadsheet files. Non-language-based information is acquired by the expert system through connections to digital or analog sensors. Output goes to electro-mechanical controls. Finally, when an expert system finds a "dead-end" in its chain of logic (i.e., a point in the line of reasoning from which no solution can be reached), a **learning system** might report this finding to the user and help the user to remedy the situation by making changes to the knowledge base during consultation (McRoberts et al., 1991).

[You will find additional information in the following sources: Articles on AI and expert systems in natural resource management began to appear in significant numbers in 1983 (Davis and Clark, 1989). The scientific journal *AI Applications*—reporting on natural resources, agriculture and environmental science—began publishing in 1987. A review article reporting on 74 AI projects in natural resources world-wide was authored by Rauscher and Hacker (1989). Another worldwide review of the literature on artificial intelligence (AI) and environmental protection was prepared by Simon et. al. (1992) in the German language.]

Benefits of Expert Systems

Many potential benefits can be derived from the application of expert systems in science and society. Expert systems can be used to synthesize and store the knowledge in a particular domain in a formulated, testable and maintainable way. This will provide an explicit record of expertise to ensure some protection from catastrophic loss and to provide a concrete record of current decision strategies. An explicit record affords the decision maker a measure of accountability—a means for, at least partially, justifying decisions. This can be especially desirable when decisions may be controversial. Also, strategies used in solving problems and making decisions become less mysterious. Artificial expertise provides a focus for upgrading and improving management strategies over time (Starfield and Bleloch, 1983). As new information is produced, it can be incorporated into our understanding (knowledge) of that field, and lead to new strategies for the application of that knowledge in the "real world."

Not every day is a good day and not everyone works in top form all the time. Subtle effects of such proficiency aberrations can have substantial consequences for critical decisions. Expert systems are consistent from one day to the next, unaffected by day-to-day bias. Human experts or other decision makers can utilize this reliability to perform checks on their own skills—a method of checks and balances. Regular consultation with an advisor program could help a decision maker avoid mistakes and also, over time, validate the performance of the system.

Knowledge is lost when human experts retire. Expert systems provide one means to ameliorate the institutional erosion of knowledge and to retain some of this hard-earned expertise for future use. One of the primary advantages of the expert system approach is the ability to create systems that fuse the knowledge of several separate disciplines [(e.g., Schmoldt and Martin (1989)]. This breadth of expertise endows them with a greater potential to solve complex problems than human experts, who are often knowledgeable only in a single domain. In a sense we can create "super experts" that have a greater breadth of knowledge and that can address a wider range of problems.

Most knowledge found in textbooks, handbooks, research papers and other "public" sources addresses general principles and recommends for the average case. Expert systems are designed to translate this general understanding into a solution for specific problems. Differences between routine situations and those requiring more analysis or data collection or attention by a human expert can be identified and treated appropriately (Luger and Stubblefield, 1989). The centuries-old gap between the producers and the consumers of knowledge may be substantially reduced. Once a system has been built, it becomes inexpensive to expose large segments of the society to this expert knowledge, making expertise more universally accessible and usable. By capturing and distributing knowledge, expert systems can be used as assistants to humans with less than expert skills in some area and as training vehicles for novices (Bowerman and Glover, 1988). The most current knowledge becomes available for many users rather than the for a few individuals who are acquainted with the human expert. Expert systems are currently available that enable employees to perform tasks that previously required close supervision or access to knowledgeable senior specialists (Harmon and King, 1985). A novice can function at a level comparable to more experienced members of the organization. New employee transition periods may be shortened and the usually concomitant loss of productivity greatly reduced. Continuity can be maintained despite changes in staff; inexperienced colleagues may see an expert system as a safety net and more experienced staff may regard it as a check-list (Starfield and Bleloch, 1986).

Limitations of Expert Systems

Expert systems, being part of and reliant on AI, have all the limitations associated with our current inability to field truly intelligent systems. As AI develops, and more sophisticated and intelligent systems are created, expert systems will improve to the point where many limitations may eventually disappear. Bowerman and Glover (1988) note several categories of expert system weaknesses:

1. **Cognitive**: Systems function in a subdomain of the full human potential, i.e., cognitive and logical. They are not adept in managing sensory input or output, or dealing with less analytical aspects of human reasoning abilities, e.g., ellipsis, intuition.
2. **Narrow**: The total knowledge available to a human expert appears in many forms, of which expert systems have no understanding, e.g., social knowledge, analogical reasoning, random memories, feelings, emotions, common sense and irrational information. Their problem-solving skills have a very narrow focus.

3. **Brittle**: Expert systems degrade almost immediately when confronted with situations outside their domain. The ability to recognize and recover from errors is non-existent.

4. **Non-learning**: Human intellectual development relies extensively on learning abilities; few current systems have any capacity for machine learning.

5. **No self-awareness**: They exhibit no sense of self; therefore, they possess no understanding of what they know—or what they do not know.

Limitations do not imply any theoretical defect, currently apparent, that would prevent expert system developers from eventually overcoming these problems. This technology remains in its infancy. Vigorous research in AI has as its goal the solution of several of these problems; in particular, machine learning, alternative logics and common-sense reasoning are receiving substantial effort. The critical breakthrough for solving many of these problems may likely involve some capacity for learning (Schank, 1988; McRoberts et al., 1991). Learning will enable systems to recognize familiar situations and to react quickly, to adapt to new and unfamiliar circumstances, and to expand their narrow focus to larger, more general problems. An expert system can only reasonably be expected to incorporate in its reasoning that part of the domain that human experts would regard as routine (Sell, 1985). Many of the current expert systems are not as good as the best human expert or even the "average" human expert. Jackson (1986) observes, however, that, like their human parents, expert systems need not be right all the time to be useful. While such inconsistency may create difficulties for system users, this and other problems are now the case, given the poor general availability of human expertise.

THE KNOWLEDGE REPRESENTATION SPECTRUM

To better explain the utility of hypertext and expert system methods, it is helpful to examine the different forms knowledge may take. Data, information and knowledge can reside in diverse forms and media. During the last 40 years, computer technologies have been developed to electronically manipulate most of the forms of non-language and language-based knowledge.

Non-Language-Based Knowledge

Non-language-based knowledge can reside in thought, sound, smell, feel and sight. We have had very little success in electronically manipulating thoughts and smells. The robotics industry has developed computer controlled arms and hands sensitive enough in touch to play a piano and handle delicate objects. Computer-interpreted vision is good enough to allow the development of rudimentary, autonomously mobile all-terrain vehicles for military and space exploration purposes. The playing of music and recorded sounds under computer control is quite advanced, whereas voice recognition is still rudimentary.

We have had much more success in manipulating graphic, photographic and video images using computers. With the recent advent of CD-ROM (compact disk-read only memory) devices into mainstream computing, knowledge represented in these formats is almost commonplace. One advantage of representing knowledge in the non-language based media is compactness, i.e., a picture is worth a thousand words. Another advantage is the ability to impart certain types of knowledge better through sound, smell, sight or feel than through language. For example, it is difficult to describe the difference in smell of a hardwood forest versus a pine forest in autumn; yet, anyone familiar with these ecosystems can instantly differentiate between them on the basis of smell alone. Thus, the sensation of smelling provides information which leads to knowledge.

Language-Based Knowledge

Language is defined as a shared system for symbolic communication whether oral or text (Ashcraft, 1989). Because meaning is not inherent in the symbols, but is attributed to them by common consent, we are free to devise any number of languages to represent knowledge. Linguistics recognizes two classes of languages: natural and artificial (Table 1) (Gal et al., 1991). **Natural languages**, such as English, are robust (they can be ungrammatical and still comprehensible), ambiguous (multiple definitions for the same word and implied meanings are allowed), and indeterminate (unable to enumerate all grammatically correct sentences ahead of time). Furthermore, natural languages, which are still spoken and written, are always changing in an uncontrolled way. The grammatical rules themselves may change radically from century to century.

Table 1. Comparison of Natural and Artificial language characteristics. [from Gal et al. (1991)]

Natural Languages	Artificial Languages
1. Not controllable	1. Completely under human control
2. Indeterminate	2. Determinate
3. Ambiguous	3. Precise
4. Multiple definitions	4. Single definitions
5. Robust	5. Brittle
6. Implicit meaning allowed	6. Implicit meaning not allowed

In contrast, the aim of **artificial languages**, such as FORTRAN, calculus and numerical logic, is to describe a very limited set of phenomena very precisely (Gal et al., 1991). Artificial languages use strictly controlled vocabulary and grammar with simple structure designed to eliminate ambiguity. Each valid word has a unique meaning and all words and structures can be enumerated (determinism). The cost of precision is lack of generality and increased brittleness (small errors in grammar cause comprehension to fail). Languages, both natural and artificial, can be used to represent knowledge. The concept of **knowledge representation** refers to methods of structuring, encoding and storing symbols so that the attributed meaning is clear. Language-based knowledge representation methods can be organized according to their problem solving attributes using the dimensions of generality, ambiguity, and power (Table 2). **Generality** refers to the size of the set of problems that the method is capable of addressing. **Ambiguity** refers to the precision with which meaning can be attributed to the symbols used by the method. **Power** refers to the effectiveness of the method in solving specific problems.

For better understanding, I will first discuss the extremes and then fill-in the middle. We have learned to manage **natural language text** (NLT) electronically with word processing software. NLT has great generality, i.e., it can be used to represent almost any knowledge but it is not very powerful in communicating because it is ambiguous and imprecise. To compensate for this ambiguity, language is bulky—it takes many words to describe something a picture or an equation, if available, could perhaps represent more succinctly.

Mathematics contrasts with NLT because it is the study of symbols which are numbers, i.e., they have magnitude, direction and formally defined operations. Mathematics was invented partly out a of desire to be precise and to solve problems infallibly. It can be argued that mathematical knowledge representation is a reaction to the ambiguity of the natural world and to the comparatively feeble problem solving power of NLT. Mathematical equations with analytical solutions are the most powerful problem-solving tools humans have yet invented. When scientifically validated analytical equations are available and applicable, they represent the most efficient and effective way to model a natural system, understand nature and communicate knowledge. Unfortunately, equations with analytical solutions are not available for most biological processes; indeed, they may never be available for some biological systems. Therefore, in contrast to NLT, analytical equations are powerful, but apply only to a few situations.

Table 2. Language-based knowledge representation methods organized according to problem solving power and generality.

Natural Language Systems		Artificial Language Systems		
		Logic	Mathematics	
NLT[1] Documents	Hypertext Documents	Expert Systems Qualitative Simulations Neural Networks	Numerical Equations	Analytical Equations

High <- Generality - - - - - - - - - - - - - - - - - -> Low
High <- Ambiguity - - - - - - - - - - - - - - - - - -> Low
Low <- - - - - - - - - - - - - - - - - - Problem-Solving Power - - - - - - - - - - - - - -> High

[1] Natural Language Text

Until electronic computers were invented, language-based knowledge could only be represented and manipulated as NLT or as analytical equations. The first knowledge representation method, totally dependent upon the availability of electronic computers, was that of iteratively solved systems of interdependent mathematical equations. Most such numerical systems are theoretically, but not practically, solvable without computers. When a system of interdependent equations cannot be solved analytically, we typically use iterative numerical methods to solve them approximately, usually in the form of **quantitative simulation models**. Being satisfied with sufficiently precise solutions rather than demanding exact solutions to systems of mathematical equations, allows us to increase the computationally solvable set of problems by orders of magnitude. Simulation models are not as compact or powerful as analytical models but they are significantly more general in their applicability to real world problems.

Expert systems (Luger and Stubblefield, 1989), qualitative simulation models (McRoberts et al., 1991) and neural network models (Wasserman, 1989) take another giant step in the direction of generality. These systems are not at all based on mathematical equations. They grew out of the science of symbolic logic and were adapted for practical problem solving and understanding by workers in the field of AI. Researchers in AI realized that much of what human experts knew could not easily be formulated into a system of mathematical equations. And yet, human experts do exist that are able to solve some problems consistently better than most non-experts. Expert systems were invented to capture the knowledge of experts into sets of IF-THEN logical statements and then to manipulate the resultant knowledge-base, to achieve some goal. Expert systems both store and apply knowledge. They trade off a gain in generality to increase the set of potentially soluble problems, with a decrease in compactness and problem solving power. Indeed, expert systems, just like their human counterparts, are not always correct; they are not even guaranteed to find a solution all the time.

The movement from analytical systems to numerical systems to expert systems all progressed from the mathematical end of the knowledge representation spectrum. In contrast, hypertext systems originated from the NLT end of the spectrum. In most NLT, the structure of the subject is more or less hidden, camouflaged by the sequential nature of the medium and the need to bridge ideas gracefully. With some notable exceptions (such as encyclopedias), structure in most NLT is subservient to the content of the material being presented. Hypertext forces the author to explicitly highlight the structure (outline or concept map) first and foremost for the user. Only secondarily is the user exposed to the content matter. It is the structure that guides the user, time-and-again, to try different paths in the hyperdocument in an order that is user-determined. The author can no longer rely on sequential reading to present material, which means that each block or chunk of the hypertext must be independently understandable, much as we demand that journal figures and tables be independently understandable. Hypertext gives up some generality to increase the power of text to communicate meaning more clearly and to increase problem solving power by reducing ambiguity.

Refining NLT text, first into hypertext, then into expert systems, and finally into mathematical models produces a significant compaction of knowledge bulk (the amount of redundant text). It also increases understanding through progressive organization, evaluation and synthesis, and promotes more effective problem solving and communication. The cost is a step-by-step reduction in the set of problems potentially solvable and in the effort of refining knowledge at each step.

A THEORY OF ROOT-SHOOT INTERACTION AND CARBON ALLOCATION

The highly fragmented state of the international scientific knowledge practically guarantees that finding, organizing and synthesizing any particular piece of it will be an arduous, time-consuming task. I found that a comprehensive and concise theory of root-shoot interaction and C allocation does not exist even in modern ecophysiological textbooks such as Kozlowski et al. (1991). This section, therefore, contains a preliminary theory of root-shoot interaction and C allocation based on a literature review. It is intended primarily for illustrative purposes and should be viewed much as a "test" data set might be viewed as a means to illustrate a new statistical procedure.

I will present the theory NLT because this chapter has been published using print technology. I actually did organize and synthesize the literature using hypertext. Unfortunately, rendering a 3-dimensional structure of chunks and links (Fig. 1) into a sequential, printed output prevents the reader from understanding the organizing and communicating power of hypertext. (**The hypertext version of this preliminary theory of root-shoot interaction and C allocation is available from me upon request.**)

Understanding the C-dynamics of plants is fundamental to understanding their relative success or failure in any given physical environment (Mooney, 1972). A plant's ability to accumulate assimilated C depends upon its photosynthetic capacity, its pattern of carbon distribution and its ability to maintain dry matter accumulation in the face of environmental stresses (Geiger and Servaites, 1991). It is well known that most plants fall far short of their genetic potential. Agricultural crops grow only 12 to 30% of their recorded maximums; forest stand production is probably even lower. A detailed account of plant C allocation processes under stress may be found in Mooney et. al. (1991).

Carbon supply to a sink is controlled by (Kramer and Kozlowski, 1979):

1. the absolute size of the C supply, i.e., currently produced photosynthate and stores of reserves.

2. the relative strength of the various "sinks."

3. the relative distance from the source to sink (sinks are supplied from the nearest source).

4. growth regulators (hormones) play a role in determining sink strength and are capable of changing relative sizes of both sources (by increasing photosynthetic rates) and sinks.

5. the existing vascular connections between sources and sinks (Leopold and Kriedemann, 1975).

6. the feedback from sink to source.

A hypothesis, termed the optimum resource allocation hypothesis, has been formulated to explain the strategy of whole-plant C allocation. The optimum resource allocation hypothesis states that: "...Plants respond homeostatically to imbalances in resource availability by allocating newly acquired C to organs that enhance acquisition of those resources most strongly limiting growth...." (Bloom et al., 1985). This theory is a formalized statement summarizing observations of plant growth behavior, but it is not sufficiently detailed for modeling purposes. Although recent studies have improved our understanding of the dynamics of C allocation in plants, much remains to be learned, and a comprehensive and sufficiently detailed theory does not exist.

The most important questions to be answered are: "What controls the amount of C delivered to the various competing organs and tissues and brings about the necessary changes at various stages of development and levels of stress...." (Kramer and Kozlowski,1979). And, if the translocating system is viewed as a system of sinks and sources, what controls relative sink strength? and how does a gradient of C supply take place so that large structural imbalances, such as a wavy tree stem, do not occur?

Unstressed Carbon Allocation

Trees exhibit a generalized pattern of C allocation under "normal" growing conditions. This generalized pattern is altered with stress and during years of heavy fruiting, when reproductive tissues monopolize the bulk of available carbohydrates. Each phenological phase of growth, or **phenophase**, during the season exhibits a unique carbohydrate allocation pattern. It is useful to discuss "normal" C allocation patterns within the framework of these phenophases. A growth-year may be classified into five major phenophases: dormant, pre-budbreak, early growth, main growth and post-bud-set. It should be emphasized that the changes from one to another phenophase are fairly gradual, not abrupt. Nonetheless, for the present purposes, it is sufficiently accurate to treat the switches between the phenophases as abrupt. It is also clear that flushing-species will cycle through these phenophases in a cyclical pattern. Once a useful general theory has been formulated, species specific variants can be more readily identified.

There are also patterns of C allocation that change with the age or size of trees. These long-term patterns, associated with tree maturity, gradually influence the seasonal C allocation patterns.

Pre-Budbreak Carbon Allocation Patterns. Root growth usually occurs before bud-break (Wilson, 1984) and is the major carbohydrate sink before buds open (Kramer and Kozlowski, 1979). The carbohydrate comes from storage in the parenchyma cells of the large woody roots. The priority for normal C allocation during this phenophase is: a) new roots, b) diameter growth in suberized roots, and c) protective chemicals

Early Growth Carbon Allocation Patterns. Early in the growing season, while the new leaf primordia are still developing, and thus have a small sink strength, the roots are the strongest sinks and large amounts of storage carbohydrates are translocated to them. After buds open, a reversal of translocation gradually takes place, with larger amounts of carbohydrates moving to the expanding shoots from storage locations. In gymnosperms, the old needles are the most important source of metabolites for early shoot growth. At this time, cambial growth forms a secondary carbohydrate sink (Kramer and Kozlowski, 1979). A priority for normal C allocation during this phenophase might be: a) new foliage and buds; b) new roots; c) canopy, stem and root storage; d) diameter growth in stems and branches; and, e) protective chemicals

Main Growth Carbon Allocation Patterns. During the main-growth phenophase, the C for shoot growth is largely supplied by current photosynthate. Shoot growth has the highest priority. One estimate of export rates from leaves of deciduous trees is: 40% of C remains in the mother leaf; 10% exported from mature to developing leaves; 25% goes to the stem; and, 25 % goes to root (Porter, 1966). In evergreen trees, greater amounts of C remain in the mature leaves because they serve as important storage organs (Mooney 1972). The C produced by the more mature leaves in the lower half of the canopy is used first for storage, then diameter growth and, finally, root growth. At this time, cambial growth forms a secondary carbohydrate sink. While the cambial meristems in trees are a relatively weak sink for carbohydrates, their large areas ensure that they consume large quantities of carbohydrates over the growing season. Use of carbohydrates in cambial growth is not uniform in time or space. In many species cambial growth occurs in only a few months of the year. Even during that time period, the use of carbohydrates is intermittent because of competition of cambial growth with stronger sinks and because of the environmental stress effects on photosynthate supply (Kramer and Kozlowski, 1979).

Table 3. Summary of photosynthate transport coefficients (%) for juvenile hybrid poplar during the main growth phenophase. [from Rauscher et al. (1990)]

Photosynthate Transported		Photosynthate Sinks		
LPI[1]	(%)	Other Leaves	Stem Internodes	Roots
(Apex of Shoot)				
1-4	0	0	0	0
5	15	73	27	0
6	30	67	31	2
7	45	54	39	7
8	60	47	43	1
9	75	36	52	12
10	90	27	60	13
11	90	17	69	14
12	90	12	64	24
13	90	13	59	28
14	90	10	60	30
15	90	10	56	34
16	90	6	53	41
17	90	3	53	44
18	90	2	53	45
19-50	100	0	55	45
(Base of Crown)				

[1]Leaf Plastochron Index, a morphological index that permits comparison of leaves at the same stage of development among plants grown at different times and under different environmental conditions (Larson and Isebrands, 1971).

C transport must be accounted for in detail to understand whole-tree ecophysiological dynamics (Isebrands et al., 1990; Dickson and Isebrands, 1991). In angiosperms, each leaf has its own C allocation pattern associated with its physiological age (Table 3). Specific, detailed data of whole tree C allocation such as presented in Table 3 and by Porter (1966) is extremely rare in the scientific literature on C allocation. A priority for normal C allocation (Waring and Schlesinger, 1985) during this phenophase is: a) new foliage and buds; b) canopy, stem and root storage; c) diameter growth in stems and branches; d) new root growth; and, e) protective chemicals.

Post-Budset Carbon Allocation Patterns. After budset, current photosynthate as well as remobilized photosynthates from senescing leaves go into branch, stem and root storage. Bud set in poplar marks the transition between photosynthate going to the stem for diameter growth and the flow of photosynthate to the stem and roots for storage (Isebrands and Nelson, 1983).

The preeminence of the lower bole and large-diameter roots as storage organs is consistent for all types of trees. Some diameter growth in stems and branches occurs before the accumulation of reserves, but the bulk comes after storage has been satisfied (Waring and Schlesinger, 1985). This is particularly true for large-root diameter growth. Strong downward translocation of C triggered by bud set was observed in poplar clones (Pregitzer et al., 1990). After bud set there is a large increase in root biomass, mostly in the form of nonstructural carbohydrates (Nguyen et al., 1990; Pregitzer et al., 1990). A 410 kg oak may have reserves equivalent to 54 kg of sugar, sufficient to support the replacement of the leaf canopy three times (McLaughlin et al., 1980).

When storage demand has been satisfied, cambial growth and renewed root growth commonly comprise the major carbohydrate sinks. Roots of temperate zone trees continue to grow later in the season than shoots and stems and are, therefore, stronger sinks than shoots and stems in the late summer and early autumn (Kramer and Kozlowski, 1979). A priority for normal C allocation for this phenophase is: a) canopy, stem and root storage; b) new root growth; and, c) diameter growth in stems, branches and suberized roots.

Age-Related Carbon Allocation Patterns. Changes in the age or size of trees have been correlated with changes in C allocation (Table 4). For poplars and aspen, the proportion of C in leaves and roots decreases with age, whereas that in the stem increases. Such net C-pool comparisons can only be used as a validation of the end result of gross C allocation patterns. C allocation coefficients cannot be directly derived from biomass pool sizes. Unfortunately, C allocation coefficients along an age sequence are seldom available in the literature.

Table 4. The percentage of carbon by age of major tree components in hybrid poplar and aspen leaves, stems and roots. [Data are a composite from unpublished and published (Ruark and Bockheim, 1987) sources.]

	Age in Years						
	1	5	10	15	20	30	60
	Percent of Whole-Tree Carbon						
Leaves	30	8	5	4	4	3	3
Stems	20	47	59	69	71	80	82
Roots	50	45	36	27	25	17	15

There is considerable evidence that as trees increase in size, and the translocation path from the shoots to the roots becomes longer, there is decreasing dependency of growing tissues in the upper crown on the carbohydrate reserves in the lower stem and roots (Kramer and Kozlowski, 1979). Utilization of carbohydrates also varies by stem position (Kramer and Kozlowski, 1979). In seedlings and in vigorous, open-grown adult trees, more carbohydrates are used in cambial growth in the lower stem than in the upper stem. In suppressed trees and very old trees, the reverse is true as shown by the absence of xylem rings in the lower stem. Leaves exhibit radically changing C allocation patterns as a function of their physiological age (Table 3). During early growth, a leaf is an importer of assimilates until the laminar surface is one-third to one-half full size (Leopold and Kriedemann, 1975). When leaves attain half size, they will for a short time simultaneously import and export C (Mooney, 1972). When assimilate export starts, it goes mostly to the nearest apex to the leaf.

Subsequently, leaves will produce enough for their own C needs and export a portion to other parts of the plant. Leaves near the apex of the plant export primarily upward; lower leaves export primarily to the root (Mooney, 1972). Branches are simply functional replicas of the main stem. Leaves near the apex export primarily upwards to the branch apex; lower leaves export primarily to the main stem and roots. As the leaf ages, the root system gradually receives more and more of the leaf's C. Ultimately, the source leaf senesces, and organic and inorganic reserves are mobilized and exported to storage compartments (Leopold and Kriedemann, 1975). Older leaves cannot import C from younger ones, thus precluding "parasitism" during senescence (Mooney, 1972).

Carbon Allocation During Stress

Plants draw on C reserves to combat the negative effects of stress (Geiger and Servaites, 1991). Because stress can occur at any time, allocation must provide C reserves throughout the plant's life. While the normal patterns of C allocation may be predictable, changes due to stress will cause significant deviations (Waring and Schlesinger 1985).

Plants respond to stress by changing C assimilation and partitioning (Dickson and Isebrands, 1991). Environmental changes are sensed primarily by leaves. C partitioning and plant growth then change to compensate for the stress. Balanced allocation of C between immediate use and storage is essential for plant growth and survival during times of stress (Geiger and Servaites, 1991). A tree's ability to withstand a particular kind of stress can be assessed by evaluating the soluble and easily mobilized carbohydrate reserves near the points of potential need. For example, current photosynthate is reallocated quickly to combat defoliation, but will not supply C to stems and roots for repair of any damage to those organs (Waring and Schlesinger, 1985). C stored in large-diameter roots is used to support spring root growth or replacement of foliage following defoliation. Plants that are acclimated to resource-rich environments allocate photosynthate flexibly in response to environmental stress (Bloom et al., 1985). In contrast, plants from resource-poor environments have more rigid allocation patterns and will tend to allocate excess C (temporary surplus) into storage. As a general pattern, light and C limitations lead to proportionately more shoots, while water and nutrient limitations lead to more roots (Sharpe and Rykiel, 1991). The impact of stressors on C allocation patterns will be reviewed below.

Light Stress. Inadequate light in parts of or the entire canopy will reduce root growth in favor of new foliage growth and terminal elongation (Mooney, 1972, Kozlowski et al., 1991, Waring and Schlesinger, 1985). Increasing light availability increases the photosynthetic rate (Wilson, 1988). As photosynthesis becomes more efficient, it places a greater demand on the root system for water and nutrients, thus requiring more C investment in the root system. Low irradiance or shortened photoperiods reduce plant C. Leaves divert newly fixed C from leaf growth to starch biosynthesis, which ensures a reserve needed during the dark period. Within two days, reserve starch increases 1.5 to 2 times normal, which decreases leaf growth by the same amount. This change is gradual and continues until the leaf reaches its compensation point, beyond which C reserves are exported and the leaf senesces. Growth of leaf area is maintained by reductions in leaf density and increases in shoot-root ratio (Geiger and Servaites, 1991).

If the photosynthetic surface area is much reduced by defoliation, root growth is decreased temporarily by the reduced supply of photosynthate, but leaf growth is stimulated and leaf area tends to increase (Wilson, 1988; Kozlowski et al., 1991). Such compensatory translocation of carbohydrates within the shoot, resulting from shading or defoliation of sections of the shoot, has been demonstrated repeatedly (Kramer and Kozlowski, 1979). This process most likely occurs in the roots as well when sections of the root zone die.

Water Stress. Under mild drought, C fixation is slightly reduced but shoot growth is inhibited (Waring and Schlesinger 1985, Wilson 1988). When the relative water content of an exporting leaf falls, C may be allocated to synthesis of osmotic agents in the leaf in order to increase the sink strength of the leaf for water (Geiger and Servaites 1991). Exportable C is transported down the phloem into the roots. Roots, being near the water source, are less water-stressed than cells in the stem and branches and are able, with improved C supply, to grow actively. Water stress will also reduce the rate at which assimilate is converted into structural carbohydrates.

There is a negative feedback between water stress and shoot growth. Mild water stress may be induced by the leaf surface area becoming too large in proportion to the root system, thus making it difficult for the root system to maintain an optimum water balance (Kozlowski et al. 1991). This feedback is a part of the poorly understood mechanism which maintains a rough proportional balance between roots and shoots. It is possible to model the negative linkage of water stress and growth by decreasing the proportion of C allocated for growth of leaves whenever the relative water content of the leaf decreases. De Wit (1968) used this method in his BACROS simulation model. The C diverted from the shoot can be used to increase the amount of C exported to the root system.

Nutrient Stress. When a nutrient, such as nitrogen (N), is deficient, construction of photosynthetic enzymes in new foliage is restricted, canopy growth slows and an increasing proportion of C moves toward the roots (Wilson, 1988). The general experience has been that if nutrients are limited, C partitioning tends to favor root growth (Schulze, 1983). Increasing root growth may increase uptake of N and permit shoot growth to gradually continue (Waring and Schlesinger, 1985). For example, nutritionally stressed Scots pine (*Pinus sylvestris*) have been measured to allocate 60% of their photosynthate to the roots whereas fertilized individuals allocate only 40% to the roots (Linder and Axelsson, 1982). Deciduous hardwoods are expected to exhibit similar responses. They have the additional flexibility to improve their photosynthetic efficiency (Waring and Schlesinger, 1985).

Although trees respond similarly to water and nutrient stress, there is, nevertheless, a significant difference between water- and nutrient-related stresses (Schulze, 1983). Storage and incorporation of water in growth processes is usually quantitatively insignificant—the mass of water simply flows through from the soil to the atmosphere—whereas in the case of nutrients, considerable mass storage occurs. Thus, the plant may be able to buffer short-term nutrient deficiencies by remobilizing needed elements from storage. At toxic levels of nutrients, plants are expected to suppress root growth more than shoot growth resulting in a larger shoot-to-root ratio (Wilson, 1988).

Temperature Stress. It appears that allocation of C to roots is negatively affected by either lower or higher than normal soil temperatures. If root temperature is increased, while keeping shoot temperature constant, the shoot-to-root ratio increases (Wilson, 1988). At high temperatures, root processes are more active and become more efficient so that the C invested in the root system can be decreased. Low soil temperature slows root growth and results in the reduction of C allocation to the roots (Geiger and Servaites, 1991).

* * * * *

In summary, research results reporting C allocation coefficients of trees growing under stress are rare in the scientific literature. Reports of periodic biomass fractions during a season (Pregitzer et al., 1990) or over the age of stands (Beets and Pollock, 1987) are more common. Unfortunately, biomass fractions are not the sensitive a measure of whole-tree carbon allocation that is necessary to develop a solid predictive understanding of tree growth and development. Furthermore, biomass partitioning fractions are an end product, not a process, and are not necessarily indicative of the whole-tree C allocation coefficients.

In the present state of theory, it is unclear what effects indirectly induced stresses have on C allocation patterns. For example, if water availability is low, can the uptake of minerals be inhibited to the point where the plant exhibits symptoms of mineral deficiency, even though the minerals may be available in the soil?

A THEORY OF ROOT-SHOOT INTERACTION AS AN EXPERT SYSTEM

The function of a plant regulatory expert system is to monitor and control tree growth and developmental processes in a mechanistic simulation model. Plants exhibit behaviors as if they have a regulatory control system. We realize that these behaviors are the outward expression of complex, genetically and environmentally determined chemical processes. It is reasonable and practical, however, to analyze and model these unknown interactions teleologically as a regulatory control system with "goals," "strategies" and "tactics?" Most probably, Yes. And, because all regulatory systems are conceptually similar, I can and will use a simple home-heating system as an analogy to guide you.

Strategic Rules

At the heart of any regulatory system is a goal-driven strategy. The goal of the home-heating system is to keep the temperature inside the house as close as possible to the temperature set on the thermostat. It is quite possible to be continually changing the thermostat setting, which continually changes the goal of the regulatory system, but at any given instant there is a goal. Plants also exhibit definite, goal-driven behavior. The goal settings change with the plant's phenophase and its age. For example, before bud-break the goal of poplar is to invest C in new root growth (see previous section). After bud-break, the goal is to invest C in new foliage and buds with a secondary emphasis on new root growth and so on. These goals are valid only under "normal" growing conditions. When stress occurs, new goals may go into effect. One can imagine the poplar "thermostat" in a continual state of change and yet, at any given hour or day, some goal or hierarchy of goals is in force. The strategic rules of the expert system express what the poplar needs to do in order to achieve its goals at any given time. An example of a strategic rule is:

IF	Current phenophase is Pre-Budbreak
AND	Growing degree-days ≥ 265.0
THEN	Current phenophase is Early Growth

Using only the variable "growing degree-days" to trigger bud-break is a simplification for the sake of this discussion. The bud-break trigger could just as easily take the form of a complex simulation submodel. In place of the "growing degree-days" rule, there might be an instruction to execute a quantitative simulation model capable of predicting the probability of bud-break. Linking the results into the expert system poses few problems.

The strategic level of the expert system must also determine the stress level of the plant:

IF	Temperature is Normal
AND	Water availability is Normal
AND	Nitrogen availability is Low
THEN	Stress is Nitrogen Stress

There would be other rules that determine whether each of these stress factors is normal, high or low, based upon environmental and plant internal statuses. It is important to realize that the hypertext system is used not only to improve the identification of the necessary rules for the expert system knowledge-base. The hypertext system is an integral part of the expert system because each rule, like the one above, is linked to a particular portion of the hyperdocument for explanatory purposes. The ability to explain its own reasoning as well as to justify the knowledge that it uses are major features that separates expert systems from mathematical models (Table 2).

Once the state of the system has been determined, the strategic level must formulate instructions to the tactical level of the expert system:

IF	Current phenophase is Early Growth
AND	Stress is Nitrogen Stress
OR	Stress is Potassium Stress
OR	Stress is Phosphorus Stress
THEN	Instruction is Reduce Storage-Carbon Allocation to Leaves
AND	Instruction is Increase Storage-Carbon Allocation to Roots

The example rules given above were extracted from the hypertext literature review. The hypertext knowledge-base is a resource that guides the development and testing of the expert system. It is possible to take the hypertext description of the theory of root-shoot interactions and C allocation, and translate it into a set of rules such as the ones above. Every possible combination of circumstances in which the growing tree could find itself must be covered by the rules. Gaps become painfully obvious. For example, the previous section, in which I discussed the effects of water stress, considered only instances of minimal water stress and did not consider the equally likely occurrence of maximal water stress. If two or more levels of water stress are important to tree behavior, then they need to be explicitly considered. As we gain a better understanding of the impact of water stress on tree growth and development, we will move closer to a quantitative rather than qualitative model. The result of the strategic level is a set of qualitative instructions that serve as input to the tactical level.

Tactical Rules

Given strategic direction, a regulatory system also needs tactical rules of behavior for implementing the strategy. For the home-heating system, these rules might govern whether to activate the heating or cooling subsystem, when and for how long to turn the oil burner on, or when to shut down operations in the event a stress to the system is detected such as C-soot buildup on the spray nozzle of the oil burner or no more fuel oil reaching the burner. Some equivalent tactical rules for the poplar regulatory system might address how, and by how much, to change C allocation rates or when to shut down various organs, i.e., senesce leaves or roots in response to stress. An example of a tactical rule that might implement the instructions in the above strategic rule is:

IF	Instruction is Increase Storage-Carbon Allocation to Roots
THEN	Current Coefficient = Current Coefficient + 10%
IF	Instruction is Reduce Storage-Carbon Allocation to Leaves
THEN	Current Coefficient = Current Coefficient - 10%

The tactical rules translate the qualitative direction of the strategic level into changes that a quantitative ecophysiological model can use. If the quantitative model executes hourly, and the expert system daily, then the C allocation coefficients might be adjusted once every 24 hours. If the status of the tree 24 hours later still requires the same adjustment as above, then another 10% increase in the allocation of C from storage to roots, and another 10% decrease to leaves, would be ordered. Other rules, becoming active under other conditions, could reverse this flow of C.

Finally, there is the mechanistic level. For the home-heating regulatory system, this equates to the continuous process of transporting fuel oil from the storage tank to the burner. A regulatory valve sets the rate of flow of the oil; but, once set, a simple flow equation can be used to quantitatively estimate the oil supply to the burner as long as there is oil in the tank and the burner is operational. The ECOPHYS process model (Rauscher et al. 1990) is an example of a model operating at the mechanistic level. In the next section, I will discuss how a qualitative expert regulatory system can be linked to a quantitative process model such as ECOPHYS.

A PLANT REGULATORY CONTROL SYSTEM FOR ECOPHYS

ECOPHYS was developed for poplar trees growing under near-optimal conditions (Host et al., 1990; Isebrands et al., 1990; Rauscher et al., 1990). Thus, the regulatory strategy and tactics in ECOPHYS are static, and they "force" the simulated tree into the same developmental strategy and tactics regardless of the weather in a simulated year. Growth can vary as a function of genetic constitution of the tree, and light and temperature, but the developmental pattern is fixed. This is a valid assumption under near-optimal growing conditions (Host and Rauscher, 1990) but becomes invalid under stressful conditions. Trees growing under natural conditions react to stress by adopting defensive developmental patterns. ECOPHYS must be responsive to these stress factors to be useful in making research or applied forest management decisions.

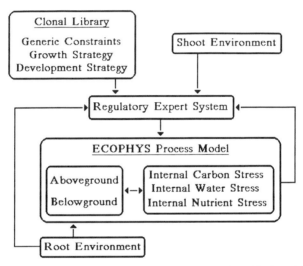

Figure 3. The conceptual diagram of ECOPHYS v.5.0 showing the logical positioning of the regulatory expert system.

There is a clear need to link the plant regulatory control expert system into the ECOPHYS process model (Fig. 3). The proposed model, ECOPHYS (version 5.0), would operate according to the following scenario. The genetic constraints and the growth and development strategy for a specific poplar clone is contained in a clonal library subsystem (Host et al., 1990). The information in the clonal library has been developed previously from research specific to a particular clone or synthesized from the literature for a generic, generalized ideotype (Dickmann, 1985). The clonal library provides ECOPHYS with the values for equation variables that apply to growth under near-optimal conditions and with the developmental goals, in the form of strategic rules, appropriate for the genetic material being simulated.

The regulatory expert system analyzes the state of the root and shoot environments and the state of the tree each day. It translates the goal strategy from the clonal library into tactical rules, which implement changes in the parameter values, which in turn govern the mass transfer rates of the quantitative process subsystem. The mechanistic process submodel operates on an hourly time scale and simulates dimensional growth of leaves, stem, branches and roots as functions of the availability of C, water and N within the tree (Rauscher et al., 1990).

Root-shoot interaction is the end result of a continual and gradual adjustment of the C allocation coefficients by the regulatory expert system, as a function of phenophase, environment and the state of the plant's resources. Thus, if we give the regulatory system control over the allocation coefficients, we can effectively change the size of the different organs of the tree over time according to some experimentally derived strategic plan. More importantly, the strategic and tactical rules are explicitly stated, clearly understandable, and they can be used to test the consequences of various theories of plant growth regulation.

The behavior of poplar trees demands that the regulatory system also exercise control over morphological coefficients such as those that control leaf shape. Other developmental behaviors are not gradual and reversible but rather more abrupt and irreversible, as seen from the "granularity" of ECOPHYS: examples are budbreak, budset and leaf initiation. These types of controls can also be implemented in an expert system, but they are beyond the scope of this paper, although they have been discussed by Rauscher and Isebrands (1990).

DISCUSSION

He that will not apply new remedies must expect new evils; For time is the greatest innovator... —Francis Bacon

Organizing and synthesizing the scientific literature to produce a comprehensive theory of root-shoot interactions and C allocation is time-consuming and labor intensive. Considerable prior knowledge and skill are needed to produce a useful product. Using hypertext technology to store and access the knowledge requires additional skills. Yet more time and still different skills are needed to refine the knowledge into an expert system capable of simulating the qualitative behavior of a real tree growing in real space. Is it worth it? Will anyone do it? In order to address these important questions, we need to expand the scope of the discussion to consider scientific knowledge management in general.

Pressure to apply new knowledge-management remedies, such as hypertext, is increasing, based on at least four "new evils." First, as I pointed out in Introduction, the 200-year-old system of scientific publishing is in crisis. Experimental justification for this belief is difficult to find although anecdotal evidence is readily available. The modern scientific literature is full of books, articles and editorials pointing to this crisis. If doubt still exists, an informal survey of scientists in almost any field should support this assertion. I ask: Do you feel overwhelmed with the volume of literature in your field? Can you keep up with it to your satisfaction? What proportion of the applicable literature do you think you actually read? The social and personal costs of lost knowledge, of misplaced knowledge, of continually recreated knowledge, of missed opportunities because we could not recognize anomalies for what they are, are unknown but probably substantial.

Second, trying to solve increasingly complex and global problems with inappropriate resources (i.e., lack of adequate knowledge and people out of their depth) is likely to produce poor decisions because of the need to oversimplify and rigidly constrain the inputs in order to make the problem manageable. As long as the problems were relatively simple (e.g., How many boardfeet of lumber will this stand yield in 50 years?) and the impact of decisions were local, we could muddle-through without concentrating a great deal of resources on ensuring that we found successful solutions. In the last 30 years, however, we have gradually come to realize that management actions such as pesticide use, agricultural fertilization, and coal and oil energy consumption, and even intensive forest harvesting, have greater local and national consequences than we previously imagined. Just within this last decade is it has become clear that human activities, previously thought to have only local or at most regional effects, are causing global climate changes. The public expects the scientific community to provide administrative and legal decision makers with scientifically sound and ecologically credible knowledge. There is too little evidence that we, the scientists, are consistently delivering useful knowledge on a timely basis to the right people.

Third, the history of science offers many examples of an enabling technology forcing major changes in science. A particularly powerful and appropriate one for our discussion is the advent of the printing press in medieval Europe (Eisenstein, 1979). Similarly, the rapid development of computer hardware capabilities and computer software suitable for knowledge management (i.e., hypertext and expert systems) enables changes in science. In a single decade, 1982 to 1992, computers have become familiar tools. The skill to use computer programs has become widespread as well. Words are processed electronically, numbers are processed electronically and pictures are processed electronically. More recently, scientific references and their abstracts are processed electronically. Multi-media capability based on CD-ROM technology is rapidly being phased-in by the computer software and hardware industry. Personal publishing of gigantic multi-media CD-ROM titles became a reality with the introduction, in 1992, of a desktop CD-ROM copying system (Udell, 1993).

Fourth and last, there is pressure to change from the competitive innovation that underlies science and technology. The best and most sought after computer software development knowledge is being written by Microsoft Corp. and delivered in CD-ROM format. Books, articles, example software code and fully executable computer programs −over 35,000 pages of well-organized and accessible knowledge in hypertext format−are available for $30. This concrete example underscores the fact that the current scientific publication process is making less and less sense. Scientists read and cite paper-text scientific literature. They produce their manuscripts electronically with word processors and submit them to journals. Journals edit the copy electronically and use text formatting software to typeset the manuscript before printing. The printing process produces paper-text, which once again is read and cited by scientists. The electronic manuscript is then discarded.

Notice from the foregoing that production and editing of manuscripts are already done electronically. Paper-text is rapidly becoming an illogical, expensive and unnecessary medium for the publication of scientific research results. Hypertext software and CD-ROM hardware deliver performance that no print-based technology can match in price or in utility. Very soon, the best and most sought-after literature reviews may be authored electronically using hypertext software, and delivered and stored on CD-ROM. Eventually, most scientific literature, including extensive data sets and full model descriptions and code, may be published electronically, disseminated electronically, stored electronically and read/used electronically.

Change causes stress and problems both personally and institutionally. Changes in the way humans are willing to interact with knowledge are certainly complex. What will be lost if we move from print to hypertext? What are the dangers inherent in electronic review articles? Will the widespread acceptance of a set of heuristics embedded in trusted expert systems somehow limit the scope of human creativity? These and many other important questions remain unanswered.

In the final analysis, however, we really have little choice but to rely upon some form of electronic management system to deal with the ever-increasing explosion of scientific knowledge. Currently, hypertext appears the most promising electronic technology for general purpose knowledge management. Hypertext will be especially attractive as it becomes more integrated with other desktop applications such as full-power wordprocessing, spreadsheet, database and graphic programs, and as the sharing of documents in national and international networks becomes more common. As hardware limitations, and user unfamiliarity and resistance, begin to disappear and as more useful hyperdocuments become available, hypertext may eventually replace printed text for many, if not all, scientific reporting purposes. The theory of root-shoot interactions and C allocation presented in this chapter, and its representation as a hypertext and expert system, may serve as a demonstration of the feasibility of electronic knowledge management in science.

ACKNOWLEDGMENTS

I thank Drs. Fred Beall, George Host, Jud Isebrands and Dale Johnson, and three anonymous reviewers, for their significant contributions to this chapter. This research was performed as part of the USDA Forest Service Northern Station's Global Climate Change research project. Mention of companies or tradenames does not constitute endorsement by the Forest Service.

REFERENCES

Allen, T.F.H., and Hoekstra, T.W., 1992. "Toward a Unified Ecology," Columbia Univ. Press, New York.

Ashcraft, M.H., 1989. "Human Memory and Cognition," Scott, Foresman and Co., Glenview.

Bauer, H.H., 1992. "Scientific Literacy and the Myth of the Scientific Method," Univ. of Illinois Press, Urbana.

Beets, P.N., and Pollock, D.S., 1987, Accumulation and partitioning of dry matter in *Pinus radiata* as related to stand age and thinning, *N.Z. J. For. Sci.* 17:246.

Berk, E., 1991, "Text-Only Hypertexts," *in*: E. Berk, and J. Devlin, eds., "Hypertext/Hypermedia Handbook," McGraw Hill, New York.

Bloom, A.J., Chapin III, F.S., and Mooney, H.A., 1985, Resource limitation in plants - an economic analogy, *Annu. Rev. Ecol. Syst.* 16:363.

Bowerman, R.G., and Glover, D.E., 1988, "Putting Expert Systems into Practice," Van Nostrand Reinhold, New York.

Bush, V., 1945, As we may think, *Atlant. Month.* 176.1:101.

Conklin, J., 1987, "Hypertext: an Introduction and Survey," *in*: *IEEE Computer*, Sept. 17.

Davis, J.R., and Clark, J.L., 1989, A selective bibliography of expert systems in natural resource management, *AI Applic.* 3:1.

de Wit, C.T., 1968, Plant production, *Misc. Pap. Landbouw. Wag.* 3:25.

Dickmann, D.I. 1985, The ideotype concept applied to forest trees, *in*: "Institute of Terrestrial Ecology," M. Cannell, and J. Jackson, eds., Huntington.

Dickson, R.E., and Isebrands, J.G., 1991, Leaves as regulators of stress response, *in*: "Response of Plants to Multiple Stresses," H.A. Mooney, W.E. Winner, and E.J. Pell, eds., Academic Press, San Diego.

Eisenstein, E. 1979, "The Printing Press as an Agent of Change," Cambridge Univ. Press, Cambridge.

Fersko-Weiss, J., 1991, 3-D reading with the hypertext edge, *PC Mag.* , May 28, p. 241.

Forscher, B.K., 1963, Chaos in the brickyard, *Science* 142:339.

Gal, A., Lapalme, G., Saint-Dizier, P., and Somers, H., 1991, "PROLOG for Natural Language Processing," John Wiley and Sons, Chichester.

Geiger, D.R., and Servaites, J.C., 1991, Carbon allocation and response to stress, *in*: "Response of Plants to Multiple Stresses," H.A. Mooney, W.E. Winner, and E.J. Pell, eds., Academic Press, San Diego.

Giere, R.N., 1984, "Understanding Scientific Reasoning," 2nd ed., Holt, Rinehart, and Winston, New York.

Harmon, P., and King, D., 1985, "Expert Systems: Artificial Intelligence in Business," Wiley and Sons, New York.

Herring, C., 1968, Distill or drown: the need for reviews. *Phys. Today* 21:27.

Horn, R.E., 1989, "Mapping Hypertext: Analysis, Linkage, and Display of Knowledge for the Next

Generation of On-line Text and Graphics," Information Mapping, Inc., Waltham.

Host, G.E., and Rauscher, H.M., 1990, Validating the regional applicability of a whole-plant ecophysiological growth process model of poplar, *in*: "Proc. IUFRO Forest Simulation Systems Conf.," L.C. Wensel, and G.S. Biging, eds., Berkeley.

Host, G.E., Rauscher, H.M., Isebrands, J.G., Dickmann, D.I., Dickson, R.E., Crow, T.R., and Michael D.A., 1990, "The Microcomputer Scientific Software Series No. 6: The ECOPHYS User's Manual," Gen. Tech. Rep. NC-141, USDA Forest Service, North Central For. Exp. Sta., St. Paul.

Isebrands, J.G., and Nelson, N.D., 1983, Distribution of ^{14}C-labeled photosynthates within intensively cultured *Populus* clones during the establishment year, *Physiol. Plant.* 59:9.

Isebrands, J.G., Rauscher, H.M., Crow, T.R., and Dickmann, D.I., 1990, Whole-tree growth process models based on structural-functional relationships, *in*: "Process Modeling of Forest Growth Responses to Envrionmental Stress," R.K. Dixon, R.S. Meldahl, G.A. Ruark, and W.G. Warren, eds, Timber Press, Portland.

Jackson, P., 1986, "Introduction to Expert Systems," Addison-Wesley Company, Inc, Reading.

James, G., 1985, "Document Databases," Van Nostrand Reinhold Co., New York.

Kozlowski, T.T., Kramer, P.J., and Pallardy, S.G., 1991, "The Physiological Ecology of Woody Plants," Academic Press, San Diego.

Kramer, P.J., and Kozlowski, T.T., 1979, "Physiology of Woody Plants," Academic Press, New York.

Kurzweil, R., 1990, "The Age of Intelligent Machines," MIT Press, Cambridge.

Larson, P.R., and Isebrands, J.G., 1971, The plastochron index as applied to developmental studies of cottonwood, *Can. J. For. Res.* 1:1.

Leopold, A.C., and Kriedemann, P.E., 1975, "Plant Growth and Development," 2nd ed., McGraw-Hill Book Co., New York.

Linder, S., and Axelsson, B., 1982, Changes in carbon uptake and allocation patterns as a result of irrigation and fertilization in a young Pinus sylvestris stand, *in*: "Carbon Uptake and Allocation: Key to Management of Subalpine Forest Ecosystems," R.H. Waring, ed., IUFRO Workshop, For. Res. Lab., Oregon State Univ., Corvallis.

Luger, G.F., and Stubblefield, W.A., 1989, "Artificial Intelligence and the Design of Expert Systems," Benjamin/Cummings Pub. Co., Inc., Redwood City.

McLaughlin, S.B., McConathy, R.K., Barnes, R.L., and Edwards, N.T., 1980, Seasonal changes in energy allocation by white oak (*Quercus alba*), *Can. J. For. Res.* 10:379.

McNeill, D., and Freiburger, P., 1993. "Fuzzy Logic," Simon and Schuster, New York.

McRoberts, R.E., Schmoldt, D.L., and Rauscher, H.M., 1991, Enhancing the scientific process with artificial intelligence: forest science applications, *AI Applic.* 5(2):5.

Mooney, H.A, 1972, The carbon balance of plants, *Annu. Rev. Ecol. Syst.* 3:315.

Mooney, H.A., Winner, W.E., and Pell, E.J., 1991, "Response of Plants to Multiple Stresses," Academic Press, San Diego.

Nielsen, J., 1989, "Hypertext and Hypermedia," Academic Press, New York.

Nguyen, P.V., Dickmann, D.I., Pregitzer, K.S., and Hendrick, R., 1990, Late-season changes in allocation of starch and sugar to shoots, coarse roots, and fine roots in two hybrid poplar clones, *Tree Physiol.* 7:95.

Parsaye, K., Chignell, M., Khoshafian, S., and Wong, H., 1989, "Intelligent Databases," John Wiley & Sons, Inc., New York.

Porter, H.K., 1966, Leaves as collecting and distributing agents of carbon, *Aust. J. Sci.* 29:31.

Pregitzer, K.S., Dickmann, D.I., Hendrick, R., and Nguyen, P.V., 1990, Whole-tree carbon and nitrogen partitioning in young hybrid poplars, *Tree Physiol.* 7:79.

Price, D.J., 1963, "Little Science, Big Science," Columbia Univ. Press, New York.

Rauscher, H.M., and Hacker, R., 1989, Overview of artificial intelligence applications in natural resource management, *J. Knowl. Eng.* 2:30.

Rauscher, H.M., and Host, G.E., 1990, Hypertext and AI: a complementary combination for knowledge management, *AI Applic.* 4:27.

Rauscher, H.M., and Isebrands, J.G., 1990, Using expert systems to model tree development, *in*: "Proc. IUFRO Forest Simulation Systems Conf.," L.C. Wensel, and G.S. Biging, eds., Berkeley.

Rauscher, H.M., Alban, D.H., Johnson, D.W., A brief overview of hypertext system development, *in*: "Proc. IUFRO Cent. Meet.," 1992 Aug. 31-Sept. 7, Berlin. (in press)

Rauscher, H.M., Isebrands, J.G., Host, G.E., Dickson, R.E., Dickmann, D.I., Crow, T.R., and Michael, D.A., 1990, ECOPHYS: an ecophysiological growth process model for juvenile poplar, *Tree Physiol.* 7:255.

Rauscher, H.M., Bartos, D.L., Davey, S.M., Downing, K., Elmes, G.A., Gertner, G., Biing, T., Stockwell, D.R.B., Twery, M.J., and Schmoldt, D.L., 1991, The encyclopedia of AI applications to forest science, (Hypertext version, computer disk), *AI Applic.* 5: insert 592,080 bytes; 235 chunks; and 449 links.

Ruark, G.A., and Bockheim, J.G., 1987, Biomass, net primary production, and nutrient distribution for an age sequence of *Populus tremuloides* ecosystems, *Can. J. For. Res.* 18:435.

Schank, R.C., 1988, "The Cognitive Computer: On Language, Learning, and Artificial Intelligence," Addison-Wesley Publishing Co., Inc., Reading.

Schlumlienzer, P.C., 1989. Shadow-fusing hypertext with AI, *IEEE Expert*, Winter, p. 65.

Schmoldt, D.L., and Martin, G.L., 1989, Construction and evaluation of an expert system for pest diagnosis of red pine in Wisconsin, *For. Sci.* 35:364.

Schulze, E.D., 1983, Root-shoot interactions and plant life forms, *Nether. J. Agric. Sci.* 4:291.

Sell, P.S., 1985, "Expert Systems: A Practical Introduction," John Wiley & Sons, Inc., New York.

Seyer, P., 1991, "Understanding Hypertext: Concepts and Applications," Windcrest Books, Blue Ridge Summit.

Sharpe, P.J.H., and Rykiel, Jr., E.J., 1991, Modelling integrated response of plants to multiple stresses, *in*: "Response of Plants to Multiple Stresses," H.A. Mooney, W.E. Winner, and E.J. Pell, eds., Academic Press, San Diego.

Shneiderman, B., and Kearsley, G., 1989, "Hypertext Hands-on! An Introduction to a New Way of Organizing and Accessing Information," Addison-Wesley, Reading.

Shneiderman, B., Kreitzberg, C., and Berk, E., 1991, Editing to structure a reader's experience, *in*: "Hypertext/Hypermedia Handbook," E. Berk, and J. Devlin, eds., McGraw-Hill, New York.

Simon, K.H., Manche, A., and Uhrmacher, A., 1992, "Expertensysteme auf dem Umweltsektor," Umweltbundesamt, Berlin .

Slatin, J.M., 1991, Composing hypertext: A discussion for writing teachers," *in*: "Hypertext/Hypermedia Handbook," E. Berk, and J. Devlin, eds., McGraw-Hill, New York.

Starfield, A.M., and Bleloch, A.L., 1983, Expert systems: an approach to problems in ecological management that are difficult to quantify, *J. of Environ. Manag.* 16:261.

Starfield, A.M., and Bleloch, A.L., 1986, "Building models for conservation and wildlife management," MacMillan Pub. Co., New York.

Udell, J., 1993, Desktop CD-ROM publishing, *BYTE* 18:217.

Thomson, A.J., Sutherland, J.R., and Carpenter, C., 1993, Computer-assisted diagnosis using expert systems- guided hypermedia, *AI Applic.* 7(1):17.

Waring, R.H., and Schlesinger, W.H., 1985, "Forest Ecosystems: Concepts and Management," Academic Press, Inc., Orlando.

Wasserman, P.D., 1989, "Neural computing," Van Nostrand Reinhold, New York.

Wilson, B.F., 1984, "The Growing Tree," rev. ed., Univ. Mass. Press, Amherst.

Wilson, J.B., 1988, Shoot competition and root competition, *J. Appl. Ecol.* 25:279.

Ziman, J., 1976, "The Force of Knowledge: The Scientific Dimension of Society," Cambridge Univ. Press, Cambridge.

CARBON ALLOCATION TO ROOT AND SHOOT SYSTEMS

OF WOODY PLANTS

Alexander L. Friend[1], Mark D. Coleman[2], and J.G. Isebrands[2]

[1]Department of Forestry
Mississippi State University
P.O. Drawer FR
MSU, MS 39762 USA

[2]USDA Forest Service
North Central Experiment Station
Forestry Sciences Laboratory
P.O. Box 898
Rhinelander, WI 54501 USA

INTRODUCTION

Carbon allocation to roots is of widespread and increasing interest due to a growing appreciation of the importance of root processes to whole-plant physiology and plant productivity. Carbon (C) allocation commonly refers to the distribution of C among plant organs (e.g., leaves, stems, roots); however, the term also applies to functional categories within organs such as defense, injury, repair and storage (Mooney, 1972). It also includes C consumed by roots in maintenance respiration and nutrient uptake (Lambers, 1987). In this paper we will use the terms "C allocation," "C partitioning," and "component biomass accumulation" (i.e., leaf, stem and root biomass) according to the process-based definitions of Dickson (1989), and Isebrands and Dickson (1991). C allocation is the process of distribution of C within the plant to different parts (i.e., source to "sink"). C partitioning is the process of C flow into and among different chemical fractions (i.e., different molecules, different storage and transport pools). Biomass component accumulation is the end product of the process of C accumulation at a specific sink. In the present review, allocation, partitioning and distribution will be relative terms (e.g., percent of total), whereas growth and accumulation will reflect absolute size (e.g., dry weight, moles of C, etc.).

There have been many reviews of C allocation and partitioning in herbaceous and crop plants (Mooney, 1972; Gifford and Evans, 1981; Gifford et al., 1984; Daie, 1985; Cronshaw et al., 1986; Ho, 1988; Stitt and Quick, 1989; Wardlaw, 1990). In addition, numerous reviews that have focused almost exclusively on woody plants (Kozlowski and

Biology of Adventitious Root Formation, Edited by
T.D. Davis and B.E. Haissig, Plenum Press, New York, 1994

Keller, 1966; Glerum, 1980; Oliveira and Priestley, 1988; Cannell, 1989; Dickson, 1989, 1991; Kozlowski, 1992), although there may be little reason to expect that results from woody plants would differ from other plants if plant ontogenetic stages and temporal factors are compared (Dickson, 1991). C allocation has been reviewed in the context of allocation to plant defense (Jones and Coleman, 1991), C storage (Chapin et al., 1990), C "costs" of nutrient and water uptake (Nobel et al., 1992), C fluxes to the rhizosphere (van Veen et al., 1991), C allocation to symbionts (Anderson and Rygiewicz, 1991), C allocation to reproduction (Chairiello and Gulman, 1991), and C allocation at the whole-plant (Ingestad and Agren, 1991) or ecosystem levels (Raich and Nadelhoffer, 1989). In each of the above cases, C allocation is recognized as being plastic and of fundamental importance to plant biology and ecology. Most of these reviews conclude with the message that more experimentation is needed to evaluate the multitude of hypotheses on C allocation in plants—especially in the case of woody plants, where experimentation has numerous logistical constraints.

The objective of this review will be to succintly but comprehensively review and interpret information on C allocation to root and shoot systems of woody plants, with emphasis on plants that have adventitiously generated root systems. In adventitiously rooted plants, C allocation research has been largely focused on two subjects: 1) the effects of carbohydrate storage and allocation of C to stems on rooting and establishment of cuttings, and 2) whether adventitiously rooted plants develop normal root systems, including balanced root/shoot C allocation (see, e.g., the chapter by Dickmann and Hendrick in this volume). Therefore, we will include some information on various subjects that has been obtained with nonadventitiously rooted plants when such information is absent or deficient for adventitiously rooted plants. After a brief introduction, we will explore the controls of C allocation, including plant factors and environmental factors. The plant factors section will include a general discussion of plant controls and root-shoot feedbacks, and will conclude with a discussion of differences between adventitiously rooted plants and those of seed origin, where those differences are known. The environmental factors section will review the influences of nutrients, water, O_2 and temperature on C allocation. The final section will be devoted to reviewing and evaluating methods and possible future research directions.

PLANT FACTORS

Intact Plants

General Patterns. In order to understand C allocation in rooted cuttings it is essential to have an appreciation of the general C allocation patterns that occur within intact woody plants. These patterns are highly integrated within the plant and are determined at the early embryonic stages of plant development. Moreover, they are regulated by a complex network of competing sources and "sinks" within plants (Dickson, 1991). Sources are defined as net exporters of photosynthates, and sinks as net importers of photosynthates (Ho, 1988). An interdependency between the shoot and root exists throughout the life of a plant (Kozlowski, 1971), with the feedback between the root and shoot playing an important role in the regulation of overall plant growth and development (Davies and Zhang, 1991).

C allocation patterns are a function of source-sink interactions. The sink-strength within woody plants varies markedly by species, genotype, shoot type, age of plant, location within the plant, season and environmental conditions (Kozlowski, 1992). Although there are exceptions to almost all generalizations made about root-shoot interactions and function (Groff and Kaplan, 1988; see the chapter by Barlow in this volume), there are three fundamental C allocation patterns in woody plants (Dickson, 1991; Kozlowski, 1992). The first pattern is associated with determinate (or fixed) shoot growth, which is characterized by a single, short burst of shoot growth in the late spring and early summer followed by a long lag period of budset. Distribution of assimilates is according to the flush-cycle with most of the assimilates (i.e., > 90%) directed upward to the flush during the flushing episode, and conversely most of the assimilates (i.e., > 95%) are directed downward to the lower stem and roots during the lag stage in between flushing episodes. An example of a genus with this pattern is *Picea*. The second pattern is associated with indeterminate (or free) shoot growth, characterized by continuous shoot growth over most of the growing season. Distribution of assimilate in these plants varies with the stage of development of each leaf. Young developing leaves are net importers of assimilate until they become fully expanded, at which time they export both acropetally to developing leaves and basipetally to the stem and roots. Mature leaves export almost exclusively to the lower stem and roots. An example of a genus exhibiting this pattern is *Populus*. The third pattern is associated with semi-determinate (or recurrent flushing) shoot growth, characterized by periodic flushes of shoot growth with intermediate lag stages. The distribution of assimilate in these plants is cyclic with transport upward to developing leaves during a flush and downward to the stem and roots during the lag; the cycle is repeated during each consecutive flushing episode. An example of a genus with this pattern is *Quercus*.

A common denominator in each of the allocation patterns is the importance of anatomy to understanding C allocation to roots. Specifically, the pathways through which source-sink interactions occur in woody plants are strongly dependent upon the vascular connections delimited during the early primary vascular development of the plants (Larson, 1983). This concept has been supported by Larson and coworkers in a wide variety of plants, including *Populus* (Larson et al., 1980), *Osmunda* (Kuehnert and Larson, 1983), *Gleditsia* (Larson, 1984), *Fraxinus* (Larson, 1985) and *Polyscias* (Larson, 1986).

The overall finding from this review of the literature is that most studies, particularly mechanistic ones, have focused on understanding C allocation to shoots rather than roots. For example, the comprehensive review of C allocation and partitioning by Wardlaw (1990) provides only minimal coverage of allocation to roots. This deficiency is not the fault of the author, but rather a reflection of our current state-of-knowledge on C allocation to roots in plants.

Mechanisms of Carbon Allocation. In order to better understand adventitious rooting in woody plants, it is not enough to know what the patterns of C allocation are in roots and shoots; it is necessary to know how they occur and their significance to plant function. Thus, an understanding of the mechanisms of C allocation and partitioning may facilitate improvements in adventitious rooting (e.g., recalcitrant plants; see Haissig et al., 1992). There have been numerous reviews of the mechanisms of C allocation and partitioning in plants (see Introduction). However, the controlling mechanisms of C allocation remain largely unknown. Some of the genetic, biochemical and physiological aspects of the regulation of transport from sources and sinks are becoming more clear (Cronshaw et al., 1986, p. xxix), yet considerable mechanistic information is still needed. Specifically, two recurring observations are important to understanding adventitious rooting in woody plants.

First, there is increasing evidence that the structural and functional diversity among plants and among tissues within a plant is greater than originally thought. Therefore, the mechanisms controlling C allocation and partitioning during rooting may not be universal and may be specific to species, genotypes and certain tissues (Cronshaw et al., 1986). This means that the specific anatomical differences in rooting between species, which have been shown by Lovell and White (1986), are probably very important in ultimately determining mechanisms and functional properties of rooting of each cutting system; it also means that genetic control is likely to be more important than once thought, as suggested by Haissig et al. (1992). Second, the main control of C allocation between roots and shoots is thought to lie in sink activity (Ho, 1988). Thus, factors affecting sink activity are probably the most important for regulating C allocation and partitioning (Gifford and Evans, 1981). These factors include the biochemistry of sucrose synthesis (Stitt and Quick, 1989; Dickson, 1991), respiration (Gifford, 1986), phloem loading and unloading (Ho, 1988), hormones (Gifford and Evans, 1981), cell turgor (Wyse, 1986) and genetic control (Wyse, 1986). Another important factor is the proximity of the source to the sink and vascular connections between them (Gifford et al., 1984; Dickson, 1989). Indeed, one key to understanding the mechanisms of C allocation in plants lies in understanding the properties of sinks. Important sink-related factors include the initial establishment of the sink at an early developmental stage, duration of sink growth and sink effects on photosynthesis (Gifford and Evans, 1981).

Rooted Cuttings

In this section we will focus on C allocation and partitioning during adventitious rooting of cuttings of woody plants at three plant developmental stages (Lovell and White, 1986): 1) root initiation, (2) root growth and development, and (3) early field performance. Representative sources with the genera studied are presented in Table 1. Readers interested in reviews of the use of carbohydrates during cambial growth of roots, root thickening and mycorrhizae are referred to Kozlowski (1992), and in rootstock C storage to Priestley (1970).

Table 1. List of carbon allocation and partitioning studies conducted on cuttings from woody plants[1].

Genus	Cutting Type[2]	Setting	Reference
Acer	Softwood	Polyhouse	Donnelly (1977)
"	"	Greenhouse	Smalley et al. (1991)
Castanea	Hardwood	"	Vieitez et al. (1980)
Juniperus	"	"	Henry et al. (1992)
Malus	Hardwood	Controlled environment	Cheffins & Howard (1982)
"	Root	Storage	Robinson & Schwabe (1977)
Morus	Softwood	Greenhouse	Satoh et al. (1977)
Pinus	Hardwood	Controlled environment	Cameron & Rook (1974)
"	Softwood	" "	Ernstsen & Hansen (1986)
"	Softwood and Seedlings	" "	Haissig (1984)
"	Softwood	" "	Haissig (1989)
"	"	" "	Hansen et al. (1978)
Populus	Softwood	Controlled environment	Dickmann et al. (1975)
"	Root	" "	Eliasson (1968)
"	Hardwood	" "	Fege (1983)
"	"	Storage	Fege & Brown (1984)
"	"	Field	Friend et al. (1991)
"	"	Field	Isebrands & Nelson (1983)
"	"	Outdoor - Pots	Nanda & Anand (1970)
"	"	Field	Nelson & Isebrands (1983)
"	"	Outdoor - Pots	Nguyen et al. (1990)
"	Softwood	Controlled environment	Okoro & Grace (1976)
"	Hardwood	Outdoor - Pots	Pregitzer et al. (1990)
"	"	Greenhouse	Tschaplinski & Blake (1989)
Olea	Softwood	Greenhouse	Rio et al. (1991)
Prunus	Softwood	Greenhouse	Breen & Muraoka (1974)
Rhododendron	"	Storage	Davis & Potter (1985)
"	"	Greenhouse	French (1990)
Triplochiton	"	Greenhouse	Leakey & Coutts (1989)

[1] Includes only example, primary, English-language references for studies with rooted cuttings; later papers by the same authors may be available. [2]"Softwood" and "hardwood" as per Dirr and Heuser (1987).

Root Initiation. Adventitious roots arise either *in situ* as induced root primordia or as preformed root primordia (Haissig, 1974a; Sutton, 1980; see the chapter by Barlow, and by Haissig and Davis in this volume). Thus, there are important anatomical differences among species: those that create a root primordial site and those with sites already present (Lovell and White, 1986). Initiation of adventitious roots is regulated by a balance of internal substances, including carbohydrates, hormones and nitrogenous compounds and, perhaps, related "cofactors" (Kozlowski, 1971; Haissig, 1974b, see also the chapter in this volume by Howard, and by Blakesley). However, genetic control of adventitious rooting is also very important (Haissig, 1986; Haissig, et al. 1992), as evidenced by the wide variation in rooting among closely-related taxa, and by the variety of adventitious root types (Kozlowski, 1992; see the chapter by Barlow in this volume).

Understanding source-sink relations during and subsequent to root initiation depends upon knowledge of anatomical differentiation (Haissig, 1986). Success or failure of rooting depends upon the establishment of a vascular link between source and the new sink (i.e., root). This linkage (or lack thereof) has been hypothesized to be dependent on anatomical characteristics of the cutting (Vieitez, 1974; Vieitez et al., 1980) and is related to presence of secondary development associated with maturation and aging. A common element among some species that have preformed root initials and those that form roots *in situ* is the association of the new root (i.e., sink) in close proximity to existing vascular tissue (i.e., source). For example, in *Salix*, preformed root initials occur in leaf gaps of nodes next to vascular traces, as described by Haissig (1970). In *Quercus rubra*, a recalcitrant semi-determinate species, adventitious roots form *in situ* in vertical rows in association with stem vascular traces (Isebrands and Crow, 1985).

Knowledge of source-sink relations between roots and shoots of intact plants may add to our understanding of C allocation in rooted cuttings. For example, in the nodal region of *Populus*, Isebrands and Larson (1977) found that there are two significant attributes of cambial development in vascular bundles within leaf gaps: 1) fusion of cambial initials between adjacent vascular bundles occurs when a critical minimum tangential distance is attained, and 2) cambial initials within bundles exhibit polarity with respect to adjacent bundles. Similar phenomena occur during the linkage of differentiating cells in tissue cultures of *Coleus* and *Daucus* (Lang, 1974), during differentiation of vascular bundles of secondary veins in *Populus* leaves (Isebrands and Larson, 1980), and in the vascular union between developing axillary buds (i.e., branches) and the stem in *Populus* (Larson and Fisher, 1983). If the analogy extends to adventitious roots, these phenomena may explain the ease of rooting cuttings in species with preformed root initials in leaf gaps. In those cases, root cambial cells are in close proximity to cambial cells of adjacent leaf traces and in a favorable position to establish the early vascular connections between the source and sink that are necessary to provide carbohydrates and hormones required for the formation of roots. Another example is the establishment of the vascular system in new germinants from *Populus* seed (Larson, 1979). In the early stages of seedling development, the procambium differentiates in ground tissue in the region of the prospective hypocotyl (root), and develops acropetally and continuously with the epicotyl (shoot). Thus, a procambial template system is formed between root and shoot that determines all subsequent vascularization of the plant. Early vascularization of the seedling, determined by phyllotaxy, is thereby responsible for integrating structural-functional relationships of the whole plant. Similar structural-functional relationships have been found in the early vascularization of the root and shoot in sugar beet (Stieber and Beringer, 1984). Therefore, successful functioning of plants developing from seed depends upon the intimate structural-functional connection between the shoot (source) and root (sink), just as in rooted cuttings.

Haissig (1986) presented a comprehensive review of C metabolism of cuttings during root initiation. Although carbohydrates are considered the principal source of energy during rooting, there is substantial variation in the use of carbohydrates among species and cutting types. There is considerable controversy over the role of carbohydrates during rooting due to differences in environment, experimental approaches and interpretation by researchers (Haissig, 1986). Stockplant factors are also known to be important in determining carbohydrate status in cuttings (Hansen et al., 1978; Ernstsen and Hansen, 1986; Leakey and Coutts, 1989; see also the chapter by Howard in this volume). Therefore, it is difficult to use one measure of carbohydrates as an independent variable for predicting rooting success. Nevertheless, some examples of carbohydrate dynamics associated with adventitious rooting will be presented below.

Total carbohydrates of cuttings are thought to be related to rooting, but not necessarily causally (see the chapter by Haissig and Davis in this volume). For example, Fege and Brown (1984) found that total carbohydrates were greatest in large-diameter *Populus* hardwood cuttings and were correlated with rooting success, but concluded that other factors may be more important than carbohydrate content for rooting success. Breen and Muraoka (1974) found that most ^{14}C transported from leaves (after a $^{14}CO_2$ pulse) of softwood cuttings of *Prunus* remained in the upper stem until rooting commenced, at which time a four-fold increase in ^{14}C concentration occurred at the cutting base. However, they concluded that an unspecified root promoting substance was probably more important than carbohydrate content for successful rooting. Haissig (1984) studied carbohydrate concentrations in *Pinus banksiana* stem cuttings, and found that total carbohydrate accumulation was correlated with basal stem callus rooting; again, no causal relationship was established. Moreover, he found that differential partitioning of carbohydrates, specifically, the ratio of reducing sugar to starch, was a good indication of rooting success. In a recent review, Veierskov (1988), concluded that an abundance of stored carbohydrates will improve the success of rooting if other physiological conditions are favorable. However, the physiological state of the plant may have an overriding influence and result in poor rooting even in plants with high carbohydrate concentrations. Clearly, several factors are important in determining the role of carbohydrates in root initiation: origin of roots (callus or preformed initials), types and availabilities of carbohydrates, autocorrelations between carbohydrate accumulations in cuttings and other growth-affecting chemicals, genetic variation in optimal carbohydrate balance for rooting, phenological stage, and other plant factors. In summary, more research is needed which incorporates the plant factors mentioned above (Haissig, 1986; Haissig et al., 1992).

Root Growth and Development. Although there has been much research on C allocation and partitioning at the initiation stage of adventitious rooting of woody plants, there are far fewer studies during the "early" stages of root growth and development. This paucity of research seems to be related to the lack of suitable methods and approaches for studying cuttings during the transition between root initiation and root growth. Carbohydrate reserves are very important during root initiation, while subsequent growth may be almost entirely dependent on current photosynthate. Haissig (1988) has suggested that radiotracers offer the most suitable techniques for studying these processes and has called for a multidisciplinary use of radiotracers to improve our understanding of C metabolism during the various stages of adventitious rooting. However, despite the availability of simple and inexpensive radiotracer methods (Isebrands and Dickson, 1991; Isebrands and Fege, 1993), few investigators have used them.

As an exception, early C allocation patterns in softwood cuttings were investigated by Breen and Muraoka (1974) using leafy softwood cuttings exposed to $^{14}CO_2$ to study the changes in carbohydrate content in softwood cuttings of *Prunus* during and after root initiation. They showed that the proportion of ^{14}C increased at the stem base after root initiation. They also found a decline in ^{14}C in sugar over time, indicating a metabolism of carbohydrate reserves in the cutting during rooting. After labeling, the percentage of total ^{14}C in reserves declined in the cutting base from 80% at the start to nearly 20% at day 28 of the chase period, while the residue fraction (i.e., structural carbohydrates) increased during the chase period mostly in cuttings with a good rooting percentage. Another study of ^{14}C allocation found that 75% of the photosynthate transported to developing roots of *Pinus radiata* softwood cuttings came from the older, mature foliage when compared with the younger, recently expanded foliage (Cameron and Rook, 1974). Results from both of these studies indicate that source-sink relations of foliage and roots are very important in affecting the allocation of C to developing roots in softwood cuttings.

The role of stored assimilates during adventitious root formation and subsequent growth after outplanting in the field was examined by Isebrands and Fege (unpublished) using hardwood cuttings of *Populus* in Wisconsin. They labelled coppice sprouts *in vivo*, harvested radiolabelled hardwood cuttings during the winter, then followed ^{14}C in the cuttings during early root development. Carbohydrate reserves played an important role in the growth of developing leaves during the first four weeks before any root growth occurred. After that time, current photosynthate was more important for subsequent leaf and stem growth. Root growth used stored photosynthate for a longer period than above-ground components (i.e., ca. six weeks after planting), thereby following the leaves and stem in dependence upon current photosynthate.

Some researchers have used biochemical methods other than radiotracers to study the early root growth and development phase of rooted cuttings. Okoro and Grace (1976) studied carbohydrate concentrations of leafless hardwood cuttings and softwood cuttings of various *Populus* species. In the leafless hardwood cuttings, concentrations of carbohydrates were initially high and fell rapidly as roots developed. However, even at the final stages of the experiment, much reserve carbohydrate remained. Leafy softwood cuttings had a different pattern in that they exhibited a steady accumulation of carbohydrate from the time of planting to the time that roots developed. The authors noted that root production in both hardwood and softwood cuttings extended leaf longevity on the cutting and increased leaf photosynthesis, confirming a positive feedback from developing roots to the developing shoot, as in conifer seedlings (van den Driessche, 1987). Using biochemical methods, Tschaplinski and Blake (1989) also studied carbohydrate metabolism in relation to subsequent growth from leafless hardwood cuttings in five *Populus* hybrid clones exhibiting a range of rooting patterns. Their carbohydrate analysis indicated that differences among early clonal performance were related to carbohydrate utilization. Glucose and *myo*-inositol levels declined more rapidly in faster-growing hybrid clones than in slower ones. They concluded that the faster-growing hybrids consumed the monosaccharide carbohydrates for early growth, while slower growers converted monosaccharides to storage carbohydrates. *Myo*-inositol concentrations were highly correlated with root growth, although starch was not. These results prompted the authors to recommend *myo*-inositol concentration as a possible early selection index for early root production in hardwood cuttings of poplar.

Field Performance. The number of C allocation and partitioning studies in woody plants reared from cuttings and grown under field conditions is limited. This general lack of information on field-based C physiology (Isebrands and Dickson, 1991) is somewhat surprising, considering the recently expanded commercial use of cuttings in forestry (Ritchie, 1991; see the chapter by Ritchie in this volume) and their long-standing use in ornamental horticulture (see the chapter by Davies and coworkers in this volume). Much of the existing work to date has been part of research programs focused on developing short- rotation woody crops for biomass and energy (Isebrands et al., 1983; Stettler et al., 1988). Moreover, much of the research has been with *Populus* because that genus exhibits outstanding biomass production under a short-rotation system (Hinckley et al., 1989). C physiology studies on horticultural trees have largely been on plants derived from rootstocks, not hardwood cuttings (Priestley, 1970).

Isebrands and Nelson (1983) studied C allocation patterns in the establishment year in two intensively cultured *Populus* clones with contrasting phenology and morphology; plants were grown in the field from hardwood cuttings. Using radiotracers, they found clonal differences in seasonal patterns of C allocation that were related to phenology. The timing of budset was the most important factor in determining allocation differences between the clones. Prior to budset, export was primarily upward to developing leaves. Following budset, mature leaf export was primarily to the lower stem, hardwood cutting and roots. Moreover, one of the clones, 'Tristis,' exported a greater proportion of carbohydrate to roots, suggesting that the sink strength of roots in *Populus* is under genetic control (see the chapter by Dickmann and Hendrick in this volume). As the season progressed, lower leaves were the primary source of photosynthates for root growth. The C allocation patterns were largely consistent with the general patterns characteristic of indeterminate woody plant seedlings (Dickson, 1986). In a companion study with a clone that exhibited late-season leaf retention, it was concluded that autumnal retention of leaves is important for root growth and the accumulation of reserve storage in the cutting and roots (Nelson and Isebrands, 1983).

Although relatively little is known about the role of hardwood cuttings in C allocation during years subsequent to establishment (Isebrands et al., 1983; Dickson, 1986), some information exists. We have used ^{14}C methods to study C allocation of *Populus* clones in the second year in the field, from hardwood cuttings (Friend et al., 1991). In that study, the hardwood cutting itself grew considerably in size from the time of planting through the second season when it became fully incorporated in the stem-root transition (see the chapter by Dickmann and Hendrick in this volume). C allocation to below-ground biomass components was evenly distributed among the cutting, coarse roots and fine roots, indicating that the cutting is still an active sink for photosynthate beyond root initiation stages and into the second growing season.

Hinckley et al. (1989) reviewed C allocation patterns during the first and second year of four hybrid poplar clones grown from hardwood cuttings in Washington state in a study conducted by Scarascia-Mugnozza (1991). They found a progressive shift of photosynthate allocation from the upper leaves and stem to the lower stem and roots as the season progressed. Clonal differences were observed in allocation to roots, with one clone (i.e., H44-136) maintaining a high allocation to the cutting, coarse roots and fine roots throughout the entire growth period. Clones also varied in the number, size and orientation of roots as they developed. It was concluded that high allocation to roots in the first year resulted in greater productivity of the clone in the second year. Similar studies have been conducted by Pregitzer et al. (1990) and Nguyen et al. (1990).

Root-Shoot Feedback

Within higher plants the specialized functioning of photosynthetic energy fixation in leaves must be coordinated with nutrient and water absorption of roots. The coordination between plant roots and shoots is apparent when considering the dynamic responses of root-shoot C allocation to environmental factors. Early establishment of root-shoot balance is especially important for rooted cuttings. Without an adequate root system to balance the existing shoot, normal plant development is not possible. A number of proposed mechanisms account for the coordination of specialized functions among the various plant organs, including resource-based functional equilibrium, metabolic control and chemical signals such as phytohormones or metabolites. These possible mechanisms for coordination of C allocation will be discussed below.

Functional Equilibrium. Brouwer (1983) emphasized that the dynamics of root-shoot interactions are dependent upon a functional equilibrium between root and shoot. Under certain environmental conditions, plants will adjust relative growth to maintain a constant root:shoot biomass ratio, even if roots or shoots are partially abscised. Further, if the ability of one organ to supply C or mineral nutrient resources is disturbed through changing environmental conditions, it will affect the growth and functioning of other dependent organs, resulting in a new equilibrium (Brouwer, 1983). Because the dynamics of functional equilibrium are exhibited by many plant species and influenced by a variety of environmental conditions, the idea of growth regulation, through supply of and demand for internal resources, continues to be viewed favorably (Dickson and Isebrands, 1991; Klepper, 1991). Bloom et al. (1985) postulated that the cost for resource acquisition, in terms of reserve C and nutrient resources used, could be weighed against the gain of having obtained such resources. Further, and in line with functional equilibrium, they speculated that plants can adjust internal supply of and "demand" for resources by allocating internal reserves and adjusting growth patterns.

The relationship between root-shoot equilibrium and internal resource balance is most apparent when considering the balance between C and nitrogen (N). Internal N concentration is strongly related to C allocation processes (Rufty et al., 1988) and root:shoot ratio (Agren and Ingestad, 1987). The correlation of root-shoot allocation with N supply is controlled by internal non-structural C to N (C:N) concentration ratios (Brouwer, 1983; Mooney and Winner, 1991). Mooney and Winner (1991) considered the C:N ratio to be at some "set-point" value which, if disturbed, would adjust root or shoot growth and activity. The C:N ratio set-point would then be reestablished. Once the C:N set-point was reestablished, a new root:shoot ratio would be achieved. Such an approach has resulted in relatively accurate predictions of root-shoot balance based on C:N ratios in simulation models (Thornley, 1972; Reynolds and Thornley, 1982). However, the failure of these models to maintain equilibrium conditions without prescribing specific C:N ratios suggests that some other control factors may be involved. Nevertheless, similar C:N set-points may be useful constructs in which to investigate the optimum carbohydrate storage that maximizes rooting of cuttings (cf. Leakey and Storeton-West, 1992).

Metabolic Control. The correlative relationship between root-shoot C allocation processes and C:N ratios suggests the importance of C and N metabolism to understanding these processes. For example, Huber (1983) examined a variety of species under various light and N treatment conditions, and concluded that root:shoot ratio is inversely related to accumulation of leaf starch. Leaf starch accumulation and sucrose formation are regulated by sucrose phosphate synthetase (SPS) (Huber et al., 1985). Because SPS activity affects the supply of sucrose for export (Kerr et al., 1984; Huber et al., 1985), a possible link between leaf carbohydrate metabolism and control of root-shoot C allocation is apparent. Regulation of enzymes known to be responsible for the partitioning between starch and sucrose, including SPS, has been extensively studied (Stitt et al., 1987; Stitt and Quick, 1989; Stitt, 1990; Huber and Huber, 1992). On the basis of this work, it is evident that a number of factors that affect C allocation to the root and shoot also affect the partitioning between starch and sucrose. For example, it is possible that the interdependence of carbohydrate metabolism and nitrate reduction in leaves (Huber and Huber 1992) is an important regulating point for C allocation processes. In this case, the N moving in the vascular system may act as a controlling factor over C allocation processes.

Nitrogenous compounds are known to move not only from root to shoot in the xylem but also in the phloem (Lambers et al., 1982; Patrick, 1988), especially under nutrient-stress conditions where shoot activity is minimal. In addition, roots growing in N-rich soil patches tend to show greater growth and respiration rates, as well as nitrate reductase activity (Lambers et al., 1982; Lambers, 1987; Granato and Raper, 1989), all of which increase the demand for carbohydrate and the amount translocated to roots (Patrick, 1988). In this way, vascular transport of nitrogenous compounds, N uptake and N metabolism in roots may have considerable control over C allocation.

Phytohormones. Plant hormones and other metabolically active compounds moving up from roots to leaves in xylem or down to roots from leaves in phloem have long been considered strong influences in root-shoot interactions (Skene, 1975). There is little doubt that signals coming from herbivore-damaged tissue are responsible for initiating gene transcription in remote locations (Ryan and Farmer, 1991). Abscisic acid (ABA) signals moving in the xylem sap clearly demonstrate that root-shoot interactions occur via hormones. Soil-drying experiments are known to affect leaf conductance and growth before affecting turgor (Davies and Zhang, 1991), and exogenous leaf application of ABA will reduce leaf conductance (Mansfield, 1987).

Hormones may also influence C allocation processes. It is known that the enzymes responsible for regulation of starch-sucrose partitioning are influenced by plant hormones (Daie, 1986; Cheikh and Brenner, 1992). Growth of roots is favored over the growth of shoots following exogenous application of ABA, and this response is similar to the effects of water-stress treatments (Watts et al., 1981). The negative effect of root or shoot excision on sucrose translocation toward that organ can be reversed by application of auxin or cytokinin to the site of the removed organ (Gersani et al., 1980). Dependence of shoot growth on root-supplied hormones is considered to be most important under conditions of limiting nutrients. Therefore, it may be that hormonal control is less important under greenhouse than field conditions where nutrients are often limiting (Wareing, 1980).

While a number of processes are affected by phytohormones, there is little consensus among physiologists that hormonal control is necessary to maintain root-shoot equilibrium. Although it is likely that plant hormones and other metabolically active compounds are participating in the fine control of feedback between roots and shoots, their role is unclear (Patrick, 1987). When considering this point, and the effective coarse-control of functional equilibrium to describe C allocation, it is understandable that researchers have not incorporated the fine control of hormones in simulation models (cf. de Wit and Penning de Vries, 1983; Wilson, 1988).

Comparison of Adventitiously Rooted Plants and Seedlings

There has been considerable research on the field performance of adventitiously rooted plants compared with seedlings, but only a few studies have investigated differences in C allocation between vegetative propagules and seedlings. One of these is the work of Haissig (1984) who compared carbohydrate accumulation and partitioning in *Pinus banksiana* seedlings and cuttings during root initiation. He found differences between the two in the amount of individual carbohydrates accumulated and in the location of accumulation. All parts of cuttings accumulated more carbohydrates when compared with seedlings. The most pronounced difference between cuttings and seedlings was in the reducing sugar:starch ratio, which was much lower in cuttings than in seedlings. Studies of *Populus* hybrids have found one distinct difference between hardwood cutting-derived plants and seedlings: the hardwood cutting is a relatively large and active sink for photosynthate compared with the stem of a seedling of similar size (Isebrands and Nelson 1983). In fact, large quantities (i.e., 20 to 30%) of ^{14}C-labelled photosynthate from mature leaves were actively incorporated into storage products and cambial derivatives in the cutting during the entire establishment year (Isebrands and Nelson, 1983). These findings are corroborated by the recent work of Scarascia-Mugnozza (1991), which indicates that the cutting is more than simply a conduit for carbohydrates and hormone transport from shoot to root, but instead plays a key role beyond root initiation in the integrated anatomical and physiological development during establishment and subsequent growth of plants from hardwood cuttings in the field. In contrast, recent work with softwood cuttings of *Populus tremuloides* has found no appreciable difference in root-shoot biomass components in response to stress treatments between cuttings and seedlings subjected to ozone-stress (Coleman and Isebrands, unpublished).

After root initiation, the ability to maintain a balance between root and shoot C allocation, and to produce an effective root system, will determine the success of vegetative propagules. Numerous studies have investigated the field performance of plantlets relative to seedlings (Gupta et al., 1991), but few have attempted to link the success, or lack thereof, to C allocation or root-shoot balance. Most effort in this area has been with coniferous tree species, which are of great economic importance but notoriously difficult to propagate vegetatively. For example, in *Pinus taeda*, allometric relations between root and shoot dry weight (DW) do not appear to differ between plantlets and seedlings. However, adventitiously generated plantlet roots are less effective at nutrient uptake due to inadequate branching and large diameters, compared with roots of seedling origin (McKeand and Allen, 1984). Later research with *Pinus taeda*, propagated in a similar manner, confirms these findings and even quantifies the negative impact of this less effective root system on field performance of young plantations (Anderson et al., 1992). This work also noted that a larger part of the early lag in growth of plantlets compared with seedlings is attributable to shoot rather than root effects. Specifically, Anderson et al. (1992) observed a mature shoot morphology in plantlets and provided evidence that this was caused by the shoot rather than factors associated with the adventitious roots. In contrast, work with *Betula* found tissue culture plantlets showed more juvenile shoot morphological traits when compared with rooted cuttings (Brand and Lineberger, 1992). In the above cases, the juvenility or maturity of the roots was not discussed, but presumably roots also have juvenile and mature characteristics, although this phenomenon is less well characterized in roots than in shoots.

Field plantings of *Pseudotsuga menziesii* have been similarly limited by survival, growth and stem form of adventitiously rooted plants compared with seedings (Ritchie and Long, 1986). In contrast to *Pinus taeda*, relative root biomass (root:shoot ratio) is reported to be greater in rooted cuttings than in seedlings of *P. menziesii* (Ritchie et al., 1992). Also, it has been hypothesized that plagiotropism is caused, in part, by abnormal root development (Timmis et al., 1992). Specifically, it is reasoned that the observed asymmetric development of adventitious roots (i.e., on only one side of the cutting) may cause an imbalance in plant growth substances of root origin and provide an erroneous message to the shoot apical meristem. In contrast, mature shoot characteristics of *Pinus taeda* were associated with signals from the shoot and not the root system (Anderson et al., 1992).

ENVIRONMENTAL FACTORS

The ability of plants to adjust C allocation in response to environmental changes is widely documented. While the role of these changes in environmental stress tolerance or acclimation may be debated, many environmentally-induced changes in C allocation are well established. These will be reviewed in the section below, where the information will apply to intact plants and maybe comparable to the responses of stock plants from which softwood or hardwood cuttings might be taken. In addition to the information on intact plants, the C allocation of cuttings in response to environmental factors will be included. The hypothesis that environmental responses of C allocation to adventitious rooting will not differ from that of intact plants will also be examined.

Nutrients

It has been widely stated that the general influence of nutrient-stress on C allocation is to increase C allocation to the roots compared with that allocated to the shoot. This change would seem both intuitive and adaptive because increased root development should, in part, compensate for a nutrient deficiency by increasing the nutrient absorption capacity of the plant. However, this hypothesis has not been rigorously tested until recently. Evidence will be provided from both ^{14}C tracer studies and whole-plant growth analyses.

Although nutrient uptake and metabolism are acknowledged to increase root C consumption (Lambers, 1987; Patrick, 1988), net allocation of ^{14}C to roots has been found to decrease with improved nutrient availability for *P. menziesii* (Friend, 1988). One cause of this pattern appears to be a shift in C partitioning from carbohydrates toward nitrogenous compounds (cf. Marx et al., 1977, Margolis and Waring 1986). Another cause appears to be an increased sink strength of actively growing foliage (high N), which consumes a disproportionate amount of C compared with low N conditions. In a study by Friend (1988) of four-month-old *P. menziesii* seedlings, net ^{14}C allocation patterns (CA, Fig. 1) were caused by N effects on both growth of tissues (i.e., sink size; see DW, Fig. 1) and on ^{14}C accumulation rate on a per-unit-tissue basis (i.e., unloading rate or sink activity; see SA, Fig. 1). In this and other studies, trends in DW distribution are usually representative of net C allocation patterns, but caution must be used in drawing inferences about C allocation from analyses of DW, especially given the seasonal variations in C storage common in woody plants and the importance of carbohydrate storage to adventitious rooting.

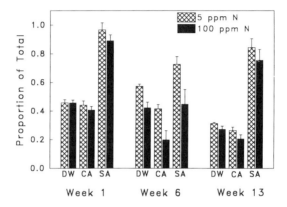

Figure 1. Three measures of carbon allocation to roots in response to nitrogen (N). Four-month-old *Pseudotsuga menziesii* seedlings grown in silica sand with N-deficient (5 ppm N as NH_4NO_3) or N-sufficient (100 ppm N as NH_4NO_3) nutrient solutions added thrice per week. Bars represent paired comparisons between 5 ppm N and 100 ppm N for relative dry weight accumulation (DW; proportion of total), net carbon allocation (CA, proportion of recovered ^{14}C in roots), and relative specific activity [SA; as defined by Mor and Halevy (1979)] at one, six and 13 weeks after treatment. Error bars depict one standard error of the mean. [after Friend (1988)]

One of the best illustrations of nutritional effects on root-shoot development at the whole-plant level is from Ingestad and Agren (1991), who support the hypothesis that nutrient-stress increases root development by showing increased proportions of dry matter in roots compared with shoots as a steady-state nutritional regime is decreased from the optimum. However, their results indicate that this phenomenon cannot be generalized to all nutrients. For example, potassium (K) appears to have no effect on root-shoot biomass distribution, while N, phosphorus (P) and sulfur (S) deficiencies each increase the proportion of biomass in the root system (Ingestad and Agren, 1991). An important additional finding from the use of steady-state nutrition techniques is that the shift in root-shoot biomass distribution occurs over a relatively short time period; subsequently, the relative growth rates of roots and shoots are equal (Ingestad and Lund, 1979). This single, rapid adjustment may apply to steady state conditions only, while a nutrient deficiency under temporally variable field conditions may result in numerous such periods of root-shoot adjustment, due to short-term fluctuations in nutrient availability.

The increase in root-shoot biomass ratio brought about by nutrient-stress may be caused by either an increase in root growth or by a larger decrease in shoot growth than in root growth. Studies of seedlings commonly report the latter pattern (change in relative not absolute growth). For example, N-stress increased root:shoot ratio in *P. menziesii* seedlings primarily through a more negative effect on foliage and stem growth than on root growth (Fig. 2) (Friend et al., 1990). From this and other related seedling studies (Squire et al., 1987; Johnson, 1990), the N-stress effect on root:shoot ratio may be viewed as caused by foliage and shoot growth being more sensitive to and limited by N than root growth. In contrast with seedlings, studies of mature trees indicate that older, larger plants may accomplish a similar shift in root-shoot biomass distribution by different mechanisms. For example, Gower et al. (1992) found that N fertilization of *P. menziesii* forests resulted in absolute decreases in fine root productivity.

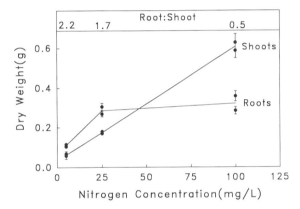

Figure 2. Response of shoot growth, root growth and root:shoot ratio to nitrogen (N) stress. *Pseudotsuga menziesii* seedlings were grown from seed at a 2.5 cm x 2.5 cm spacing in boxes of silica sand for four months. Weekly, they received a complete and optimum nutrient solution with the exception of N, which was supplied at 5, 25 or 100 mg L^{-1} as NH$_4$NO$_3$. Each symbol represents the mean dry weight of the 25 seedlings in each box; error bars represent ± one S.E. of the mean. [Friend and Coleman, unpublished]

Nutrient-stress does not always result in shifts in biomass distribution to roots. Substantial nutrient-induced growth reductions are commonly required before such shifts are evident (Dewald et al., 1992). An additional source of uncertainty is the change in root:shoot ratio that occurs as a function of plant size rather than nutrient addition *per se* (cf. Carlson and Preisig, 1981). These uncertainties about nutrient-stress and DW accumulation may have fueled previous skepticism about the nutrient-stress-induced shift in root-shoot biomass distribution (e.g., Carlson and Preisig, 1981).

Most of the studies of C:N ratios and adventitious rooting have been oriented toward C partitioning, given the potential role of carbohydrate storage in promotion of rooting. Under conditions of high irradiance, addition of N to stock plants may improve rooting of associated cuttings (Henry et al., 1992; Leakey and Storeton-West, 1992). This response is not always positive, however. The rooting of *Juniperus* declined as N levels added to the stock plant approached optimum for whole-plant growth (Henry et al. 1992) and, at low irradiance, NPK addition to *Triplochiton* stock plants had a negative effect on rooting. In both studies, sugars were better correlated with rooting than was N, but with opposite effects. Specifically, rooting of *Juniperus* cuttings was positively correlated with sucrose content, while that of *Triplochiton* was negatively associated with reflux-soluble carbohydrates. Of additional interest was the very strong correlation between K and boron (B) concentration at the time of excision and the percent rooting of *Juniperus* cuttings (Henry et al., 1992). Collectively, these findings suggest two important differences between the response of root development in intact plants (stock plants) to nutrients and the rooting response of cuttings collected from intact plants. First, severe N-stress increases the root-shoot biomass ratio of intact plants but may not encourage rooting of cuttings. Second, K and B may encourage rooting of cuttings but do not appear to affect biomass distribution between roots and shoots of intact plants.

Water

Water plays a fundamental solute role in the translocation of sucrose from source to sink, in "pressure flow" and in root-shoot "communication." A water deficiency has a cascade of effects on C allocation and partitioning, from those that occur over seconds/minutes to those that occur over days/weeks [reviewed by Geiger and Servaites (1991)]. Fundamental effects of water-stress include an increased C partitioning to osmotica (Geiger and Serviates, 1991) and altered source-sink relations. For example, water-stress caused decreased export rates of ^{11}C from the source leaf and decreased ^{11}C import rates to sink leaves and roots, relative to unstressed conditions. However, root import under water-stress was decreased less than leaf import (Schurr and Jahnke, 1991), resulting in relative increase in C allocation to roots. Similar results have been reported for *Theobroma* but the change from unstressed conditions was not continuous: net ^{14}C allocation to roots decreased at moderate water-stress before it increased to above-control levels at severe water-stress (Deng et al., 1990). In the same study, the relative accumulation of biomass in roots responded differently than that of ^{14}C because relative root biomass decreased continuously with increasing water-stress. This emphasizes the potential differences between C allocation and DW distribution. In general, however, the proportion of C distributed to roots increases with increasing water-stress, as with nutrient-stress, but the absolute amount does not. As a result, root systems under water-stress increase in size compared to the shoot.

The effects of water-stress and nutrient stress on root-shoot biomass distribution are related in two important ways. The first is a similarity: the general effect of both is to increase the proportion of biomass accumulated in roots. However, this increase is caused by a different mechanism in water-stress than nutrient-stress. Water-stress promotes root development because root growth can be maintained at soil water potentials that slow or stop foliage growth (Sharpe and Davies, 1979). In drying soil, roots are able to maintain a greater turgor pressure than foliage and thus continue to grow (Kozlowski et al., 1991, p. 253). The ability of net photosynthesis to continue under water-stress conditions that stop or substantially reduce leaf growth (Kozlowski et al., 1991, p. 265) also contributes to the water-stress-induced increase in root growth; C fixed under these conditions is allocated to roots. In contrast, nutrient-stress effects on root biomass accumulation appear to be caused by inadequate substrate availability for shoot growth, resulting in C accumulation in the roots, which require a lower nutrient supply for growth. The second water- and nutrient-stress relation is an interaction: the effects of drought and water-stress may cause changes in C allocation through the negative indirect effects of water availability on plant nutrient availability. Such indirect effects include the negative effects of dry soil on nutrient release

258

from organic matter (Alexander, 1977, p. 137) and the reduction of ionic mass flow and diffusion through the soil to the root surface (Mengel and Kirkby, 1982, p. 69).

Several experimental studies have investigated the influence of water and nutrients on biomass distribution between roots and shoots with combined irrigation-fertilization experiments (Table 2). Results from these experiments indicate that the effect of nutrient-stress is stronger and more consistent than that of water-stress on C allocation to roots. The common supposition that improved water regime results in less C allocation to roots of woody plants is generally true. However, irrigation does not appear to have an additional effect on fertilization treatments. Thus, while the combination of fertilization and irrigation may improve tree growth and productivity, their combination does not appear to alter C allocation more than fertilization alone. This observation may reflect inherent limits to the degree of plasticity in plants or it may reflect interactions between plant water relations and mineral nutrition.

Table 2. Effects of water (W), nutrients (N) and water plus nutrients (W+N) on distribution of biomass to root systems, expressed as percentage changes relative to control (stressful) conditions.

Plant Type	W	N	W+N	Measure	Author(s)
PSME[1] stand	-27	-34	---	BNPP[2]	Gower et al. (1992)
PITA[3] seedlings	-6	-24	-21	k (R:S)[4]	Johnson (1990)
PIEL[5] seedlings	-12	+2	-3	k (R:S)	Johnson (1990)
PIRA[6] seedlings	+20	-50	-25	R/TTL[7]	Squire et. al. (1987)
PISY[8] seedlings	-15	-22	-37	BNPP/TNPP[9]	Axelsson (1985)

[1]*Pseudotsuga menziesii*; [2]Below-ground net primary production; [3]*Pinus taeda;* Allometric coefficient, k, for total root versus total shoot biomass; [5]*Pinus elliottii*; [6]*Pinus radiata*; [7]Root biomass as a proportion of total biomass; [8]*Pinus sylvestris*; [9]Below-ground net productivity as a proportion of total net primary productivity

In contrast to rooting of intact plants, it is generally agreed that water-stress of stock plants or cuttings decreases root development (Loach, 1988). Although certainly negative for the rooting of cuttings, mild water-stress in stock plants may have a positive influence on rooting in cuttings collected from such a plant (cf. Rajagopal and Andersen, 1980).

Oxygen

The term "water-stress" is commonly associated with water deficits, but excess water, or flooding, also results in plant stress. One of the principal mechanisms of injury from flooding is through O_2 deficit, or hypoxia (Drew and Stolzy, 1991). Removal of O_2 from the root atmosphere results in two significant chemical changes. First, the root is in an energy deficient state due to the low-yielding anaerobic metabolism of glucose (Kramer and Kozlowski, 1979, p. 235). Thus, energy-requiring processes such as nutrient assimilation and growth are decreased. Second, the by-products of anaerobic respiration may be toxic in themselves; [e.g., ethanol may injure cell membranes (Glinski and Lipiec, 1990, p. 135)]. The net effect on partitioning is a rapid consumption of sugar by inefficient respiration and the potential for accumulation of ethanol as a byproduct of anaerobic decomposition. Few studies have investigated the role of hypoxia on net allocation of C between shoot and root, but C allocation patterns can be inferred to a limited extent from knowledge of root growth responses.

The most commonly observed response of plants to hypoxia is a decrease in root growth which has been shown for a variety of plants [reviewed by Glinski and Lipiec (1990), p. 139]. Although leaf growth can be very sensitive to short-term reductions in soil O_2 (Smit et al., 1989), root growth is usually more sensitive than shoot growth to soil O_2

deficit, leading to a decrease in root-shoot biomass ratio (Kozlowski et al., 1991, p. 321). For example, in hypoxia-resistant woody plants, shoot growth can remain unchanged after 30 days of hypoxic treatments, while root growth and root-shoot ratio may be reduced by more than 30% (Topa and McLeod, 1986). Because root growth is severely limited by O_2 supply, and because growth activity is an important component of sink strength, C allocation to roots is likely to be less under hypoxic conditions than aerobic conditions for most plant species.

Apart from net allocation of C between root and shoot systems, hypoxia may cause allocation to change within the root system. The most notable change is the formation of hypertrophied lenticels and adventitious roots at the soil surface, which may enable partial compensation for the O_2 deficit and impaired functioning of the original root system (Kozlowski et al., 1991, p. 331). Although this response to hypoxia is not evident in all species, and the specific role of hypoxia in the process of adventitious root formation is not yet fully resolved (Haissig, 1986), the phenomenon of adventitious rooting under flooded, hypoxic conditions is clearly of importance as an adaptive mechanism and as a potential key to a basic physiological understanding of the causes of adventitious root formation; both lines of investigation warrant further study.

Other root-oriented adaptations to hypoxia include anatomical changes, such as the formation of structures that improve root access to O_2 from above the soil surface. These structures include air channels within the roots (aerenchyma) and air ports on the exterior of the root (lenticels), as well as structures that emerge from the soil and obtain O_2 for the roots enduring hypoxic conditions (pneumatophores) (Drew and Stolzy, 1991). The ability of roots to maintain metabolic activity under hypoxic conditions is further assisted by a favorable anaerobic metabolism and by the ability to store carbohydrate reserves to fuel this process (Drew and Stolzy, 1991). Although C allocation to roots is generally reduced under hypoxic conditions, the C costs of inefficient anaerobic metabolism and specialized root structures could conceivably result in greater net C allocation when considered over the growing season in intermittently flooded conditions (for related information, see the chapter by Barlow in this volume).

Light

In general, C allocation to roots of woody plants decreases with decreasing irradiance. This response has been shown from analyses of relative biomass accumulation (Loach, 1970; Kolb and Steiner, 1990a,b) and ^{14}C distribution (Lockhart, 1992). Within this general pattern, however, much complexity exists. Plant species may be more or less plastic in their C allocation patterns in response to light, with variations in successional status and age. For example, greater effects of irradiance on relative biomass accumulation by roots were observed in two-year-old *Quercus* and *Liriodendron* seedlings compared with one-year-old seedlings (Kolb and Steiner, 1990a). Few studies have investigated the effects of light quality on C allocation in woody plants, but light quality may also affect C allocation, as evidenced by changes in relative dry matter accumulation in *Pinus* cuttings (Morgan et al., 1983).

In contrast to intact plant root development, decreased irradiance may result in increased rooting in the collected cuttings, although this varies considerably with species (Maynard and Bassuk, 1988; Moe and Andersen, 1988; see the chapter by Howard, and by Murray and coworkers in this volume). For example, *Triplochiton* cuttings taken from shaded plants rooted more when the stock plants were exposed to low rather than high irradiance (Leakey and Storeton-West, 1992). Reports of increased rooting of cuttings grown at low irradiance exist, but may be confounded by other environmental factors such as evaporative demand. In a manner similar to intact plants, it appears that better rooting of cuttings is associated with increased irradiance up to a saturation level, due to the importance of carbohydrate supply for developing roots [see Loach (1988), Haissig

(1990)]. The light quality environment of a stock plant may also affect the rooting of cuttings collected. For example, *Triplochiton* cuttings from stock plants grown with a low red:far-red ratio had a better rooting percentage than those from a high red:far-red ratio (Leakey and Storeton-West, 1992). However, light quality effects on rooting cuttings has received little emphasis in the literature.

Temperature

As the net result of enzymatically catalyzed processes, C allocation to a particular sink is likely to decrease when that sink is cooled (cf. Klepper, 1991). In fact, root cooling has been shown to decrease the rate of carbohydrate translocation from source leaves [see Bowen (1991)]. As straightforward as this seems, the role of temperature in controlling C allocation in the plant is still uncertain. In a recent review, Bowen (1991) was unable to find any reports of complete C-balance studies that adequately address the effect of temperature on C fluxes in the plant. Particularly uncertain is the effect of temperature on the rate of C loss in exudation. Instead of comprehensive studies, most existing literature focuses on either DW accumulation in roots and shoots, or on dimensional growth, in response to temperature.

In general, root cooling causes shoots to accumulate a greater proportion of biomass than roots, when compared with optimal root temperatures (Bowen, 1991). However, this is a relative increase in shoots. Suboptimal root temperatures actually decrease shoot growth and photosynthesis, even when shoots are growing at optimal temperatures (Vapaavuori et al., 1992). C allocation at supraoptimal root temperatures has received even less attention than that at suboptimal root temperatures. Bowen's recent (1991) review notes that root-shoot biomass ratios are reported to increase or decrease at non-optimal temperatures; some variation is associated with plant age and species. Such generalizations must be made with caution, however, because studies with agricultural crops indicate that relatively complex patterns of root-shoot growth occur in response to temperature when both root and shoot temperature are varied and when plant ontogeny and acclimation are considered (Reddy et al. 1992).

Another root response to temperature, perhaps even more important to plant functioning than C allocation, is root morphology. Root length commonly increases with increasing temperature, with little or no change in root DW (discussed in Bowen, 1991). Thus, as temperature increases, root diameter decreases. Given the importance of root length density (cm root per cm^3 soil) to water and nutrient uptake, the role of temperature in controlling root morphology may be important in causing stress interactions (e.g., cold temperature-nutrient-stress). The similarity between the effects of cold temperatures on root diameter and the observation of larger than normal diameters of roots in adventitiously rooted trees, as previously discussed, suggests that investigations are needed on the physiological controls of root morphology. An additional consideration in the control of rooting is the temperature difference between day and night periods, which may be more important than absolute temperature in affecting certain growth processes (cf. Erwin et al., 1989).

Temperature considerations in adventitious root formation generally agree with intact plant responses in that soil heating to maintain a particular optimum generally improves rooting (Loach, 1988).

TECHNIQUES AND EXPERIMENTAL APPROACHES

Techniques

Two of the more important questions concerning C allocation to roots and shoots are: how much of the plant's C is allocated to roots or shoots? and what are the dynamics of C fluxes within the plant? These questions are especially relevant for the root system, given the inaccessibility of roots for growth and gas exchange measurements. With improvements in the availability of technology for the analysis of radioactive [reviewed by Isebrands and Dickson (1991)] and stable isotopes [reviewed by Ehleringer and Osmond (1991), and Caldwell and Virginia (1991)], such methods can be increasingly relied upon for quantification of C allocation and partitioning in root systems. Isotopes have been used to refine estimates of translocation rates between shoot and roots (Spence et al., 1990) and from roots to the rhizosphere (van Veen et al., 1991); to estimate root turnover and production, N fixation and nutrient acquisition (Caldwell and Virginia, 1991); and to infer patterns of root distribution (Friend et al., 1991). One of the principal advantages of isotopes is that the plant system can be studied in its natural state rather than in an artificial environment, thus avoiding the artifacts associated with most methods of root quantification. At the level of the adventitiously developing root, existing carbon-isotope labelling techniques hold promise for answering many of the persistent questions on the role of stored assimilates in rooting, and about source-sink relations between the shoot and newly developing roots. Particularly, short-term tracing of C within and between roots and shoots may be accomplished with ^{11}C labeling (cf. Schurr and Jahnke, 1991; Spence et al., 1990), and mid- to long-term analysis of C allocation and partitioning may be accomplished by ^{13}C labeling (cf. Cliquet et al. 1990) or ^{14}C labeling (Isebrands and Dickson, 1991).

At the whole-plant level, it has become increasingly apparent that nonstructural C allocation to root systems, including the respiratory costs of growth and maintenance, nutrient uptake, allocation to mycorrhizal fungi and exudation of C from roots into the rhizosphere, may be quite substantial (Buwalda et al., 1992). Recent reviews of methods for quantifying the allocation of C to nonstructural root functions are presented by Lambers et al. (1983), Lambers (1987) and Vogt et al. (1989). An important aspect of this question is the importance of realistic soil environmental conditions for the estimation of respiratory C allocation (cf. Naganawa and Kyuma, 1991). Such techniques for evaluating C fluxes to nonstructural uses may be directly and productively applied to adventitiously developed roots to evaluate their functioning relative to those developing from normal ontogenetic processes.

At the ecosystem or stand level, C allocated to roots is frequently estimated from below-ground net primary production (NPP), which is driven largely by fine root production and estimated from periodic collections of soil cores, careful separation of roots from soil, and analysis of patterns to estimate net C input [reviewed by Vogt and Persson, 1991, and Caldwell and Virgina, 1991; see also the chapter by Dickmann and Hendrick, and by Morita in this volume). Although absorbing or fine roots are the most dynamic and appear to consume the most C, larger, structural roots may serve important storage functions. Quantification of the storage role of the roots of woody plants has also been recently reviewed by Loescher et al. (1990). There is little reason to expect that fully developed stands of adventitiously rooted plants will differ from those of seedling origin in NPP or storage. However, application of these tools in developing stands may provide useful insights, especially in cases where the growth of clonal stands lags behind that of seedlings [e.g., Anderson et al. (1992)].

Another technology for quantifying C allocation to roots in stands is through nondestructive minirhizotron techniques (see the chapter by Dickmann and Hendrick in this volume). Initial efforts in this area relied upon glass or plastic tubes inserted in the soil, with root activity observed with a periscope device and quantified according to root intersections, with scribed lines on the tubes or intersections viewed with an ocular cross-hair (Richards, 1984). The current and most sophisticated technology uses similar tubes with data collected with miniature video cameras, stored on magnetic tape and analyzed with advanced image analysis systems (Brown and Upchurch, 1987; Hendrick and Pregitzer, 1992; see the chapter by Dickmann and Hendrick in this volume). While the magnitude of C consumed by root turnover (root death and subsequent replacement, often with little net change in living root biomass) has received considerable emphasis (Fogel, 1990), new nondestructive techniques using minirhizotron tubes offer great potential for improving estimates of turnover with direct observation of root death and disappearance. Minirhizotron tubes or modifications of this principle might find useful application in the characterization of root development from cuttings in the field. The dynamics and locations of root initiation are usually inferred from destructive harvest, but nondestructive observation of cuttings as they root under field conditions, as could be seen with a rhizotron or minirhizotron, would provide better information on successful rooting characteristics.

In addition to answering specific experimental questions, the output from these root quantification techniques is an important source of data for simulation modelling efforts. Root system dynamics is one of the greatest uncertainties in most current growth simulation models. The techniques reviewed above will enable essential root system parameters such as: absolute and relative elongation rate, vertical distribution, root length density, turnover rate and net primary productivity to be determined and incorporated into models. Sensitive techniques will also be essential in refining models to the point that temporal and spatial heterogeneity can be included in modelling root function (cf. Fitter, 1992).

Process Simulation Models

Use of computer simulation models by plant scientists has increased in recent years as the availability and speed of high powered computers has expanded (for related information see the chapter by Rauscher in this volume). Most models are used either as a predictive tool or as a research tool for formulating hypotheses, interpreting results and understanding biochemical and physiological processes. Two types of simulation models are generally recognized, empirical and mechanistic, or process-based. Empirical models describe relationships among variables without reference to the underlying principles; process models are based upon underlying biochemical and physiological processes of the system (Isebrands and Burk, 1992).

There is a multitude of mechanistic process models available for agronomic, herbaceous and forest plants (Thornley, 1976; Whisler et al, 1986; Graham et al., 1985; Isebrands et al., 1990). However, few if any of the models deal adequately with C allocation processes, because the mechanisms of C allocation and partitioning are not fully understood (as discussed previously). Most simulation models allocate C empirically through predetermined allocation coefficients based upon biomass components (Wilson, 1988). One notable exception is ECOPHYS, a physiological process model of juvenile poplar trees growing from hardwood cuttings (Rauscher et al., 1990; see the chapter by Rauscher in this volume), where C is allocated from sources to sinks over the season via dynamic transport coefficients determined from radiotracer studies. However, one of the current limitations to wider applications of the ECOPHYS model results from the lack of knowledge of adventitious root development. For example, we do not understand the factors controlling root initiation and location (i.e., siting) on cuttings, vascular connections between roots and shoots, or the mechanisms of root-shoot feedback in the developing plant. Again, these gaps in knowledge will no doubt require an interdisciplinary team effort, including molecular biology, in order to achieve a solution (Haissig et al., 1992).

In the future, the best use of mechanistic simulation models for understanding adventitious rooting will probably be as an interactive tool for knowledge synthesis in association with experimentation. In parallel modelling and experimentation, process simulation models are valuable for integrating a series of experiments or for testing simpler, empirical simulation models (Isebrands and Burk, 1992). A good example of this approach is the work of Dick and Dewar (1992), who developed a model of carbohydrate dynamics during adventitious root development in leafy cuttings of *Triplochiton*. They expanded the concepts of Thornley and coworkers concerning balanced root-shoot development (Thornley, 1972, 1976; Johnson and Thornley, 1987) to adventitious rooting in cuttings of woody plants. The processes represented in their model were leaf photosynthesis, starch mobilization, sugar transport and sugar utilization for root growth—a sensitivity analysis was also conducted on model parameters. Although the authors concluded that further parallel experimental work is needed to parameterize and verify their simulation model, this work is a useful new approach for use in conjunction with such experimentation for the eventual understanding of C metabolism during adventitious rooting of cuttings of woody plants.

Woody Plant Model Experimental Systems

Throughout this chapter we have emphasized the need for an integrated multidisciplinary research approach to understanding the mechanisms of C metabolism during adventitious rooting of woody plants. In this regard, certain plant materials represent ideal experimental systems for employing this integrated approach (see the chapter by Ernst, and by Riemenschneider in this volume). Without question, the genus *Populus* stands-out among woody plants as a representative with mostly induced or mostly preformed root initials, depending on species. As evidenced by our literature citations, *Populus* has been studied in detail by many scientific disciplines at all scales of plant organization. There is a wealth of information on the molecular biology, genetics, biochemistry, anatomy, physiology, propagation and silviculture of the genus. A few of the reasons for selecting *Populus* for such a model experimental system are: 1) abundant natural genetic variation in rooting capability, 2) traditional genetic improvement programs that yield pedigreed lines, 3) ease of hybridization and genetic manipulation, 4) molecular genetics research programs (Stettler et al., 1992), 5) fundamental anatomical, biochemical and physiological databases (Isebrands and Dickson, 1991), (6) interdisciplinary research programs (Stettler and Ceulemans, 1993), (7) process simulation models (Isebrands et al., 1990), (8) worldwide economic importance and utilization, and (9) much silvicultural information (Hinckley et al, 1989).

Another candidate for an ideal woody plant experimental system is jack pine (*Pinus banksiana*) as a representative of a recalcitrant conifer that forms only induced (also termed post-formed) adventitious roots. Jack pine does not have all the "credentials" of *Populus*, but it has the advantage of having the most comprehensive fundamental biochemical data base with respect to adventitious rooting of any conifer [see Haissig (1984, 1986, 1988), Haissig et al. (1992)]. In addition, extensive jack pine genetically unique materials exist, and the species is genetically highly variable. Moreover, jack pine flowers at a very young age (i.e., ca. age 3 years) compared to most conifers, and several generations of inbred lines are available for study. Unfortunately, the companion molecular genetics programs needed for integration with the anatomical and biochemical studies of the controls of adventitious rooting as called for by Haissig et al. (1992) are just beginning.

SYNTHESIS AND FUTURE RESEARCH DIRECTIONS

Synthesis

C allocation in woody plants has been addressed in the fields of plant physiology, ecology, horticulture and forestry. In broad terms, C allocation is controlled by plant and environmental factors. Plant factors include growth patterns (i.e., determinate, indeterminate, semideterminate), and species-specific factors such as those that determine root morphology and the ability to produce adventitious roots. Environmental factors affecting C allocation include but are not limited to nutrients, water, O_2, light and temperature. At a more mechanistic level, C allocation is controlled by source-sink relations. All of these factors cause a shift in allocation by altering the rate at which carbohydrates are loaded, unloaded and consumed by a particular sink. Simple as this may seem, the physiological processes underlying source-sink relations are poorly understood. C allocation is of key importance to adventitiously rooted plant biology because it may help explain poor rooting success and poor field performance, and it may be used to improve the productivity and yield of adventitiously rooted species.

Future Research

Clearly, more research is needed in all aspects of C allocation in woody plants and especially important is research on C allocation to root systems. We have identified six areas that require urgent attention. The first area involves examining the differences in C allocation between adventitiously rooted plants and those of seedling origin. These differences are most important in species that are difficult to root, because inadequate allocation to roots may offer some explanation for poor field performance. A critical issue is the control of juvenility and maturity in affecting C allocation. Carefully designed and controlled experiments are needed to establish the seasonal C dynamics of seedlings of varying ages in comparison with cuttings of similar sizes.

The second area is in the role of C partitioning in successful adventitious rooting. Much correlative evidence presently exists, including relationships with rooting success and various carbohydrates or environmental factors that are believed to be important because of their effect on carbohydrates. However, other factors may ultimately be more limiting and difficult to control. Thus, it is suggested that researchers attempt to control or at least monitor carbohydrate balance in adventitious rooting experiments. One approach would be to develop optimum carbohydrate balances for cuttings and other vegetative propagules, then use these optima as a baseline in subsequent investigations of environmental and plant effects. Another approach, if carbohydrate allocation or partitioning is believed to be a causal and limiting factor in rooting success, would be to vary carbohydrate balance by genetic rather than environmental means, and thus remove the other potentially confounding effects of environment on morphogenesis (see the chapter by Riemenschneider in this volume). This could be accomplished, for example, by screening genotypes or mutants for C partitioning patterns. Such standardization would greatly improve the mechanistic understanding of processes affecting adventitious rooting.

The third and perhaps most ambitious area is the mechanistic explanation of C allocation to adventitious roots. While evidence exists for the role of newly developing roots as a sink for carbohydrates, the magnitude to which roots can direct assimilates of leaf origin is uncertain. Also, resolving the roles (if any) of key enzymes, known to be involved in C partitioning (e.g., SPS) by mediating sink strength, will further a mechanistic understanding of plant and environmental factors that control C allocation to roots. This research direction will necessitate an integrated approach that includes anatomical, biochemical and molecular biological investigations, and will result in basic knowledge useful to all plant scientists.

The fourth area concerns whole-tree C-budgets. Too much of the current information on C allocation is estimated by difference or other indirect means. Experimentation is needed that quantifies structural and functional costs of all major plant organs throughout a growing season. Such research should result in specific C influx and efflux rates for individual tissues that can be used in plant growth simulation models. This approach could also provide useful information in determining relative importance of stored C reserves to rooting and plant establishment for species with different adaptations, growth patterns and life histories.

The fifth area is related to the fourth, but is directed only to the root system. Roots may consume more than one-half of the plant's C, yet good data on the magnitude and fate of C allocated to the root system are infrequent in the literature. Included under this topic are the C costs of nutrient uptake, symbionts and the potential for "wasteful" respiration and exudation. For such information to be useful to simulation modelling efforts, more information on root dynamics is also needed. This effort will include variation in C allocation and physiological activity over spatial and temporal scales. Much of the inaccuracy of current simulation models results from the overuse of physiological parameters collected from a single time during the growing season.

In the sixth and final area, more whole-plant simulation modelling efforts are needed in conjunction with whole-plant C allocation experimentation. Such efforts will provide guidance in the processes that are most important for study, and in the appropriate level for examination of the processes. Simulation models can thus help identify gaps in knowledge and provide feedback to experimentation. A successful illustration of this approach to solving problems associated with adventitious rooting is provided by Dick and Dewar (1992).

ACKNOWLEDGMENTS

We thank L.J Frampton, S.B. Land, and G.A. Ritchie for sharing unpublished data and personal observations with us. Review comments and suggestions on earlier drafts of the manuscript were generously supplied by H.F. Hodges, J.D. Hodges, S.B. Land and three anonymous reviewers. The assistance of J.D. Friend, R.L. Harty, J.A. Mobley, R. Schultz and P.C. Smith with graphics and manuscript preparation is gratefully acknowledged. ALF acknowledges support under the Mississippi Agricultural and Forestry Experiment Station Stress Physiology Project. JGI acknowledges support under US Department of Energy Interagency Agreement No. DE A105-800R20763. This paper is contribution PS8241 of the Mississippi Agricultural and Forestry Experiment Station.

REFERENCES

Agren, G.I., and Ingestad T., 1987, Root : shoot ratio as a balance between nitrogen productivity and photosynthesis, *Plant Cell Environ.* 10:579.

Alexander, M., 1977, "Introduction to Soil Microbiology," John Wiley & Sons, New York.

Andersen, C.P., and Rygiewicz, P.T., 1991, Stress interactions and mycorrhizal plant-response: understanding carbon allocation priorities, *Environ. Pollut.* 73:217.

Anderson, A.B., Frampton, L.J., Jr., McKeand, S.E., and Hodges, J.F., 1992, Tissue-culture shoot and root system effects on field performance of loblolly pine, *Can. J. For. Res.* 22:56.

Axelsson, B., 1985, Increasing forest productivity and value by manipulating nutrient availability, *in*: "Weyerhaeuser Sci. Symp.: Forest Potentials, Productivity and Value," R. Ballard, P. Farnum, G.A. Ritchie, and J.K. Winjum, eds., Weyerhaeuser Co., Tacoma.

Bloom, A.J., Chapin, F.S., III., and Mooney, H.A., 1985, Resource limitation in plants - an economic analogy, *Annu. Rev. Ecol. Syst.* 16:363.

Bowen, G.D., 1991, Soil temperature, root growth, and plant function, *in:* "Plant Roots: The Hidden Half," Y. Waisel, A. Eshel, and U. Kafkafi, eds, Marcel Dekker, Inc., New York.

Breen, P.J., and Muraoka, T., 1974, Effect of leaves on carbohydrate content and movement of ^{14}C assimilate in plum cuttings., *J. Amer. Soc. Hortic. Sci.* 99:326.

Brand, M.H., and Lineberger, R.D., 1992, *In vitro* rejuvenation of *Betula* (*Betulaceae*): morphological evaluation, *Amer. J. Bot.*, 79:618.

Brouwer, R., 1983, Functional equilibrium: sense or nonsense?, *Neth. J. Agric. Sci.* 31:335.

Brown, D.A., and Upchurch, D.R., 1987, Minirhizotrons: a summary of methods and instruments in current use, *in* "Minirhizotron Observation Tubes: Methods and Applications for Measuring Rhizosphere Dynamics," H.M. Taylor, ed., ASA Special Pub. No. 50, Amer. Soc. Agron., Madison, p. 15.

Buwalda, J.G., Fossen, M., and Lenz, F., 1992, Carbon dioxide efflux from roots of calamodin and apple, *Tree Physiol.* 10:391.

Caldwell, M.M., and Virginia, R.A., 1991, Root systems, *in* "Plant Physiological Ecology: Field Methods and Instrumentation," R.W. Pearcy, J.R. Ehleringer, H.A. Mooney, and P.W. Rundel, eds., Chapman and Hall, New York.

Cameron, R.J., and Rook, D.A., 1974, Rooting stem cuttings of Radiata pine: environmental and physiological aspects, *N.Z. J. For. Sci.* 4:291.

Cannell, M.G.R., 1989, Physiological basis of wood production: A review, *Scand. J. For. Res.* 4:459.

Carlson, W.C., and Preisig, C.L., 1981, Effects of controlled-release fertilizers on the shoot and root development of Douglas-fir seedlings, *Can. J. For. Res.* 11:230.

Chairiello, N.R., and Gulmon, S.L., 1991, Stress effects on plant reproduction, *in:* "Response of Plants to Multiple Stresses," H.A. Mooney, W.E. Winner, E.J. Pell, and E. Chu, eds., Academic Press, Inc., San Diego.

Chapin, F.S. III, Schulze, E.D., and Mooney, H.A., 1990, The ecology and economics of storage in plants, *Annu. Rev. Ecol. Syst.* 21:423.

Cheffins, N.J., and Howard, B.H., 1982, Carbohydrate changes in leafless winter apple cuttings, II. Effects of ambient air temperature during rooting, *J. Hortic. Sci.* 57:9.

Cheikh, N., and Brenner, M.L., 1992, Regulation of key enzymes of sucrose biosynthesis in soybeans leaves, *Plant Physiol.* 100:1230.

Cliquet, J., Deleens, E., Bousser, A., Martin, M., Lescure, J., Prioul, J., Mariotti, A., and Morot-Gaudry, J., 1990, Estimation of carbon and nitrogen allocation during stalk elongation by ^{13}C and ^{15}N tracing in *Zea mays* L., *Plant Physiol.* 92:79-87.

Cronshaw, J., Lucas, W.J., and Giaquinta, R.T., 1986, "Phloem Transport," Alan Liss, Inc., New York.

Daie, J., 1985, Carbohydrate partitioning and metabolism in crops, *Hortic. Rev.* 7:69.

Daie, J., 1986, Hormone-mediated enzyme activity in source leaves, *Plant Growth Reg.* 4:287.

Davies, W.J., and Zhang, J., 1991, Root signals and the regulation of growth and development of plants in drying soil, *Annu. Rev. Plant Physiol. Mol. Biol.* 42:55.

Davis, T.D., and Potter, J.R., 1985, Carbohydrates, water potential, and subsequent rooting of stored *Rhododendron* cuttings, *HortSci.* 20:292.

Deng, X., Joly, R.J., and Hahn, D.T., 1990, The influence of plant water deficit on distribution of ^{14}C-labelled assimilates in cacao seedlings, *Ann. Bot.* 66:211.

de Wit, C.T., and Penning de Vries, F.W.T., 1983, Crop growth models without hormones, *Neth. J. Agric. Sci.* 31:313.

Dewald, L., White, T.L., and Duryea, M.L. 1992, Growth and phenology of seedlings of four contrasting slash pine families in ten nitrogen regimes, *Tree Physiol.* 11:255.

Dick, J.M., and Dewar, R.C., 1992, A mechanistic model of carbohydrate dynamics during adventitious root development in leafy cuttings, *Ann Bot.* 70:371.

Dickmann, D.I., Gjerstad, D.H., and Gordon, J.C., 1975, Developmental patterns of CO_2 exchange, diffusion resistance and protein synthesis in leaves of *Populus* x *Euramericana, in:* "Environmental and Biological Control of Photosynthesis," R. Marcelle, ed., Dr. W. Junk, The Hague.

Dickson, R.E., 1986, Carbon fixation and distribution in young *Populus* trees, *in:* "Crown and canopy structure in relation to productivity," T. Fujimori, and D. Whitehead, eds., For. and For. Prod. Res. Inst., reprinted from the proceedings, Ibaraki.

Dickson, R.E., 1989, Carbon and nitrogen allocation in trees, *Ann. Sci. For.* 46 (suppl.):631s.

Dickson, R.E., 1991, Assimilate distribution and storage, *in:* "Physiology of trees," A.S. Raghavendra, ed., John Wiley and Sons, Inc., New York.

Dickson, R.E., and Isebrands, J.G., 1991, Leaves as regulators of stress response, *in:* "Response of Plants to Multiple Stresses," H.A. Mooney, W.E. Winner, E.J. Pell, and E. Chu, eds., Academic Press, San Diego.

Dirr, M.A., and Heuser, C.W., 1987, "The Reference Manual of Woody Plant Propagation," Varsity Press, Inc., Athens.

Donnelly, J.R., 1977, "Morphological and Physiological Factors Affecting Formation of Adventitious Roots on Sugar Maple Stem Cuttings, *USDA Forest Service Res. Pap.* NE-365.

Drew, M.C., and Stolzy, L.H., 1991, Growth under oxygen stress, *in:* "Plant Roots: The Hidden Half," Y. Waisel, A. Eshel, and U. Kafkafi, eds., Marcel Dekker, Inc., New York.

Ehleringer, J.R., and Osmond, C.B., 1991, Stable isotopes, *in:* "Plant Physiological Ecology: Field Methods and Instrumentation," R.W. Pearcy, J.R. Ehleringer, H.A. Mooney, and P.W. Rundel, eds., Chapman and Hall, New York.

Eliasson, L., 1968, Dependence of root growth on photosynthesis in *Populus tremula, Physiol. Plant.* 21:806.

Ernstsen, A., and Hansen, J., 1986, Influence of gibberellic acid and stock plant irradiance on carbohydrate content and rooting in cuttings of Scots pine seedlings (*Pinus sylvestris* L.), *Tree Physiol.* 1:115.

Erwin, J.E., Heins, R.D., and Karlsson, M.G., 1989, Thermomorphogenesis in *Lilium longiflorum, Amer. J. Bot.,* 76:47.

Fege, A.S., 1983, "Changes in *Populus* Carbohydrate Reserves During Induction of Dormancy, Cold Storage of Cuttings, and Development of Young Plants," Ph.D. thesis, Univ. of Minnesota, St. Paul.

Fege, A., and Brown, G., 1984, Carbohydrate distribution in dormant *Populus* shoots and hardwood cuttings, *For. Sci.* 30:999.

Fitter, A.H., 1992, Architecture and biomass allocation as components of the plastic response of roots to soil heterogeneity, *in:* "Exploitation of Environmental Heterogeneity by Plants," M.M. Caldwell, and R.W. Pearcy, eds., Academic Press, New York. (in press)

Fogel, R., 1990, Root turnover and production in forest trees, *HortSci.* 25:270.

French, C.J., 1990, Rooting of *Rhododendron* 'Anna Rose Whitney' cuttings as related to stem carbohydrate concentration, *HortSci.* 25:409.

Friend, A.L., 1988, "Nitrogen Stress and Fine Root Growth of Douglas-fir," Ph.D. thesis, Univ. of Washington, Seattle.

Friend, A.L., Eide, M.R., and Hinckley, T.M., 1990, Nitrogen stress alters root proliferation in Douglas-fir seedlings, *Can. J. For. Res.* 20:1524.

Friend, A.L., Scarascia-Mugnozza, G., Isebrands, J.G., and Heilman, P.E., 1991, Quantification of two-year-old hybrid poplar root systems: morphology, biomass, and ^{14}C distribution, *Tree Physiol.* 8:109.

Geiger, D.R., and Serviates, J.C., 1991, Carbon allocation and response to stress, *in:* "Response of Plants to Multiple Stresses," H.A. Mooney, W.E. Winner, E.J. Pell, and E. Chu, eds., Academic Press, Inc., San Diego.

Gersani, M., Lips, S.H., and Sachs, T., 1980, The influence of shoots, roots and hormones on sucrose distribution, *J. Exp. Bot.* 31:177.

Gifford, R.M., 1986, Partitioning of photoassimilate in the development of crop yield, *in:* "Phloem Transport," J. Cronshaw, W.J. Lucas, and R.T. Giaquinta, eds., Alan R. Liss, Inc., New York.

Gifford, R.M., and Evans, L.T., 1981, Photosynthesis, carbon partitioning and yield, *Annu. Rev. Plant Physiol.* 32:485.

Gifford, R.M., Thorne, J.M., Hitz, W.D., and Giaquinta, R.T., 1984, Crop productivity and photoassimilate partitioning, *Sci.* 225:801.

Glerum, C., 1980, Food sinks and food reserves of trees in temperate climates, *N.Z. J. For. Sci.* 10:176.

Glinski, J. and Lipiec, J., 1990, "Soil Physical Conditions and Plant Roots," CRC Press, Inc., Boca Raton.

Gower, S.T., Vogt, K.A., and Grier, C.C., 1992, Carbon dynamics of Rocky Mountain Douglas-fir: influence of water and nutrient availability, *Ecol. Monog.* 62:43.

Graham, R.L., Farnum, P., Timmis, R., and Ritchie, G.A., 1985, Using modeling as a tool to increase forest productivity and value, *in:* "Weyerhaeuser Science Symposium: Forest Potentials, Productivity and Value," R. Ballard, P. Farnum, G.A. Ritchie, and J.K. Winjum, eds., Weyerhaeuser Co., Tacoma.

Granato, T.C., and Raper, C.D., 1989, Proliferation of maize (*Zea mays* L.) roots in response to localized supply of nitrate, *J. Exp. Bot.* 40:263.

Groff, P.A., and Kaplan, D.R., 1988, The relation of root systems to shoot systems in vascular plants, *Bot. Rev.* 54:387.

Gupta, P.K., Timmis, R., and Mascarenhas, A.F., 1991, Field performance of micropropagated forestry species, *In Vitro Cell Devel. Biol.* 27P:159.

Haissig, B.E., 1970, Preformed adventitious root initiation in brittle willows grown in a controlled environment, *Can. J. Bot.* 48:2309.

Haissig, B.E., 1974a, Origins of adventitious roots, *N.Z. J. For. Sci.* 4:299.

Haissig, B.E., 1974b, Metabolism during adventitious root primordium initiation and development, *N.Z. J. For. Sci.* 4:324.

Haissig, B.E., 1984, Carbohydrate accumulation and partitioning in *Pinus banksiana* seedlings and seedling cuttings, *Physiol. Plant.* 61:13.

Haissig, B.E., 1986, Metabolic processes in adventitious rooting of cuttings, *in:* "New Root Formation in Plants and Cuttings," M.B. Jackson, ed., Martinus Nijhoff Pubs., Dordrecht.

Haissig, B.E., 1988, Future directions in adventitious rooting research, *in:* "Adventitious Root Formation in Cuttings," T.D. Davis, B.E. Haissig, and N. Sankhla, eds., Dioscorides Press, Portland.

Haissig, B.E., 1989, Removal of the stem terminal and application of auxin change carbohydrates in *Pinus banksiana* cuttings during propagation, *Physiol. Plant.* 77:179.

Haissig, B.E., 1990, Reduced irradiance and applied auxin influence carbohydrate relations in *Pinus banksiana* cuttings during propagation, *Physiol. Plant.* 78:455.

Haissig, B.E., Davis, T.D., and Riemenschneider, D.E., 1992, Researching the controls of adventitious rooting, *Physiol. Plant.* 84:310.

Hansen, J., Strömquist, L., and Ericsson, A., 1978, Influence of the irradiance on carbohydrate content and rooting of cuttings of pine seedlings (*Pinus sylvestris* L.), *Plant Physiol.* 61:975.

Hendrick, R.L., and Pregitzer, K.S., 1992, The demography of fine roots in a northern hardwood forest, *Ecol.* 73:1094.

Henry, P.H., Blazich, F.A., and Hinesley, L.E., 1992, Nitrogen nutrition of containerized eastern redcedar,

II. Influence of stock plant fertility on adventitious rooting of stem cuttings, *J. Amer. Soc. Hortic. Sci.* 117:568.

Hinckley, T.M., Ceulemans, R., Dunlap, J.M., Figliola, A., Heilman, P.E., Isebrands, J.G., Scarascia-Mugnozza, G., Schulte, P.J., Smit, B., Stettler, R.F., van Volkenburgh, E., and Waird, B.W., 1989, Physiological, morphological and anatomical components of hybrid vigor in *Populus*, *in:* "Structural and Functional Responses to Environmental Stresses: Water Shortage," K.H. Kreeb, H. Richter, and T.M. Hinckley, eds., SPB Academic Pubs., The Hague.

Ho, L.C., 1988, Metabolism and compartmentation of imported sugars in sink organs in relation to sink strength, *Annu. Rev. Plant Physiol. Mol. Biol.* 39:355.

Huber, S.C., 1983, Relation between photosynthetic starch formation and dry-weight partitioning between the shoot and root, *Can. J. Bot.* 61:2709.

Huber, S.C., and Huber, J.L., 1992, Role of sucrose-phosphate synthase in sucrose metabolism in leaves, *Plant Physiol.* 99:1275.

Huber, S.C., Kerr, P.S., and Kalt-Torres, W., 1985, Regulation of sucrose formation and movement, *in:* "Regulation of Carbon Partitioning in Photosynthetic Tissue," R.L. Heath, and J. Preiss, eds., Waverly Press, Baltimore.

Ingestad, T., and Agren, G.I., 1991, The influence of plant nutrition on biomass allocation, *Ecol. Appl.* 1:168.

Ingestad, T., and Lund, A.B., 1979, Nitrogen stress in birch seedlings, I. Growth technique and growth, *Physiol. Plant.* 45:137.

Isebrands, J.G., and Burk, T.E., 1992, Ecophysiological growth process models of short rotation forest crops, *in:* "Ecophysiology of Short Rotation Forest Crops," C.P. Mitchell, J.B. Ford-Robertson, T. Hinckley, and L. Sennerby-Forsse, eds., Elsevier, New York.

Isebrands, J.G., and Crow, T.R., 1985, Techniques for rooting juvenile softwood cuttings of northern red oak, *in:* "Proc. 5th Central Hardwood For. Conf.," J.O. Dawson, and K.A. Majerus, eds., Univ. of Illinois, Urbana-Champaign.

Isebrands, J.G., and Dickson, R.E., 1991, Measuring carbohydrate production and distribution--radiotracer techniques and applications, *in:* "Techniques and Approaches in Forest Tree Ecophysiology," J.P. Lassoie, and T.M. Hinckley, eds., CRC Press, Inc., Boca Raton.

Isebrands, J.G., and Fege, A.S., 1993, Applications of ^{14}C methods for the study of metabolism in hardwood cuttings, *Tree Physiol.* (in press)

Isebrands, J.G., and Larson, P.R., 1977, Vascular anatomy of the nodal region in *Populus deltoides* Bartr., *Amer. J. Bot.* 64:1066.

Isebrands, J.G., and Larson, P.R., 1980, Ontogeny of major veins in the lamina of *Populus deltoides* Bartr., *Amer. J. Bot.* 67:23.

Isebrands, J.G., and Nelson, N.D., 1983, Distribution of [14C]-labeled photosynthates within intensively cultured *Populus* clones during the establishment year, *Physiol. Plant.* 59:9.

Isebrands, J.G., Nelson, N.D., Dickmann, D.I., and Michael, D.A., 1983, "Yield Physiology of Short Rotation Intensively Cultured Poplars," USDA Forest Service, Gen. Tech. Rep. NC-91.

Isebrands, J.G., Rauscher, H.M., Crow, T.R., and Dickmann, D.I., 1990, Whole-tree growth process models based on structural-functional relationships, *in:* "Process Modeling of Forest Growth Responses to Environmental Stress," R.K. Dixon, R.S. Meldahl, G.A. Ruark, and W.G. Warren, eds., Timber Press, Portland.

Johnson, I.R., and Thornley, J.H.M., 1987, A model of shoot:root partitioning with optimal growth, *Ann. Bot.* 60:133.

Johnson, J.D. 1990. Dry-matter partitioning in loblolly and slash pine: effects of fertilization and irrigation, *For. Ecol. Mgmt.* 30:147.

Jones, C.G., and Coleman, J.S., 1991, Plant stress and insect herbivory: toward an integrated perspective, *in:* "Response of Plants to Multiple Stresses," H.A. Mooney, W.E. Winner, E.J. Pell, and E. Chu, eds., Academic Press, Inc., San Diego.

Kerr, P.S., Huber, S.C., and Israel, D.W., 1984, Effect of N-source on soybean leaf sucrose phosphate synthase, starch formation, and whole-plant growth, *Plant Physiol.* 75:483.

Klepper, B., 1991, Root-shoot relationships, *in:* "Plant Roots: The Hidden Half," Y. Waisel, A. Eshel, and U. Kafkafi, eds., Marcel Dekker, Inc., New York.

Kolb, T.E., and Steiner, K.C., 1990a, Growth and biomass partitioning of northern red oak and yellow-poplar seedlings: effects of shading and grass root competition, *For. Sci.* 36:34.

Kolb, T.E., and Steiner, K.C., 1990b, Growth and biomass partitioning response of northern red oak genotypes to shading and grass root competition, *For. Sci.* 36:293.

Kozlowski, T.T., 1971, "Growth and Development of Trees; Cambial Growth, Root Growth, and Reproductive Growth," vol. 2, Academic Press, New York.

Kozlowski, T.T., 1992, Carbohydrate sources and sinks in woody plants, *Bot. Rev.* 58:107.

Kozlowski, T.T., and Keller, T., 1966, Food relations of woody plants, *Bot. Rev.* 32:293.

Kozlowski, T.T., Kramer, P.J., and Pallardy, S.G., 1991, "The Physiological Ecology of Woody Plants," Academic Press, Inc., San Diego.

Kramer, P.J., and Kozlowski, T.T., 1979, Physiology of Woody Plants, Academic Press, Inc., Orlando.

Kuehnert, C.C., and Larson, P.R., 1983, Development and organization of the primary vascular system in the phase II leaf and bud of *Osmunda cinnamomea* L., *Bot. Gaz.* 144:310.

Lambers, H., 1987, Growth, respiration, exudation, and symbiotic associations: the fate of carbon translocated to roots, *in:* "Root Development and Function," P.J. Gregory, J.V. Lake, and D.A. Rose, eds., Cambridge Univ. Press, Cambridge.

Lambers, H., Simpson, R.J., Beilharz, V.C., and Dalling, M.J., 1982, Growth and translocation of C and N in wheat (*Triticum aestivum*) grown with a split root system, *Physiol. Plant.* 56:421.

Lambers, H., Szaniawski, R.C., and de Visser, R., 1983, Respiration for growth, maintenance and ion uptake: an evaluation of concepts, methods, values and their significance, *Physiol. Plant.* 58: 556.

Lang, A., 1974, Inductive phenomena in plant development, *in:* "Basic Mechanisms in Plant Morphogenesis," P.S. Carlson, ed., Brookhaven Symp. Biol. No.25, Nat. Tech. Info. Serv., Springfield.

Larson, P.R., 1979, Establishment of the vascular system in seedlings of *Populus deltoides* Bartr., *Amer. J. Bot.* 66:452.

Larson, P.R., 1983, Primary vascularization and the siting of primordia, *in:* "Growth and Functioning of Leaves," J.E. Dale, and F.L. Milthorpe, eds., Cambridge Univ. Press, Cambridge.

Larson, P.R., 1984, Vascularization of developing leaves of *Gleditsia triacanthos* L., II. Leaflet initiation and early vascularization, *Amer. J. Bot.* 71:1211.

Larson, P.R., 1985, Rachis vascularization and leaflet venation in developing leaves of *Fraxinus pennsylvanica*, *Can. J. Bot.* 63:2383.

Larson, P.R., 1986, Vascularization of multilacunar species: *Polyscias quilfoylei* (*Araliaceae*), II. The leaf base and rachis, *Amer. J. Bot.* 73:1632.

Larson, P.R., and G.D. Fisher, 1983, Xylary union between elongating lateral branches and the main stem in *Populus deltoides*, *Can. J. Bot.* 61:1040.

Larson, P.R., Isebrands, J.G., and Dickson, R.E., 1980, Sink to source transition of *Populus* leaves, *Ber. Deutsch. Bot. Ges.* 93:79.

Leakey, R.R.B., and Coutts, M.P., 1989, The dynamics of rooting in *Triplochiton scleroxylon* cuttings: their relation to leaf area, node position, dry weight accumulation, leaf water potential and carbohydrate composition, *Tree Physiol.* 5:135.

Leakey, R.R.B., and Storeton-West, R., 1992, The rooting ability of *Triplochiton scleroxylon* cuttings: the interactions between stockplant irradiance, light quality and nutrients, *For. Ecol. Mgmt.* 49:133.

Loach, K., 1970, Shade tolerance in tree seedlings, II. Growth analysis of plants raised under artificial shade, *New Phytol.* 69:273.

Loach, K., 1988, Water relations and adventitious rooting, *in:* "Adventitious Root Formation in Cuttings," T.D. Davis, B.E. Haissig, and N. Sankhla, eds., Dioscorides Press, Portland.

Lockhart, B.R., 1992, "Morphology, Gas-Exchange, and ^{14}C-Photosynthate Allocation Patterns in Advanced Cherrybark Oak Reproduction," Ph.D. theses, Mississippi State Univ.

Loescher, W.H., McCamant, T., and Keller, J.D., 1990, Carbohydrate reserves, translocation, and storage in woody plant roots, *HortSci.* 25:274.

Lovell, P.H., and White, J., 1986, Anatomical changes during adventitious root formation, *in:* "New Root Formation in Plants and Cuttings," M.B. Jackson, ed., Martinus Nijhoff Pubs., Dordrecht.

Mansfield, T.A., 1987, Hormones as regulators of water balance, *in:* "Plant Hormones and Their Role in Plant Growth and Development," P.J. Davies, ed., Martinus Nijhoff, Dordrecht.

Margolis, H.A., and Waring, R.H., 1986, Carbon and nitrogen allocation patterns of Douglas-fir seedlings fertilized with nitrogen in autumn, I. Overwinter metabolism, *Can. J. For. Res.* 16:897.

Marx, D., Hatch, A.B., and Mendicino, J.F., 1977, High soil fertility decreases sucrose content and susceptibility of loblolly pine roots to ectomycorrhizal infection by *Pisolithus tinctorius*, *Can. J. Bot.* 55:1569.

Maynard, B.K., and Bassuk, N.L., 1988, Etiolation and banding effects on adventitious root formation, *in:* "Adventitious Root Formation in Cuttings," T.D. Davis, B.E. Haissig, and N. Sankhla, eds., Dioscorides Press, Portland.

McKeand, S.E., and Allen, H.L., 1984, Nutritional and root development factors affecting growth of tissue culture plantlets of loblolly pine, *Physiol. Plant.* 61: 523.

Mengel, K., and Kirkby, E.A., 1982, "Principles of Plant Nutrition," Int. Potash Inst., Worblaufen-Bern.

Moe, R., and Andersen, A.S., 1988, Stock plant environment and subsequent adventitious rooting, *in:* "Adventitious Root Formation in Cuttings," T.D. Davis, B.E. Haissig, and N. Sankhla, eds., Dioscorides Press, Portland.

Mooney, H.A., 1972, The carbon balance of plants, *Ann. Rev. Ecol. Syst.* 3:315.

Mooney, H.A., and Winner, W.E., 1991, Partitioning response of plants to stress, *in:* "Response of Plants to Multiple Stresses," H.A. Mooney, W.E. Winner, E.J. Pell, and E. Chu, eds., Academic Press, Inc., San Diego.

Mor, Y., and Halevy, A.H., 1979, Translocation of ^{14}C-assimilates in roses, I. The effect of the age of the shoot and the location of the source leaf, *Physiol. Plant.* 45:177.

Morgan, D.C., Rook, D.A., Warrington, I.J., and Turnbull, H.L., 1983, Growth and development of *Pinus radiata* D. Don.: The effect of light quality, *Plant Cell Environ.* 6:691.

Naganawa, T., and Kyuma, K., 1991, Concentration-dependence of CO_2 evolution from soil in chamber with low CO_2 concentration (< 2,000 ppm), and CO_2 diffusion/sorption model in soil, *Soil Sci. Plant Nutr.* 37:381.

Nanda, K.K., and Anand, V.K., 1970, Seasonal changes in auxin effects on rooting of stem cuttings of *Populus nigra* and its relationship with mobilization of starch, *Physiol.Plant.* 23:99.

Nelson, N.D., and Isebrands, J.G., 1983, Late-season photosynthesis and photosynthate distribution in an intensively-cultured *Populus nigra* x *laurifolia* clone, *Photosynthetica* 17:537.

Nguyen, P.V., Dickmann, D.I., Pregitzer, K.S., and Hendrick, R., 1990, Late-season changes in allocation of starch and sugar to shoots, coarse roots, and fine roots in two hybrid poplar clones, *Tree Physiol.* 7:95.

Nobel, P.S., Alm, D.M., and Cavelier, J., 1992, Growth respiration, maintenance respiration and structural-carbon costs for roots of 3 desert succulents, *Funct. Ecol.* 6:79.

Okoro, O.O., and Grace, J., 1976, The physiology of rooting populus cuttings, I. Carbohydrates and photosynthesis, *Physiol. Plant.* 36:133.

Oliveira, C.M., and Priestley, C.A., 1988, Carbohydrate reserves in deciduous fruit trees, *Hortic. Rev.* 10:403.

Patrick, J.W., 1987, Are hormones involved in assimilate transport?, *in:* "Hormone Action in Plant Development, A Critical Appraisal," G.V. Hoad, J.R. Lenton, M.B. Jackson, and R.K. Atkin, eds., Butterworths, London.

Patrick, J.W. 1988, Assimilate partitioning in relation to crop productivity, *HortSci.* 23:33.

Pregitzer, K.S., Dickmann, D.I., Hendrick, R., and Nguyen, P.V., 1990, Whole-tree carbon and nitrogen partitioning in young hybrid poplars, *Tree Physiol.* 7:79.

Priestley, C.A., 1970, Carbohydrate storage and utilization, *in:* "Physiology of Tree Crops," L.C. Luckwill and C.V. Cutting, eds., Academic Press, London.

Raich, J.W., and Nadelhoffer, K.J., 1989, Belowground carbon allocation in forest ecosystems: global trends, *Ecol.* 70:1346.

Rajagopal, V., and Andersen, A.S., 1980, Water stress and root formation in pea cuttings, I. Influence of the degree and duration of water stress on stock plants grown under two levels of irradiance, *Physiol. Plant.* 48:144.

Rauscher, H.M., Isebrands, J.G., Host, G.E., Dickson, R.E., Dickmann, D.I., Crow, T.R., and Michael, D.A., 1990, ECOPHYS: an ecophysiological growth process model for juvenile poplar, *Tree Physiol.* 7:255.

Reddy, K.R., Hodges, H.F., McKinion, J.M., and Wall, G.W., 1992, Temperature effects on pima cotton growth and development, *Agron. J.* 84:237.

Reynolds, J.F., and Thornley, J.H.M., 1982, A shoot:root partitioning model, *Ann. Bot.* 49:585.

Richards, J.H., 1984, Root growth in response to defoliation in two *Agropyron* bunchgrasses: Field observations with an improved root periscope, *Oecologia* 64:21.

Rio, C. del, Rallo, L., and Caballero, J.M., 1991, Effects of carbohydrate content on the seasonal rooting of vegetative and reproductive cuttings of olive, *J. Hortic. Sci.* 66:301.

Ritchie, G.A., 1991, The commercial use of conifer rooted cuttings in forestry: a world overview, *New For.* 5:247.

Ritchie, G.A., and Long, A.J., 1986, Field performance of micropropagated Douglas fir, *N.Z. J. For. Sci.* 16:343.

Ritchie, G.A., Tanaka, Y., and Duke, S.D., 1992, Physiology and morphology of Douglas-fir rooted cuttings compared to seedlings and transplants, *Tree Physiol.* 10:179.

Robinson, J.C., and Schwabe, W.W., 1977, Studies on the regeneration of apple cultivars from root cuttings, II. Carbohydrate and auxin relations, *J. Hortic. Sci.* 52:221.

Rufty, T.W., Huber, S.C., and Volk, R.J., 1988, Alterations in leaf carbohydrate metabolism in response to nitrogen stress, *Plant Physiol.* 88:725.

Ryan, C.A., and Farmer, E.E., 1991, Oligosaccharide signals in plants: a current assessment, *Annu. Rev. Plant Physiol. Mol. Bio.* 42:651.

Satoh, M., Kriedemann, P.E., and Loveys, B.R., 1977, Changes in photosynthetic activity and related processes following decapitation in mulberry trees, *Physiol. Plant.* 41:203.

Scarascia-Mugnozza, G., 1991, "Physiological and Morphological Determinants of Yield in Intensively Cultured Poplars (*Populus* spp.)," Ph.D. thesis, Univ. of Washington, Seattle.

Schurr, U., and Jahnke, S., 1991, Effects of water stresses and rapid changes in sink water potential on phloem transport in *Ricinus*, *in:* "Recent Advances in Phloem Transport and Assimilate Compartmentation", J.L. Bonnemain, S. Delrot, W.J. Lucas, and J. Dainty, eds., Ouest Editions Presses Academiques, Nantes.

Sharp, R.E., and Davies, W.J, 1979, Solute regulation and growth by roots and shoots of water-stressed maize plants, *Planta* 147:43.

Skene, K.G.M., 1975, Cytokinin production by roots as a factor in the control of plant growth, *in:* "The Development and Function of Roots," J.G. Torrey, and D.T. Clarkson, eds., Academic Press, London.

Smalley, T.J., Dirr, M.A., Armitage, A.M., Wood, B.W., Teskey, R.O., and Severson, R.F., 1991, Photosynthesis and leaf water, carbohydrate, and hormone status during rooting of stem cuttings of *Acer rubrum*, *J. Amer. Soc. Hortic. Sci.* 11:1052.

Smit, B., Stachowiak, M., and van Volkenburgh, E., 1989, Cellular processes limiting leaf growth in plants under hypoxic root stress, *J. Exp. Bot.* 40:89.

Spence, R.D., Rykiel, E.J., Jr., and Sharpe, P.J.H., 1990, Ozone alters carbon allocation in loblolly pine: assessment with carbon-11 labeling, *Environ. Pollut.* 64:93.

Squire, R.O., Attiwill, P.M., and Neales, T.F., 1987, Effects of changes of available water and nutrients on growth, root development and water use in *Pinus radiata* seedlings, *Aust. For. Res.* 17:99.

Stettler, R.F., and Ceulemans, R., 1993, Clonal material as a focus for genetic and physiological research in forest trees, *in:* "Clonal Forestry: Genetics, Biotechnology and Application," M.R. Ahuja, and W.J. Libby, eds., Springer-Verlag, Berlin. (in press)

Stettler, R.F., Bradshaw, H.D., and Zsuffa, L., 1992, The role of genetic improvement in short rotation forestry, *in:* "Ecophysiology of Short Rotation Forest Crops," C.P. Mitchell, J.B. Ford-Robertson, T. Hinckley, and L. Sennerby-Forsse, eds. Elsevier Appl. Sci., London.

Stettler, R.F., Fenn, R.C., Heilman, P.E., and Stanton, B.J., 1988, *Populus trichocarpa* x *Populus deltoides* hybrids for short-rotation culture: variation patterns and 4-year field performance, *Can. J. For. Res.* 18:745.

Stieber, J., and Beringer, H., 1984, Dynamic and structural relationships among leaves, roots, and storage tissue in the sugar beet, *Bot. Gaz.* 145:465.

Stitt, M., 1990, Fructose-2,6-bisphosphate as a regulatory molecule in plants, *Annu. Rev. Plant Physiol. Mol. Bio.* 41:153.

Stitt, M., and Quick, W.P., 1989, Photosynthetic carbon partitioning: its regulation and possibilities for manipulation, *Physiol. Plant.* 77:633.

Stitt, M., Gerhardt, R., Wilke, I., and Heldt, H.W., 1987, The contribution of fructose 2,6-bisphosphate to the regulation of sucrose synthesis during photosynthesis, *Physiol. Plant.* 69:377.

Sutton, R.F., 1980, Root system morphogenesis, *N.Z. J. For. Sci.* 10:264.

Thornley, J.H.M. 1972, A balanced quantitative model for root:shoot ratios in vegetative plants, *Ann. Bot.* 36:431.

Thornley, J.H.M., 1976, "Mathematical Models in Plant Physiology," Academic Press, London.

Timmis, R., Ritchie, G.A., and Pullman, G.S., 1992, Age- and position-of-origin and rootstock effects in Douglas-fir plantlet growth and plagiotropism, *Plant Cell Tissue Organ Cult.* 29:179.

Topa, M.A., and McLeod, K.W., 1986, Responses of *Pinus clausa*, *Pinus serotina* and *Pinus taeda* seedlings to anaerobic solution culture, I. Changes in growth and root morphology, *Physiol. Plant.* 68:523.

Tschaplinski, T.J., and Blake, T.J., 1989, Correlation between early root production, carbohydrate metabolism, and subsequent biomass production in hybrid poplar, *Can. J. Bot.* 67:2168.

van den Driessche, R., 1987, Importance of current photosynthate to new root growth in planted conifer seedlings, *Can. J. For. Res.* 17:776.

van Veen, J.A., Liljeroth, E., Lekkerkerk, L.J.A., and van de Geijn, S.C., 1991, Carbon fluxes in plant-soil systems at elevated atmospheric CO_2 levels, *Ecol. Appl.* 1:175.

Vapaavuori, E.M., Rikala, R., and Ryyppo, A., 1992, Effects of root temperature on growth and photosynthesis in conifer seedlings during shoot elongation, *Tree Physiol.* 10: 217-230.

Veierskov, B., 1988, Relations between carbohydrates and adventitious root formation, *in:* "Adventitious Root Formation in Cuttings," T.D. Davis, B.E. Haissig, and N. Sankhla, eds., Dioscorides Press, Portland.

Vieitez, E., 1974, Vegetative propagation of chestnut, *N.Z. J. For. Sci.* 4:242.

Vieitez, A.M., Ballester, A., Garcia, M.T., and Vieitez, E., 1980, Starch depletion and anatomical changes during the rooting of *Castanea sativa* Mill. cuttings, *Sci. Hortic.* 13:261.

Vogt, K.A., and Persson, H., 1991, Measuring growth and development of roots, *in:* "Techniques and

Approaches in Forest Tree Ecophysiology," J.P. Lassoie, and T.M. Hinckley, eds., CRC Press, Boca Raton.

Vogt, K.A., Vogt, D.J., Moore, E.E., and Sprugel, D.G., 1989, Methodological considerations in measuring biomass, production, respiration and nutrient resorption for tree roots in natural ecosystems. *in:* "Applications of Continuous and Steady-State Methods to Root Biology," J.G. Torrey, and L.J. Winship, eds., Kluwer Academic Pubs., Dordrecht.

Wardlaw, I.F., 1990, Tansley Review No. 27, The control of carbon partitioning in plants, *New Phytol.* 116:341.

Wareing, P.F., 1980, Root hormones and shoot growth, *in:* "Control of Shoot Growth in Trees," Proc. Joint Workshop of IUFRO Working Parties on Xylem and Shoot Growth Physiol.," Frederiction.

Watts, S., Rodriguez, J.L., Evans, S.E., and Davies, W.J., 1981, Root and shoot growth of plants treated with abscisic acid, *Ann. Bot.* 47:595.

Whisler, F.D., Acock, B., Baker, D.N., Fye, R.E., Hodges, H.F., Lambert, J.R., Lemmon, H.E., McKinion, and Reddy, V.R., 1986, Crop simulation models in agronomic systems. *Adv. Agron.* 40:141.

Wilson, J.B., 1988, A review of evidence on the control of shoot:root ratio in relation to models, *Ann. Bot.* 61:433.

Wyse, R.E., 1986, Sinks as determinants of assimilate partitioning: possible sites for regulation, *in:* "Phloem Transport," J. Cronshaw, W.J. Lucas, and R.T. Giaquinta, eds., Alan R. Liss, Inc., New York.

A HISTORICAL EVALUATION OF

ADVENTITIOUS ROOTING RESEARCH TO 1993

Bruce E. Haissig[1] and Tim D. Davis[2]

[1]USDA Forest Service
North Central Forest Experiment Station
Forestry Sciences Laboratory
P.O. Box 898
Rhinelander, WI 54501 USA

[2]Texas A&M University
Research and Extension Center
17360 Coit Road
Dallas, TX 75252-6599 USA

But my course and method, as I have often clearly stated and would wish to state again, is this—not to extract works from works or experiments from experiments (as an empiric), but from works and experiments to extract causes and axioms, and again from those causes and axioms new works and experiments, as a legitimate interpreter of nature... —Francis Bacon, 1620

INTRODUCTION

Vegetative propagation of plants by rooting of cuttings (cuttage) was successfully used hundreds of years before there was any study, much less understanding, of the underlying biological processes. For some species, cuttage was old practice even in antiquity, as evidenced in the writings of Aristotle (384-322 B.C.), Theophrastus (371-287 B.C.) and Pliny the Elder (23-79 A.D.). But cuttage was never successful enough to fulfill all then-current public and commercial demands and it still is not [e.g, see chapter by Howard in this volume]. In addition, organ formation has long been a study area within plant morphogenesis (Went and Thimann, 1937), which has made adventitious rooting of academic botanical interest. Hence research on the fundamental biology of adventitious rooting began and continues.

In this chapter we will provide a brief interpretive history of research that has sought knowledge about what causes plants and their parts to *sometimes* initiate adventitious roots and about what *sometimes* prevents such rooting. Hence, our primary subject will be the control of adventitious rooting. Within the biology of adventitious rooting, control is only one of many worthy study subjects, but its study has tended to dominate because control governs initiation; and, initiation necessarily precedes study of adventitious roots and their biological significances [see chapter by Dickmann and Hendrick, and by Friend and coworkers in this volume]. Even for writers with more catholic intentions than ours it would be difficult to treat the whole of the biology of adventitious rooting because its natural history is so incompletely described. We lack cohesive knowledge of principal facts and characteristics of adventitious rooting and roots; therefore, there are large gaps in our information and knowledge of origins, evolution, ecological interactions of all sorts, etc., etc. We know almost nothing, for example, about the roles of adventitious rooting and roots in the structure and function of ecosystems of various sorts. As a result of our lacks of information and knowledge, and the proper assembly of information that is available, experimentation about the biology of adventitious rooting has had limited bases upon which to build and progress.

Our presentation will mainly concern the last 125 years. We will summarize and evaluate observations and experiments before 1900, because they resulted in philosophical bases that influenced research about the control of adventitious rooting throughout this century—to a greater extent than is commonly realized. Many of the earlier experiments and observations that we will recount dealt with, sometimes simultaneously, animals and plants. Only as the 20th century progressed did plant regeneration in general and, specifically adventitious rooting, become specialty subjects within plant science. During the past two centuries alone there have been thousands of publications that have treated some aspect of the biology of adventitious rooting; therefore, we will of necessity limit references for discussion to representative examples. In doing so, we will try to adhere to Bacon's (1620, p. 274) advice: "Never cite an author except in a matter of doubtful credit; never introduce a controversy unless in a matter of great moment...." Nonetheless, we will attempt to faithfully depict this era of research and to use these cameos as bases for suggesting what the next era holds. The research and opinions that we will discuss were selected as typifying an era, rather than as excellent models of research or thinking, although some are both.

For an additional understanding of various aspects of the history of adventitious rooting research and related subjects, we refer the reader to the 20th century authors Goebel (1898-1901, 1905a), Morgan (1901), Klebs (1903), Balfour (1913), Loeb (1924), Priestley (1926), Büsgen and Münch (1929), Hatton (1930), Aucher (1930), Krenke (1933), van der Lek (1924, 1934), Went and Thimann (1937), Anon. (1941), Avery and Johnson (1947), Pledge (1947), Marston (1955), Schaffalitzky de Muckadell (1959), Fernqvist (1966), Hinds (1974), Zimmerman (1976), Jackson (1986), Davis et al. (1988) and Hartmann et al. (1990).

TERMINOLOGY

Adventitious roots and their biology are difficult subjects to precisely describe and discuss because much of the related terminology is like a mirage—distinct from some locations but nonexistent or fleeting from others, which causes wonderment as to whether the subject was or was not there. The terminology is imprecise partly because it has arisen on account of needs to simply communicate about types of rooting that are ecologically, morphologically, anatomically, physiologically, etc., diverse and complex [see Barlow (1986)]. And, terminology and usages are evolving as we learn more about all types of rooting [see Barlow (1986), Groff and Kaplan (1988), and the chapter by Barlow in this volume]. However, the vagueness in terminology has tended to persist because there has not been a concerted effort by interested scientists to create and support, and continue to develop, effective terminology. We have seemed to be too engaged by the research itself, perhaps because there are no rewards, and often criticisms, for those who attempt to develop terminologies. All-in-all, literature about adventitious rooting is subject to ambiguous and differential interpretations, depending on the reader or, as in our case, the readers and writers. We cannot discuss all of the difficulties with the terminology of rooting, but will point-out a few major ones.

Plants are capable of the two so-called vegetative restitution phenomena, regeneration and reparation (Harder et al., 1965, p. 319). Reparation, such as the renewed meristematic activity of surface cells that exactly replaces a damaged root tip, is less common in the plant kingdom, especially in higher plants, than regeneration. Regeneration is the regrowth of lost or destroyed parts or organs, but not as same-number, same-location replacements for the lost organs:

> ...By the outgrowth of a resting primordium already present, or by completely adventitious and new formation. ... [due to] the awakening of a predisposition ... [and] regeneration merges so insensibly into ordinary vegetative growth [that] the necessary limitations as to the use of the term must be entirely artificial....[Respectively from Harder et al. (1965, p. 319), Goebel (1903a, p. 198) and McCallum (1905a, p. 99)].

The questions arise, then, when is adventitious rooting regeneration? when is it something else? what is that something else to be termed? and, very importantly from our viewpoint, what are adventitious roots? As the term "adventitious rooting" has been commonly used it may have included more than the "regrowth" of regeneration. For example, with regard to what qualifies as an adventitious root, Esau's (1977, p. 253) stated that:

> The term adventitious root has somewhat varied meanings. In this book it is used to broadly designate roots that arise on aerial plant parts, on underground stems, and on more or less old root parts, especially those having undergone secondary growth. Adventitious roots may develop in intact plants growing under natural conditions, or arise in connection with infections by disease agents ... or after experimental surgery or other injury. They may develop on excised plant parts or in tissue cultures. Some adventitious roots develop from preformed dormant primordia that require an additional stimulus to resume growth. Sometimes the distinction between adventitious and lateral roots is not sharp....

Esau, in other publications, has characterized adventitious roots by using different descriptions than those quoted above, but all of them contain the general same meanings about what types of roots qualify as adventitious, in her view.

Most scientists agree that the term adventitious properly applies to roots that arise de novo (meaning "post-formed") from any detached aerial part of a plant, from intact plant parts as a result of trauma or from the ontogenetically older portions of severed roots. But we lack clear definition of what might qualify in terms of type or magnitude as special root-inducing traumas, as opposed to the traumas of everyday plant life. Most scientists would probably agree that physical injury is a qualifying, root-inducing trauma, but most might not agree that sphagnum moss slowly overgrowing a lower branch of a spruce tree, to produce a rooted "ground-layer," is a qualifying trauma.

There is no general agreement about the classification of roots (or their primordia) that arise in stems of intact plants in ontogenetically—morphologically and temporally—predictable manners. In Esau's terminology such roots are adventitious, so her terminology can be considered to be "broad-sense." However, the sense of what constitutes an adventitious root can be narrowed by exclusion of the previously described roots that normally arise in intact stems of some species. Although the original authorship is uncertain, such roots have been termed "shoot-borne" and not necessarily to be considered as adventitious (see chapter by Barlow in this volume). By analogy, roots that arise in intact, ontogenetically older parts of roots might then be "root-borne" roots. The "-borne" terminology is appealing because adventitious roots are then only the roots of regeneration. But only time will tell whether the borne terminology becomes common. If so, we will still have to deal with terminology for the "odd" situation, such as when a root primordium-containing branch becomes detached and again becomes a plant by developing a root system, due to outgrowth of the preformed primordia that were contained in the originally detached branch—which would clearly be regeneration.

The previous discussion indicates that whether or not rooting qualifies as adventitious depends on where it is located in the plant and it also depends on what has happened during the life of the plant that may have changed ontogenetically expected root-root and root-shoot relations. But we lack definitions of "ontogenetically expected." For example, is it to be ontogenetically expected that lower branches of spruces, firs and many other species may ground-layer as the result of the upwelling of moss or duff? Many individuals in a population of such plant species may do so, often depending upon the site, but many more may not do so [see Barlow (1986) for other examples].

Terminology concerning the development of adventitious roots is also vague and confusing. Terms such as "rhizogenesis," and adventitious root "formation," "initiation," "induction" and "development" have been used, mostly without definition, and sometimes interchangeably within the same paper. We as writers have no choice except also to be vague, in keeping with the literature that we will report to you and discuss for you.

In this chapter, we will use the term "adventitious root," and variations thereof, in Esau's broad-sense—which is consistent with usage of the terms in most of the literature that we will discuss. We will use the term "adventitious rooting" as a generalization of all the processes of adventitious root primordial initiation, primordial development, root-stem vascularization and root elongation growth. Finally, we will attempt to be as specific as were the authors whose research we will cite.

OBSERVATIONS AND RESEARCH BEFORE 1900

Regeneration

Academic study of plant adventitious regeneration seems to have at least partly awaited the discovery that animals, too, are capable of somatic regeneration, so that regeneration became known as a universal biological phenomenon. Luckily, in 1740, Abbé Trembley observed that bisecting a fresh-water polyp resulted in the production of two polyps. Then-current dogma stated that only plants were capable of somatic regeneration. However, the locomotion and browsing behaviors of polyps suggested to Trembley that they were animals, whereupon he concluded that polyps and other animals, like plants, were capable of somatic regeneration. Confirming experiments were soon performed by Réaumur in 1742 and Bonnet in 1745 and, later and more elegantly, by Spallanzani in 1768 [see Morgan (1901)].

The relatively straight-forward organ *reparation* in animals made it difficult for developmental biologists to rationalize and accept the multifarious vegetative *regeneration* exhibited by higher and lower plants. For instance, Thomas Hunt Morgan (1901, pp. 15-16) made convoluted excuse for the regeneration capabilities of plants and the discussion of them in his book, *Regeneration*:

> In the higher plants the production of a new plant from a piece takes place in a different way from that by which in animals a new individual is formed. The piece does not complete itself at the cut ends, nor does it change its form into that of a new plant, but the leaf-buds that are present on the piece begin to develop, especially those near the distal end of the piece ... and roots appear near the basal end of the piece. The changes that take place in the piece are different from those taking place in animals, but as the principal difference is the development of the new part near the end, rather than over the end, and as in some cases the new part may even appear in new tissue that covers the end, and, further, since the process seems to include many factors that appear also in animals, we are justified, I think, in including the process in plants under the general term regeneration....

Morgan's statements, which may appear unseemly to us, were appropriate to the time, mainly as a diplomatic hedge against conflict with some of his peers. In the school of thought subscribed to by Morgan, Vöchting, Goebel and Klebs, regeneration encompassed even shoot formation from existing but previously dormant axillary buds (i.e., not only *de novo* formations) (Kupfer, 1907, p. 195). However, a second and equally formidable school (Němec, Pfeffer, Prantl, Frank) felt—at least at times—that regeneration should only apply to organs that replaced exactly in number and position the lost organs (i.e., as in animals) (Kupfer, 1907, p. 195). Both schools had critics, such as Elsie Kupfer of the New York Botanical Garden (Kupfer 1907), but we will leave their views in the past for the reader to independently extract and explore.

Morphogenesis

Morgan's statements about regeneration become more meaningful when they are evaluated from the perspective of 19th century morphogenesis research, of which regeneration research was an integral part, but only a part. During that century, adventitious organ formation was certainly well described and subjected to cursory experimentation, as we will describe below, but those pursuits were mainly related to developing an understanding of morphogenesis overall ["causal morphology" (Goebel, 1903a, p. 197)] rather than to specifically understanding regeneration of vegetative organs and, even less, to propagating plants. In the 19th century, plant morphogenesis research was defined by the four major areas of organ formation, correlation, tropisms and growth (Went and Thimann, 1937, p. 16) (Fig. 1). Morgan (1903, p. 206) described correlations as: "...Curious, connected changes in different parts of the plant...." Organ formation as a subject was important, but not singularly important. It was, for example: "...Of special interest in the study of morphogenesis, since it often involves fully developed and differentiated cells becoming once again totipotent...." (Harder et al., 1965, p. 319). The concept of totipotency was introduced in 1839 by Turpin (Goebel, 1905b).

ORGAN FORMATION	CORRELATIONS	TROPISMS	GROWTH	YEAR
BONNET (Animals)				1745
DUHAMEL	DUHAMEL			1758
		SACHS	SACHS	
VÖCHTING	VÖCHTING			
SACHS	SACHS	DARWIN		1880
BEIJERINCK	BEIJERINCK	ROTHERT		1890
KLEBS, GOEBEL		PFEFFER	PFEFFER	1900

Figure 1. Key scientists in the four apparently independent but soon to be united areas of plant physiological research. [after Went and Thimann (1937)]

Strict focus on adventitious roots, especially with a view to improve rooting for practical purposes, was more the concern of the 18th or 19th century gardener, forester or horticulturist than the botanist (and it continues so today) [see, e.g., Ott (1763), pp. 12-25]. To make excuse, 19th century botanists faced the complex task of explaining the correlated growth and development of the plant body. Botanists of that period—many of whom were, literally, born in the world's finest botanical gardens—had vast knowledge of and comprehensive interest in the plant kingdom, so manifold types of plant bodies were paramount to them, which complicated their jobs. Then, toward 1900, scientists were further pressed to explain how morphogenesis caused and/or resulted from cell, tissue and organ "polarity," a hallmark but enigmatic concept introduced into animal science in 1864 by G.J. Allman and into plant science in 1878 by Herman Vöchting. In Vöchting's case, polarity described certain seemingly immutable morphogenetic characteristics of plant parts, such as adventitious rooting from the morphologically distal (basal) portions of cuttings, regardless of their physical orientation. Hence, polarity was thought to exist at even the smallest organizational subdivisions in plants (however, it is still unexplained at any organizational level in higher plants).

Overall, then, much of what was learned about vegetative regeneration, including adventitious rooting, during the 1800s came about because newly developed organs were important indicators of the operation or non-operation of internal factors (e.g., polarity, electromotive forces, organ-forming substances, wound stimuli, traumatotropisms and "plastic", "building" or "constructive" materials) and external factors (e.g., light, gravity, water, pressure or contact) that were observed to be or hypothesized to be very influential in, and perhaps causal of, whole plant morphogenesis. Adventitious rooting was a manifestation of organ polarity and, therefore, was of fundamental botanical interest.

Organogenesis

The morphology and anatomy of adventitious rooting were thoroughly, often very accurately, described throughout the 19th century (Knight, 1809; Trécul, 1846; De Bary 1884; Vöchting, 1884; Lemaire, 1886; van Tieghem and Douliot, 1888; Corbett, 1897), so an understanding developed that two types of adventitious root primordia were possible, one type "preformed" during normal shoot development, the other type "induced" (de novo, post-formed) only after a stem or root was wholly (a cutting) or partially (a layer) severed from the stock plant, or even if stem or root was not wholly or partially severed, but rather somehow perturbed or traumatized. It was also found that cuttings of a given plant species might contain preformed primordia but also form induced primordia and that development among adventitious root primordia was somehow correlated such that growth of one primordium might hinder growth of adjacent primordia. Development of adventitious root primordia was found to be "endogenous," meaning that the histologic origins were near the vascular system, although the exact histologic origin depended on the age of the stem or root and the species [De Candolle (1825), De Bary (1884), Lemaire (1886), van Tieghem and Douliot (1888); reviewed by Girouard (1967), Haissig (1974a), and Lovell and White (1986)].

Of 19th century investigators, Vöchting, who lived from 1847 to 1917, conducted the most notable tests, made the most notable observations and drew the most notable inferences about vegetative restitution by diverse plant species, ranging from thallus reparation in liverworts (e.g., Lunularia vulgaris) to regeneration of preformed adventitious root primordia in willow stems (Salix) (Vöchting, 1878, 1884, 1892, 1906). Vöchting exhaustively experimented with and reported on growth of shoots from existing and/or adventitious buds and growth of roots from preformed and/or induced primordia, using stem, leaf and root cuttings (and numerous other types of propagules) from various herbaceous species (e.g., Begonia discolor, Heterocentron diversifolium) and woody species (e.g., *Populus dilatata, Salix viminalis*). His perspective on restitution phenomena was holistic. He wanted to understand and compare correlated restitution of all species of plants from any of their parts, and he wanted to identify whatever internal and external factors significantly conditioned restitution. Vöchting learned about adventitious root regeneration because, and probably only because, it fell within his holistic purview.

Vöchting's experiments indicated that internal factors, rather then external ones, were dominant in determining differences in organs that conditioned formation and/or development of buds and roots. His studies suggested that neither "foods" nor gravity were primary factors in regeneration, although each played a demonstrable role, and that availability of water influenced regeneration [i.e., desiccation meant death, too much water meant lack of necessary oxygen for respiration (cf. Devaux, 1899)], and that white light often inhibited rooting. Vöchting found that regeneration was polar (i.e., roots tended to form in those portions of a propagule that had been closest to the root collar, with the converse true for buds) in stems and roots but not in organs that exhibited limited growth (e.g., leaves, thorns). He also found that ontogenetically younger organ pieces better demonstrated polar regeneration, compared to relatively older pieces, and that such polarity existed equally in stem pieces that contained multiple- or single-internodes, which suggested that polarity was not organ or tissue size-related. Vöchting's greatest overall contribution was his vivid demonstration that diverse aspects of regeneration could be studied systematically by combinations of experiments using different techniques and species.

Two other "regeneration giants" of the late 19th century, Julius Sachs (the Father of Plant Physiology, according to some) and his colleague Karl Goebel also contributed significantly to thinking about vegetative regeneration, generally, and adventitious rooting, specifically. As might be expected, the thinking of the giants was close but not coincident, probably because the results of regeneration experiments could not then, as now, be unambiguously interpreted, which caused each scientist to vacillate between philosophical positions. Thus the giants bickered among themselves, often with encouragement from sideline scientists, and mostly about plant developmental questions that are as yet unanswered. Their uniform complaints were that their colleagues misinterpreted their writings or, worse yet, did not read them fully or at all. As a consequence, the defender usually performed remarkable experiments to rebuff the attacker [e.g., Vöchting (1906)].

Sachs (1880, 1882) proposed that there were special endogenous substances in plants that caused the specific formation of buds or roots or flowers. Sachs's postulate is closely related to the much earlier theories of Bonnet in 1745 [cited from Morgan (1903)] and Duhamel du Monceau in 1758. Bonnet proposed that a piece of a worm, *Lumbriculus*, regenerated a head at one end and a tail at the other because "fluids" flowed forward to activate "head-germs" and rearward to activate "tail-germs." Duhamel (1758) based his theories about organ development on a similar "two-moving-opposing-saps" theory. In Sachs's version, separation of the cutting from its root system caused a build-up of the shoot-produced, root-forming substance in the cutting, resulting in rooting by the cutting and (polarly) at the base of the cutting because of, for instance, recent or historic gravitational influences on the root-forming substance or tissues that contained it. Sachs (1882) proposed that these sometimes but not always (e.g., in tubers) polarly transported substances were present in minute quantities and that they were special, rather than, for instance, the ubiquitous carbohydrate, protein or lipid plastic or constructive or building stuffs, as they were variously termed. Goebel (1903a,c) also postulated that polar organ formation was due to formative substances but only or primarily nutritional—constructive or building—ones, although sometimes he relied on Sachs's formative substances.

The basic substances of plants—the building, plastic, constructive, food, etc., stuffs as referred to in 19th century literature—had long been of interest [(e.g., De Candolle (1832)]. Beijerinck (1886) had previously observed that leaves, apparently as sources of nutrition, were important support organs for adventitious rooting. In one of Goebel's views—his major one—polar organ formation was explained in terms of, in our contemporary jargon, "source-sink" relations [see chapter by Barlow, by Dickmann and Hendrick, and by Friend and coworkers in this volume]. Supposedly, damaging the whole plant initially changed source-sink relations; then the relations were further changed by the newly developing organs. It is hard to understand, based on the difficult-to-translate German-language scientific literature of the late 19th and early 20th centuries, what various scientists thought constituted the non-nutrient, non-constructive stuffs. Sachs (1882) suggested that these special substances were passed from sperm to egg during fertilization, as the "*Nuclein*" "fertilization stuff," and that they might be enzymes ("*die Fermente*"). Almost coincidentally, Kühne (1878) had coined the term "enzyme" for the type of organic catalyst involved in fermentations that was first clearly recognized by Payen and Persoz in 1833. Slightly later, Fitting (1909, 1910), who was one of the first, if not the first, to test the biological activity of higher plant extracts, introduced the concept of "hormone(s)" (as localized stimulators) into plant science. Then, too, vitamins were newly discovered. So the non-food stuffs were variously interpreted to be enzyme-, hormone-, vitamin-like or something(s) else.

"Stuff," which originated from Old French, had a much more precise meaning when used in the older scientific literature than it has today in American English. The German-language "*der Stoff*" and "*die Stoffe*" referred to, and still refer to, fundamental, concrete, distinct, identified or identifiable material or materials, respectively. Hence, "*der Stoffe-wechsel*" is metabolism. As the reader will learn, many stuffs that have been considered in rooting research did not actually qualify in a strict sense.

At this point the reader may think, erroneously, that late 19th century plant physiologists were ill-prepared for the complex problems of rooting because "we know so much more nowadays." However, in 1887, Sachs published the second edition of his plant physiology lectures (884 pages, 391 figures), as an abridgement of the first edition that was requested by the publisher to reduce costs. Indeed, Sachs wanted a larger and better illustrated book, supported by more than his personal research. Nonetheless, his seven subject categories, totaling 43 lectures, ranging from organ morphology to enzymes, are comprehensive even by today's standards. Sachs lacked some of our information or more nearly correct information on specific subjects—advances in knowledge in the 20th century have certainly honed intuitions about plant biochemistry and physiology, and genetics was introduced—but Sachs's teachings made up for any informational deficiencies by instilling scientific zeal for openly exploring life's mysteries in the plant kingdom, of which adventitious rooting was and is one [see also De Candolle (1832) and earlier]. In general, plant physiologists of the period 1850-1900 had excellent fundamental understandings of biology, physics, chemistry and mathematics—because knowledge in each of these disciplines had increased markedly between 1700 and 1850, in some aspects to nearly modern levels [see Pledge (1947)]. And, those same physiologists had access to powerful technologies, many of them still in use, to aid observation and experimentation [see Pledge (1947)].

In summary, at entry into the 20th century, vegetative regeneration was thought to be caused by internal "forces" and "stuffs," with the forces emanating from polarity and the stuffs embodied in nutrients and/or more specific, postulated growth substances or enzymes, which perhaps caused polarity. Overall, scientists of the 19th century bequeathed to those of the 20th century what came to be known as The Hypothesis of Formative Stuffs (nicknamed The Stuff-Hypothesis), including the inherent polarizing and activating forces [see Morgan (1903)]. Vöchting's discerning mental and experimental approaches accompanied some scientists into the 20th century. Thus the stuff-hypothesis was severely tested, at least to the extent possible for any particular day, to this day, as we will recount in subsequent sections.

EXPERIMENTAL PROTOCOLS

Before we discuss 20th century rooting research *per se*, we will briefly address how relevant experiments have been conducted and interpreted, and how results may have been thereby influenced. This section necessarily intrudes into our historical progression because the reliability of results, and the scope and fidelity of inferences obtained in rooting research, as in all research, depends upon how tests were conducted and what was evaluated. To quote Bacon (1620, p. 282): "...In any new and more subtle experiment the manner in which the experiment was conducted should be added, that men may be free to judge for themselves whether the information obtained from that experiment be trustworthy or fallacious, and also that men's industry may be aroused to discover, if possible, methods more exact...."

Materials

As is probably clear to the reader, there have been abundant plant species for research subjects, at least since the times of the cryptogam- and gymnosperm-eating early dinosaurs and the angiosperm-eating later dinosaurs. Adventitious rooting was enough of a curiosity that rooting tests were commonly conducted to compare performances between species, often disparate species, so knowledge about general rooting characteristics was readily available when botanical experimentation on adventitious rooting began. In one remarkable survey, mainly with leaf cuttings but also with other plant parts, Hagemann (1932) studied adventitious shoot and root formation by 1204 species from 135 families, 711 of the species for the first time. His study included 21 gymnospermous, 141 monocotyledous and 1042 dicotyledonous species. In general, Hagemann found that leaves of the dicotyledonous species rooted easier than the monocotyledonous ones. He also rooted roots, internodes, inflorescences, fruits, embryonic leaves, scale leaves, rhizomes, sepals, carpels and tendrils; whereas, for example, stamens did not root. Finally, Hagemann observed that adventitious buds and roots did not necessarily form simultaneously or even equally well, if at all, in the same species or plant part. As in Hagemann's test, dicotyledonous species have predominanted in most investigations of rooting fundamentals. Interestingly, however, *Nicotiana* spp., which have been ubiquitous test materials elsewhere in plant physiological research, including even the earliest research with plant growth regulators, did not receive a high grade for rooting research (Hitchcock, 1935a) and they have not been much used in *ex vitro* rooting tests. Indeed, the first commonly used rooting bioassay employed *Pisum sativum* [Went (1934a); see also Went (1934b)], as discussed below.

Basically, the chosen test species were handy and easily used, and consistently ready-rooters. F.W. Went (1934a), for example, reported the approximate numbers of roots developed per untreated cutting for some commonly used species: *Helianthus annuus*, 50; *Impatiens balsamina*, 30; *Phaseolus radiatus*, 10. Implicit in the foregoing is the assumption that the fundamentals of adventitious rooting throughout the life of the plant are about the same in all species, especially between monocotyledons, dicotyledons and gymnosperms, so that if rooting of pea is effectively studied, rooting in a bristle cone pine or an oil palm will be better understood. In this regard, study of easily rooted plants has provided information that most probably also pertains to adventitious rooting of woody plants during their early weeks, months or years of life, before their propagules have lost the predisposition to initiate adventitious roots, as usually happens (see section below on Chronological and Physiological Aging). However, we also need information about the effects of aging and maturation on the predisposition for adventitious rooting by the usually recalcitrant, long-lived perennial species—and those needs are not new.

It has become increasingly apparent [e.g., Haissig et al. (1992)] that available genotypes for species that have been commonly used for rooting research do not allow for discerning, comparative biochemical and physiological investigations; for example, between "good" and "poor" rooters within a species. Few, if any, suitable rooting mutants have been found, nor have suitable pedigreed lines been produced and used for selection of differently rooting types. Genetic comparisons within a species cannot be very discerning unless isogenic lines are available because hypothesized effects (e.g., of a treatment) are confounded by indirect and correlated effects, with reference to adventitious rooting. The foregoing indicates that so-called genetic approaches within a species provide meaningful inferences only to the extent that genetically suitable experimental materials are available (see chapters by Riemenschneder and by Ernst in this volume). Genetic approaches between species will seemingly advance only once genetic approaches within species have been better mastered. But, overall, genetic approaches should prove useful because substantial evidence for a genetic basis of adventitious rooting has been obtained [see Haissig and Riemenschneider (1988)], as we will discuss later. Indeed, even many investigators of the late 19th and early 20th centuries alluded to a genetic basis for adventitious rooting. These early investigators based their assumptions on voluminous visual observations, which they no doubt coupled to the "particulate" genetic theories that were developing during the 19th century [see Pledge (1947)].

As previously stated, horticulturists, foresters and botanists had distinguished between induced and preformed root primordia at least as early as the 19th century [e.g., Knight (1809)]. However, van der Lek (1934) was the first to clearly show that results obtained from experimentation might depend, at least quantitatively, upon the type of root primordium being considered. So little physiological research has been done with preformed root primordia that it is uncertain whether the control of their initiation is the same or nearly the same as that for induced primordia. In addition, it even is unclear for induced primordia whether the control is the same or nearly the same when they form in already existing tissues (e.g., adjacent to vascular bundles) compared to newly developing tissues (e.g., callus). Similarly, we do not know whether fundamental studies of lateral rooting or roots also tell us about adventitious rooting. And, the influences of preformed primordia on the subsequent formation and development of induced primordia, and the converse, is unclear, although such interactions were reported long ago [e.g., Loeb (1919a,b)]. Many species that are commonly used in rooting research contain at least some preformed primordia, which many investigators have not realized.

Methods with Cuttings

Frits W. Went, son of a famous botanist and, appropriately, born in a botanical garden in the Netherlands (Went, 1974), identified the need for testing and assessing rooting under reproducible conditions, and for setting reasonably high standards for such tests and assessments. As a result, he developed the now-classic pea test (Went, 1934a). Went's idea for a rooting bioassay, to be used in studying the hypothetical rhizocaline, logically followed his development of the so-called *Avena* (Went, 1928) and split-pea tests (Went, 1934b) to "quantitatively" determine concentrations of what later proved to be the auxin, indole-3-acetic acid (IAA). (We will discuss both rhizocaline and auxin in later sections.) Went had previously conducted tests with defoliated cuttings of the woody shrub *Acalypha* (which develops many preformed root primordia), to test root-promoting effects of extracts from leaves and rice polishings. But he concluded that *Acalypha* was unsuitable for "quantitative" experimentation because rooting was too variable among cuttings due to their different original positions in stock plants, which were themselves subject to the environmental vagaries of the outdoors. Went would not likely have chosen a woody species if he had read Curtis's (1918, p. 77) earlier lament about using woody cuttings to study rooting fundamentals:

> *It was important that cuttings should be used which would root in as short a time as possible. The field of research is new, and if one were forced to wait two months*

*or more for results on preliminary tests the experiments would be long drawn out
or it would be too late to obtain cuttings for a second set. The writer experienced
this trouble for three consecutive years, even with comparatively quick-rooting*
Ligustrum....

After *Acalypha*, Went tried several herbaceous species, deciding upon *Pisum
sativum*, whose untreated cuttings produced the fewest roots (1.43 per cutting, based on
1000 observations), which he felt was desirable as a low baseline in the comparison of
treatments, as embodied in his standards (Went, 1934a, pp. 446-447):

> *A good test method must be simple, reproducible and must give mathematically
> significant results ... As a criterion for the suitability of a certain treatment in
> regard to the test method the following rules were adopted: 1. The untreated
> controls must have as few roots as possible. 2. Plants treated with a given amount
> of rhizocaline must have as many roots as possible. 3. The range over which an
> increase in the amount of rhizocaline gives an increased number of roots must be
> as large as possible. The described method is a compromise between all these
> different factors, involving in some respects a sacrifice of accuracy in favour of
> simplicity....*

Went's pea test incorporated standardized sample sizes, number of replications,
environmental conditions (temperature, light quantity and quality, humidity), genotype and
origin of seed, cutting size range, statistical analysis, etc. He recognized and attempted to
counter possible effects of contaminant chemicals by using oxidant-free glassware, ether-
purified sucrose, and of microbes by using surface sterilization, solution sterilization and
$KMnO_4$ treatment of cuttings. Possible influences of test solution pH were examined, and
found to be unimportant. However, Went's test included several possible sources of error,
such as: presence of a carbon source tended to promote microbial contamination, cuttings
were exposed to aqueous $KMnO_4$ before treatment, mineral-containing tap water was the
solute for treatment chemicals, lower internodes contained preformed root primordia
(Schmidt, 1956) and leaf-scales remained on cuttings. $KMnO_4$, which may influence
respiration rates, reduced rooting of the controls (cf. Curtis, 1918) and leaf-scales promoted
rooting. Either or both $KMnO_4$ and chemicals in tap water might have interacted with test
chemicals (cf. Curtis, 1918). Only macroscopic evidence of rooting was gathered, so
treatment results were generalized across phases of rooting, as we will discuss later. Finally,
selection for poor-rooting controls may have biased the test. But, as stated above, Went
admitted that the test was a simplicity favoring compromise, designed, we surmise, to
provide straight-forward data by reducing potential sources of variation, because Went
(1974), on reviewing his career, admitted that he did not believe in statistical inferences.
Subsequent macroscopic rooting bioassays that were developed with herbaceous [see
reviews by Fernqvist (1966) and Heuser (1988)] and, less so, with woody species were
often simpler than the pea test but were less, or at least no more, rigorous.

Methods with Tissue Cultures

Establishment of long-term tissues cultures, which was reported independently in the
1930s by Roger Gautheret, Pierre Nobécourt and Philip White, provided new perspectives
on improving plant propagation [see Haissig (1965)]. Tissue cultures might be used directly
for propagation (now termed "micropropagation") or such cultures might be used in
physiological investigations of rooting [see Haissig (1965)]. Tissue culture as a method
offered the potentials for achieving physically and chemically controlled environments and,
therefore, experimentation that would be otherwise difficult to conduct at reasonable cost.
Then, too, tissue culture was developing as a useful technology during the very period when
it was becoming eminently clear that little progress was being made in understanding the
control of adventitious rooting in general and especially for recalcitrant rooters, which are
almost invariably the most aboriginal woody species, specifically gymnosperms and

determinate-growth angiosperms, or the tree-like monocotyledons. Simplistically but accurately stated, tissue culture offered the possibility of using factorial experimental designs to "throw the chemical cabinet" at rooting problems just when plant growth regulator research was blossoming, and when cytokinin/auxin ratios offered the promise of controlling organogenesis [e.g., Geissbühler and Skoog (1957) and Skoog and Miller (1957)]. However, tissue culture of woody plants and monocotyledons has often proven difficult because, in the absence of adequate physiological knowledge, we have not been lucky in overcoming obvious deficiencies in technique, as evidenced by repeated failures; and, plant growth regulators have not compensated for those deficiencies (see later sections). Hence there has not been much progress in using tissue culture to help us understand the control of adventitious rooting compared to, for instance, somatic embryogenesis [see Haissig et al. (1992)], lateral rooting and root growth. But *in vitro* cultures may help us to learn about predisposition or competence for rooting (see chapter by Mohnen in this volume).

Tissue culture has, however, provided a stable environmental platform for developing rooting bioassays, including some with woody species. As one excellent example, Wesley Hackett (1970) used stem apices from the juvenile and mature phases of *Hedera helix*, which were originated from the same plant, in various tests. In one test, he determined the rooting response of juvenile apices to chromatographically fractionated extracts of tissues from juvenile- and adult-phase plants. (We will specifically discuss phase-change in a later section.) Hackett tested the hypothesis that good rooting of the juvenile-phase, compared to the mature, was related to a greater content of promotive substances, as previously postulated by Charles Hess [see Hackett (1970)]. Hess's rooting bioassay was based on application of *Hedera* extracts to mung bean. Hackett's results refuted the hypothesis and in doing so pointed-out: "...A danger in using easy-to-root tissue as a rooting assay in studies of root initiation in difficult-to-root plants...." (Hackett, 1970). Indeed, use of other than the species whose physiology was being researched has been and continues to be a potential pitfall of most rooting bioassays, going back at least as far as Went's *Acalypha* test [see also Linser (1948)].

Assessments

In early tests, the number of roots that formed per propagule was determined; for example, roots per cutting or roots per rooted cutting, which may lead to different conclusions. F.W. Went (1929) is sometimes credited as being the first to assess adventitious rooting by measuring root elongation. However, as might be expected, Vöchting (1906) had much earlier recognized that counting roots and measuring root lengths provided different inferences about rooting, although he opted for counting because it was easier and supplied the desired information (Vöchting, 1906). Vöchting (1906) also recognized that his earlier research would have benefited if he had used "statistical" as well as visual observational approaches. He (1906, p. 104) clearly stated the potential pitfall of visual observations alone, as in his earlier research: "The physical eye sees objects as the mind's eye previously imagined...." Hence, Vöchting (1906) standardized on 10 cuttings per treatment and counted the number of roots produced in each of three sections per cutting, with the results expressed as means (i.e., the statistic). He noted that counting the number of roots in each of four rather than three sections per cutting would have yielded more illustrative data, by increasing the means for the better-rooting basal sections, compared to the other sections. He also would have liked to use more than 10 cuttings per treatment to reduce variability in the data. Knowingly or not, Curtis (1918) used Vöchting's protocols.

Assessments have tended to be numerically rather than functionally oriented; initial profuseness of rooting has most often been evaluated—because more roots have been assumed to be somehow "better" than fewer roots—but we have little information, much less knowledge, about how initial root mass may relate to subsequent plant vigor. And, we do not know how chemical treatments that may increase initial root numbers and/or lengths may positively or negatively influence the functioning of roots and, again, plant vigor.

Like us and many of his other successors in rooting research, Vöchting found that for one reason or another it was almost impossible to conduct rooting experiments with an optimal experimental design. Unfortunately, many of Vöchting's successors have not clearly and uniformly understood that proper use of biometrics would help to obtain or more nearly obtain optimal experimental designs. Only recently has a physiological hypothesis about rooting been clearly framed and tested in biometric terms (Haissig and Riemenschneider, 1992), and that research further demonstrated that assumptions about so-called "scale" may greatly influence the interpretation of experimental results, to the extent of causing rejection of a likely true hypothesis or acceptance of a likely false one. In this regard, the investigator must consider the assessment variables (e.g., proportion rooted, number of roots per cutting, dry weight, etc.) but, additionally and importantly, the fundamental relations (e.g., linear, curvilinear, logrithmic, etc.) between those variables and the hypothesized control.

In an integration of Vöchting's quantitative ideas, and predating Went's research, dry and/or fresh weights of new roots have also been used as bases for assessment [e.g., Loeb (1919a,b)]. But, especially given the high state of present day technology, no assessment method has been developed that can be considered to even border on the elegant, by which we mean a method that would give one or a set of numerical observations or even visual observations for any given treatment that accurately estimated how rooting had progressed up to any chosen point during an experiment (e.g., preprimordium, putative primordium, identifiable primordium, growing root, functional root).

As previously discussed, studies of root initiation (de novo formation) and of root protrusion (i.e., from preformed primordia) might lead to different conclusions about possible causal (controlling) internal and external factors. Macroscopic rooting bioassays have often been used to provide data that supposedly related only or mostly to root initiation, but only microscopically based assays can actually tell about the earliest stages of rooting. There is no clear point in history when root initiation and root elongation growth, however either was then defined, were treated separately as experimental subjects. It did not happen on one particular day or in one particular year; indeed, it is still happening. Likely, however, the distinction began to arise after the discovery that, whereas auxin treatment would promote the formation of root primordia, it might also inhibit their growth into roots, especially if the treatment dosage was physiologically high. The observation that auxin treatment would promote and inhibit "rooting" would have favored dividing adventitious rooting into "initiation" and "growth," although arbitrarily because the processes obviously form anatomical and physiological continua until death of the plant. Of course, a third phase, "pre-initiation" had also long been recognized, at least as a mental baseline.

There have been further and anatomically justified attempts to further subdivide the rooting processes [see Lovell and White (1986), pp. 114-116, and chapter by Blakesley in this volume]. There have also been attempts to describe rooting as consisting of physiological phases, linked to at least the anatomical phases of preinitiation, initiation and growth. Went (1939) concluded that adventitious rooting of Pisum sativum consisted of an initial phase, which could be satisfied by several chemical substances (including auxin), during which the hypothetical rhizocaline was redistributed to the cutting's base. Then there was an auxin-requiring phase during which occurred: "...The activation of the accumulated rhizocaline..." (Went 1939, p. 28). Similarly, in experiments with epicotyl cuttings of Azukia angularis (Phaseolus angularis), Mitsuhashi and coworkers (1969a,b) observed that there was an early, IAA-insensitive phase of rooting and a later, IAA-sensitive phase. Three phases were subsequently defined for Azukia by Shibaoka (1971), as Ruge (1957, 1960) had previously done for *Tradescantia*. Gaspar and coworkers (1977) found evidence which suggested to them that adventitious rooting consisted of an "inductive" and an "initiative" phase. However, except for the phases of preinitiation, initiation and growth, interested scientists have not achieved any consensus about further, and perhaps more meaningful, subdivisions of adventitious rooting into anatomical or physiological or anatomophysiological phases. Achieving consensus on definition of phases, however defined, has been impossible because no standardized terminology or standardization in uses of terminology have arisen for the developmental anatomy and physiology of adventitious rooting, as previously discussed.

Two basic methods, whole stem "staining" and tissue sectioning, have been used to specifically study adventitious root initiation. Succulent cuttings, such as bean cuttings, have been stained to show, perhaps using low power magnification, developing primordia [e.g., Fernqvist (1966)]. Sectioning has mostly been limited to herbaceous stems that form readily located ranks of many primordia or to woody species that form preformed primordia, which are also readily located but are often few in number [e.g., Carlson (1938, 1950) and Haissig (1970)]. Neither whole organ staining nor tissue sectioning have been used much to assess the progress of rooting in biochemical and physiological studies, with the consequence that: "In many studies it is not clear which developmental stage of 'rooting' was sampled and, therefore, what aspect of anatomical differentiation was chemically described...." (Haissig, 1986, p. 143).

Early in the 20th century anatomical and chemical research concerning adventitious rooting were sometimes combined, for reasons that will become clear to the reader in the next section. Successful attempts to study adventitious root initiation, for example, by using histochemical "staining" (Smith, 1928; Carlson, 1929; Molnar and LaCroix, 1972a,b) or microautoradiography (Haissig 1970, 1971; Tripepi et al. 1983) or macroautoradiography (Breen and Muraoka, 1973, 1974; Altman and Wareing, 1975) to locate specific metabolites, applied auxin, or anabolic or catabolic activities, or to assess structural development, have not been much emulated. Similarly, the newer methods of immunocyto-localization are yet to be employed in adventitious rooting research. The previously named techniques have probably not been used much because they are tedious. The consequence of not using tedious technologies has been lesser than might have been gains in specific knowledge about the fundamentals of root primordium initiation, which the reader should keep in mind as we progress in our presentation into the 20th century [see Haissig (1986), Davis and Haissig (1990), and chapter by Hand in this volume].

THE HYPOTHESIS OF FORMATIVE STUFFS

In 1902, a fourth regeneration giant, T.H. Morgan, presented an invited lecture, *The Hypothesis of Formative Stuffs*, before the Botanical Society of America (Morgan, 1903). In that lecture Morgan (1903) summarized the stuff-hypothesis, at least as used by Goebel (1898-1901) in *Organographie der Pflanzen*, and resoundingly criticized it as encompassing too much in terms of explaining polarized regeneration in animals and plants, and encompassing too little in terms of unambiguously defining "regeneration-triggering" components. In this instance, Morgan was apparently directing himself at the hypothesis, not at Goebel, who had clearly stated (1898-1901, pp. 173-174) that biological processes should be "causally understandable" through research. Whatever the intent, Morgan (1901; 1903, pp. 212-213) concluded, based on his and other scientists' regeneration experiments with the tubularian hydroid, *Tubularia mesembryanthemum*, that the stuff-hypothesis was erroneous and deficient, and suggested that:

> ...*Some physical factor enters into the problem ... in the form of different tensions in the living tissues. ...* [perhaps the] *expression of osmotic changes, which are themselves in turn the result of the presence of certain chemical substances present, or being produced in the organism. ...* [or maybe, as just then suggested by

Matthews] *an expression of the electrical conditions of the piece; in short, that 'polarity' is an electrical phenomenon; possibly, I may add, due to the movement of ions....*

Thus, although critical of the stuff-hypothesis, Morgan did not defeat it: "Morgan has strongly objected to this theory, but his evidence against it does not seem to me to be necessarily fatal...." (McCallum, 1905b, p. 244). Indeed, Morgan expanded the hypothesis's complexity and, hence, uncertainty, which was soon pointed-out by Goebel [see Morgan (1903)]. In retrospect, Morgan's attempt to defeat the stuff-hypothesis actually enshrined it because his 1903 paper was so widely read and discussed and cited.

Like Morgan, it is doubtful that other plant or animal developmental biologists fully accepted the stuff-hypothesis, as evidenced by the repeated attempts after Morgan's to debunk its intricate and manifold assumptions. Debunking would have allowed development of new, cogent and researchable theories, meaning theories that were rationally partitioned by testable hypotheses. Two such attempts are discussed in some detail below.

McCallum (1905a, pp. 98, 100, 102, 113), in testing the stuff-hypothesis, identified three primary classes of plant regeneration phenomena, summarized the then-current knowledge, presented his experimental approach and explained his bases for evaluating observations:

We have in these cases at least three seemingly diverse phenomena: (1) the part removed is entirely restored by growth of the cells at the cut surface [e.g., thallus of *Marchantia* and *Lumularia*]; *(2) there is no growth of embryonic tissue at the wounded surface, but at a greater or less distance from it the organization of entirely new primordia which develop organs that replace those removed* [e.g., adventitious shoots from roots of *Taraxacum* and leaves of *Begonia*]; *(3) the organ removed, e.g., the shoot, is restored by the development of already existing dormant buds* [e.g., preformed root primordia and axillary buds of *Salix*]. *Between these no hard and fast lines can be drawn, for they all exhibit intergradations ... We know little enough of the external factors concerned and almost nothing at all definite about the internal ones. ... Various theories have been proposed, but none have yet been supported by experimental evidence. ... In conducting some investigations in this subject it soon became evident* [to me] *that the best method of attack would be to take all the possible factors and work on them separately, subjecting each, one at a time, to a more exact physiological analysis. ... The various theories and possibilities suggested fall naturally into a few general classes: (1) wound stimulus* [e.g., Goebel (1903a,b,c), Wiesner, cited from McCallum (1905b)]; *(2) disturbance in nutritive relations* [e.g., Goebel (1902), Klebs (1903)]; *(3) changes in water content* [e.g., Klebs (1903)]; *(4) accumulation at certain places of definite formative substances* [Sachs (1887) but sometimes also Goebel]; *(5) correlation* [various scientists but see Noll (1900) and Driesch (1901)]; *(6) relative age and degree of maturity of the different parts of a member* [e.g., Vöchting (1878)]; *and (7) growth tensions* [e.g., Morgan (1903)]. *... Before we can attribute a result to any factor it is necessary to show (1) that that factor is always present when the result occurs, and (2) that when it is absent the result will not occur....*

McCallum (1905a,b) tested various parts of the stuff-hypothesis, as described above, and even in expanded context, with reference to shoot formation from axillary buds and rooting from preformed and induced primordia. In doing so, he conducted 47 controlled, replicated, often intricate experiments (e.g., hydrogen gas to cause anoxia in stem terminals; an ether solution as an "anaesthetic" for "girdling" stems), mostly with *Phaseolus* but also with a host of other herbaceous and woody species for verification. He concluded that organ forming stimuli (assuming separate stimuli for shoot and root formation) did not result from: "food" from cotyledons or the young shoot, moisture or water content of cells or transpiration rate, wounding, irradiance or photosynthesis, aeration or respiration (maybe), gravity or specific formative substances. McCallum did not specifically research Morgan's growth-tension hypothesis, probably because he categorized it with other hypotheses, such as Noll's (1900) "body-forming stimulus," that could not be demonstrated or refuted.

Unfortunately, McCallum did not come to any definite conclusions about what did provoke shoot development or adventitious rooting. In various places in his second (1905b) paper he suggested that organ regeneration occurred when correlation was upset by something, like removal (in the extreme case) of the shoot apex or roots, which prevented or slowed "usual" growth processes. In turn, such interference with growth processes caused "new conditions" or positive "influences," or eliminated negative influences. The conditions and/or influences were local but also at times (and incongruously) polarly translocated in the vascular system, and they specifically promoted either shoot or root development. McCallum wanted to divorce himself—with a scientific basis, of course—from all aspects of the stuff-hypothesis, but he did not succeed in all cases (e.g., polarity-based, specific stimuli and correlations); nor did McCallum define any clear, defensible, alternative theory. Time and again (1905a,b) he expressed the view that removal of growing points or slowing their growth were the only two consistent conditions that provoked organogenesis. Perhaps not coincidentally, it would later be observed that auxin treatment of cuttings to promote rooting was also likely to inhibit shoot elongation, as we will discuss in a subsequent section.

Elsie Kupfer, who roundly criticized McCallum's research and conclusions (Kupfer, 1907, pp. 234-235), also tested the stuff-hypothesis. She conducted 55 experiments with species ranging in morphological complexity from an alga, *Penicullus capitatus*, to a pine tree, *Pinus laricio*. Her experiments primarily concerned *de novo* (i.e., adventitious) formation of buds and roots on roots, stems, leaves, inflorescences and fruits. In view of her results and her interpretations of the results of others, Kupfer (p. 236) concluded that "The only recourse, therefore, seems to be to go back to the old formative stuffs idea in perhaps a somewhat altered form...." But Kupfer did not alter its basic form; she used somewhat newer words for the old stuff. Kupfer proposed [pp. 237-238; cf. Sachs (1882)], without any direct evidence, that there was a localized (e.g., nodes, stem growing points) shoot-forming enzyme and a more generally distributed root-forming enzyme, perhaps existing as "pro-enzymes" or "zymogens" that were activated by, for example, high relative humidity, undefined "pressure relations," or aggregated food (i.e., by the stuffs and forces inherent in The Hypothesis):

> The root-forming enzyme which is ... present in nearly all cambial cells is normally
> prevented from acting by the fact that the growth is hindered by the surrounding
> cells and perhaps also by the absence of a static food material upon which to act.
> When a cut is made, however, the stimulus of the wound rouses the enzyme into
> activity, the food aggregating here supplies the material, and the organ, no longer
> hindered by encompassing tissue, makes its appearance. ... The presence of a
> substance called a kinase is known to enhance the action of an enzyme, and it may
> be that such a substance, in addition to the enzyme is present at the nodes and
> vegetative points ... Cases are [also] known where, in company with an enzyme, a
> compound of opposing nature, termed an 'anti-enzyme' exists, and according as one
> or the other gains the ascendant the particular effect of the enzyme is manifest or
> obscured....

McCallum's and Kupfer's work, and similar work by others in the late 19th and early 20th centuries, has been repeated in one way or another, again and again, throughout the 20th century, to this day, for obvious reasons: Like McCallum and Kupfer, other scientists have not been able to definitively experiment on *all* aspects of the stuff-hypothesis with *diverse* species of plants, such that interpretations of results provided clear guidance for further experimentation. The resulting inferences provided too narrow a scope for validating or rejecting old hypotheses and constructing new hypotheses that could be embodied in one or more scientifically helpful theories. As a major complication, exemplified by Morgan's and Kupfer's previously stated conclusions, the stuff-hypothesis gathered new components as biochemical, biophysical and physiological knowledge increased during the 20th century. And, when damaged or stressed, the stuff-hypothesis itself underwent regeneration, which repaired and expanded its body, and prolonged its life. It is still alive and robust today in many laboratories, including ours.

POLARITY

After Vöchting's introduction of the concept of polarity into plant science, polarity has been variously defined, but Bloch's (1943a,b, 1965) definitions have been commonly used:

> ...Any internal asymmetrical state or any situation where two ends or surfaces in a living system are different. ... [1943a, p. 297] [or] any internal asymmetrical axiate condition or any situation where two ends or surfaces in a living system react unequally or are substantially different. ... [1943b, p. 262] [or, in a broad sense] any internal assymmetrical state within a living system as expressed in differentness between two ends or two surfaces, and also between the ends or surfaces and the central region. [Hence] Polarity is part of the general process of orderly differentiation and one of its most fundamental, early expressions.... [1965, p. 235]

As previously described, polarity was a force that superimposed itself on the stuff-hypothesis as a vexing complication. As with the stuff-hypothesis, there were repeated attempts by notable botanists, such as Strasburger and Kny and Klebs, to challenge polarity theories by, for example, showing that polarity actually did not exist or was "reversible." For example, in 1903, Georg Klebs experimented with some of the willow species used by Vöchting (*Salix laurina, S. pentandra*) and confirmed Vöchting's conclusions about polarity [see Vöchting (1906)]. However, Klebs's additional tests with *Salix alba vitellina pendula*, a species that Vöchting had not previously used, led Klebs to conclude that polarity was "easily and certainly removable." Klebs's conclusions so infuriated Vöchting that he retaliated (1906) with quantitative experiments that refuted Klebs's assertions about reversing polarity with *Salix alba vitellina pendula*.

Goebel (1905b) stated that polarity could be suppressed but not removed. He reasoned that regenerants should not be polarized if they arose from organs that had been depolarized, but the existence of non-polarized regenerants was inconsistent with fact. Polarity was also studied and discussed by other investigators during the early 20th century [Winkler (1900), Küster (1903), Loeb (1917a,b; 1919a,b), Massart (1918), van der Lek (1924), De Haan (1936); see also references in Bloch (1943a, 1965)]. Of the classical treatments that were tried to modify polarized responses, changing correlated structural development (e.g., slowing growth, wounding), changing water relations, varying irradiance, and/or changing contact pressures, with reference to the organ or whole plant, have most often and consistently modified polarized rooting and budding, but without known reasons. And, there have been many exceptions.

By way of explaining polarity in higher plants, F.W. Went (1932) proposed a theory based on his research with movement of supposedly nontoxic dyes in etiolated hypocotyls of *Impatiens balsamina*, which demonstrated pronounced polarities. He concluded that differences in electrical potentials within the plant caused acidic substances (like IAA) to move (electrophoretically) basipetally, with the reverse being true for alkaline substances. According to Went (1932), polarity could theoretically only be due to pressure differentials, concentration gradients or electrical potential differentials within plants. He discounted pressures and concentrations because, respectively, polarity was not sensitive to external factors and it was not reversible (and other reasons); but, by this time it was known that polarity was respiration-dependent.

Went's interest in electrically based polarity was in keeping with the times: In the 1800s and early 1900s there was substantial and significant research on electrophysiology [see Went (1932)], presumably because of the then-new technology of galvanometers, which were rapidly developed and improved in the early 1800s—see Pledge (1947)]. As we have previously described, Morgan had in about 1900 suggested that the force inherent in polar plant organ regeneration might be electrical. Most specifically, however, Went was extending the research and reasoning of Kunkel (1878, 1879) and Lund (1923, 1924, 1928, 1930, 1931; Lund and Kenyon, 1927). Kunkel (1879) in his paper, *About a few characteristics of the electrical transport potentialities of living plant parts*, had found that an electric current moved with less resistance from the root to shoot tip, compared to the reverse electron flow; hence, cations would move downward and anions would move upward. But, Kunkel did not relate his observations to either polarity or the movement of hormonal stuffs.

Lund (1923) initially found that polarity (axis of symmetry) in a *Fucus* egg and the future plant body, which was known to be influenced by applied unilateral light, was similarly controlled even in darkness by suitably sized, applied electrical potential differences. However, he also found that *Fucus*'s development was, as would be expected, inhibited if too much electromotive force was applied. Lund (1923, p. 300) concluded that: "The establishment of an electrochemical polarity of some sort may possibly be a fundamentally associated condition for the development of morphological polarity, because the physiological mechanism in the *Fucus* egg which determines morphological polarity can be controlled and directed by the application to the egg of a difference of electrical potential of external origin....," which he had found to be, as a minimum, 25 ± 2 millivolts. Thus, he continued research on electrical polarity theories with a marine hydroid (*Obelia*), roots of various plant species and the whole aerial part of the Douglas-fir tree (Lund 1924, 1928, 1930, 1931; Lund and Kenyon, 1927). Subsequent research [see Bloch (1965)] has cast doubts on the electrical polarity theory—are electrical changes the cause or the manifestation? But such research has not much improved our knowledge about the causes of polarity in higher plants beyond Morgan's (1904, p. 227) comment that: "...The result is not so much the outcome of the polarity of the piece, acting at the time of regeneration, as of the preexisting conditions in the piece at the time of its removal from the plant...." Some recent research has (again) suggested that the wound stimulus is an electrical phenomenon (Wildon et al., 1992).

The possible relations between polarity, wounding and adventitious organ formation are still unclear. Wiesner proposed that an "original relation" existed between cellular death, as it occurred after wounding, and the new formation of tissues and organs [cited from Goebel (1903c)]. However, Goebel (1903c), based on his and other investigators' experiments and observations with *Bryophyllum* leaves, stated that: "The changes that occur in severed plant parts—either apparent in new formations or otherwise—are mainly the result of the severance and not of the wounding...." We interpreted Goebel's declaration to mean that the wound stimulus was sufficient to incite regeneration (i.e., it became operable) only after polarity was disrupted, and that the wound stimulus did not necessarily arise *de novo* at the time of injury but might have been ever-present: polarity change was the only critical factor in triggering regeneration [see Wilson and van Staden (1990), Table 2, p. 484, for a listing of metabolic responses to wounding].

It is still unclear whether polarity in higher plant regeneration is truly reversible because the possibility for existence of an immutable polarized state depends upon how polarity is defined—it is as much semantics as physiology. Bloch's commonly accepted definitions, as quoted above, indicated to us that reversing polarity or otherwise modifying it would simply produce a new polarized state, so that polarity may be immutable only in the sense that its flux cannot be prevented, or reversible only in the sense that flux results. Block (1965, p. 235) offered the following opinion about tests that were designed to alter or reverse polarity:

> They mean as a rule that by factorial manipulation we can alter, re-orient or mask
> the expression of the polar tendency of the organism. It remains doubtful, however,

whether we can really change, by ordinary experimental-morphological means, the basic nature of specific polarity which appears to be rooted in some unknown property of the protoplasm and cell. In fact, it would be difficult to single out a clear-cut case in which this fundamental property, this innate tendency toward differential orientation and development in space, has been experimentally modified or reverted....

Rooting research now readily accepts the existence of polarity and sometimes uses polarity to obtain specific treatment effects (i.e., basal vs apical auxin treatment of cuttings), but has not significantly contributed to understanding its causes. But, research on polarity theories advanced knowledge about the control of adventitious rooting in at least one major way, as we will discuss later. Unfortunately, there has been little interest in polarity theories in adventitious rooting research since World War II, apart from polar auxin transport and chemical effects thereon [e.g., Warmke and Warmke (1950), Niedergang-Kamien and Skoog (1956) and Niedergang-Kamien and Leopold (1957)].

CHRONOLOGICAL AND PHYSIOLOGICAL AGING

As we have previously described, Vöchting (1878) observed the influences of organ age on regeneration and McCallum (1903a) classified "relative age and degree of maturity of the different parts of a member" as germane to the stuff-hypothesis. By-and-large such maturation phenomena have been most obvious in the long-lived woody perennial plants [see reviews by Moorby and Wareing (1963), Zimmerman (1976), Hackett (1988) and Haffner et al. (1991); see chapter by Murray and coworkers, and by Howard in this volume]. Vöchting's previous observations about advanced tissue age deterring rooting were verified for several woody species by Hitchcock and Zimmerman (1931, 1932), but the reasons for tissue age-rooting relations remain unknown. With regard to causes, Zimmerman and Hitchcock (1946, p. 31) suggested that: "...Change in structure or [change in] protoplasm with age is a major controlling factor...."

Deleterious maturation effects on adventitious rooting due to aging of the whole stock plant ("meristematic" or "physiological" aging) have proven equally enigmatic [see review by Zimmerman (1976)]. Whole organism maturation effects were known long before Vöchting. For example, morphological equality among embryos of species that have distinctly different mature forms were discussed by Charles Darwin (1859, pp. 337-346) in *The Origin of Species*. Goebel (1898-1901, pp. 38-39 and 132-133) observed for members of the Cupressaceae that branches of juvenile (morphology) plants rooted easier than those of mature plants, and that juvenile and mature morphologies, and intermediate morphologies, could be "fixed" by rooting cuttings from side shoots or from basal shoots of plants whose tops had been removed. He (Goebel, 1898-1901, pp. 132-133; 1889, p. 36) also reported his inability to root cuttings of juvenile pines, but that (1889, p. 36) Hochstetter rooted cuttings of 2- and 3-year-old *Pinus caneriensis*, which then retained their juvenile morphological characteristics. Goebel (1889) concluded that plants in the juvenile phase lacked the special organ-modifying (e.g., leaf morphology-modifying) substances that he supposed were developed during maturation. According to Goebel (1898-1901, p. 134), Beijerinck felt that "diminished nutrition" favored a plant's retention of juvenile characteristics. Goebel (1903a) also stated that seedlings often had greater capacity for adventitious organ regeneration than older plants.

Nonetheless, we have not been able to determine when the chronological age of a stock plant was *first* observed to influence adventitious rooting of its parts or when it was *first* observed that some woody plants, such as *Hedera helix*, had morphologically and physiologically different juvenile and mature phases with which adventitious rooting potential was correlated. However, there is no doubt that the often deleterious effect of stock (perennial) plant age on rooting of cuttings was clearly demonstrated by Gardner (1929) [see also Schaffalitzky de Muckadell (1959)] (Fig. 2). Gardner propagated cuttings from 21 woody species of ages ranging from one up to 12 years, at least for some species. Cuttings of most species showed reduced and slower rooting with increasing stock plant age. As exceptions, cuttings of *Amorpha fruticosa* did not show a decline in rooting when taken from stock plants from one to five years old; *Prunus persica* cuttings rooted poorly even when taken from one-year-old stock plants. Gardner's tests indicated that rooting ability was not determined by some (chemical) factor that was contained in the seed, and that cutting-back seedling stock plant shoots helped to retain rooting ability of their cuttings that was lost due to aging. Hence, since Gardner's demonstration, there have been repeated attempts to physiologically "rejuvenate" older plants or to prevent maturation of juvenile-phase plants, by physical and/or chemical treatments, so that adventitious rooting potential

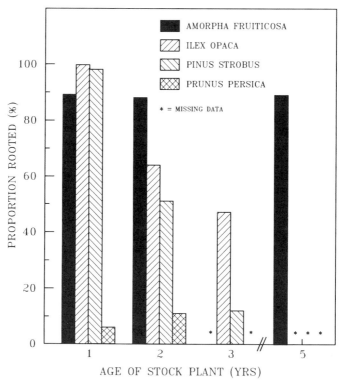

Figure 2. Species-dependent effects of stock plant age, or degree of maturation, on adventitious rooting of cuttings. [data from Gardner (1929)]

would, respectively, be restored or retained (see chapter by Howard in this volume). But, neither empirical trial to modify maturation nor physiological research to understand maturation have made much progress to date, even though the ability to control maturation of woody species would be of immense economic benefit in forestry and horticulture. As Hitchcock and Zimmerman (1932) had previously suggested for tissue age effects, Trippi (1963, p. 165) opined that the effects of plant maturation on rooting were: "...Correlated with rather stable structural changes in... [meristem] cells...". Much earlier, Diels (1906) had suggested that differences he observed between juvenile and mature forms were due at least somewhat to heritable (genetic) processes, although his interests were in the general morphological manifestations of phase change, not physiological predisposition for adventitious rooting.

CAUSAL ANATOMY

There was renewed research into the cytology and anatomy of adventitious rooting in the first half of the 20th century [see Priestley and Swingle (1929)], partly to better understand remarkable types of adventitious rooting such as burr-knots of apple (Swingle, 1927) [see also Knight (1809)] and "air-roots" or preformed root primordia of many species [e.g., Carlson (1938); see chapter by Barlow in this volume]. In addition, however, the research of the foregoing scientists and many of their colleagues was also designed to explore possible relations among plant structural (anatomical) characteristics, tissue aging and plant maturation, polarity, the supposed wound stimulus and adventitious rooting potential or expression—what Priestley and coworkers (Priestley and Ewing, 1923; Priestley and Swingle, 1929) termed "causal anatomy." From a wholly practical viewpoint, anatomical characteristics might predict a predisposition to initiate adventitious roots, for example, Lugovoy's (1937) observed relations between lenticel anatomy and adventitious rooting ability of cuttings. Then, too, anatomical characteristics might inhibit rooting [e.g., Smith (1928) and Mahlstede and Watson (1952)], so that also needed to be examined. The capabilities for exploring such anatomical and cytological scientific interests had been greatly improved in the last half of the 19th century through development of durable, specific "stains" for tissues and through development of the substage-illuminated Abbé microscope and, first water-immersion, then oil-immersion lenses [see Pledge (1947)].

Goebel (1903a) had previously noted relations between anatomical organizational complexity and regeneration responses. He noted that organ regeneration in fungi was much more sensitive to external conditions, compared to higher plants, and, even more interestingly, that (1903a, p. 205):

> The leaves of the foliaceous liverworts do not show any tendency to limit regeneration to the bases. This however is easily explainable in my point of view. These leaves are composed of only one or two layers of cells and so have at the time of separation neither a large amount of constructive materials, nor definite conducting systems to carry them....

Goebel's reference was adventitious bud formation but he clearly points out a supposed general relation between the stuff-hypothesis and cellular complexity—particularly with regard to vasculature—and adventitious organ formation. Vöchting (1906) more generally concluded that the location of root primordia was determined by "internal laws," laws that are still eluding us [see Haissig (1988)].

During this time period, much of what we know about the developmental cytology, anatomy and histochemistry of preformed and induced primordia was learned through tedious and laborious research by obviously dedicated scientists. Priestley and Swingle (1929, p. 84) concluded that polarity did not have an anatomical basis and that the developmental performance of cells, for instance in forming or not forming a root primordium, was determined by their relative positions in the plant. However, there have been very few studies that have critically scrutinized ontogenetic development of plant architecture, especially the vasculature with which adventitious root primordia are so often associated, in order to possibly relate physiological with physical situation of primordia.

Somewhat before Priestley's and Swingle's (1929) observations, Annie Hartsema (1924) conducted a remarkably detailed cytological and anatomical study of adventitious budding and rooting by leaves of *Begonia rex*, based on observations of living tissue sections. Her research was conducted from 1922 to 1924 in the elegantly equipped (Went, 1974) laboratory of F.A.F.C. Went at the Botanical Institute of the University of Utrecht, the Netherlands. She compared her observations with other relevant research from the 19th and early 20th centuries, which makes her contribution particularly valuable to us today. Hartsema found that root primordia formed near the phloem poles of vascular bundles. She painstakingly described the complex protoplasmic streaming, nuclear movements and mitoses in root-forming cells and non-root-forming cells in the wounded leaf. She noted two kinds of responses to an apparent wound stimulus that was propagated in the phloem, upward from the wound: The primary response triggered division of cells immediately adjacent to the injury; the secondary response triggered division of root primordium-forming parenchyma cells, at a distance from the wound, adjacent to the phloem poles. Hartsema looked for but did not observe starch accumulation above the injury and concluded that starch accumulation could not be the cause of rooting. Past research aside, starch anabolism and catabolism during adventitious rooting are still anatomical and physiological curiosities [e.g., Seago and Marsh (1990)].

Carbohydrates in general and starch in particular have been objects for observation during adventitious rooting, probably because of the long-observed positive relations between CO_2 assimilation by leaves and rooting. Such observations became connected with anatomical studies because etiolation of stock plants or organs, which might profoundly change anatomy and decrease starch content of cells, often increased adventitious rooting. In one early, detailed study, Smith (1928) employed partial etiolation of *Clematis* shoots from which cuttings were made. In that instance, etiolation had notable anatomical consequences and resulted in increased starch anabolism. Smith concluded that: "This partial etiolation acts partly by restoring the necessary carbon:nitrogen ratio for protoplasmic synthesis, and partly by softening (delignifying) the hard tissues of the stem, thus rendering the egress of the root-primordium easier...." (p. 663). The cell "softening" occurred because, as Smith observed: "...The process of polysaccharide deposition on a cell-wall is reversible so long as the protoplast remains alive, and that one means of reversing the direction of the reaction is by etiolation...." (p. 662). We are unaware of subsequent research on Smith's interesting hypothesis about etiolation-driven cell wall "rejuvenation" (except see chapter by Murray and coworkers in this volume).

Stoutemyer (1937, p. 347) concluded that anatomical changes were not responsible for reduced rooting of cuttings taken from apple stock plants that had undergone phase-change to the mature form, as indicated by their development of lobed leaves and shoots with less anthocyanin:

> ...Rooting is dependent on some unknown biochemical differences in the meristematic cells of stems in the various stages of growth rather than upon any feature of anatomical structure. ... the causes for the change of growth phase are not related to changes in anatomical structure within the plant. A twig may have only primary wood and be in the mature growth phase...

Stoutemyer's conclusion cannot be considered as definitive because the term "primary wood" has had manifold meanings in the literature (P.R. Larson, personal communication). The foregoing aside, causal anatomy as a possible control in rooting has lived on and on [e.g., Mahlstede and Watson (1952)], and was especially vitalized after the discovery of auxins because of the promotive influences of auxin treatment on cell proliferation [e.g., Kraus et al. (1936)]. Causal anatomy has remained of interest as one of the possible controls of adventitious rooting because it has not been so deeply and widely studied that it can be rejected as at least a partially valid hypothesis [see chapter by Barlow, and by Friend and coworkers in this volume].

Like Stoutemyer (1937), other investigators have observed a possible relation between anthocyanin content of organs and their adventitious rooting ability [see, e.g., Lee (1971)]. This possible relation is one of the most intriguing forms of often suggested causal relations between phenylpropanoids and adventitious rooting potential—phenylpropanoids have been of interest because some are lignin precursors and, therefore, related to vascularization. Recent research by Hackett and coworkers (see chapter by Murray and coworkers in this volume) has shown that differences in anthocyanin content between the good-rooting, juvenile-phase *Hedera helix*, compared to the mature-phase, is due to differential transcription of the gene that codes for dihydroflavonol reductase, which is catalytically involved late in the anthocyanin pathway [see also Claudot et al (1993)].

In summary, there is enough direct and circumstantial evidence to suggest that the development of adventitious organs in higher plants is causally connected to vascularization, its progression and its control. In that context, studying vascularization-adventitious rooting relations would most probably provide us with strong inferences about the control of rooting, and, perhaps, even what the controllers are. The research of Edith Philip Smith (1928) with the genus *Clematis* and of Philip R. Larson and coworkers with *Populus deltoides* [e.g., Larson (1975, 1979, 1980, and many others)] would be good starting models. Such research should include detailed explorations of possible relations between development of procambial (primary) versus cambial (secondary) tissues in woody plants and the initiation of induced and preformed adventitious root primordia.

CONSTRUCTIVE STUFFS

Stoutemyer's (1937) "biochemical differences" were, of course, not new. As previously discussed, the constructive materials of biochemistry, and the related enzymes, are implicit in the stuff-hypothesis. Perhaps for that reason, the 20th century was hardly open when biochemically related experimentation about adventitious rooting began. As one example, Curtis (1918) observed that cuttage needed to be improved, because often it was too uncertain, and that the underlying propagation practices were unfounded in physiological facts. Thus, he (1918, p. 75) proposed further exploration of the biology of adventitious rooting: "...Definitely directed research in the physiology of root formation in cuttings would be of value in this field...." And, Curtis's research basis was essentially biochemical. His experiments are too lengthy to recount here, but they generally concerned the effects of applied mineral nutrients and sugar on rooting of, mostly, *Ligustrum* but also tomato cuttings—research that arose based on an earlier observation at his Cornell University Laboratory of Plant Physiology (USA) that $KMnO_4$ solution-treatment of twigs of *Ligustrum ovalifolium* promoted rooting. The research is especially interesting because it called attention to rooting as an aerobic process [see Devaux (1899), Zimmerman (1930) and Goodwin and Goddard (1940)]; it indicated that modified, specific mineral and/or carbohydrate nutrition might or might not improve rooting; and it demonstrated the as yet unexplained, somewhat consistently observed promotion of rooting by boron-containing salts. Curiously, Curtis's research has often been missed by subsequent investigators of boron effects on adventitious rooting [e.g., Nussey (1948)].

Temporally coincident with Curtis, Kraus and Krabill (1918) suggested that high carbohydrate and low nitrogen levels (to the point of stock plant foliar yellowing) favored adventitious rooting in tomato. Kraus's and Krabill's thesis of causal "carbon/nitrogen ratios" became instantly enduring, perhaps because it offered simple, *measurable* means to assess a predisposition to root and other plant developmental potentials. But, interest in possible relations between so-called C/N ratios (of stock plants and cuttings) and adventitious rooting response may also have been spurred by the exuberant conclusions of other investigators. For example, Phyllis A. Hicks (1928, p. 208), based on her research with *Salix viminalis* cuttings, firmly concluded that:

> Loeb's hormone hypothesis [of course, it was not exclusively Loeb's] *is explained as a gradation of the C/N ratio throughout the cutting. Shoots grow at the area of the lowest C/N ratio, roots at the higher. The effect of a leaf at the apex or base of the shoot is also interpreted on this basis....*

For whatever the reason, the greatest amount of rooting research since Kraus and Krabill, apart from plant hormones, has been directed at the possible influences of carbohydrates and/or non-hormone nitrogenous compounds on adventitious rooting [see Carlson (1929) for early references]. So, for example, Margery Carlson's (1929) microchemical studies of good-rooting Dorothy Perkins and poor-rooting American Pillar roses included measurements of starch, fructose, glucose, reducing substances, asparagine and nitrates. However, one could never be certain what to test, given the all-encompassing nature of the stuff-hypothesis, so Carlson also measured calcium, potassium, magnesium, phosphorus and tannin. Of the stuffs measured, Dorothy Perkins contained more reserve starch, compared to American Pillar; but, of course, starch content was not proven causal of rooting. Indeed, subsequent experiments by others concerning endogenous or applied amino acids, purines, pyrimidines, carbohydrates, alcohols and sugar alcohols, vitamins, steroids, mineral nutrients, etc., etc., have not yielded any consistent story about the control of adventitious rooting—except that the underlying processes are easier to inhibit than to promote. Hence, the results have generally agreed with McCallum's (1905a,b) and earlier investigators' theses that nutrients and other constructive stuffs were necessary to but not *primarily* causal of adventitious rooting, which was verified for regenerative phenomena in animals by Morgan (1906) at about the same time. Again, however, the discovery of auxin treatment effects breathed new life into examinations of stuff-rooting relations, especially regarding carbohydrates and/or nitrogen [e.g., Stuart (1938)] that continue to this day [e.g., Dick and Dewar (1992) and Orlikowska (1992)].

SPECIAL STUFFS

Leaf and Bud Effects

If plastic and building stuffs were not causal in rooting, that left for study the hypothetical special substances or their like, as originally proposed by Sachs. Beijerinck (1886) had focused attention on leaves as sources of substances, presumably nutritious ones, that were important for rooting. However, Jacques Loeb's (1917a,b) research with *Bryophyllum* on polarity theories contributed to fundamental theories about the control of adventitious rooting, as we previously intimated. Whether Loeb was particularly astute or exceedingly lucky or both is unclear but he essentially predicted the discovery of IAA as a companion to the hypothetical rhizocaline (Loeb, 1917b, pp. 550-551):

> The experiments, therefore, seem to prove that axial polarity in the regeneration of the stem is due to the fact that the apical bud (as well as the apical leaf) send out substances toward the base of the stem which inhibit the buds from growing out. These inhibitory substances may be identical with or may accompany the root-forming hormones....

Loeb (1919b, 1923) later proposed a polarity and regeneration theory based on mass action, rather than special substances, so perhaps his 1917b paper amounted to serendipitous casting-about or perhaps he had a falling-out with geneticists: "The assumption of the existence of specific inhibitory substances which the writer had used tentatively in two preceding papers, and which was based on the experiments of geneticists, is not needed for the explanation of the results published in this paper...." (Loeb, 1919b, pp. 713-714). Nonetheless, his (1919a) studies confirmed and expanded upon Beijerinck's contributions. Loeb (1917a) observed that rooting decreased if leaves of cuttings were removed. And he later (1919b) found that keeping cuttings in darkness or covering leaves with tin foil reduced CO_2 assimilation and, concomitantly, adventitious rooting [see Newton et al. (1992) for one recent verification].

Thus early in the 20th century the arena was opened for scientific combat with possible causal bud and leaf affects upon adventitious rooting. In 1924, van der Lek began reporting about the studies that he conducted as part of his Ph.D. dissertation. He (1924) performed the first detailed investigations of internal factors that may influence adventitious rooting from induced primordia (*Salix caprea, Salix aurita, Populus alba, Vitis vinifera*) and, separately, from preformed primordia (*Ribes nigrum*, most species of *Salix*, many species of *Populus*) [see also van der Lek (1934)]. Van der Lek observed that adventitious rooting, especially from induced primordia, was promoted by buds and mostly by developing buds, even if the buds were unilluminated or encased in plaster. His experiments (1934, p. 228) indicated that the bud-related stimulatory effect, which resembled Haberlandt's (1914) cell-division hormone from crushed cells, was propagated downward to the base of the cutting: "...The endogenous, strongly polar root-formation is connected with a hormone moving downward regularly through the phloem...." Finally, van der Lek (1924, 1934) observed that buds of *Populus* inhibited rooting if the cuttings were collected when the stock plants were dormant, in December and January, but that the buds again promoted rooting during the next months: "...The influence of the bud depends also in a large degree upon the time of the year in connection with the normal periodicity of the plant...." [(1924), p. 218; cf. Curtis (1918)]—whereupon quietly entered the research subject of "seasonal variations" in rooting ability [see Anon. (1941)], which had been recorded much earlier [e.g., Ott (1763)]. Subsequently, F.W. Went (1929) observed that buds and, to some extent, leaves promoted rooting of *Acalypha* cuttings, and that diffusate from *Carica* leaves stimulated rooting. F.A.F.C. Went (1930) concluded that primarily older leaves produced the rooting stimulus, based on experiments with *Bryophyllum*. However, the root-promoting substance was not limited to higher plants because Němec observed that *Bacterium tumefaciens* (Němec, 1930) and *Pseudomonas rhizogenes* (Němec, 1934) cultures contained root-promoting substances. Finally, based on research with *Impatiens*, Bouillenne and Went (1933) concluded that the rooting stimulus (rhizocaline) was acidic, non-nutritive, thermostable, produced in illuminated leaves, stored in cotyledons and buds, and polarly and basipetally transported—about as Sachs had proposed 50 years earlier, and very difficult to distinguish physiologically from the about to be discovered IAA, as we will discuss later. Němec (1934) reached conclusions like Went's and Bouillenne's at about the same time concerning his postulated root-producing "*hormons*" or "*organogens.*"

Rooting Inhibitors

The existence of rooting inhibitors has sometimes been postulated as a controlling factor of adventitious rooting, especially when promotive substances have failed to explain differential adventitious rooting capacities between species or age classes of stock plants. The concept of rooting inhibitors is old. For example, Goebel (1898-1901) suggested that poor rooting in difficult-to-root plants might be due to endogenous inhibitors. But the possibility of endogenous rooting inhibitors was not explored in detail until about mid 20th century. One of the first reports correlating lack of rooting capacity with inhibitor content was published in German [Linser (1948); see also Linser (1940)] and is not widely cited in subsequent literature. Linser's rationale was that if endogenous growth substances promoted shoot elongation (e.g., auxin in *Avena* coleoptiles) they should also promote rooting, and the converse should also be true: substances that inhibited shoot elongation should also inhibit rooting. His studies, however, did not confirm that hypothesis because he obtained alcoholic extracts that promoted cell elongation and shoot growth but inhibited adventitious rooting. Thus he postulated the existence of endogenous rooting inhibitors.

One of the first reports in English dealing with rooting inhibitors was presented by the Israeli scientist, P. Spiegel, at the 14th International Horticultural Congress held in the Netherlands (Spiegel, 1955). He reported that difficult-to-root *Vitis Berlandieri* hybrids possessed a rooting inhibitor which, when crudely extracted, reduced rooting of the easy-to-root *Vitis vinifera*. As we will describe in more detail later, Spiegel's observation inspired Hess (1959) to test whether difficult-to-root mature *Hedera helix* contained similar rooting inhibitors. Hess's results failed to provide any evidence for such inhibitors. His findings have since been corroborated by others who grafted mature-phase leaf lamina onto juvenile-phase petioles, whose rooting was not thereby inhibited [Geneve et al. (1991); see also Hackett (1988)]. Mullins (1985) likewise found no evidence of rooting inhibitors in apple.

Perhaps the most convincing evidence of rooting inhibitors as controlling factors came from Australian scientists working with *Eucalyptus* in the late 1960s and the 1970s. These scientists reported in a series of 1970s papers that concentrations of several cyclic ß-triketones from *E. grandis* were positively correlated with reduced rooting capacity of cuttings taken from mature stock plants [Paton et al. (1970); see also Hackett (1988)]. The substances inhibited rooting when applied to cuttings from juvenile stock plants. In addition, the substances inhibited rooting of the easy-to-root species, *E. deglupta*. These compounds have subsequently been referred to as "G inhibitors" (after *grandis*, the species from which they were first obtained). During the early stages of their work, however, the Australians realized that *Eucalyptus* might be an unusual case. In comparing their work with previous rooting studies using other species, Paton and coworkers (1970, p. 182) noted that:

> *Eucalyptus appears to be the exception. ... The concept of cofactors proposed by Hess (1959) may be applicable in Eucalyptus to some extent, but does not explain the apparent overriding influence of the rooting inhibitor in adult tissue of E. grandis....*

Table 1. Some research reports about endogenous rooting inhibitors in species other than *Eucalyptus*.

Authors	Species
Biran & Halevy, 1973	*Dahlia variabilis*
Fadl & Hartmann, 1967a	Pear
Gesto & coworkers, 1981	Chestnut
Reuveni & Adato, 1974	*Phoenix dactylifera*
Richards, 1964	*Camellia reticulata*
Taylor & Odom, 1970	*Carya illinoensis*
Vieitez & coworkers, 1966a, 1987	Chestnut

Putative rooting inhibitors have been extracted from a variety of other species during the past 20 to 30 years (Table 1). Most of these, however, do not meet the criteria for a true endogenous rooting inhibitor as proposed by Hackett (1988, p. 21):

> *...1) the active substance must be identified; 2) a correlation must be established between endogenous levels of the identified substance and rooting capacity; and 3) the extracted substance must be effective when applied to the species from which it was extracted....*

Thus, there is no generally accepted evidence for the existence of universal rooting inhibitors, even though inhibitors may have a role in controlling rooting in some species, under some circumstances. It is possible that some endogenous compounds are indirectly involved in controlling rooting because they counteract the effects of rooting promoters. For example, evidence has accumulated that endogenous cytokinins inhibit rooting (Bollmark et al., 1988; van Staden and Harty, 1988; Bollmark and Eliasson, 1990) and, therefore, might be broadly defined as rooting inhibitors.

The "Root-Forming Hormone": Discovery of Auxin

The discovery in the mid 1930s that auxin could promote adventitious rooting was a major milestone in the history of adventitious rooting research because it seemed to fulfill Julius Sachs's postulate of specific, mobile substances that, in minute amounts, controlled organ formation [see chapter by Blakesley in this volume]. As summarized by Hess (1961, p. 3): "At this point it appeared that root initiation was controlled by a single, specific substance which was produced in the leaves and buds, transported in the phloem, and accumulated at the base of the cutting...." Moore (1979) has stated that F.W. Went made "the first definitive discovery of auxin" in higher plants (Moore, 1979). Actually, Went should be credited with the first *physiological* discovery of auxin. Went undertook his M.S. (equivalent) research under close supervision in his father's laboratory at Utrecht by continuing the experimentation of others on phototropic responses of *Avena* coleoptiles [see Went (1974)]. He then continued this "auxin research" for his doctoral dissertation, which was published in 1928 under the simple title *Wuchsstoff und Wachstum* [Growth Substances and Growth, (Went, 1928)]. Went's thesis reported, for example, the discovery of polar auxin transport, which amazed Went, even in his advanced years (Went, 1974). But Went was unable to isolate and purify IAA.

After the time of Went's doctoral research, the literature concerning putative plant hormones, then including auxin, became voluminous and very confusing. The German-language papers are particularly difficult to understand. Regardless of language, however, in most papers the chemical procedures were much more clearly described than the physiological tests.

During the 1930s, Kögl and coworkers at the *Organisch-chemischen Institut der Rijksuniversität*, Utrecht, conducted numerous diligent studies—for which those scientists have seldom been given due credit—to isolate and characterize, from a variety of sources, compounds that had auxin-like activity [references in Went and Thimann (1937)]. Before they discovered IAA, Kögl and coworkers had isolated the non-indole auxins, auxin a (auxentriolic acid) and auxin b (auxenolonic acid); Went's *Avena* test was used to estimate physiological activity. The starting material was human urine. Auxins a and b readily became physiologically inactive, probably due to spontaneous isomerization, thereby auxin a was converted to "pseudo auxin" (Went and Thimann, 1937, p. 109). Subsequently, Kögl, Haagen-Smit and Erxleben [(1934), the 11th paper in the series] isolated nitrogen-containing organic crystals from human urine that, after final purification to remove an optically active contaminant, were determined by a number of tests to be chemically identical to the IAA that they had chemically synthesized by the method of Majima and Hoshino (1925). The existence of IAA in urine made biochemical sense, based on the likelihood that ingested tryptophane would be converted to IAA by intestinal bacteria. Indeed, Kögl and coworkers found that the amount of IAA contained in urine depended on, for example, the source's nutritional status, such that the well-nourished, and certain individuals, excreted the most IAA. Of course, plant propagators had been using manures in their propagating media to improve rooting of cuttings long before the urine-auxin link was forged. And, of even greater coincidence, in 1880, E. and H. Salkowski had isolated IAA from putrefying meat, and, according to Went and Thimann (1937, p. 111), from fermentations and from urine in 1885—but the Salkowski's research did not deal with putative plant hormones and, apart from that, was probably unknown by plant physiologists.

Kögl and coworkers (1934, p. 94), with the agreement of F.W. Went, named the new auxin "hetero-auxin." Went (1974) later attributed to Kögl alone the invention of the name hetero-auxin. Kögl and coworkers (1934) also determined that, for example, neither *purified* tryptophane nor indolepropionic acid had auxin-like activity in the *Avena* test, but they noted that there were many other possible precursors and analogues of IAA that should be tested for auxin activity. Everything considered, it was yet to be determined which auxins were the naturally-occurring ones in higher plants. In that regard, as late as 1937, Went and Thimann (p. 113) stated that:

> *Now that it has been shown that such chemically different compounds* [e.g., auxins a and b, and hetero-auxin], *all active on* Avena, *are widely distributed in nature, it becomes of interest to know which of these is the native growth hormone in the various higher plants. While final proof can only be given by isolations, Kögl, Haagen Smit, and Erxleben (1934) have given good evidence by indirect methods that the active substance of the* Avena *coleoptile is auxin a....*

The foregoing erroneous assumption by Went and Thimann appears to have been based on the too high molecular weight that Went (1928) had determined for *Avena* auxin, based on its rate of diffusion in agar. By way of possible explanation, the IAA molecules were probably diffusing in pairs or their diffusion was otherwise slowed by intermolecular attractions, so that Went estimated the molecular weight of *Avena* auxin at about twice the value obtained by Kögl and coworkers (1934) for IAA, which we now accept as the *Avena* auxin. We have observed a similar molecular association phenomenon during purifications of indole auxins by high performance liquid chromatography.

The first papers describing the ability of auxin to promote adventitious rooting appeared in the mid 1930s (Thimann and Went, 1934; Fischnich, 1935; Hitchcock, 1935b; Thimann and Koepfli, 1935; Went, 1935; Zimmerman and Wilcoxon, 1935). Papers such as the foregoing were the basis for references to auxin as the "root-forming hormone." Interestingly, however, it was evident to Went that buds seemed to contain a root-forming substance that was different than auxin even while he was discovering auxin (Went, 1934a; Went and Thimann, 1937). However, knowledge about the apparent non-auxin rooting stimulus became subservient for many years to the wonder of having discovered auxin as the root-forming hormone. Accordingly, research during the mid to late 1930s was focused on gaining more knowledge about the root-inducing properties of auxin.

Research on auxins at the Boyce Thompson Institute (Ithaca, New York, USA) resulted in the significant findings that indole-3-butyric acid (IBA) and naphthaleneacetic acid (NAA) were "effective root-forming substances" (Zimmerman and Wilcoxon, 1935). IBA and NAA are today still the most widely used auxins for stimulating adventitious rooting *in vitro* and *ex vitro* [see Blazich (1988)]. The work of Zimmerman and Wilcoxon was apparently stimulated by the findings of their colleague A.E. Hitchcock who reported that application of various indoles caused initiation of roots on stems of several plant species (Hitchcock, 1935b). Hitchcock's tests, in turn, had been inspired by the observations of Kögl and coworkers (1934) on the hormonal properties of auxin. Interestingly, Hitchcock (1935b) did not cite any of F.W. Went's work, which perhaps indicated that he was unaware of it or felt that it was irrelevant.

The finding that IBA was a strong promoter of adventitious rooting was clearly a pleasant surprise to Zimmerman and Wilcoxon (1935, p. 235). In discussing their findings they stated that: "...From the standpoint of molecular makeup, it is hard to believe that the slight difference in structure [of IBA compared to IAA] could account for the vast difference in effectiveness... [as a root-forming substance]." Thus, as has frequently occurred in science in the past, an unexpected outcome led to the development of a product that proved to be of considerable commercial utility. To this day, however, there is no definitive explanation as to why IBA usually so effectively promotes rooting in many plant species, compared to IAA and other auxins.

In contrast to IBA, it is not clear what led Zimmerman and Wilcoxon (1935) to test naphthalene derivatives for their ability to promote rooting. The best explanation appears to be that the group at Boyce Thompson had for several years been empirically screening chemicals for their ability to alter developmental responses in plants. In the process, they had discovered, for example, the first chemically characterized compound (carbon monoxide) that somewhat regularly induced rooting (Zimmerman et al., 1933); at the same time they discovered the root-promoting effects of ethylene (Zimmerman and Hitchcock, 1933), which we will discuss later. Zimmerman and Wilcoxon apparently included NAA in their screening tests because it could be prepared by using a synthesis described by Mayer and Oppenheimer (1916). It is interesting to note that Zimmerman and Wilcoxon (1935, p. 225) felt that:

> ...Of the many thousands of known chemical compounds, there are probably many others which would be equally effective [in inducing adventitious rooting] with those known to date. Considered from that angle, the recent disclosures mitigate against the idea of a specificity for a particular 'growth substance' or 'hormone' unless we choose to put a new meaning to those terms. It would seem more logical to speak simply of the response of plants to certain chemical compounds. If the plants manufacture their own growth-regulating substances, it is not likely that any one plant would make all of those known to be effective. Neither is it logical to assume that all plants naturally make and use one and the same growth substance. ... [Our] results suggest that the raw materials at hand may determine the kind of substances the tissue manufactures. Perhaps the growth of a given species is not always regulated by the same substance but it may vary with the environmental conditions of the plant....

The foregoing statements aid our understanding of the chemical "screening" approach at Boyce Thompson. Zimmerman's and Wilcoxon's statements also signal an early departure from the hypothesis of the hormonal control of organ formation. The statements also opened the concept of plant growth regulators, so that The Plant Growth Regulatory Society of America is now dedicated to learning about the physiological and biochemical functioning of both naturally occurring and wholly synthetic chemicals.

Very soon after the discovery that auxin promoted rooting, investigations were undertaken to determine possible horticultural applications. For example, W.C. Cooper, a Ph.D. candidate working at the California Institute of Technology, Pasadena (USA), applied IAA to cuttings from a variety of horticultural plants: "...With the purpose of determining its practical value in propagation by cuttings...." [Cooper (1935), p.789; see also Cooper (1936, 1938)]. Interestingly, *Plant Physiology* published Cooper's paper, but today the editors would almost certainly not accept a paper with applied horticultural objectives. In a subsequent paper in *Botanical Gazette*, Cooper (1938) reported additional research, including what are some of the first, if not the first, measurements of endogenous IAA concentrations in IAA-treated cuttings.

Research by Grace (1937) provided additional insights into the stimulation of rooting by auxins. He worked in the same laboratory (National Research Laboratories, Ottawa, Canada) as R.H. Manske who synthesized and provided the IBA used in Zimmerman's and Wilcoxon's (1935) landmark studies. Grace was one of the first to recognize the importance of auxin concentration as influential in adventitious root initiation *and* growth. His findings, and Bouillenne's and Bouillenne-Walrand's (1939), are the bases for many of the auxin dose-response curves presented in plant physiology textbooks. Grace also found that "dust treatment" of cuttings with auxin (probably in talc) induced rooting, and he recognized the potential commercial value of the method. Indeed, several modern auxin-based rooting agents are formulated as powders [see Blazich (1988)].

Subsequent studies on auxin's role in rooting dealt with absorption, transport and metabolism of applied compounds [e.g., Niedergang-Kamien and Skoog (1956) and Strydom and Hartmann (1960); see chapter by Blakesley in this volume]. However, little new fundamental knowledge on the role of auxin in rooting has been obtained since around 1940, except that in 1970 Haissig published reasonably good evidence that applied auxin promotes adventitious rooting by influencing the earliest stages of primordial development. Before that time one could only conclude that: "...It remains uncertain whether and to what degree the hormone is acting during the earliest state of root formation (appearance of the first primordium cells) or is important in subsequent stages (growth of the primordium in cell number, differentiation of root tissues, and root emergence)...." (Haissig, 1970, p. 27). Also, a significant change in thinking occurred in the early 1980s because of ideas championed—but not originated—by Trewavas (1981), who challenged then-current dogma regarding the role of plant hormones in growth and development. He proposed that tissue "sensitivity" was important in explaining developmental phenomena in plants, which meant that presence of a hormone was necessary for but not sufficient to influence a growth response. Such conceptualization fueled the search for auxin receptors that had begun in the 1950s and continues today [e.g., Jones (1990); see chapter by Palme and coworkers in this volume]. It is still too soon to determine whether or not the concept of receptivity will help us to gain a better understanding of the role of auxins in rooting.

The discovery of auxin propelled the search for other endogenous chemicals that might be involved in the control of adventitious rooting because it provided evidence that such substances might exist and, equally so, because it was evident to many by about 1940 that auxin was not the sole hormone-like factor, or what seemed to be a hormone-like factor, involved in the control of rooting. That is not to state that opinions, even among close colleagues, were uniform. For example, of scientists who worked closely together, F.W. Went held that auxin was a specific requirement for adventitious rooting, but Cooper (1938) and Bouillenne and Bouillenne-Walrand (1939) opined that auxin was one of the non-specific but special supportive stuffs for rooting. We will discuss the confusion that arose about the possible influences of hormone-like substances on rooting below.

The Hypothetical Non-Auxin Rooting Stimulus

As we have briefly touched upon above, over the years there have been numerous proposals that specific, endogenous, non-auxin substances exist in plant tissues that alone or in combination with an auxin promote the formation of adventitious roots. Somewhat recently, Haissig coined the terms non-auxin "endogenous rooting stimulus" (Haissig, 1982a) or "endogenous root-forming stimulus" (Haissig, 1982b) to describe these, as yet, hypothetical substances or conditions. Haissig used the term "stimulus" because there is no evidence that the hypothetical rooting stimulus is a chemical. And, even though Haissig used the word stimulus, his mental reference was the entire body of candidates of every type that "evidence" or, more often, supposition have suggested [see Haissig (1974b); and Wilson and van Staden (1990), especially Table 1, p. 481]. Rhizocaline was the first non-auxin rooting stimulus to be proposed, as will be described in the following paragraphs.

Of first importance, in 1929, F.W. Went wrote in a research paper that:

...I proved [in my Ph.D. research (Went, 1928)] *the existence of a substance* [i.e., auxin] *that causes the growth in the coleoptile of* Avena sativa *...There are several reasons to assume the existence of an allied substance, causing the development of roots. ...* [p 35] *The results of this* [present] *paper can be summarized as follows. A special rootproducing* [sic.] *substance (hormone according to van der Lek), not specific, heat-resisting, is shown to be extractable from leaves* [Acalypha, Carica] *and germinating barley, and to have the effect of starting the development of new roots. It seems to be transported by the phloem and is formed in the branches....* [p. 39]

At about the same time, his father (Went, 1930) found that leaves of *Bryophyllum* contained a similar root-promoting substance. During this period, F.W. Went was director of the Treub laboratory, which was surrounded by the botanical garden at Buitenzorg. There he met R. Bouillenne, who was *Chargé de Mission* of the Dutch Indies for the Ministry of Belgian Colonies at the University of Liège. Bouillenne and Went discovered that they had been separately researching the chemical bases for adventitious rooting and thus decided to publish a joint paper about their results (Bouillenne and Went, 1933). Bouillenne wrote the first part of the paper, Went wrote the second part, and together they wrote a summary and conclusion—for the lengthy paper. In Bouillenne's part of that paper (p. 52), he provided a rationale for the probable existence of specific root-forming substances, and how he and F.W. Went *jointly* named *them* rhizocaline:

We will have the opportunity to bring to light that the problem [of explaining adventitious rooting] *cannot wait for a solution by the theory of wound hormones. As a matter of fact, it is hypocotyls that although wounded do not give rise to any newly formed roots; and, besides that, the law of mass action by Loeb does not suffice to explain the appearance of roots by the interplay of nutritive substances. It is* [thus] *necessary to interpose the existence of a special factor: root-forming substances, which we will call by the collective term rhizocaline, and the following experiments will lead to the proof thereof. This term is formed from* ριξα [rhiza] *: root, and from* χαλεῶ [chaleō], *I summon, I make come....*

Bouillenne's and Went's (1933) conclusions about the existence of the root-forming rhizocaline have often been misinterpreted in the subsequent literature, because their work has not been evaluated from the proper temporal and physiological perspectives. In one major way, Bouillenne's and Went's findings have been misrepresented by the continued (to this day) use of the term rhizocaline: Usage of the term implies that rhizocaline, or at least evidence for it, was discovered. On the contrary, Bouillenne and Went should be credited with having obtained extensive physiological evidence for, and only for, the root-promoting properties of endogenous auxin, for the following reasons. First, it was not known in 1933 that auxin was IAA, and IAA had not then been chemically (re)discovered. Hence, crystalline IAA was not available for use as a root-promoting treatment, to compare its effects with those of the supposed (and known to be acidic) rhizocaline. Second, Went's (1928) physiological discovery of auxin did not include tests of adventitious rooting. Bouillenne and Went (1933, p. 80) fleetingly mentioned auxin, but did not associate it with an ability to promote adventitious rooting, nor, previously, did either Went (1929) or his father (Went, 1930). The discovery that IAA was the "root-forming hormone" was to occur, however, soon thereafter, as we have previously discussed. Third, it is unlikely that Bouillenne and Went could have distinguished physiologically between the effects of rhizocaline and endogenous auxin, because making that distinction has proven difficult (impossible?) to this day, and it has never been made with easily rooted species such as those used by Bouillenne and Went. Finally, Bouillenne and Went would not have looked for that difference because, as previously stated, auxin was not then associated with adventitious rooting. Even the nature of rhizocaline has often been misrepresented, as being a single substance, whereas Bouillenne and Went (1933) clearly postulated that rhizocaline was a complex of substances, including sugar and oxygen, based on their tests—more factors were to be added to the hypothetical complex later (see below).

Went distanced himself from the rhizocaline concept as soon as it became apparent that auxin promoted adventitious rooting (Table 2). Hence, Went did not focus his research on the identification of rhizocaline or its role in controlling adventitious rooting. But, in 1938, he reported additional research on the physiological effects of rhizocaline and, additionally, invented the term "caulocaline" (p. 60), to describe a root-originated substance that moved upward in the stem where, in conjunction with auxin, it promoted cell elongation growth (Went, 1938). Perhaps Went's interest in the calines was renewed because of the contemporary research of W.C. Cooper (see below). In summary, however, the final sentence of Went's (1938, p. 78) paper is as appropriate for the calines today as it was then: "So far it has been impossible to handle them outside living tissue...." In the following paragraphs we will illustrate that many attempts have been made to isolate and study the non-auxin rooting stimulus, but that abundant evidence for its existence, and any evidence as to its identity, are still lacking.

Cooper (1936, 1938) reached conclusions like Bouillenne's and Went's (1933) about the existence of a non-auxin rooting stimulus. Although we know of no papers coauthored by Went and Cooper, the two had an obvious connection because they both worked at the California Institute of Technology, and Cooper [e.g., (1938), p. 599] wrote of his discussions with Went. Cooper suggested, based on work with several horticultural species, that a substance other than auxin was required for rooting. He specifically referred to that substance as rhizocaline. Further, he concluded that basal treatment of cuttings with auxin caused a rapid basipetal movement of rhizocaline, which was putatively produced in leaves and upper stems. [Recently, Haissig and Riemenschneider (1992) confirmed what might be termed "Cooper's phenomenon," using *Pinus banksiana* cuttings.] It seems that Cooper and Went generally agreed about the rooting stimulus. But, a footnote in Cooper's (1936, p. 791) paper indicated friction between Cooper and the Boyce Thompson scientists. In essence the footnote stated Cooper's exception to Hitchcock's and Zimmerman's (1936) criticisms of his previous paper about the rooting stimulus (Cooper, 1935). Specifically, the footnote suggested disagreement between the parties as to what type of data constituted evidence or "proof" of treatment effects on rooting, which is a point of contention among rooting researchers that has persisted to the present.

The possible existence of a non-auxin rooting stimulus came under attack in the 1940s. Studies done in Puerto Rico with hibiscus cuttings by van Overbeek and coworkers confirmed that something from the leaves besides auxin was required for rooting of difficult-to-root cultivars (van Overbeek and Gregory, 1945; Gregory and van Overbeek, 1945). However, in a subsequent study (van Overbeek et al., 1946) those authors demonstrated that applied sucrose and nitrogenous compounds mimicked the promotive influence of leaves on rooting of an easily rooted hibiscus cultivar, which led to their (p. 107) conclusion that:

> *...The main function of the leaves in the process of root initiation is to supply the cutting with sugars and nitrogenous substances, factors which may be termed nutritional. There is no a priori reason to postulate the production of a special root-forming substance such as 'rhizocaline' in the leaves of red hibiscus....*

Despite the foregoing conclusion, the idea of a non-auxin rooting stimulus was not abandoned by many subsequent investigators, probably because no other controlling factors were identifiable. And, there simply was no generally acceptable alternative hypothesis upon which to base research.

Table 2. Viewpoints about the roles of auxin and the hypothetical rhizocaline in adventitious rooting. [copied and/or translated from Bouillenne and Bouillenne-Walrand (1947), pp. 794-795]

1933: Bouillenne & Went	Root formation = rhizocaline photosynthesized. in green leaves in light or stored in seed.
1934: Thimann & Went	Auxin = rhizocaline.
1934: Kögl & coworkers	Heteroauxin = rhizocaline.
1934: P. White	*In vitro* culture of root meristems: The unlimited growth of root tips in nutritive medium (containing yeast extract) invalidates the theory of a specific hormone provided by the leaves and necessary for root growth and differentiation.
1935: Zimmerman & coworkers	The discovery of numerous "root-forming substances" or root-forming hormones.
1935: Bouillenne	Auxin = metabolic activator. Rhizocaline = rhizogenic hormone.
1935: Snow	Heteroauxin = cambial growth stimulus.
1936: Cooper	No rhizocaline, no roots.
1936: Delarge	About *in vitro* culture of root meristems: Auxin is not a factor in differentiation of lateral roots. Factor of differentiation = rhizocaline.
1938: Went	Rhizocaline + auxin = preparatory reaction (specific). No auxin, no roots. No rhizocaline, no roots.
1939: Moreau	1) Like Snow (1935). 2) Heteroauxin = activator and amplifier of rhizogenic processes, which normally occur later or slower. 3) Heteroauxin = rhizogenic action is variable in proportion to number of leaves *left in light*.
1939 & 1941: Delarge	Auxin is not a root-forming factor in itself.
1942 & 1944: Gautheret	It is hardly possible to call into question the existence of a rhizogenic hormone.

Grace (1945) published an isolated but interesting report on the production of physiologically active substances by *Salix*. As previously described, Grace was the Canadian scientist who in the 1930s investigated the influences of auxin concentration and formulation on rooting of cuttings. In the interim, he apparently had received a clue from

H.T. Gussow, who Grace referred to as a "botanist," which suggested that willow cuttings produced physiologically active materials. Grace's subsequent work supported that contention, although virtually nothing was learned about the chemical nature of the substances that promoted rooting. Grace did not suggest that the substances were non-auxin rooting stimuli, but later work by Makoto Kawase in the 1960s attempted that connection, as we will describe later in this section.

Unlike Went, R. Bouillenne's interest in rhizocaline never seemed to flag. Hence, after World War II, Bouillenne and M. Bouillenne-Walrand revived the concept of rhizocaline. Despite little corroborating evidence, the previous authors (1955, p. 231) suggested that rhizocaline was composed of three parts:

1. A specific factor, circulating from the leaves and characterized chemically by ortho-diphenol groups.

2. A non-specific factor, which is also translocated, and is auxin itself, biologically circulating at low concentration.

3. An enzyme factor, specifically located in particular cells or tissues....

Bouillenne and Bouillenne-Walrand proposed that the enzyme catalyzed a reaction between the o-diphenol group and auxin, giving rise to the "rhizocaline complex." The primary contribution of Bouillenne and Bouillenne-Walrand was, however, theoretical because few supporting data were published for critical review. In fact, one of their most oft-cited papers dealing with rhizocaline appeared in *Proceedings of the 14th International Horticultural Congress*, which was held in the Netherlands in 1955. That French-language paper provided almost no experimental data. Nonetheless Bouillenne's and Bouillenne-Walrand's concepts of rhizocaline have prevailed, and thereby earned a place in the premier modern plant propagation textbook by Hartmann et al. (1990, p. 214). The 1955 paper of Bouillenne and Bouillenne-Walrand seems to have been frequently cited in the rooting literature for at least two reasons: 1) they provided a seemingly plausible hypothesis for the elusive control of rooting, and 2) the Congress where the paper was presented was attended by many prominent "rooting" scientists whose attention was apparently captured by the presentation. Recently, the roles, if any, of phenolics in the control of adventitious rooting have been seriously challenged (Wilson and van Staden, 1990).

At about the same time, Libbert (1956) also proposed that rhizocaline was a complex, which consisted of two mobile components and an immobile cellular factor. IAA and an unidentified factor X were the mobile components, and physiological age was the immobile cellular factor [cf. Went (1974)]. Libbert suggested that factor X was produced in buds, basipetally transported and united with IAA by mass action to form dissociable IAA-X. Libbert further proposed that IAA-X and IAA both promoted rooting, but at different stages of the overall process. Libbert's work was not extended.

Apparently undaunted by van Overbeek's previously discussed conclusions, Charles E. Hess of Purdue University (USA) began searching for the non-auxin rooting stimulus in woody species during the mid 1950s. His work was inspired by a paper presented by P. Spiegel at the same congress where Bouillenne and Bouillenne-Walrand explained their concept of rhizocaline (Hess, 1959). Spiegel essentially reported that rooting inhibitors could be leached from grape cuttings by soaking them in water. His observation led Hess to test whether inhibitors could likewise be leached from the difficult-to-root cuttings of mature-phase *Hedera helix*. However, Hess was unable to demonstrate that the difference in rooting capacity between juvenile- and mature-phase forms was related to inhibitors. Thereafter, Hess (1959, p. 42) began working with etiolated, defoliated mung bean cuttings because:

...These cuttings do not respond to auxin or hormones. The Mung bean cutting is, then, very similar to the mature ivy cutting, which does not respond to the addition of root-promoting substances....

Rooting of mung bean cuttings was dramatically promoted after Hess treated them with auxin plus an extract from juvenile-phase *Hedera*. Hence, Hess concluded that the juvenile phase contained non-auxin rooting stimuli that he termed "cofactors." He derived the term cofactor by way of enzymological analogy—enzyme activities and thus chemical reactions were then known to be influenced by non-protein "cofactors" (C.E. Hess, personal communication).

Subsequent research by Hess focused on determining the chemical nature of the rooting cofactors. In contrast to van Overbeek's early conclusions regarding hibiscus cuttings (van Overbeek et al., 1946), Hess concluded that four specific cofactors were present in an easy-to-root hibiscus cultivar whereas only one cofactor existed in a difficult-to-root cultivar (Hess, 1959). As a consequence, Hess (p. 43) expressed optimism about potentials for progress in understanding adventitious rooting: "...We eventually hope to be able to make a difficult-to-root cutting, easy to root by feeding it with the cofactors that are missing...." His optimism was later echoed by a session moderator at the 10th Annual Meeting of the International Plant Propagators' Society, who suffixed Hess's presentation (1960, p. 123) by remarking, albeit somewhat tongue-in-cheek that: "...We look forward to the day, Charlie, that you will give us a small black box that is readily portable [and] that we can take with us and take a small piece of juniper cutting, close the door and it will light up either green or red [,] saying now or wait another week...."

Hess (1962) subsequently numbered the cofactors, based on their relative positions on a chromatogram (Hess, 1962). Cofactor 3 was putatively identified as chlorogenic acid (C.E. Hess, personal communication). Hess's research on cofactors ended when he began a successful administrative career, which led to positions as Dean of the College of Agriculture at the University of California at Davis and as U.S. Assistant Secretary of Agriculture for Science and Education.

Table 3. Some research reports about the non-auxin rooting stimulus (1963-1992).

Authors	Species
Becker & coworkers, 1990	Avocado
Bojarczuk, 1978	*Syringa vulgaris*
Challenger & coworkers, 1965	Apple, plum
Choong & coworkers, 1969	*Rhododendron*
Fadl & Hartmann, 1967b	Pear
Girouard, 1969	*Hedera helix*
Haissig & Riemenschneider, 1992	*Pinus banksiana*
Jackson & Harney, 1970	*Pelargonium*
Lee & coworkers, 1969	*Rhododendron*
Lipecki & Dennis, 1972	Apple
Raviv & coworkers, 1986	Avocado
Sagee & coworkers, 1992	*Citrus* spp.
Vieitez & coworkers, 1966b	*Ribes rubrum*
Zimmerman, 1963	*Pinus* spp.

Makoto Kawase, then in Canada, initiated research on rhizocaline at about the time that Hess's research was ending in the early to mid 1960s. Kawase chose *Salix* cuttings as his primary experimental material, although, surprisingly, he did not cite Grace's previously discussed earlier work in Canada with *Salix*. Apparently Kawase was unaware of, or unimpressed by, Grace's results. Kawase also did not seem aware that Küster in 1904 and Vöchting, even earlier, had used a centrifuge and a clinostat, respectively, to influence polarized regeneration in twigs [cited from Vöchting (1906)]. Kawase aimed to use centrifugation to force rhizocaline to the base of the cutting, where rooting occurred. Indeed, centrifugation of the cuttings promoted rooting (Kawase 1964). In addition, the water in which the cuttings were stored reportedly contained four factors that promoted rooting of mung bean cuttings (Kawase, 1970). Kawase's findings led him to remarkable optimism (Kawase, 1981, p.8): "A naturally-occurring compound isolated from shoots of the common willow, may prove to be the 'dream chemical' long sought by nurserymen for stimulating rooting of plant species which are difficult to propagate by cuttings...." In concluding his report, Kawase (p.10) issued an important caveat:

Only one difficult step remains before the dream becomes reality. [An] *intensive research effort is needed to chemically identify the willow rooting substance so that it may be synthesized for commercial use....*

Unfortunately Kawase could not substantiate his bold assertions because he died at age 57 in 1983. Since then, no substantive research on the *Salix* rooting substance(s) has been reported.

Reports on the non-auxin rooting stimulus have appeared sporadically since the mid 1960s (Table 3). Those papers collectively demonstrated the strong tendency of plant scientists to explain adventitious rooting on the basis of a chemical stimulus. But, as pointed out by Haissig and Riemenschneider (1992), no study has shown that the non-auxin rooting stimulus is a chemical as opposed to, for example, a biophysical or anatomical one. Furthermore, it has not been demonstrated that, even if these substances do exist, they are any more controlling factors in adventitious rooting than were all chemical substances investigated thus far.

The Intrusion of Non-Auxin Plant Hormones

In this section we will briefly summarize past rooting research that considered the effects of each major group of non-auxin plant hormone. For detailed history about the discovery of non-auxin plant hormones, we refer the reader to *Biochemistry and Physiology of Plant Hormones,* published in 1979 by Thomas C. Moore of Oregon State University (USA) or *Plant Hormones and Plant Development*, published in 1979 by William P. Jacobs of Princeton University (USA). Use of the term "hormone" in plant science has been justifiably criticized in recent years, but we have used it above and will continue to use it below, simply for consistency with the majority of original research reports. Davies (1988) defined a plant hormone as a: "...Natural compound in plants with an ability to affect physiological processes at concentrations far below those where either nutrients or vitamins would affect these processes...."

The discoveries of the major groups of non-auxin hormones were not linked to rooting research. Soon afterward, however, the effects of each group on adventitious rooting were tested—and those tests continue [e.g., Selby et al. (1992)]. The scientific objectives of these tests have not always been clear; perhaps many investigators hoped for a repeat of the auxin story, whereby a new, generally or often-efficacious root-promoting hormone would be discovered. Such discoveries, however, did not happen, so the roles of the non-auxin hormones in rooting, if any, remain obscure; and, worse yet, very confusing. However, rooting-related hormone research has unintentionally but clearly demonstrated the tenacity of the stuff-hypothesis and its fusion with Sachs's previously discussed postulates about organ-forming substances. Unfortunately, much of the research dealing with hormones and rooting has been based upon exogenous treatments. In contrast, little work has critically tested the roles of endogenous hormones and their interactions, and their interactions with applied hormones. Particularly lacking is work aimed at determining how hormones might regulate gene expression and thereby influence rooting, directly and indirectly. For the foregoing reasons, it is difficult to distinguish between possible direct controlling roles of hormones compared to non-specific effects on supportive physiology, both in reference to adventitious rooting.

Abscisic Acid. The definitive discovery of abscisic acid (ABA) as a plant hormone is credited to scientists who, at two different laboratories, had been working on independent investigations during the early and mid 1960s. Addicott's group at the University of California, Davis, had been studying the role of substances that stimulated leaf abscission. The term "abscisin" was coined in 1961 to designate these abscission-causing compounds. Wareing's group in Wales had been studying substances that appeared to induce bud dormancy in woody plants. By about 1965, it became apparent that these two research groups were studying the same substance. In 1967, "abscisic acid" was adopted as the name of that substance.

Shortly thereafter, scientists in Calcutta, India, claimed to have isolated ABA from mango shoots (Basu et al., 1968). Their substance(s) stimulated rooting of mung bean cuttings, but the authors did not mention whether it also stimulated rooting by mango cuttings. In retrospect, the methods that the Indian scientists used to putatively isolate ABA (paper chromatography and bioassays) were insufficient to ensure that ABA was the active component that stimulated rooting of the mung bean cuttings. Nonetheless, the endeavors of Basu and coworkers (1970) temporarily resulted in the classification of ABA as a "rooting factor." At about the same time, Chin and coworkers (1969) at the University of Illinois (USA), who were apparently unaware of Basu's research, discovered that applied ABA promoted adventitious rooting of cuttings of mung bean and *Hedera helix*. They (Chin et al., 1969) tested ABA because it was then known to antagonize the effects of gibberellins and cytokinins, both of which inhibited rooting, as we will describe below.

Despite the promising results of early work, subsequent studies in the 1970s failed to clarify the role of ABA in rooting. Research during that period was aptly summarized by Rasmussen and Andersen (1980, p.152) who concluded that: "...Provided the experimental procedure is suitable, it is possible to obtain almost any rooting response to ABA...." Indeed the literature contains a conglomeration of reports that indicate that ABA can promote, inhibit or not affect rooting. Accordingly, the role of ABA in rooting, if any, is no better understood now than it was 25 years ago. And, during that time period, ABA's proposed roles in plant growth and development have changed considerably: ABA is now often considered to be an important mediator of plant stress tolerances. Thus adventitious rooting-ABA studies are bound to continue because cuttings under propagation are highly stressed.

Cytokinins. The existence of natural substances that control cell division was postulated in the late 1800s by J. Wiesner, a German physiologist [see Moore (1979)], and observed by Haberlandt (1914) as arising from crushed cells. But it was not until the mid 1950s that cytokinins were found to be active in stimulating cell division in plant tissue. A group at the University of Wisconsin at Madison (USA) including Carlos Miller, Folke Skoog, M.H. von Saltza and F.M. Strong found that kinetin isolated from herring sperm DNA promoted cell division in tobacco callus. It was not until 1963, however, that the New Zealander D.S. Letham reported the first definitive isolation of a naturally occurring cytokinin, zeatin, from plant tissue. Letham conducted his research in the Fruit Research Division of the Department of Scientific and Industrial Research, Auckland.

Initially, substances, like kinetin, that caused kinetin-like cell division responses were termed "kinins." However, various polypeptide hormones of animals, which influence smooth muscle function, had also been termed kinins (from the Greek *kinein*, to move); and the animal physiologists prevailed. As a result, several new terms were coined and tried for the plant cell division hormones (e.g., phytokinins) but no name became standard. Finally, Skoog and coworkers (1965) suggested "cytokinin," in the prestigious journal *Science*, and the lasting term was born.

The effect of kinetin on rooting was tested before the existence of cytokinins in plants was confirmed. Humphries (1960), working at Rothamsted Experiment Station in England, reported that kinetin inhibited rooting of dwarf pea petioles and hypocotyls. His research was probably inspired by the many reports in the late 1950s on the growth regulatory properties of kinetin. In 1963, Bachelard and Stowe likewise reported that cytokinins inhibited rooting when applied to the base of *Acer rubrum* cuttings, but that kinetin applied to leaves promoted rooting. It is unclear why Bachelard and Stowe (1963, p. 752) tested kinetin for its ability to influence rooting because the stated objective of their work was very general: "...To investigate the physiology of root initiation on cuttings...". However, those authors were aware of reports that indicate that another factor besides auxin was required for rooting, so perhaps they postulated that cytokinin was the other factor.

Little additional research on possible cytokinin-rooting relations was performed in the 1960s, probably because of Humphries' initial negative report and because of, for example, Heide's (1965) subsequent observations that applied cytokinins favored shoot formation *in vitro*. Then, for reasons that are unclear, several reports appeared in the 1970s concerning various and variable rooting responses to applied cytokinins [see review by van Staden and Harty (1988)]. Collectively, those papers failed to convey any clear message regarding the role of cytokinins in rooting, as evidenced by van Staden's and Harty's (1988, p. 196) conclusion that:

> The role of CKs [cytokinins] in root formation is clearly a topic encompassing more questions than answers. What is equally apparent is that many of the 'answers' derived from previous research have been misleading...."

Van Staden and Harty recognized two major and specific areas in need of further study: 1) quantitative determination of endogenous cytokinins at various stages of rooting, and 2) the importance of site of cytokinin application on rooting. The extent to which detailed studies in those areas will foster our understanding of rooting remains to be learned.

Ethylene. Neljubow, who worked in Russia in the late 1800s and early 1900s, is credited with being the first to recognize the effects of ethylene on plants [see Moore (1979)]. He observed that applied ethylene inhibited stem elongation, increased radial stem growth and altered shoot orientation. About 1910, ethylene was observed to hasten fruit ripening and fruits themselves were suspected of producing ethylene. Subsequent tests from about 1910 until the early 1930s focused on the fruit-ripening aspects of ethylene biology. Not until the late 1960s was ethylene widely accepted as a plant hormone.

However, as previously described, in the early 1930s researchers at the Boyce Thompson Institute began screening various chemicals for their effects on plants because they felt that: "Hormonal studies in plants will no doubt be benefited by a broad survey of the field of organic chemicals and their effects upon plant development...." (Crocker et al., 1935, p. 245). Among the chemicals tested was ethylene gas. Applied ethylene, as we have previously stated, was found to promote adventitious rooting in a variety of species (Zimmerman and Hitchcock, 1933)—before the discovery that auxin promoted rooting. Hence, ethylene and auxin were compared for root-forming activity by the Boyce Thompson group as soon as it was recognized that auxin might be a rooting hormone. The Boyce Thompson scientists concluded that ethylene and auxin had "similar root-initiating power" (Crocker et al., 1935). Thereafter, however, interest in ethylene subsided, probably because then-current technology was insufficient to adequately detect the small quantities of gas that were active in altering plant development. Furthermore, ethylene dosages administered during experiments were difficult to accurately control.

Interest in ethylene as a plant hormone surged after development of flame ionization gas chromatography in the late 1950s. Accordingly, more studies on ethylene and rooting were also conducted [see review by Mudge (1988)], especially studies aimed at determining whether ethylene was involved in auxin-induced rooting. The collective results of such studies led Mudge (1988, p. 158) to conclude that: "...A preponderance of evidence suggests that endogenous ethylene is not involved in auxin induced rooting...." As with other non-auxin hormones, we can only conclude that the direct role, if any, of ethylene in adventitious rooting is unclear. It is clear, however, that the effects of ethylene on rooting are not as predictable or consistent as those of auxin, despite the promising initial results of the 1930s.

Gibberellins. The initial discovery of gibberellins as plant hormones is attributed to E. Kurosawa, a Japanese plant pathologist, who in 1926 found that gibberellins produced by fungi dramatically increased shoot elongation in rice [cited from Moore (1979)]. Hence, gibberellins were named after the genus of the fungus of origin. It was not until the mid 1950s that the definitive existence of gibberellins in higher plants was demonstrated. The lag between Kurosawa's initial discovery and the recognition of gibberellins as a class of plant hormones was largely due to the poor political relationship, and thus poor scientific communications, between Japan and western countries.

In the 1950s, scientists at Imperial Chemical Industries in England undertook a project dealing with the mass production of gibberellins from fungal cultures. Included in that group was P.W. Brian, who compared the biological properties of gibberellins and auxins. His rationale was that gibberellins, like auxin, stimulated shoot growth and might be a new type of auxin. By extension, it was logical to test gibberellins in rooting trials because auxins promoted adventitious rooting. Initial results, however, clearly indicated that gibberellins inhibited rooting in *Pisum sativum* cuttings (Brian et al., 1955). Even so, Brian and coworkers (1955, p. 911) concluded that:

> GA [gibberellic acid] *must be considered to be an auxin in the sense of the recent definition of Tukey et al. ... It is considered premature to conclude that its mode of action in promoting shoot cell growth is distinct from that of IAA and other auxins....*

Brian's research prompted J. Kato, a Japanese scientist, to conduct further comparative studies with auxin and gibberellin. Like Brian's group, he found that gibberellin inhibited rooting, but nonetheless correctly concluded that: "...Gibberellin acts on some point of the growth mechanism different from that on which the known auxin acts...." (Kato, 1958, p. 15). Thereafter, it became apparent that auxins and gibberellins were indeed distinct classes of plant hormones.

Many subsequent studies have confirmed that gibberellins generally inhibited rooting, depending on cutting developmental stage at the time of application [see Hansen (1988)]. Several hypotheses have been put forth to explain the inhibition of rooting by gibberellins. Chief among these, Brian et al. (1960) concluded that gibberellins inhibited early cell division related to rooting. Haissig (1972) later confirmed Brian's hypothesis, but the biochemical and physiological mechanisms by which applied gibberellins inhibit adventitious rooting remains unknown.

Many investigators have tested the effects on adventitious rooting of inhibitors of gibberellin biosynthesis, with the idea of determining whether or not endogenous gibberellins, like applied gibberellins, inhibited rooting [reviewed by Davis and Sankhla (1988)]. The rationale behind most of these tests was that lesser concentrations of endogenous gibberellins might yield greater rooting. Indeed various gibberellin biosynthesis inhibitors have been found to promote rooting. Unfortunately, such inhibitors were not specific enough to influence only gibberellin biosynthesis without influencing untargeted metabolic pathways. As a consequence, lessened endogenous gibberellin and enhanced rooting would not necessarily have been causally related. Overall, the effects of gibberellin biosynthesis inhibitors on rooting have been less consistent than effects of applied auxin, as indicated by lack of development of the inhibitors as proprietary root-promoting agents [see Davis and Haissig (1990)].

THE HYPOTHESIS OF "BALANCED" FORMATIVE STUFFS

The stuff-hypothesis has, since its inception, become more complex whenever study of one or another subject, such as carbon/nitrogen ratios or rooting cofactors, failed to reveal much about the control of adventitious rooting. Each modification of the stuff-hypothesis resulted in an ever-greater complex of interacting factors. And, each of those factors has usually had different postulated roles, depending on stock plant physiology, cutting physiology, stage of rooting, etc. Furthermore, the postulated roles have differed, depending on how any particular developmental physiological state was influenced by the then-current and historical environments. There is a definite time when this complexity began to infiltrate the stuff-hypothesis: immediately after its proposal, early in the 20th century, as single or few factors (stuffs) were ruled out as controllers of adventitious rooting. Hence, Went and Thimann (1937) and Went (1939) reported rooting experiments from the 1930s that tested effects of higher and lower plant tissue extracts, carotin, nucleotides, vitamins, human sex hormones, etc., etc. However, the rate of increase of complexity in the stuff-hypothesis elevated as more scientists began studying adventitious rooting, thereby more quickly discarding specific single- or few-factor hypotheses. In addition, capabilities for biochemical and physiological detections and measurements increased, thereby offering the possibility to study more factors, "in balance." We cannot, on account of space limitations, dwell on all the factors or stuffs that have joined The Hypothesis in the last 75 years [see instead Jackson (1986) and Davis et al. (1988)]. Our primary purpose in this section will be to examine in some detail *why* complexity, and as a result "balance," were added to the stuff-hypothesis.

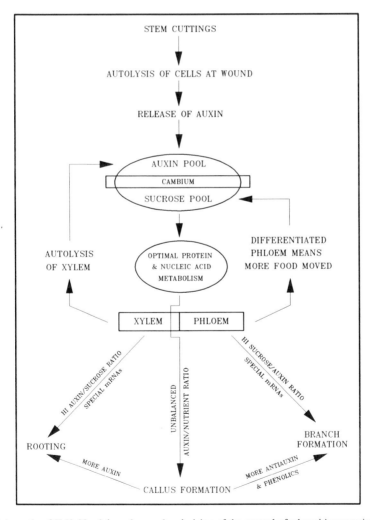

Figure 3. Schematic of K.K. Nanda's and coworkers' vision of the control of adventitious rooting in *Populus* cuttings. [redrawn from Nanda et al. (1974), p. 345]

313

Many authors have referred to the importance of balanced factors in adventitious rooting. For example, Balasimha and Subramonian (1983, p. 65)—who worked in Vittal, India, at a small Central Plantation Crops Research Institute laboratory—began one of their papers by stating that: "The balance of various internal and external factors influence the rhizogenesis in stem cuttings of plants...." Another group of Indian scientists, led by K.K. Nanda at Punjab University, Chandigarh, had likewise envisioned rooting as a result of a complex balance of physiological factors. They concluded one of their numerous papers (Nanda et al., 1971, p. 391) by stating that: "...The ability of stem cuttings to root is determined by a proper balance between the nutritional factors and regulating substances and that rooting may not occur even when the concentration of one of these factors is very high...." (Fig. 3). Other noteworthy scientists, who studied relations between adventitious rooting and interacting physiological factors, included a Danish group in Copenhagen led by A. Skytt Andersen and a Swedish group in Stockholm led by Lennart Eliasson. These productive scientists published numerous papers about adventitious rooting in the 1970s and 1980s.

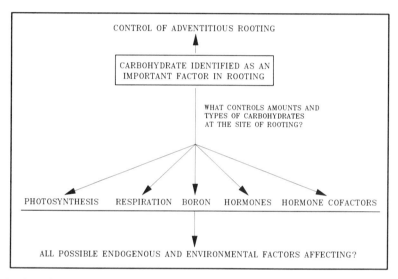

Figure 4. A schematic example of how physiological studies have expanded, rather than reduced, the number of factors supposedly related to adventitious rooting. As an example, when carbohydrates were proposed as important in rooting, investigators studied the many factors that could influence carbohydrates in the rooting zone. As a consequence, attention was diverted to supportive physiology from controllers, which likely reside at or near genes.

There are several reasons why researchers in the 1960s and beyond tended to view rooting from the balanced factor perspective. For instance, in the early and mid 1960s, several papers in the then-emerging field of plant tissue culture demonstrated that organ formation *in vitro* could be controlled by auxin-cytokinin ratios in the culture medium [e.g., Heide (1965); reviewed by Haissig (1965)]. Such results obtained with tissue cultures undoubtedly influenced the thinking of adventitious rooting researchers. Unfortunately, however, the effect of exogenous hormones *in vitro* has not necessarily reflected endogenous roles in cuttings. Another basis for the balanced factor perspective was the generally held belief that rooting could be comprehensively understood at only the level of physiology. For example, Bachelard and Stowe (1963, p. 751) stated that: "...An understanding of the physiology of root initiation is the key to the successful rooting of cuttings...." [essentially paraphrasing Curtis's (1918, p. 75) comments of 45 years earlier]. The assumption that physiological research alone would define the controllers of rooting was false. But earlier investigators could not have known that because they could not apply the "genius" to their subject that retrospect always provides. Physiological studies have in fact expanded rather than reduced the number of factors that could be (but likely are not) involved in the control of adventitious rooting (Fig. 4).

The physiological approaches used to study rooting in the 1960s and beyond were also a reflection of theoretical and methodological limitations. Compared to the present, there was little understanding of plant biology at the molecular level. Conversely, methodology for measuring "physiology-level" factors (e.g., many of the factors listed in Table 4) was becoming widely available. To some extent, then, experiments were designed around available technology and tended to be descriptive, which is often characteristic of early experimentation on any subject. Many *influential* factors were correlated with rooting capacity but no generally accepted *controlling* factors could be identified.

Table 4. Some chemical factors ("stuffs") that have been implicated in adventitious rooting.

abscisic acid	acid/bases	amino acids	auxins	anthocyanins
carbohydrates	carbon dioxide	cofactors	cytokinins	enzymes
ethylene	gibberellins	histones	inhibitors	minerals
nitrogen	nucleotides	nucleosides	nucleic acids	oxygen
phenolics	polyamines	vitamins	water	

QUANTITATIVE AND MOLECULAR GENETICS

The concept of heredity was of considerable interest to biological scientists in the 19th century. Thus, the possibility that genetics played an important role in organ regeneration was recognized many years ago. For example, McCallum (1905b, p. 245), in referring to the earlier writings by Noll (1900), stated that: "Somewhat analogous to this is Noll's [1900] idea of a body forming stimulus (*Körperformreizen*), by which he implies that there is an innate impulse in the organism toward a definite form, and when a part is removed the resulting disturbance (*Formstörung*) acts as a stimulus to reconstruction of the whole...." Kupfer (1907, p. 237) likewise hinted that genetics might play a role in rooting:

Inasmuch as the polar tendency in a plant may be regarded as one of its hereditary qualities, the power to form roots or shoots at any point might be ascribed to the presence at such a place of particular enzymes, which are responsible for the formation of the organ in question. ... The root-forming enzyme ... apparently remains a permanent constituent of nearly all cambial cells throughout the plant....

The suggestions about possible genetics-rooting relations that were made by McCallum, Noll, Kupfer and others were remarkable given their time period. But, no experimental evidence for a genetic control of rooting was obtained at the time because the hypotheses were considered to be untestable. For example, McCallum (1905b, p. 245) stated that:

Such hypotheses as these are at present incapable of demonstration as they are of refutation, and can only serve a useful purpose if they form the starting point for experimentation. Unfortunately they can scarcely be said to do that....

Despite the suppositions of the early 1900s, virtually no attention was given to the genetic control of rooting until the 1960s. This lack of attention is surprising because genetics is the only identifiable link between the stock plant and propagules taken therefrom [see Haissig (1988)]. With regard to genetics-rooting relations, clonal differences in rooting of cuttings had been observed early in the 20th century (Zimmerman and Hitchcock, 1929; Snow, 1939; Deuber, 1940; Dunn and Townsend, 1954). Theoretical and methodological constraints seem to have suppressed interest in the role of genetics until genetic regulatory mechanisms began to be understood based upon classical work by scientists such as James Watson and Francis Crick in 1953, François Jacob and Jacques Monod (1961), and James Bonner and coworkers (1963).

Several papers appeared during the mid 1960s and early 1970s that described research aimed at describing processes of adventitious rooting at or near the gene level. For example, the German scientists Böttger and Lüdemann (1964) studied changes in RNA fractions and protein synthesis during adventitious rooting. Guillot (1965, 1971, 1972), working in France, and Fellenberg (1965, 1966, 1967, 1969a,b,c), working in Germany, used various inhibitors of protein and nucleic acid metabolism to study adventitious rooting. These papers were not widely appreciated for at least two reasons: 1) they were not written in English, and 2) the results were unfortunately difficult to interpret, except to verify that protein and nucleic acid metabolism were involved in adventitious rooting, as expected. Moreover, such research was not very attractive at the time because available technologies for use with higher plant tissues were not equal to the molecular tasks expected of them. Accordingly, the brief surge of interest in genetic regulatory mechanisms involved in adventitious rooting subsided in the early 1970s.

Interest in the genetics of rooting has increased in recent years because of the significant progress that has been made in molecular biology and because the relevant technologies are easier to use with higher plants. But molecular genetic studies of rooting are too new to have an interpretable history, even though available molecular genetic technologies now hold considerable promise for elucidating the control of rooting [see Haissig et al. (1992)]. However, the new technologies have largely been used so far only to test the old stuff-hypothesis. For example, *Agrobacterium*-mediated genetic transformation has been used to evaluate the role of auxin and/or auxin sensitivity in adventitious rooting [e.g., Schmülling et al. (1989), Maurel et al. (1990, 1991), Julliard et al. (1992); see also chapter by Hamill and Chandler, by Palme and coworkers, and by Tepfer and coworkers in this volume]. It is still too early to tell how valuable molecular approaches will be in unravelling the control of rooting. But history suggests that new hypotheses and theories may be needed to fully capitalize on new molecular genetic technologies, as we will discuss in the next two sections.

THE EVOLUTION OF HYPOTHETICO-DEDUCTIVE REASONING

How scientists view, approach and use science has slowly changed over the period of years that our chapter treats, as we will discuss in this section. The reader may initially feel that this section is somewhat far-afield of our subject, but in interpreting history it is imperative that we consider two main points that are related to scientific change. First, different uses of the so-called scientific method (defined below) have differently influenced the types and values of results that have been obtained. Second, we cannot draw conclusions from previous research without addressing the issue of how research was done, compared to how it would perhaps have better been done—error can be a helpful teacher.

Thus, to properly conclude our chapter, we must achieve a common understanding with the reader about the development and use of the scientific method for the last 400+ years, based first on a common terminology. Unfortunately, there are no commonly accepted, commonly used definitions of the scientific method and attendant terminology; indeed, books have been written on scientific method-terminology subjects [see, e.g., Bunge (1967)]. However, for our simple purposes we have accepted the following terms and their definitions from Webster's Third New International Dictionary:

> **Scientific Method** (unknown date of origin): The principles and procedures used in the systematic pursuit of intersubjectively accessible knowledge and involving as necessary conditions the recognition and formulation of a problem, the collection of data through observation and if possible experiment, the formulation of hypotheses, and the testing and confirmation of the hypotheses formulated ...
>
> **Hypothetico-deductive** (originated 1912): *Of or relating to scientific method in which hypotheses suggested by the facts or observation are proposed and*

consequently deduced from them so as to test the hypothesis and evaluate the consequences. ... **Law** (unknown date of origin, from Middle English): *A statement of an order or relation of phenomena that so far as is known is invariable under given conditions. ...* **Theory** (originated 1597): *The coherent set of hypothetical, conceptual, and pragmatic principles forming a general form of reference for a field of inquiry (as for deducing principles, formulating hypotheses for testing, undertaking actions). ...* **Hypothesis** (originated 1596): *A proposition tentatively assumed in order to draw out its logical or empirical consequences and so test its accord with facts that are known or may be determined....*

Our presentation will be brief and limited to three scientists who have made *remarkable, original* contributions to our scientific thinking since the 16th century: Francis Bacon, T.C. Chamberlin and John R. Platt; but the interested reader should also review, for example, the contributions of all early scientific philosophers [see Pledge (1947)], Mario Bunge (1967) and Karl Popper (see Bondi, 1992).

Bacon was born in London in 1561 and entered Trinity College at age 12 years. A short time later, at Cambridge, he developed a dislike for Aristotle's philosophy; he also formed notable personal aspirations: "I have as vast contemplative ends as I have moderate civil ends, for I have taken all knowledge as my province...." (Hutchins 1953, p. v). Bacon's career included philosophy, science, politics and law—and some less estimable activities (Hutchins, 1953, pp. v-vi). Our specific interest, however, is that Bacon is credited with inventing the conditional inductive tree, i.e., the dichotomous test scheme whereby hypotheses can be tested, and the likely false ones left behind, so that the investigator can advance to formulating and testing a new hypothesis. Viewed from another perspective, Bacon in a way invented the "If, Then, Else" logic of computer programming.

Chamberlin (1897), a brilliant geologist and, at one time, President of the University of Wisconsin, expanded upon so-called Baconian thinking by developing the concept of multiple hypotheses. Before its recent reprinting, Chamberlin's paper—now considered a hallmark—was essentially unknown, except to a few academics who somehow obtained copies for classroom use. Chamberlin first presented his ideas before the Society of Western Naturalists in 1892. A few years later he published a revised version in *The Journal of Geology*. In his paper, Chamberlin proposed the formulation and testing of multiple hypotheses, partly for the following reasons:

> *Three phases of mental procedure have been prominent in the history of intellectual evolution thus far. ... The method of the ruling theory, the method of the working hypothesis, and the method of multiple working hypotheses. ... The habit of precipitate explanation leads rapidly on to the birth of general theories. When once an explanation or special theory has been offered for a given phenomenon, self-consistency prompts to the offering of the same explanation or theory for like phenomena when they present themselves and there is soon developed a general theory explanatory of a large class of phenomena similar to the original one. In support of the general theory there may not be any further evidence or investigation than was involved in the first hasty conclusion. But the repetition of its application to new phenomena, though of the same kind, leads the mind insidiously to the delusion that the theory has been strengthened by additional facts. ...* [p. 839] *A premature explanation passes first into a tentative theory, then into an adopted theory, and lastly into a ruling theory. ... The method of the ruling theory occupied a chief place during the infancy of investigation. It is an expression of a more or less infantile condition of the mind. ...* [p. 841] *Conscientiously followed, the method of the working hypothesis is an incalculable advance upon the method of the ruling theory ...* [but lest the hypothesis become a controlling idea] *the method of multiple working hypotheses is urged. ...* [to bring] *into view every rational explanation of the phenomena in hand and to develop every tenable hypothesis relative to its nature, cause or origin, and to give to all of these as impartially as possible the working form and a due place in the investigation....* [p. 843]

In brief, Chamberlin suggested that scientific progress would be optimal when scientists, to the best of their individual abilities, framed and tested every conceivable hypothesis that was consistent with the underlying theory; and, as necessary, modified theories based on their on observations. He reasoned that no individual hypothesis would likely tell much about the nature of a phenomenon, and that individual hypotheses were likely to become ruling and wrong, general or special theories. Presumably Chamberlin followed his own advice because he and Moulton, about 1900, demonstrated that Laplace's hypothesis could not account for the solar system [see Pledge (1947), pp. 291-292]. Chamberlin's multiple hypothesis concept has been questioned [e.g., Beveridge (1957), p. 50] because, for example, it demands the personal development and exercise of extreme intellectual capacity. Beveridge (1957) suggested that more often hypotheses were tested serially, and Chamberlin would probably have approved of that approach, too, for the less agile-minded. With regard to hypotheses, Karl Popper, an acclaimed scientific philosopher, espoused the idea that working premises should be vulnerable to disproof, so that the more rigid, at-risk postulate was preferable to the supple one [see Bondi (1992) for a brief overview of Popper's ideas].

Platt built upon Bacon's and Chamberlin's thoughts in developing the final concept of experimentation that we will discuss, that of Strong Inference. Platt, at the time he published his proposal, was Professor of Biophysics and Physics at the University of Chicago (USA). Like Chamberlin, Platt's initial dissertation took place before a society, the Division of Physical Chemistry of the American Chemical Society, in 1963. Platt (1964) then published his ideas in *Science*. Platt (1964) viewed intellectual advancement, such as Bacon's and Chamberlin's, as the primary factor in scientific advancement. He pointed out (p. 347) that in certain fields, such as molecular biology and high-energy physics, the advancement of knowledge has been comparatively faster than in other fields because the scientists applied:

> *...An accumulative method of inductive inference that is so effective that I think it should be given the name of 'strong inference'. ... Strong inference consists of applying the following steps to every problem in science, formally and explicitly and regularly:*
> *1) Devising alternative hypotheses;*
> *2) Devising a crucial experiment (or several of them), with alternative possible outcomes, each of which will, as nearly as possible, exclude one or more of the hypotheses;*
> *3) Carrying out the experiment so as to get a clean result;*
> *1') Recycling the procedure, making subhypotheses or sequential hypotheses to refine the possibilities that remain; and so on....*

Platt further opined that strong inference was difficult to embrace, because its use caused penetration of the naturally dreaded unknown, that which lies beyond our scientific comfort zone. Also, strong inference places on the investigator the burden of deep thought in developing many dichotomous test schemes. But, Platt also noted that strong inference was the surest and fastest way to rapid progress, and he provided many concrete scientific examples that validated his viewpoints, which are now commonly accepted [see, e.g., the Publisher's Note that prefaces Chamberlin (1897)]. Finally, Platt suggested that the application of strong inference would help an investigator to maintain problem-orientation rather than method-orientation, the latter of which would more likely achieve "busy work" rather than scientific progress.

The reader may ask, have developmental plant biologists other than molecular biologists demonstrated the results of strong inference? The answer is, yes, and we will briefly describe one case, that of Hans Fitting's initiation of the concept of plant hormones. Fitting reported his experiments in two clearly and simply written, fascinating papers (Fitting, 1909, 1910). Fitting had become curious as to why pollinated orchid flowers quickly died. He was mulling it over one New Year's Day during a walk at the Buitenzorg botanical garden when he observed a magnificent orchid that had over 100 flowers. On

January 2nd, Fitting used that orchid as one of his many test specimens, in a running application of strong inference. He applied sand to the stigmas of some flowers, pollen to others and left some untouched. The flowers that were treated with sand or pollen promptly wilted; the controls survived. Fitting also observed that the flowers died, for example, if wounded or treated with saliva, or even if ants crawled upon them. Instead of giving up because of the complexity of his observations, and the number of possibly relevant hypotheses, Fitting conducted 96 experiments to key his way through the problem. He learned that application of even purified, sterile sand caused the flowers to die. Hence, the effect of sand was only physical—but he reasoned that pollen's influence could still be different. He observed that even ungerminated pollen resulted in death of flowers. He postulated that, unlike sand, pollen contained a readily diffusable substance that quickly passed to the flower, causing death. For verification, Fitting extracted pollen with water and found that the applied extract caused death of the flowers. Fitting concluded that orchid pollen contained a substance (*der Reizstoff*), similar to Starling's animal hormones (Fitting, 1910, pp. 264-299)—although Fitting did not develop the full modern-day plant hormone concept. Fitting suggested that the term hormone be applied only to such plant substances that were found to influence development. He did not learn why physical stimulation caused orchid flowers to die, but he did learn that physical stimulation killed orchid flowers only if they were susceptible to death from pollen.

We conclude that Hans Fitting's problem was no more or less complex than the problems that we face in learning about the control of adventitious rooting. As Fitting stated, his research took an amazing turn through observation, and, we add, through strong inference. Unlike Fitting's studies, adventitious rooting research has explored only one major hypothesis, the stuff-hypothesis, with variable consequences, as discussed immediately below.

CONCLUSION

The history of science since 1500 indicates that centuries seem to alternate between rapid advancement of scientific knowledge and slow assimilation of that which has been previously learned or postulated [see Pledge (1947)]. In terms of the biology, especially the control of adventitious rooting, the 20th century has primarily ingested, ruminated upon and is very slowly digesting the observations, experiments and hypotheses *cum* theories of the 19th century—and as we researchers have expanded them. Our major research efforts have explored the stuff-hypothesis, primarily the constructive and special stuffs, because we have longed for the simplicity of adventitious rooting controlled by chemicals.

Perhaps the stuff-hypothesis was really a hypothesis at one time in the 19th century, shortly after its inception. But, even if so, the stuff-hypothesis quickly became the ruling general theory, of the type that Chamberlin so clearly described and cautioned us to avoid. Ruling theories conflict with a main purpose of science, namely, to determine the bounds of applicability of various theories, thereby exposing voids in knowledge that can be filled only by the formulation and testing of hypotheses whose conjunction results in a new theory or theories. In the case of the stuff-theory there have been ample demonstrations that it applies—and applies well—within the bounds of practical plant propagation and rooting-supportive physiology, at the least. There is also substantial evidence which suggests that the stuff-theory is out of bounds, or inadequately formulated, for application in studies of the control of adventitious rooting. Hence, as our ruler, the stuff-theory has been both beneficial and detrimental.

As a benefit, testing the multifarious aspects of stuff-theory has markedly improved practical plant propagation of many species, which has been a contribution to plant commerce. Compared to 1900, we have better knowledge of stock plant manipulations to achieve greater predisposition to root. We have better *ex vitro* and *in vitro* propagating environments. We have more chemicals and more efficacious chemical treatments for propagules. We know something of selecting stock plants for adventitious rooting ability. But, we also still have a plethora of species, varieties or age-classes of stock plants whose propagules cannot be rooted by any way. Thus, as a detriment, exploring stuff-theory has taught us biological sleight of hand and mind, in place of teaching us about fundamental biology, of which we do not know anything with *certainty* concerning the *control* of adventitious rooting. However, that biology contains knowledge of the control of rooting that we have wanted for so long to understand, for purely scientific reasons and because the new scientific knowledge would bolster commerce, as has been true of all great gains in scientific knowledge in the past [see Pledge (1947)]. It is time to leave what can somewhat incongruously, but also accurately, be described as the "complex oversimplification" of the old stuff-theory. Alternatively, new, testable hypotheses must be derived from stuff-theory, but, if that were possible, would it not have already happened? Stuff-theory is simply not very researchable because it tends to lead to circular rather than dichotomous testing, so nothing can be negated through the use of strong inference.

It is time to develop new and original theories based on hypotheses that replace—at least in the domain of control—the aged and flabby stuff-theory, so that we can justify, to ourselves and our publics, further experimentation to learn about the control of rooting. In the end, stuffs may prove to be of paramount importance in the control of adventitious rooting—as retinoids have been shown to be of major importance in differentiation phenomena in animals [see, e.g., Kraft et al. (1993) and Ross (1993)]—but somehow there seems to be a bigger part of the picture that is sorely wanting. New theories about the control of rooting are required because hypotheses that are not couched in an accepted theory invariably suffer from scientific disdain because they appear to be "illogical" [see, e.g., Bunge (1967)]. The time is right for new theories because our hypothetico-deductive reasoning about how we should experiment has improved during the 20th century; we now need to exercise our capabilities for using strong inference. In developing theories, we must not be sidetracked with the "busy work" of method-driven experimentation. Hence, "the new approaches," whatever they may then be, should not limit the number and diversity of hypotheses that we test. Why? During the last two centuries no new technology has improved our understanding of the control of adventitious rooting. It will be more logical for us to accept the truism that technology is reason's adjunct, not its replacement.

With regard to new technologies, as we previously stated, molecular genetic approaches may help us to understand the control of adventitious rooting. But we may also find that such approaches alone are insufficient. Molecular approaches are powerful and popular, but they focus attention only at the cellular level, just when the classic "cell theory" of Schleiden and Schwann is being seriously questioned: "Reductionistic exaggerations of this kind cannot be defended today...." (Sitte, 1992, p. S1). In discussing the cell theory, Peter Sitte (1992, p. S1) [see also Stebbins (1992)] has also pointed out that:

> The formulation of a unifying concept appears to be hindered not only by obsolete
> historical reasons but also by conceptual differences and also by methodological
> barriers. Concerning the balance of genetic and epigenic factors in morphogenesis,
> it appears unlikely that information stored in one-dimensional nucleic acid
> molecules could suffice to direct ordered three-dimensional development....

Sitte's comments apply to the specific subject of adventitious rooting, even though that was not his primary intent. Judith Verbeke (1992, p. S88) has also suggested the need for technolological and hypothetical balance in how we proceed in studies of differentiation: "Although cell and tissue differentiation must certainly be the result of selective gene expression, our inability to make the connection between phenotype and genotype remains a critical question that must be addressed by a combination of the 'new' technologies with the 'old' structural approaches...." One role of the "new technology" of molecular biology may be, because it is qualitative, to give us the opportunity to quickly make the type of observations that Hans Fitting found to cause an amazing turn in his research progress. New observations will certainly be needed to help us develop new theories about the control of adventitious rooting.

In developing new theories we might profitably embody some of the old. As previously stated, control-related rooting research of the 20th century has been badly manipulated by—paraphrasing Chamberlin (1898)—the unruly child of exclusively chemical explanations of the physiology of rooting. In comparison, as we have made amply clear in previous sections of this chapter, the original stuff-theory was infused with nuances of anatomical structure and of osmotic and electrical forces that have been ignored for the last 50 or more years. Our ignorance of the possible and likely biophysical effects on rooting of osmotic, electrical or unknown forces—interacting with the three-dimensional structures of organs, tissues and cells—is especially curious because nearly all of the stuffs that we have been researching are osmotica or electrolytes, or both, as Morgan suggested so long ago. Stuffs alone have thus far proven insufficient, for example, to explain stock plant maturity and organ polarity effects on adventitious rooting.

We would like to end this chapter by providing the reader with new, concise, researchable hypotheses, even theories, for exploring the control of adventitious rooting. However, we are incapable of doing so, perhaps because we have been influenced by stuff-theory for too long, which makes it difficult and disturbing to think differently. Then, too, theoretical development takes time, and is not necessarily the responsibility of one or a few scientists. Most likely, improved hypotheses for researching "non-conventional" roots—whether described as adventitious, air, shoot-borne, etc.—will arise from bases in Natural History and, especially therein, ecology. Indeed, meaningful definitions of non-conventional rooting may only arise from our further understanding of germane Natural History. Unfortunately, the Natural History of non-conventional roots is fragmentary at best.

At present, before additional natural historical study, it seems that improved hypotheses for adventitious rooting research would include:
- primordium initiation.
- structural (anatomical) levels that are relevant to all *imaginable* aspects of adventitious rooting in *diverse species*, with plants of reproducible pedigree.
- the many different types of adventitious, shoot-borne, air, etc., rooting.
- the plant and propagule as physiological *and* thermodynamic units.
- the predisposition (competence) to root, including polarities, plant maturation, organ and tissue age, and other "immobile" components.
- relations between predisposition to root, stock plant maturation and developmental anatomy.
- how plant species, especially perennials, "learn" and "remember" their daily, monthly, yearly, etc., "appropriate" physiological functions.
- dichotomous test schemes for multiple hypotheses.
- quantitative *and* qualitative research.
- observational and experimental data.
- only tests for which there are available and discerning methods, or for which such methods can soon be developed, for the chosen organizational level of the study.
- communicating purpose and intent to other scientists.

The foregoing research will not be easy, but great discovery never has been, as Christopher Columbus learned [Beveridge (1957, p. 41); for the whole story, see Lunenfeld (1991)]:

> ...(a) He [Columbus] *was obsessed with an idea—that since the world was round he could reach the Orient by sailing west, (b) the idea was by no means original, but evidently he had obtained some additional evidence from a sailor blown off his course who claimed to have reached land in the west and returned, (c) he met great difficulties in getting someone to provide the money to enable him to test his idea, as well as in the actual carrying out of the experimental voyage, (d) when finally he succeeded he did not find the expected new route, but instead found a whole new*

world, (e) despite all evidence to the contrary he clung to the bitter end to his hypothesis and believed that he had found the route to the Orient, (f) he got little credit or reward during his lifetime and neither he nor others realized the full implications of his discovery, (g) since his time evidence has been brought forward showing that he was by no means the first European to reach America.... [For a rather complete confirmation of the above truths, see, e.g., Fitting (1909, 1910)].

We encourage the reader to embark or continue on a voyage of great discoveries about the fundamental biology, especially the control, of adventitious rooting. The fastest vessel to that New World of The First Law of Adventitious Rooting is the stout ship Strong Inference; it could depart from your laboratory, momentarily.

ACKNOWLEDGMENTS

We thank Ms. K. Haissig for her exacting translations and interpretations of non-English-language articles—and B.E.H. thanks her for doing so for 33 years. We thank Ms. S.K. Haissig for help in checking our literature citations. We also thank, for their critical reviews and discussions of the manuscript: Prof. L. Eliasson, Prof. W.P. Hackett, Ms. K. Haissig, Dr. P.R. Larson, Dr. J.R. Potter, Dr. D.E. Riemenschneider, Prof. J. van Staden and Dr. J. Zasada. Finally, we thank Prof. P.W. Barlow for reviewing the Terminology section.

REFERENCES

Anon., 1941, Factors affecting the vegetative propagation of forest trees, *For. Abs.* 3:2.

Altman, A., and Wareing, P.F., 1975, The effect of IAA on sugar accumulation and basipetal transport of ^{14}C-labelled assimilates in relation to root formation in *Phaseolus vulgaris* cuttings, *Physiol. Plant.* 33:32.

Aucher, E.C., 1930, American experiments in propagating deciduous fruit trees by stem and root cuttings, *in*: "Rept. and Proc. 9th Int. Hortic. Cong.," p. 287.

Avery, G.S., and Johnson, E.B., 1947, "Hormones and Horticulture," McGraw-Hill Book Co., New York.

Bachelard, E.P., and Stowe, B.B., 1963, Rooting of cuttings of *Acer rubrum* L. and *Eucalyptus camaldulensis* Dehn, *Aust. J. Biol. Sci.* 16:751.

Bacon, F., 1620, "The New Organon," F.H. Anderson, ed., The Library of Library of Liberal Arts, The Bobbs-Merrill Co., Inc., New York.

Balasimha, D., and Subramonian, N., 1983, Roles of phenolics in auxin induced rhizogenesis & isoperoxidases in cacao (*Theobroma cacao* L.) stem cuttings, *Indian J. Exp. Biol.* 21:65.

Balfour, I.B., 1913, Problems of propagation, *J. Roy. Hortic. Soc.* 38:447.

Barlow, P.W., 1986, Adventitious roots of whole plants: their forms, functions, and evolution, *in*: "New Root Formation in Plants and Cuttings," M.B. Jackson, ed., Martinus Nijhoff Pubs., Dordrecht.

Basu, R.N., Ghosh, B., and Sen, P.K., 1968, Naturally occurring rooting factors in mango *(Mangifera indica* L.), *Indian Agric.* 12:194.

Basu, R.N., Roy, B.N., and Bose, T.K., 1970. Interaction of abscisic acid and auxins in rooting of cuttings, *Plant Cell Physiol.* 11:681.

Becker, D., Sahali, Y., and Raviv, M., 1990, The absolute configuration effect on the activity of the avocado rooting promoter, *Phytochem.* 29:2065

Beijerinck, M.W., 1886, Beobachtungen und Betrachtungen über Wurzelknospen und Nebenwurzeln, *Verz. Geschr.* II:7.

Beveridge, W.I.B., 1957, "The Art of Scientific Investigation," W.W. Norton & Co., Inc., New York.

Biran, I., and Halevy, A.H., 1973, Endogenous levels of growth regulators and their relationship to the rooting of *Dahlia* cuttings, *Physiol. Plant.* 28:436.

Blazich, F.A., 1988, Chemicals and formulations used to promote adventitious rooting, *in*: "Adventitious Root Formation by Cuttings," T.D. Davis, B.E. Haissig, and N. Sankhla, eds., Adv. in Plant Sci. Ser., vol. 2, Dioscorides Press, Portland, p. 132.

Bloch, R., 1943a, Polarity in plants, *Bot. Rev.* 9:261.

Bloch, R., 1943b, The problem of polarity in plant morphogenesis, *Chron. Bot.* 7:297.

Bloch, R., 1965, Polarity and gradients in plants: A survey, *Handb. D. Pflanzenphysiol.* 15:234.

Bojarczuk, K., 1978, Studies on endogenous rhizogenic substances during the process of rooting lilac *Syringa vulgaris* L.) cuttings, *Plant Prop.* 24:3.

Bollmark, M, and Eliasson, L., 1990, A rooting inhibitor present in Norway spruce seedlings grown at high irradiance-a putative cytokinin, Physiol. Plant. 80:527.

Bollmark, M., Kubat, B., and Eliasson, L., 1988, Variation in endogenous cytokinin content during adventitious root formation in pea cuttings, J. Plant Physiol. 132:262.

Bondi, H., 1992, The philosopher of science, Nature 358:363.

Bonner, J., Huang, R.C., and Gilden, R.V., 1963, Chromosomally directed protein synthesis, Proc. Nat. Acad. Sci. USA 50:893.

Böttger, I., and Lüdemann, I., 1964, Über die Bildung einer stoffwechsel-aktiven Ribonucleinsäurefraktion in isolierten Blättern von Euphorbia pulcherrima zu Beginn der Wurzelregeneration, Flora 155:331.

Bouillenne, R., and Bouillenne-Walrand, M., 1939, Teneur en auxines des plantules et hypocotyles inanitiés de "Impatiens Balsamina" L. en rapport avec l'organogénèse des racines, Bull. Acad. Roy. Belg. 16:473

Bouillenne, R. and Bouillenne-Walrand, M., 1947, Détermination des facteurs de la rhizogénèse. Bull. Acad. Roy. Belg. 33:790.

Bouillenne, R., and Bouillenne-Walrand, M., 1955, Auxins et bouturage, in: "Proc. 14th Int. Hortic. Cong.," 1:231.

Bouillenne, R., and Went, F., 1933, Recherches expérimentales sur la néoformation des racines dans les plantules et les boutures des plantes supérieures, Ann. Jard. Bot. Buitenzorg 43:25.

Brian, P.W., Hemming, H.G., and Lowe, D., 1960, Inhibition of rooting of cuttings by gibberellic acid, Ann. Bot. 24:407.

Brian, P.W., Hemming, H., and Radley, M., 1955, A physiological comparison of gibberellic acid with some auxins, Plant Physiol. 8:899.

Breen, P.J., and Muraoka, T., 1973, Effect of indolebutryic acid on distribution of ^{14}C-photosynthate in softwood cuttings of 'Marianna 2624' plum, J. Amer. Soc. Hortic. Sci. 98:436.

Breen, P.J., and Muraoka, T., 1974, Effect of leaves on carbohydrate content and movement of ^{14}C-assimilate in plum cuttings, J. Amer. Soc. Hortic. Sci. 99:326.

Bunge, M., 1967, "Scientific Research I, The Search for System," and "II, The Search for Truth," Springer-Verlag, New York.

Büsgen, M., and Münch, E., 1929, "The Structure and Life of Forest Trees," 3rd ed., English trans. by T. Thomson, John Wiley & Sons, Inc., New York.

Carlson, M.C., 1929, Microchemical studies of rooting and non-rooting rose cuttings, Bot. Gaz. 87:64.

Carlson, M.C., 1938, The formation of nodal adventitious roots in Salix cordata, Amer. J. Bot. 25:721.

Carlson, M.C., 1950, Nodal adventitious roots in willow stems of different ages, Amer. J. Bot.37:555.

Challenger, S., Lacey, H.J., and Howard, B.H., 1965, The demonstration of root promoting substances in apple and plum rootstocks, in: "Rep. East Malling Res. Sta. for 1964," p. 124.

Chamberlin, T.C., 1897, Studies for students, The method of multiple working hypotheses, J. Geol. 5(8):837; reprinted as "Multiple Hypotheses, A Method for Research, Teaching, and Creative Thinking," Inst. for Humane Studies, Inc., Stanford.

Chin, T., Meyer, M.M, Jr., and Beevers, L., 1969, Abscisic acid-stimulated rooting of stem cuttings, Planta 88:192.

Choong, L.I., McGuire, J.L., and Kitchin, J.T., 1969, The relationship between rooting cofactors and easy and difficult-to-root cuttings of three clones of Rhododendron, J. Amer. Soc. Hortic. Sci. 94:45.

Claudot, A.C., Jay-Allemand, C., Magel, E.A., and Drouet, A., 1993, Phenylalanine ammonia-lyase, chalcone synthase and polyphenolic compounds in adult and rejuvenated hybrid walnut tree [sic.], Trees 7:92-97.

Cooper, W.C., 1935, Hormones in relation to root formation on stem cuttings, Plant Physiol. 10:789.

Cooper, W.C., 1936, Transport of root-forming hormone in woody cuttings, Plant Physiol. 11:779.

Cooper, W.C., 1938, Hormones and root formation, Bot. Gaz. 99:599.

Corbett, L.C., 1897, The development of roots from cuttings, Meehans' Monthly 7:32.

Crocker, W., Hitchcock, A.E., and Zimmerman, P.W., 1935, Similarities in the effects of ethylene and the plant auxins, Contrib. Boyce Thomp. Inst. 7:231.

Curtis, O.F., 1918, "Stimulation of Root Growth in Cuttings by Treatment with Chemical Compounds," Cornell Univ. Agric. Exp. Sta. Mem. 14, New York.

Darwin, C., 1859, "The Origin of Species," reprinted by The Modern Library, New York.

Davies, P.J., 1988, The plant hormones: their nature, occurrence, and functions, in: "Plant Hormones and Their Role in Plant Growth and Development," P.J. Davies, ed., Academic Pubs., Dordrecht.

Davis, T.D., and Haissig, B.E., 1990, Chemical control of adventitious root formation in cuttings, Plant Growth Reg. Soc. Amer. Quart. 18(1):1.

Davis, T.D., and Sankhla, N., 1988, Effect of shoot growth retardants and inhibitors on adventitious rooting, in: "Adventitious Root Formation by Cuttings," T.D. Davis, B.E. Haissig, and N. Sankhla, eds., Adv. in Plant Sci. Ser., vol. 2, Dioscorides Press, Portland, p. 174.

Davis, T.D., Haissig, B.E., and Sankhla, N., eds., 1988, "Adventitious Root Formation in Cuttings," Dioscorides Press, Portland.

De Bary, A., 1884, "Comparative anatomy of phanerogams and ferns," English trans. by F.O. Bower, and D.H. Scott, Clarendon Press, Oxford.

De Candolle, A.-P., 1825, Premier mémoire sur les lenticelles des arbres et le développement des racines qui en sortent, *Ann. Sci. Nat.*, p. 5.

De Candolle, A.-P., 1832, "Physiologie Végétale, ou Exposition des Forces et des Fonctions Vitales des Végétaux," vols. 1 and 2, Béchet Jeune, Lib. Fac. Med., Paris.

De Haan, I., 1936, Polar root formation, *Rec. Trav. Bot. Néerl.* 33:292.

Deuber, C.G., 1940, Vegetative propagation of conifers, *Trans. Conn. Acad. Arts and Sci.* 43:1.

Devaux, H., 1899, Asphixie spontanée et production d'alcool dans les tissus profonds des tiges ligneuses poussant dans les conditions naturelles, *C.R. Acad. Sci.* Paris 128:1346.

Dick, J. McP., and Dewar, R.C., 1992, A mechanistic model of carbohydrate dynamics during adventitious root development in leafy cuttings, *Ann. Bot.* 70:371.

Diels, L., 1906, "Jugendformen und Blütenreife," Gebrüder Borntraeger Verlag, Berlin.

Driesch, H., 1901, "Die organisechen Regulationen," Leipzig.

Duhamel du Monceau, H.L., 1758, "La Physique des Arbres," Vols. I and II, Guerin and Delatour, Paris.

Dunn, S., and Townsend, R.J., 1954, Propagation of sugar maple from vegetative cuttings, *J. For.* 52:678.

Esau, K., 1977, "Anatomy of Seed Plants," John Wiley & Sons, Inc., New York.

Fadl, M.S., and Hartmann, H.T., 1967a, Relationship between seasonal changes in endogenous promoters and inhibitors in pear buds and cutting bases and the rooting of pear hardwood cuttings, *Proc. Amer. Soc. Hortic. Sci.* 91:96-112.

Fadl, M.S., and Hartmann, H.T., 1967b, Isolation, purification, and characterization of an endogenous root-promoting factor obtained from basal sections of pear hardwood cuttings, *Plant Physiol.* 42:541.

Fellenberg, G., 1965, Hemmung der Wurzelbildung an etiolierten Erbsenepikotylen durch Bromuracil und Histon, *Planta* 64:287.

Fellenberg, G., 1966, Die Hemmung auxininduzierter Wurzelbildung an etiolierten Erbsenepikotylen mit Histon und Antimetaboliten der RNS- und Proteinsynthese, *Planta* 71:27.

Fellenberg, G., 1967, Möglichkeiten der Regulierung differentieller DNS-Aktivitäten bei höheren Pflanzen durch Histon, *Planta* 76:252.

Fellenberg, G., 1969a, Veränderungen des Nucleoproteids von Erbsenepikotylen durch synthetische Auxine bei der Induktion der Wurzelneubildung, *Planta* 84:195.

Fellenberg, G., 1969b, Veränderungen des Nucleoproteids unter dem Einfluss von Auxine und Ascorbinsäure bei der Wurzelneubildung an Erbsenepikotylen, *Planta* 84:324.

Fellenberg, G., 1969c, Hemmung der Wurzelneubildung durch saure und neutrale Kernproteine, *Planta* 86:165.

Fernqvist, I., 1966, Studies on factors in adventitious root formation, *Ann. Agric. Coll. Sweden* 32:109.

Fischnich, O., 1935, Über den Einfluss von ß-Indolylessigsäure auf die Blattbewegungen und die Adventivwurzelbildung von *Coleus*, *Planta* 24:552.

Fitting, H., 1909, Die Beeinflussung der Orchideenblüten durch die Bestäubung und durch andere Umstände, *Zeit. Planzenphysiol.* 1:1.

Fitting, H., 1910, Weitere entwicklungsphysiologische Untersuchungen an Orchideenblüten, *Zeit. Pflanzenphysiol.* 2:225.

Gardner, F.E., 1929, The relationship between tree age and the rooting of cuttings, *Proc. Amer. Soc. Hortic. Sci.* 26:101.

Gaspar, T., Smith, D., and Thorpe, T., 1977, Arguments supplémentaires en faveur d'une variation inverse du niveau auxinique endogène au cours des deux premières phases de la rhizogénèse, *C.R. Acad. Sci. Paris* 285:327.

Geissbühler, H., and Skoog, F., 1957, Comments on the application of plant tissue cultivation to propagation of forest trees, *TAPPI* 40:258.

Geneve, R.L., Mokhtari, M., and Hackett, W.P. 1991. Adventitious root initiation in reciprocally grafted leaf cuttings from the juvenile and mature phase of *Hedera helix* L., *J. Exp. Bot.* 42:65.

Gesto, M.D.V., Vazques, A., and Vieitez, E., 1981, Changes in the rooting inhibitory effect of chestnut extracts during cold storage of the cuttings, *Physiol. Plant.* 51:365.

Girouard, R.M., 1967, Anatomy of adventitious root formation in stem cuttings, *in*: "Proc. Int. Plant Prop. Soc. Annu. Meeting," 1967:289.

Girouard, R.M., 1969, Physiological and biochemical studies of adventitious root formation, Extractable rooting cofactors from *Hedera helix*, *Can. J. Bot.* 47:687.

Goebel, K., 1889, Ueber die Jugenzustände der Pflanzen, *Flora* 72:1.

Goebel, K., 1898-1901, "Organographie der Pflanzen," Gustav Fischer Verlag, Jena.

Goebel, K., 1902, Ueber Regeneration im Pflanzenreich, *Biol. Centralbl.* 22:385.

Goebel, K., 1903a, Regeneration in plants, *Bull. Torrey Bot. Club* 30:197.

Goebel, K., 1903b, Morphologische und biologische Bemerkungen, 14. Weitere Studien über Regeneration, *Flora* 92:132.

Goebel, K., 1903c, Studien ueber Regeneration, *Flora* 92:132.

Goebel, K., 1905a, "Organography of Plants Especially of the Archegoniatae and Spermatophyta," English edition by I.B. Balfour, Clarendon Press, Oxford.

Goebel, K., 1905b, Allgemeine Regenerationsprobleme, *Flora* 95:384.

Goodwin, R.H., and Goddard, D.R., 1940, The oxygen consumption of isolated woody tissues, *Amer. J. Bot.* 27:234.

Grace, N.H., 1937, Physiologic curve of response to phytohormones by seeds, growing plants, cuttings, and lower plant forms, *Can. J. Res.* 15(C):538.

Grace, N.H., 1945, Liberation of growth stimulating materials by rooting *Salix* cuttings, *Can. J. Res.* 23(C):85.

Gregory, L.E., and van Overbeek, J., 1945, An analysis of the process of root formation on cuttings of a difficult hibiscus variety, *Proc. Amer. Soc. Hortic. Sci.* 46:427.

Groff, P.A., and Kaplan, D.R., 1988, The relation of root systems to shoot systems in vascular plants, *Bot. Rev.* 54:387.

Guillot, A., 1965, Action de la 2-thiouracile sur la rhizogénèse dans les boutures de plantules étiolées de tomate, *Planta* 67:96.

Guillot, A., 1971, Action de la 5-bromouracile et de ses nucléosides sur la morphogénèse des boutures de plantules étiolées 2-thiouracile sur la rhizogénèse dan les boutures de plantules étiolées de tomate, *Planta* 67:96.

Guillot, A., 1972, Action de la désoxyuridine sur l'inhibition de la rhizogénèse et de la croissance de l'hypocotyle observée en présence de 5-bromodésoxyuridine chez le boutures de plantules étiolées de tomate, *Planta* 102:127.

Haagen-Smit, A.J., Dandliker, W.B., Wittwer, S.H., and Murneek, A.E., 1946, Isolation of 3-indoleacetic acid from immature corn kernels, *Amer. J. Bot.* 33:118.

Haberlandt, G., 1914, Zur Physiologie der Zellteilung, *Sitz. Ber. K. Preuss. Akad. Wiss.* 1914:1096.

Hackett, W.P., 1970, The influence of auxin, catechol and methanolic tissue extracts on root initiation in aseptically cultured shoot apices of the juvenile and adult forms of *Hedera helix*, *J. Amer. Soc. Hortic. Sci.* 95:398.

Hackett, W.P., 1988, Donor plant maturation and adventitious root formation, *in:* "Adventitious Root Formation by Cuttings," T.D. Davis, B.E. Haissig, and N. Sankhla, eds., Advances in Plant Sciences Series, vol. 2, Dioscorides Press, Portland, p. 11.

Hagemann, A., 1932, Untersuchungen an Blattstecklingen. *Gartenbauwiss.* 6:69.

Haffner, V., Enjalric, F., Lardet, L., and Carron, M.P., 1991, Maturation of woody plants: a review of metabolic and genomic aspects, *Ann. Sci. For.* 48:615.

Haissig, B.E., 1965, Organ formation *in vitro* as applicable to forest tree propagation, *Bot. Rev.* 31:607.

Haissig, B.E., 1970, Influence of indole-3-acetic acid on adventitious root primordia of brittle willow, *Planta* 95:27.

Haissig, B.E., 1971, Influences of indole-3-acetic acid on incorporation of ^{14}C-uridine by adventitious root primordia of brittle willow, *Bot. Gaz.* 132:262.

Haissig, B.E., 1972, Meristematic activity during adventitious root primordium development, Influences of endogenous auxin and applied gibberellic acid, *Plant Physiol.* 49:886.

Haissig, B.E., 1974a, Origins of adventitious roots, *N.Z. J. For. Sci.* 4:299.

Haissig, B.E., 1974b, Influences of auxins and auxin synergists on adventitious root primordium initiation and development, *N.Z. J. For. Sci.* 4:311.

Haissig, B.E., 1982a, Carbohydrate and amino acid concentrations during adventitious root primordium development in *Pinus banksiana* Lamb. cuttings, *For. Sci.* 28:813.

Haissig, B.E., 1982b, Activity of some glycolytic and pentose phosphate pathway enzymes during the development of adventitious roots, *Physiol. Plant.* 55:261.

Haissig, B.E., 1986, Metabolic processes in adventitious rooting, *in:* "New Root Formation in Plants and Cuttings," M.B. Jackson, ed., Martinus Nijhoff Pubs., Dordrecht, p. 141.

Haissig, B.E., 1988, Future directions in adventitious rooting research, *in:*"Adventitious Root Formation by Cuttings," T.D. Davis, B.E. Haissig, and N. Sankhla, eds., Adv. in Plant Sci. Ser., vol. 2, Dioscorides Press, Portland, p. 303.

Haissig, B.E., and Riemenschneider, D.E., 1988, Genetics of adventitious rooting, *in:* "Adventitious Root Formation by Cuttings," T.D. Davis, B.E. Haissig, and N. Sankhla, eds., Adv. in Plant Sci. Ser., vol. 2, Dioscorides Press, Portland, p. 47.

Haissig, B.E., and Riemenschneider, D.E., 1992, The original basal stem section influences rooting in *Pinus banksiana*, *Physiol. Plant.* 86:1.

Haissig, B.E., Davis, T.D., and Riemenschneider, D.E., 1992, Researching the controls of adventitious rooting, *Physiol. Plant.* 84:310.

Hansen, J., 1988, Influence of gibberellins on adventitious root formation, *in*: "Adventitious Root Formation by Cuttings," T.D. Davis, B.E. Haissig, and N. Sankhla, eds., Adv. in Plant Sci. Ser., vol. 2, Dioscorides Press, Portland, p. 162.

Harder, R., Schumacher, W., Firbas, F., and Denffer, D. von, 1965, "Strasburger's Textbook of Botany," English trans. by P. Bell, and D. Coombe, Longmans, Green and Co., Ltd., London.

Hartmann, H.T., Kester, D.E., and Davies, F.T., Jr., 1990, "Plant Propagation, Principles and Practices," Prentice-Hall, Inc., Englewood Cliffs.

Hartsema, A.M., 1924, Anatomische und experimentelle Untersuchungen über das Auftreten von Neubildungen an Blättern von *Begonia rex*, *Rec. Trav. Bot. Néerl.* 23:305.

Hatton, R.G., 1930, Stock : scion relationships, *J. Roy. Soc. Hortic. Sci.* 55(II):169.

Heide, O.M., 1965, Interaction of temperature, auxins, and kinins in the regeneration ability of *Begonia* leaf cuttings, *Physiol. Plant.* 18:891.

Hess, C.E., 1959, A study of plant growth substances in easy and difficult-to-root cuttings, *in*: "Proc. 9th Annu. Intl. Plant Prop. Soc.," p. 39.

Hess, C.E., 1960, Research in root initiation - a progress report, *in*: "Proc. 10th Annu. Meeting Plant Prop. Soc.," p. 118.

Hess, C.E., 1961, The physiology of root initiation in easy- and difficult-to-root cuttings, *Hormolog* 3:3.

Hess, C.E., 1962, Characterization of the rooting cofactors extracted from *Hedera helix* L. and *Hibiscus rosa-sinensis* L., *in*: "Proc. Int. Plant Prop. Soc.," 1961:51.

Heuser, C.W., 1988, Bioassay, immunoassay, and verification of adventitious root promoting substances," *in*: "Adventitious Root Formation by Cuttings," T.D. Davis, B.E. Haissig, and N. Sankhla, eds., Adv. in Plant Sci. Ser., vol. 2, Dioscorides Press, Portland, p. 274.

Hicks, P.A., 1928, Chemistry of growth as represented by carbon/nitrogen ratio, Regeneration of willow cuttings, *Bot. Gaz.* 86:193.

Hinds, H.V., ed., 1974, "Special Issue on Vegetative Propagation," *N.Z. J. For.*, vol. 4, no. 2.

Hitchcock, A.E., 1935a, Tobacco as a test plant for comparing the effectiveness of preparations containing growth substances, *Contrib. Boyce Thomp. Inst.* 7:349.

Hitchcock, A.E., 1935b, Indole-3-*n*-propionic acid as a growth hormone and the quantitative measurement of plant response, *Contrib. Boyce Thomp. Inst.* 7:87.

Hitchcock, A.E., and Zimmerman, P.W., 1931, Rooting of greenwood cuttings as influenced by the age of tissue at the base, Proc. Amer. Soc. Hortic. Sci. 27:136.

Hitchcock, A.E., and Zimmerman, P.W., 1932, Relation of rooting response to age of tissue at the base of greenwood cuttings, *Contrib. Boyce Thomp. Inst.* 4:85.

Hitchcock, A.E., and Zimmerman, P.W., 1936, Effect of growth substances on the rooting response of cuttings, *Contrib. Boyce Thomp. Inst.* 8:63.

Humphries, E.C., 1960, Inhibition of root development on petioles and hypocotyls of dwarf bean (*Phaseolus vulgaris*) by kinetin, *Physiol. Plant.* 13:659.

Hutchins, R.M., ed., 1953, "Francis Bacon. Advancement of Learning, Novum Organum, New Atlantis," in: "Great Books of the Western World," Encyclopedia Britannica, Inc., Chicago.

Jackson, M.B., ed., 1986, "New Root Formation in Plants and Cuttings," Martinus Nijhoff Pubs., Dordrecht.

Jackson, M.B., and Harney, P.M., 1970, Rooting cofactors, indoleacetic acid and adventitious root initiation in mung bean cuttings (*Phaseolus aureus*), *Can. J. Bot.* 48:943.

Jacob, F., and Monod, J., 1961, Genetic regulatory mechanisms in the synthesis of proteins, *J. Mol. Biol.* 3:318.

Jacobs, W.P., 1979, "Plant Hormones and Plant Development," Cambridge Univ. Press, Cambridge.

Jones, A.M., 1990, Do we have the auxin receptor yet?, *Physiol. Plant.* 80:154.

Julliard, J., Sotta, B., Pelletier, G., and Miginiac, E., 1992, Enhancement of naphthaleneacetic acid-induced rhizogenesis in T_L-DNA-transformed *Brassica napus* without significant modification of auxin levels and auxin sensitivity, *Plant Physiol.* 100:1277.

Kato, J., 1958, Studies on the physiological effect of gibberellin, II. On the interaction of gibberellin with auxins and growth inhibitors, *Physiol. Plant.* 11:10.

Kawase, M., 1964, Centrifugation, rhizocaline and rooting in *Salix alba* L., *Physiol. Plant.* 17:855.

Kawase, M., 1970, Root-promoting substances in *Salix alba*, *Physiol. Plant.* 23:159.

Kawase, M., 1981, A "dream" chemical to aid propagation of woody plants, *Ohio Rept.* 66(1):8.

Klebs, G., 1903, "Willkürliche Entwicklungsänderungen bei Pflanzen," Gustav Fischer Verlag, Jena.

Knight, T.A., 1809, On the origin and formation of roots, Roy. Soc. Philos., pt. 1, p. 16., Reprinted: Sq. Q. London, 1809.

Kögl, F., Haagen-Smit, J.A., and Erxleben, H., 1934, Über ein neues Auxin ("Hetero-auxin") aus Harn, 11. Mitteilung über planzliche Wachstumsstoffe, *Hoppe-Seylers Zeit. Physiol. Chem.* 228:90.

Kraft, J.C., Shepard, T., and Juchau, M.R., 1993, Tissue levels of retinoids in human embryos/fetuses, *Reprod. Toxicol.* 7:11-15.

Kraus, E.J., and Krabill, H.R., 1918, "Vegetation and Reproduction with Special Reference to Tomato," Oregon Agric. Coll. Exp. Sta. Bull. 149.

Kraus, E.J., Brown, N.A., and Hamner, K.C., 1936, Histological reactions of bean plants to indoleacetic acid, *Bot. Gaz.* 98:370.

Krenke, N.P., 1933, "Wundkompensation, Transplantation und Chimären bei Pflanzen. Monographie aus dem Gesamtgebiet der Physiologie der Pflanzen und der Tiere," trans. from Russian, J. Springer-Verlag, Berlin.

Kühne, W. von, 1878, Erfahrungen und Bemerkungen über Enzyme und Fermente, *Heidelberg Univ. Physiol. Inst. Unters.* 1:291.

Kunkel, A., 1878, Ueber elektromotorische Wirkungen an unverletzten lebenden Pflanzentheilen, *Arb. Deutsh Bot. Inst. Würzburg* 2:2.

Kunkel, A., 1879, Ueber einige Eigenthümlichkeiten des elektrishen Leitungsvermögens lebender Pflanzentheile, Arb. Deutsch Bot. Inst. Würzburg 2:333.

Kupfer, E., 1907, Studies of plant regeneration, *Mem. Torrey Bot. Club.* 12:195.

Küster, E., 1903, Beobachtungen über Regenerationserscheinungen an Pflanzen, *Beih. Bot. Centralbl.* 14:316.

Larson, P.R., 1975, Development and organization of the primary vascular system in *Populus deltoides* according to phyllotaxy, *Amer. J. Bot.* 62:1084.

Larson, P.R., 1979, Establishment of the vascular system in seedling of *Populus deltoides* Bartr., *Amer. J. Bot.* 66:452.

Larson, P.R., 1980, Interrelations between phyllotaxis, leaf development and the primary-secondary transition in *Populus deltoides*, *Ann. Bot.* 46:757.

Lee, C.I., 1971, "Influence of Intermittent Mist on the Development of Anthocyanins and Root-Inducing Substances in *Euonymus alatus* (Sieb.) 'Compactus,' Ph.D. thesis, Cornell Univ., No. 72-8957, Univ. Microfilms Int., Ann Arbor.

Lee, C.I., McGuire, J.J., and Kitchin, J.T., 1969, The relationship between rooting cofactors of easy and difficult-to-root cuttings of three clones of *Rhododendron*, *J. Amer. Soc. Hortic. Sci.* 94:45.

Lemaire, A., 1886, Recherches sur l'origine et le developpement des racines laterales chez les dicotyledones, *Ann. Sci. Nat. Ser. VII Bot.* 3:163.

Letham, D.S., 1963, Zeatin, a factor inducing cell division isolated from *Zea mays*, *Life Sci.* No. 8, p. 569.

Libbert, E., 1956, Untersuchungen über die Physiologie der Adventivwurzelbildung, I. Die Wirkungsweise einiger Komponenten des Rhizocalinkomplexes, *Flora* 144:121.

Linser, H., 1940, Über das Vorkommen von Hemmstoff in Pflanzenextrakten, sowie über das Verhältnis von Wuchsstoffgehalt und Wuchsstoffabgabe bei Pflanzen oder Planzenteilen, *Planta* 31:32.

Linser, H., 1948, Über den Einfluß von Pflanzenextrakten auf das Streckungswachstum, Wurzel- und Sproßbildung bei Planzen, *Ost. Bot. Zeit.* 95:95.

Lipecki, J., and Dennis, F.G., 1972, Growth inhibitors and rooting cofactors in relation to rooting response of softwood apple cuttings, *HortSci.* 7:136.

Loeb, J., 1917a, Influence of the leaf upon root formation and geotropic curvature in the stem of *Bryophyllum calycinum* and the possibility of a hormone theory of these processes, *Bot. Gaz.* 63:25.

Loeb, J., 1917b, The chemical basis of axial polarity in regeneration, *Science* 46:547.

Loeb, J., 1919a, The physiological basis of morphological polarity in regeneration. I., *J. Gen. Physiol.* 1:337.

Loeb, J., 1919b, The physiological basis of morphological polarity in regeneration. II., *J. Gen. Physiol.* 1:687.

Loeb, J., 1923, Theory of regeneration based on mass action, *J. Gen. Physiol.* 5:831.

Loeb, J., 1924, "Regeneration from a Physico-Chemical Viewpoint," McGraw-Hill Book Co., Inc., New York.

Lovell, P.H., and White, J., 1986, Anatomical changes during adventitious root formation, *in:* "New Root Formation in Plants and Cuttings," M.B. Jackson, ed., Martinus Nijhoff Pubs., Dordrecht, p. 111.

Lugovoy, M., 1937, The rooting and non-rooting of tree species in connection with the anatomical structure of lenticels, *Ukrain. Akad. Sci., Inst. Bot. J.*, No. 23, p. 239.

Lund, E.J., 1923, Electrical control of organic polarity in the egg of *Fucus*, *Bot. Gaz.* 76:288.

Lund, E.J., 1924, Experimental control of organic polarity by the electric current, V. The nature of the control of organic polarity by the electric current, *J. Exp. Zool.* 41:155.

Lund, E.J., 1928, Relation between continuous bio-electric currents and cell respiration II., *J. Exp. Zool.* 51:265.

Lund, E.J., 1930, Internal distribution of the electric correlation potentials in the Douglas fir, *Pub. Puget Sound Biol. Sta.* 7:259.

Lund, E.J., 1931, The unequal effect of O_2 concentration on the velocity of oxydation on loci of different electric potential, and glutathione content, *Protoplasma* 13:236.

Lund, E.J., and Kenyon, W.A., 1927, Relation between continuous bio-electric currents and cell respiration I. Electric correlation potentials in growing root tips, *J. Exp. Zool.* 48:333.

Lunenfeld, M., ed., 1991, "1492, Discovery, Invasion, Encounter," D.C. Heath and Co., Lexington.

Mahlstede, J.P., and Watson, D.P., 1952, An anatomical study of adventitious root development in stems of *Vaccinium corymbosum, Bot. Gaz.* 113:279.

Majima, R., and Hoshino, T., 1925, Synthetische Versuche in der Indol-Gruppe, VI. Eine neue Synthese von β-Indolyl-alkylaminen, *Ber. Deutsch. Chem. Gesell.* 58:2042.

Marston, M.E., 1955, The history of vegetative propagation, *in*: "Proc. 14th Int. Hortic. Cong.," p. 1157.

Massart, J., 1918, Sur la polarité des organes végétaux, *Bull. Biol. Fr. Belg.* 51:475.

Maurel, C., Barbier-Brygoo, H., Spena, A., Tempé, J., and Guern, J., 1991, Single *rol* genes from the *Agrobacterium rhizogenes* TL-DNA alter some of the cellular responses to auxin in *Nicotiana tabacum, Plant Physiol.* 97:212.

Maurel, C., Brevet, J., Barbier-Brygoo, H., Guern, J., and Tempé, J., 1990, Auxin regulates the promoter of the root-inducing *rol*B gene of *Agrobacterium rhizogenes* in transgenic tobacco, *Mol. Gen. Genet.* 223:58.

Mayer F., and Oppenheimer, T., 1916, Über Naphthyl-essigsäuren, *Ber. Deutsh. Chem. Ges.* 49:2137.

McCallum, W.B., 1905a, Regeneration in plants. I., *Bot. Gaz.* 40:97.

McCallum, W.B., 1905b,. Regeneration in plants. II., *Bot. Gaz.* 40:241.

Mitsuhashi, M., Shibaoka, H., and Shimokoriyama, M., 1969a, Portual: A root promoting substance in *Portulaca* leaves, *Plant Cell Physiol.* 10:715.

Mitsuhashi, M., Shibaoka, H., and Shimokoriyama, M., 1969b, Morphological and physiological characterization of IAA-less-sensitive and IAA-sensitive phases in rooting of *Azukia* cuttings, *Plant Cell Physiol.* 10:867.

Molnar, J.M., and LaCroix, L.J., 1972a, Studies of the rooting of cuttings of *Hydrangea macrophylla*: enzyme changes, *Can. J. Bot.* 30:315.

Molnar, J.M., and LaCroix, L.J., 1972b, Studies on the rooting of cuttings of *Hydrangea macrophylla*: DNA and protein changes, *Can. J. Bot.* 50:387.

Moorby, J., and Wareing, P.F., 1963, Ageing in woody plants, *Ann. Bot.* 27:291.

Moore, T.C., 1979, "Biochemistry and Physiology of Plant Hormones," Springer-Verlag, New York.

Morgan, T.H., 1901, "Regeneration," Columbia Univ, Biol, Ser. VII, The MacMillan Co., New York.

Morgan, T.H., 1903, The hypothesis of formative stuffs, *Bull. Torrey Bot. Club* 30:206.

Morgan, T.H., 1904, Polarity and regeneration in plants, *Bull. Torrey Bot. Club* 31:227.

Morgan, T.H., 1906, The physiology of regeneration, *J. Exp. Zool.* 3:457.

Mudge, K.W., 1988, Effect of ethylene on rooting, *in*: "Adventitious Root Formation by Cuttings," T.D. Davis, B.E. Haissig, and N. Sankhla, eds., Adv. in Plant Sci. Ser., vol. 2, Dioscorides Press, Portland, p. 150.

Mullins, M.G., 1985, Regulation of adventitious root formation in microcuttings, *Acta Hortic.* 166:53.

Nanda, K.K., Jain, M.K., and Malhotra, S., 1971. Effect of glucose and auxins in rooting etiolated stem segments of *Populus nigra, Physiol. Plant.* 24:387.

Nanda, K.K., Kumar, P., and Kochhar, V.K., 1974, Role of auxins, antiauxin and phenol in the production and differentiation of callus on stem cuttings of *Populus robusta, N.Z. J. For. Sci.* 4:338.

Němec, B., 1930, Bakterielle Wuchsstoffe, *Ber. Deutsch. Bot. Gesell.* 48:72.

Němec, B., 1934, Ernährung, Organogene und Regeneration, *Vest. Kral. Ces. Spol. Nauk. Tr.* 7:1.

Newton, A.C., Muthoka, P.N., and Dick, McP., 1992, The influence of leaf area on the rooting physiology of leafy stem cuttings of *Terminalia spinosa* Engl., *Trees* 6:210.

Niedergang-Kamien, E., and Leopold, A.C., 1957, Inhibitors of polar auxin transport, *Physiol. Plant.* 10:29.

Niedergang-Kamien, E., and Skoog, F., 1956, Studies on polarity and auxin transport in plants, I. Modification of polarity and auxin transport by triiodobenzoic acid, *Plant. Physiol.* 9:60.

Noll, F., 1900, Über den bestimmenden Einfluss von Wurzelkrümmungen auf Entstehung und Anordnung der Seitenwurzeln, *Landw. Jahrb.* 29:361.

Nussey, A.N., 1948, "Some Effects of Boron on the Rooting of Softwood cuttings," M.S. thesis, McGill Univ., Canada.

Orlikowska, T., 1992, Influence of arginine on *in vitro* rooting of dwarf apple rootstock, *Plant Cell Tissue Organ Cult.* 31:9.

Ott, J.J., 1763, "Dendrologie Europæ Mediæ, oder: Saat, Pflanzung, und Gebrauch des Holzes, nach den Grundsätzen des Herrn Duhamel," Heidegger und Compagnie, Zürich.

Paton, D.M., Willing, R.R., Nicholls, W., and Pryor, L.D., 1970, Rooting of stem cuttings of *Eucalyptus*: A rooting inhibitor in adult tissue, *Aust. J. Bot.* 18:175.

Payen, A., and Persoz, J.F., 1833, Memoire sur la diastase, les principaux produits de ses réactions, et leurs applications aux arts industriels, *Ann. Chim. Phys.* 53:73.

Platt, J.R., 1964, Strong inference, *Science* 146:347.

Pledge, H.T., 1947, "Science Since 1500, A Short History of Mathematics, Physics, Chemistry, Biology," Min. Educ., Sci. Museum.

Pliny (C. Plinius Secundus), ca. 77 A.D., "Natural History," English trans. by T.E. Page, E. Capps, L.A. Post, W.H.D. Rouse, and E.H. Warmington, eds., Loeb Classical Lib., Harvard Univ. Press, Cambridge.

Priestley, J.H., and Ewing, J., 1923, Physiological studies in plant anatomy, *New Phytol.* 22:30.

Priestley, J.H., 1926, Problems of vegetative propagation, *J. Roy. Hortic. Soc.* 51(I):1.

Priestley, J.H., and Swingle, C.F., 1929, "Vegetative Propagation from the Standpoint of Plant Anatomy," U.S. Dept. Agric. Tech. Bull. No.151.

Rasmussen, A., and Andersen, A.S., 1980, Water stress and root formation in pea cuttings, II. Effects of abscisic acid treatment of cuttings from stock plants grown under two levels of irradiance, *Physiol. Plant.* 48:150.

Raviv, M., Becker, D., and Sahali, Y., 1986, The chemical identification of root promoters extracted from avocado tissues, *Plant Growth Regul.* 4:371.

Reuveni, O., and Adato, I., 1974, Endogenous carbohydrates, root promoters, and root inhibitors in easy- and difficult-to-root date palm (*Phoenix dactylifera* L.) offshoots, *J. Amer. Soc. Hortic. Sci.* 99:361.

Richards, M., 1964, Root formation on cuttings of *Camellia reticulata* var. 'Capt. Rawes,' *Nature* 204:601.

Ross, A.C., 1993, Overview of retinoid metabolism, *Nutrition* 123:2 suppl., p. 346.

Ruge, U., 1957, Zur Wirkstoff-Analyse des Rhizokalin-Komplexes I, *Zeit. Bot.* 45:273.

Ruge, U., 1960, Zur Wirkstoff-Analyse des Rhizokalin-Komplexes II, *Zeit. Bot.*48:292

Sachs, J., 1880 and 1882, Stoff und Form der Pflanzenorgane, I. and II., *Arb. Bot. Inst. Würzburg* 2:452 and 689.

Sachs, J., 1887, "Vorlesungen über Pflanzen-Physiologie," Wilhelm Engelmann Verlag, Leipzig.

Sagee, O., Raviv, M., Medina, Sh., Becker, D., and Cosse, A., 1992, Involvement of rooting factors and free IAA in the rootability of citrus species stem cuttings, *Sci. Hortic.* 51:187.

Salkowski, E., and Salkowski, H., 1880, Ueber die skatolbildende Substanz, *Ber. Deutsch. Chem. Gesell.* 13:2217.

Schaffalitzky de Muckadell, M., 1959, Investigations on aging of apical meristems in woody plants and its importance in silviculture, reprinted from *Det forstlige Forsφgsvæsen i Danmark* 25:309, Kandrup & Wunsch's Bogtrykkeri, Kφbenhavn.

Schmidt, E., 1956, Anatomische Untersuchungen über das Vorkommen von Wurzelanlagen in verschiedenen Internodien von *Pisum sativum*, *Flora* 144:151.

Schmülling, T., Schell, J., and Spena, A., 1989, Promoters of the *rolA*, *B*, and *C* genes of *Agrobacterium rhizogenes* are differentially regulated in transgenic plants, *The Plant Cell* 1:665.

Seago, J.L., Jr., and Marsh, L.C., 1990, Origin and development of lateral roots in *Typha glauca*, *Amer. J. Bot.* 77:713.

Selby, C., Kennedy, S.J., and Harvey, B.M.R., 1992, Adventitious root formation in hypocotyl cuttings of *Picea sitchensis* (Bong.) Carr. - the influence of plant growth regulators, *New Phytol.* 120:453.

Shibaoka, H., 1971, Effects of indoleacetic, p-chlorophenoxyisobutyric and 2,4,6-trichlorophenoxyacetic acids on three phases of rooting in *Azukia* cuttings, *Plant and Cell Physiol.* 12:193.

Sitte, P., 1992, A modern concept of the "cell theory," A perspective on competing hypotheses of structure, *Int. J. Plant Sci.* 153(3):S1.

Skoog, F., and Miller, C.O., 1957, Chemical regulation of growth and organ formation in plant tissues cultured *in vitro*, *in*: "Biol. Action of Growth Sub. 11th Symp. Soc. Exp. Biol.," Cambridge Univ. Press, Cambridge, p. 118.

Skoog, F., Strong, F.M., and Miller, C.O., 1965, Cytokinins, *Science* 148:532.

Smith, E.P., 1928, A comparative study of the stem structure of the genus *Clematis* with special reference to anatomical changes induced by vegetative propagation, *Trans. Roy. Soc. Edinburgh* 55:643.

Snow, A.G., Jr., 1939, "Clonal Variation in Rooting Response of Red Maple Cuttings," USDA Forest Service, Northeastern For. Exp. Sta. Tech. Note. No. 29.

Spiegel, P., 1955, Some internal factors affecting rooting of cuttings, *in*: "Rept. 14th Int. Hortic. Cong.," vol. 1, p. 239.

Stebbins, G.L., 1992, Comparative aspects of plant morphogenesis: A cellular, molecular, and evolutionary approach, *Amer. J. Bot.* 79:589.

Stoutemyer, V.T., 1937, "Regeneration in Various Types of Apple Wood," Res. Bull. Iowa Agric. Exp. Sta., No. 220, p. 308.

Strydom, D.K., and Hartmann, H.T., 1960, Absorption, distribution, and destruction of indoleacetic acid in plant stem cuttings, *Plant Physiol.* 35:435.

Stuart, N.W., 1938, Nitrogen and carbohydrate of kidney bean cuttings as affected by treatment with indoleacetic acid, *Bot. Gaz.* 100:298.

Swingle, C.F., 1927, Burrknot formations in relation to the vascular system of the apple stem, *J. Agric. Res.* 34:533.

Taylor, G.G., and Odom, R.E., 1970, Some biochemical compounds associated with rooting of *Carya illinoensis* stem cuttings, *J. Amer. Soc. Hortic. Sci.* 95:146.

Theophrastus, ca. 300 B.C., "Enquiry into Plants," English trans. by A. Hort, G.P. Putnam's Sons, New York (1916).

Thimann, K.V., and Koepfli, J.B., 1935, Identity of the growth-promoting and root-forming substances of plants, *Nature* 135:101.

Thimann, K.V., and Went, F.W., 1934, On the chemical nature of the rootforming [sic.] hormone, *Proc. Kon. Akad. Wetensch. Amst.* 37:456.

Trécul, A., 1846, Sur l'origine des racines, *Ann. Sci. Nat. Bot.* 6:303.

Trewavas, A., 1981, How do plant growth substances work?, *Plant Cell and Environ.* 4:203.

Tripepi, R.R., Heuser, C.W., and Shannon, J.C., 1983, Incorporation of tritiated thymidine and uridine into adventitious-root initial cells of *Vigna radiata*, *J. Amer. Soc. Hortic. Sci.* 108:469.

Trippi, V.S., 1963, Studies on ontogeny and senility in plants, IV. Activity of some enzymes at different stages of ontogeny and in clones from juvenile and adult zones of *Robinia pseudoacacia*, *Phyton* 20:160, XI-1963.

van der Lek, H.A.A., 1924, Over de wortelvorming van houtige stekken, *Meded. Landbouwhoogeschool Wageningen*, 28:1.

van der Lek, H.A.A., 1934, Over den invloed der knoppen op de wortelvorming der stekken, *Meded. Landbouwhoogeschool Wageningen* 38(2):1.

van Overbeek, J, and Gregory, L.E., 1945, A physiological separation of two factors necessary for the formation of roots on cuttings, *Amer. J. Bot.* 32:336.

van Overbeek, J., Gordon, S.A., and Gregory, L.E., 1946, An analysis of the function of the leaf in the process of root formation in cuttings, *Amer. J. Bot.* 33:100.

van Staden, J., and Harty, A.R., 1988, Cytokinins and adventitious root formation, *in*: "Adventitious Root Formation by Cuttings," T.D. Davis, B.E. Haissig, and N. Sankhla, eds., Adv. in Plant Sci. Ser., vol. 2, Dioscorides Press, Portland, p. 185.

van Tieghem, P., and Douliot, H., 1888, Recherches comparatives sur l'origine des membres endogènes dans les plantes vasculaires, *Ann. Sci. Nat. Bot.* 8:1.

Verbeke, J.A., 1992, Developmental principles of cell and tissue differentiation: Cell-cell communication and induction, *Int. J. Plant Sci.* 153(3):S86.

Vieitez, E., Vazquez, A., and Areses, M.L., 1966a, Rooting problems of chestnut cuttings, *Cong. Colloq. Univ. Liège* 38:115.

Vieitez, E., Gesto, M.D.V., Mato, M.C., Vazquez, A., and Carnicer, A., 1966b, p-hydroxybenzoic acid, a growth regulator, isolated from woody cuttings of *Ribes rubrum*, *Physiol. Plant.* 19:294.

Vieitez, J., Kingston, D.G.I., Ballester, A., and Vieitez, E., 1987, Identification of two compounds correlated with lack of rooting capacity of chestnut cuttings, *Tree Physiol.* 3:247.

Vöchting, H., 1878 and 1884, "Über Organbildung im Pflanzenreich, Physiologische Untersuchungen über Wachsthumsursachen und Lebenseinheiten," I., Max Cohen & Sohn (FR Cohen) Verlag, Bonn; II., Emil Strauss Verlag, Bonn.

Vöchting, H., 1892, "Über Transplantation am Pflanzenkörper," Tübingen.

Vöchting, H., 1906, Über Regeneration und Polarität bei höheren Pflanzen, *Bot. Zeit.* 64:101.

Warmke, H.E., and Warmke, G.L., 1950, The role of auxin in the differentiation of root and shoot primordia from rooting cuttings of *Taraxacum* and *Cichorium*, *Amer. J. Bot.* 37:272-280.

Went, F.A.F.C., 1930, Über wurzelbildende Substanzen bei *Bryophyllum calycinum* Salisb., *Zeit. Bot.* 23:19.

Went, F.W., 1928, Wuchsstoff und Wachstum, *Rec. Trav. Bot. Néerl.* 25:1.

Went, F.W., 1929, On a substance causing root formation, *Proc. Kon. Akad. Wetensch. Amst.* 32:35.

Went, F.W., 1932, Eine botanische Polaritätstheorie, *Jahrb. Wiss. Bot.* 76:528.

Went, F.W., 1934a, A test method for rhizocaline, the root-forming substance, *Proc. Kon. Akad. Wetensch. Amst.* 37:445.

Went, F.W., 1934b, On the pea test method for auxin, the plant growth hormone, *Proc. Kon. Akad. Wetensch. Amst.* 37:547.

Went, F.W., 1935, Hormones involved in rootformation [sic.], The phenomenon of inhibition, *in*: "Proc. 6th Int. Bot. Cong.," vol. 2, p. 267.

Went, F.W., 1938, Specific factors other than auxin affecting growth and root formation, *Plant Physiol.* 13:55.

Went, F.W., 1939, The dual effect of auxin on root formation, *Amer. J. Bot.* 26:24.

Went, F.W., 1974, Reflections and speculations, *Annu. Rev. Plant Physiol.* 25:1.

Went, F.W., and Thimann, K.V., 1937, "Phytohormones," MacMillan Co., New York.

Wildon, D.C., Thain, J.F., Minchin, P.E.H., Gubb, I.R., Reilly, A.J., Skipper, Y.D., Doherty, H.M., O'Connell, P.J., and Bowles, D.J., 1992, Electrical signalling and systemic proteinase inhibitor induction in the wounded plant, *Nature* 360:62.

Wilson, P.J., and van Staden, J., 1990, Rhizocaline, rooting co-factors, and the concept of promoters and inhibitors of adventitious rooting - a review, *Ann. Bot.* 66:479.

Winkler, H., 1900, Ueber Polarität, Regeneration und Heteromorphose bei *Bryopsis*, *Jahrb. Wiss. Bot.* 36:449.

Zimmerman, P.W., 1930, Oxygen requirements for root growth of cuttings in water, *Amer. J. Bot.* 17:842.

Zimmerman, P.W., and Hitchcock, A.E., 1929, Vegetative propagation of holly, *Amer. J. Bot.* 16:570.

Zimmerman, P.W., and Hitchcock, A.E., 1933, Initiation and stimulation of adventitious roots caused by unsaturated hydrocarbon gases, *Contr. Boyce Thomp. Inst.* 5:351.

Zimmerman, P.W., and Hitchcock, A.E., 1946, The relation between age of stem tissue and the capacity to form roots, *J. Gerontol.* 1:27.

Zimmerman, P.W., and Wilcoxon, F. 1935, Several chemical growth substances which cause initiation of roots and other responses in plants, *Contrib. Boyce Thomp. Inst.* 7:209.

Zimmerman, P.W., Crocker, W., and Hitchcock, A.E., 1933, Initiation and stimulation of roots from exposure of plants to carbon monoxide gas, *Contr. Boyce Thomp. Inst.* 7:209.

Zimmerman, R., 1963, Rooting cofactors in some southern pines, *in*: "Proc. Int. Plant Prop. Soc.," 13:71.

Zimmerman, R.H., ed., 1976, "Symposium on Juvenility in Woody Perennials," *Acta Hortic.*, No. 56.

RESEARCH ON ADVENTITIOUS ROOTING:

WHERE DO WE GO FROM HERE?

Wesley P. Hackett and John R. Murray

Department of Horticultural Science
University of Minnesota
St. Paul, Minnesota 55108

Agronomists, horticulturists and silviculturists have been eminently successful in manipulating rooting of cuttings of many species of plants using auxin treatments and various environmental variables in conjunction with a knowledge of the optimum stage of development of the mother plant for root initiation. Much of this success has come as a result of empirical research. However, even with this successful research effort, there are still many species, particularly in their mature phase of development, which cannot be successfully propagated from cuttings on a commercial scale. In these cases, a better understanding of the mechanisms that control the root initiation process might be useful in efforts to improve rooting.

Presentations made at the Symposium illustrated that our understanding of the mechanisms controlling root initiation in shoot tissues of plants is relatively meager and fragmentary. We do have knowledge that auxin is an inductive stimulus for root initiation in some species and phases of development, and may therefore, be limiting. In other species or phases of development, there is not an induction of root initiation by auxin, therefore, we can conclude that auxin is not the only limiting control point in the initiation process. Although, much research continues to focus on auxin as the control point, we can say very little about other controls which may limit the root initiation process in cases in which auxin does not induce root initiation. Because root initiation is a complex developmental process, perhaps we should assume that it is controlled at a number of points in the process. If we assume that there are a number of control points, then the factor or control limiting the process may differ for various species or developmental phases of a mother plant. If these assumptions are correct, dissection of root initiation as a developmental process becomes a high priority objective in order to begin to understand what limits the root initiation process in different species.

What is required to dissect root initiation as a developmental process? First, it is important to choose an experimental system or systems in which there is a large contrast phenotypically in root initiation response with little or no difference in genotype. Some examples that come to mind are: 1) A genotype or inbred line that readily forms adventitious roots, with single gene mutants which result in the complete loss of ability to initiate adventitious roots without change in other wild type phenotypic characteristics; 2) etiolated and light-grown tissue of the same clone which has little or no ability to root when grown in light, but high ability when etiolated; 3) juvenile and mature phase tissue of the same clone in which the mature phase tissue has little or no ability, and the juvenile phase tissue has high ability to initiate adventitious roots.

Secondly, it is important to dissect the root initiation process in developmental terms. When is the tissue competent to respond to auxin, and when are the cells irreversibly committed to root morphogenesis? If the tissue is incompetent to initiate roots in response to auxin, what morphogenic process limits root initiation? Is a transmissible inductive stimulus, other than auxin, or an antagonist of root initiation involved? Are these limitations related to anatomical and/or cytological changes? Some of the questions regarding the developmental stage which is limiting the root initiation process can be answered with transfer experiments using inductive and non-inductive treatments and others can be answered with grafting experiments using tissues of one of the systems with contrasting root initiation phenotypes. Use of mutants with lesions at different stages of the root initiation process would be very valuable in dissecting the control mechanisms for root initiation.

Thirdly, we need bioassays to detect root initiation inductive stimuli or antagonistic stimuli. These bioassays should involve use of tissues with the contrasting root initiation phenotypes as described above. To detect inductive stimuli that are limiting in tissues with low ability to root, we need to use as bioassays, tissues with low ability to root. To detect antagonistic stimuli that limit rooting in tissues with low ability to root, we need to use tissues with high ability to root. The bioassays should utilize tissues from the same species being used for isolation of the positive inductive and antagonistic stimuli since we do not know whether factors that limit root initiation in one species will be the same as those that limit root initiation in another.

With developmentally defined experimental systems with which to work, what are the important questions about control of the root initiation process that we want to answer? The following are some questions about control of root initiation that were raised or inferred in the oral presentations, posters and discussions at the Symposium:

1.   Do all cell/tissue types have the ability to initiate roots?

2.   If so, do all cell/tissue types require the same endogenous inductive stimulus to initiate roots?

3.   How many different endogenous stimuli are needed for induction of root initiation?

4.   Are there endogenous stimuli that are antagonistic to induction of root initiation?

5.   Is lack of an adventitious rooting response due to lack of an inductive stimuli, presence of an antagonist or lack of competence to respond to an inductive signal?

6.   Are there endogenous transmissible substances that influence competence to respond to an inductive signal?

7.   What is the nature of competence to respond to an inductive signal for root initiation, and are there useful molecular or morphological markers of competence, such as peroxidases or stem diameter, respectively?

8.   For shoot tissue that responds to inductive signals, is the tissue competent initially, or is there acquisition of competence after detachment from the mother plant?

9.   Why is wounding (detachment) important for initiation of roots in some tissues but not others?

10.   Are differences in metabolism of auxin involved in differences in competence to respond to auxin for root initiation?

11.   Are differences in levels or kinds of auxin receptors involved in differences in competence to respond to auxin for root initiation?

12.   Is competence or incompetence for root initiation maintained in isolated, homogenous cell lines?

13.   What molecular genetic mechanisms are involved in control of adventitious root initiation?

14.   How can genes involved in control of root initiation be identified and isolated?

15.   Are there genes that are negatively involved in control of root initiation as well as those that are positively involved?

16.   Are *rol* genes from *Agrobacterium rhizogenes* useful in understanding the nature of competence for root initiation?

17.   Does control of adventitious root initiation differ on a developmental, physiological or molecular level from shoot-borne-root initiation?

18.   Does lateral root initiation on roots differ on a developmental, physiological or molecular level from adventitious and shoot-borne-root initiation?

Perhaps by 1996 when the Second Symposium on the Biology of Adventitious Root Formation will be held in Israel, we will be able to provide answers to some of the questions posed. At the very least we should be able to re-phrase them to be more specific and/or ask more discerning questions.

ACKNOWLEDGEMENT

We thank Steve Chandler, Calgene Pacific Pty Ltd, for providing ideas that were used as the basis for some of the questions posed.

INDEX

Natural language text, 228-229
Net primary production, 259, 262
Nitrate reductase, 173, 253
Nitrate reductase gene, 173
Nitrogen, 30, 112, 203-204, 212, 235-237,
    249, 253, 256-257, 297-298, 301, 306,
    315
Nitrogenous compounds, 203, 249, 253, 256,
    297, 306
Nodal roots, 10-11, 26-31, 33
Node, 7-8, 14-15, 26-28, 30-31, 91,
    126-127, 132, 221, 249, 290
Non-auxin plant hormones, 309
Non-auxin rooting stimulus, 302, 304-309
Nonstructural carbohydrate, 214-215, 233
Norway spruce, 40, 44
Nucleic acid, 115
Nucleic acid markers, 115
Nursery, 41-42, 44, 49, 54, 59, 132, 207,
    309
Nutrients, 9, 15, 30, 33, 83-84, 169, 181,
    200, 206, 208-209, 211-213, 234-235,
    246, 254-256, 258-259, 264, 282,
    297-298, 309
    absorption, 29-31, 211, 255
    stress, 235, 258

Oats, 27
Octopine synthase, 171
Open reading frames, 183
Orchard seed, 40, 44
Orchid, 4, 17, 318-319
Organogenesis, 77-78, 87-88, 90, 92-94,
    106, 150, 280, 286, 290
Ornamental horticulture, 54, 251
Ornithine decarboxylase, 112, 185
Oxidation, 112-113, 146-147
Oxygen, 15, 17, 112, 259, 281, 305, 315

Parenchyma, 17-18, 92-94, 103-108, 115,
    130, 147, 149, 231, 296
Partitioning, 33, 194, 215, 234-235, 245,
    247-248, 250-251, 253-254, 256-259,
    262-263, 265
Pathway flux, 64-68, 74
Pea test, 284-285
Peroxidase, 112-113, 146-147
Phaseolus aureus, 144, 146-147, 151
Phenolic compounds, 113, 131
Phenophase, 231-233, 236-237, 239
Phenotypic plasticity, 5
Phenylalanine ammonia lyase, 114
Phenylpropanoid pathway, 114
Phenylpropanoids, 114, 296
Phosphorus, 30, 237, 256, 298
Photoaffinity labelling, 148, 155
Photosynthate, 6, 128, 132, 230, 232-235,
    246, 250-252, 254
Photosynthesis, 5-6, 129, 234, 248, 251,
    258, 261, 264, 289
Photosynthetic roots, 17

Phylogeny, 10, 19
Physicochemical assays, 148
Phytohormone, 88-89, 93, 155-156, 163,
    165, 252-254
Pigmentation, 106-108
Pinyon pine, 58
Pisum sativum, 150, 283, 285, 287, 312
Plant propagation, 69, 155, 285, 307, 319
Plantations, 43-46, 48, 50, 168, 207, 209,
    220, 255
Plasma membrane, 18, 84, 156, 158-159
Plasmid, 7, 149-150
Platt, J.R., 317
Pneumatorhizae, 12, 15
Polarity, 249, 280-282, 289-293, 295, 298,
    321
Pole-borne root, 1, 3, 5-6, 10, 13, 17-18,
    20
Pollen, 38, 171, 319
Polyamine metabolism, 112, 181, 185, 187
Polyamines, 112, 184-185, 187, 315
Polymerase chain reaction, 117
Polyphenol oxidase, 113
Poplars, 37, 45-46, 48, 50, 203, 205-207,
    209-215, 220, 233
Populus, 14-15, 37, 45, 113, 144, 203,
    206-208, 213-215, 247-252, 254, 264,
    281, 297-299, 313
Potassium, 30, 237, 256, 297
Preconditioning, 114, 124-125, 128
Predisposition, 80, 207, 277, 283, 286,
    294-295, 297, 319, 321
Predisposition to root, 297, 319, 321
Preformed root primordia, 143, 207, 249,
    284-285, 289, 295
Primordia, 7, 14, 26, 32, 88, 93, 104-105,
    111, 115, 143-145, 147, 164-166, 193,
    207, 231, 249, 277-278, 280, 284-285,
    287-289, 295, 297-298, 303
Proline-rich protein, 104-105, 115
Promoters, 95, 158, 167, 169, 171-172,
    174, 183, 187, 300
Propagation technology, 123
Protein, 62, 84, 104-106, 115-117,
    155-159, 171-172, 183-184, 215, 281, 316
Provenance trials, 38
Prunus insititia, 132-139
Pseudotsuga, 40, 255-257, 259
Putrescine, 112, 185, 187

Quantitative genetic theory, 63
Quantitative genetics, 62, 65, 81, 315-316
Quercetin, 114

Radiata pine, 38, 41-42, 48, 50
Radiotracers, 250-251
Reaction rates, 65
Reasoning engine, 224-225
Recalcitrant species, 55, 57-59, 62
Receptors, 90, 139, 149, 155, 303, 335
Reforestation, 38, 41, 43, 48-49

Regeneration, 2, 46, 53, 55, 62, 77, 82, 92, 126, 165, 183, 186, 276-282, 288-293, 295, 298, 308, 315
Regulatory sequences, 117, 173
Rejuvenation, 116, 124-125, 131-132, 135-136, 138, 296
Relative growth coefficient, 192
Reparation, 277-278, 280
Reporter genes, 171
Republic of Congo, 46
Research infrastructure, 78, 85
Respiration, 6, 15, 112, 129-130, 204-205, 209, 211, 213, 245, 248, 253, 259, 265, 281, 285, 289, 291
Restitution, 277, 280-281
Retinoic acid, 78
Rhizocaline, 284-285, 287, 298-299, 304-308
Rhizoids, 5-6
Ri plasmids, 149
Ri tl-DNA, 183-185, 187
Rice, 25-27, 32, 173-174, 187, 191-195, 199-200, 284, 312
Ringing, 124, 133, 138-139
RNA, 104-107, 115, 118, 173, 183, 316
rol genes, 115, 167-168, 174, 183, 335
    rolb gene, 150-151, 166, 172-173
    rolc gene, 166
    rold gene, 171
Root growth, 9, 25-26, 30-31, 33, 81, 112, 164, 166-167, 187, 191, 204, 206, 209-211, 213-214, 231-236, 248, 250-252, 257-260, 264, 286, 306
Root initiation, 8, 54, 64, 91, 95-96, 104, 106, 113, 115, 126, 131, 138, 143-151, 165, 167-168, 209, 248-250, 252, 254, 263, 286-288, 300, 303, 306, 311, 314, 333-335
Root length density, 193-194, 196-200, 261, 263
Root metabolism, 163, 169
Root reserves, 213, 215
Root system architecture, 181, 183-184, 187
Root system classification, 2-4
Root system dynamics, 263
Root turnover, 18, 193, 205, 215, 262
Root-inducing factor, 88-90, 95
Root-inducing medium, 88-90, 92
Root-shoot interactions, 2, 219-220, 223, 237, 239, 241, 246, 252-254
Root-specific gene expression, 163
Rooted cuttings, 40-42, 44-46, 49, 57, 73, 246, 248-249, 251-252, 255, 293
Rooting bioassay, 283-287
Rooting cofactors, 102, 113, 308, 312
Rooting inhibitors, 299-300, 307
Rooting potential, 91, 95, 100-102, 115, 123-127, 131-133, 135-139, 167, 293-296
Rooting stimulus, 129, 299, 302, 304-309
Rooting types, 2, 11-18, 284
Rooting-related genes, 61, 100, 103
Ruling theories, 319

Rye, 25, 27

Sachs, J., 281, 300
Salients, 127
Salinity, 33
Scientific method, 63, 316
Sclerenchyma, 14, 127, 130
Sclerification, 103
Secondary metabolism, 169
Secondary metabolite biosynthesis, 169
Sectioning, 288
Seed based production systems, 38
Selective breeding, 61-64, 69, 72-75
Seminal roots, 26-27, 29-31, 33
Sensitivity, 58, 68, 83, 116-117, 148-151, 166, 184, 264, 303, 316
Setts, 26-27
Severe pruning, 124, 131
Shading, 101, 234
Shoot morphology, 132-133, 135, 137, 139, 165, 255
Shoot thickness, 133-135, 137-139
Shoot-borne roots, 1-2, 4-9, 11-13, 15, 17-20
Shootless orchids, 17
Shrubs, 8-9, 55
Silvicultural practices, 215
Simulation models, 223, 225, 229, 253-254, 263-266
Simulation program, 198
Sinks, 214-215, 230-233, 246-248, 263
Sitka spruce, 38, 40
Soil temperatures, 33, 235
Somaclonal variation, 82, 135
Source-sink relations, 249, 251, 258, 262, 264, 282
Special stuffs, 298, 319
Spermidine, 112
Spruces, 44, 48, 278
Staining, 173, 288
Starch, 57, 213-215, 234, 250-251, 253-254, 264, 296-298
Stock plants, 40-41, 49, 57, 63, 123-127, 131-132, 135, 255, 257-261, 284, 294, 296-297, 299-300, 320
Stolons, 20, 26, 28-29, 32
Stoolbeds, 127, 131-132
Storage roots, 17-18
Storage subsystem, 12, 17
Strategic rules, 236, 238
Strictosidine synthase, 169, 172
Strong inference, 74, 297, 318-320, 322
Stuff-hypothesis, 282, 288-291, 293, 295, 297, 309, 312-313, 316, 319
Stuffs, 281-282, 288, 290, 292, 297-298, 304, 312-313, 315, 319-321
Sucrose, 107, 114, 253, 258, 306
Sucrose phosphate synthetase, 253
Sugar, 25-27, 33, 144-146, 206, 211, 213-215, 233, 249-250, 254, 259, 264, 297-298, 305